T0122370

Cognitive Technologies

More information about this series at http://www.springer.com/series/5216

Susanne Biundo • Andreas Wendemuth
Editors

Companion Technology

A Paradigm Shift in Human-Technology Interaction

Springer

Editors
Susanne Biundo
Institute of Artificial Intelligence
Universität Ulm
Ulm, Germany

Andreas Wendemuth
Cognitive Systems Group,
Institute for Information Technology
and Communications (IIKT)
and
Center for Behavioral Brain Sciences
(CBBS)
Otto-von-Guericke Universität Magdeburg
Magdeburg, Germany

ISSN 1611-2482 ISSN 2197-6635 (electronic)
Cognitive Technologies
ISBN 978-3-319-82880-0 ISBN 978-3-319-43665-4 (eBook)
DOI 10.1007/978-3-319-43665-4

Softcover reprint of the hardcover 1st edition 2017

Printed on acid-free paper

This Springer imprint is published by Springer Nature
The registered company is Springer International Publishing AG
The registered company address is: Gewerbestrasse 11, 6330 Cham, Switzerland

Preface

Any technical system is a *Companion*—a competent, adaptable, responsive and reliable assistant that provides its functionality individually tailored to the specific needs of each user. This vision lies behind the endeavor to open up a novel cross-disciplinary field of research. *Companion-technology* aims to overcome the discrepancy between the sometimes almost unmanageably complex technical functionality of modern systems on the one hand and the users' demand for an easy-to-use, customized deployment of this functionality on the other hand. It does so by enabling the realization of technical systems as cognitive systems with *Companion*-characteristics: competence, individuality, adaptability, availability, cooperativeness, and trustworthiness. These characteristics emerge from the controlled and well-orchestrated interplay of cognitive processes that implement perception, reasoning, planning, dialog, and interaction capabilities within technical systems.

Research and development towards *Companion*-technology needs to investigate two subjects in equal measure: cognitive technical systems, and human users of such systems. Thus, the field is cross-disciplinary by nature and involves methodological basic research across the academic disciplines of informatics, engineering, and the life sciences. As for the system side we investigate how cognitive processes could be implemented and interweaved to provide technical systems with *Companion*-characteristics. On the user side expectations towards and effects of system behavior are investigated empirically by relying on psychological behavioral models as well as on analyses of brain activity during human-system interaction.

This book reviews research results on *Companion*-technology over a wide range of topics. It is structured along four strands. The first group of articles covers knowledge modeling aspects and the high-level cognitive abilities of planning, reasoning, and decision making from various points of view. The second strand examines user-system dialogs and multi-modal interaction, while the third is devoted to affect and environment recognition from a variety of sources and input modalities. Finally the fourth strand introduces a reference architecture for *Companion*-systems and includes three articles presenting different prototypical applications of the technology. They are concerned with worker training in motor vehicle production, the assistance of users when setting up a home theater, and user-adaptive ticket purchase.

The research results reported in this book originate from the Transregional Collaborative Research Centre SFB/TRR 62 "*Companion*-Technology for Cognitive Technical Systems" (www.sfb-trr-62.de), where more than 70 researchers from various disciplines work together. The centre was founded and supported by the German Research Foundation (DFG).

We would like to thank all authors for their comprehensive contributions and the collaborative effort that brought this book into existence. We are also greatly indebted to Thomas Geier for his technical and organizational support throughout the entire development process of the book.

Ulm, Germany Susanne Biundo
Magdeburg, Germany Andreas Wendemuth
April 2017

Contents

Contributors

Simon Adler Fraunhofer-Institut für Fabrikbetrieb und-automatisierung (IFF), Magdeburg, Germany

Ayoub Al-Hamadi Institute for Information Technology and Communications (IIKT), Otto-von-Guericke Universität Magdeburg, Magdeburg, Germany

Rico Andrich Institut für Wissens- und Sprachverarbeitung (IWS), Otto-von-Guericke Universität Magdeburg, Magdeburg, Germany

Thomas Bauer Institut für Wissens- und Sprachverarbeitung (IWS), Otto-von-Guericke Universität Magdeburg, Magdeburg, Germany

Gregor Behnke Faculty of Engineering, Computer Science and Psychology, Institute of Artificial Intelligence, Ulm University, Ulm, Germany

Pascal Bercher Faculty of Engineering, Computer Science and Psychology, Institute of Artificial Intelligence, Ulm University, Ulm, Germany

Susanne Biundo Faculty of Engineering, Computer Science and Psychology, Institute of Artificial Intelligence, Ulm University, Ulm, Germany

Ronald Böck Cognitive Systems Group, Institute for Information Technology and Communications (IIKT), Otto von Guericke Universität Magdeburg, Magdeburg, Germany

Nikola Bubalo General Psychology, Institute of Psychology and Education, Ulm University, Ulm, Germany

Klaus Dietmayer Faculty of Engineering, Computer Science and Psychology, Institute of Measurement, Control, and Microtechnology, Ulm University, Ulm, Germany

Rafael Friesen Institut für Wissens- und Sprachverarbeitung (IWS), Otto-von-Guericke Universität Magdeburg, Magdeburg, Germany

Jörg Frommer Universitätsklinik für Psychosomatische Medizin und Psychotherapie, Otto-von-Guericke-Universität Magdeburg, Magdeburg, Germany

Thomas Geier Faculty of Engineering, Computer Science and Psychology, Institute of Artificial Intelligence, Ulm University, Ulm, Germany

Birte Glimm Faculty of Engineering, Computer Science and Psychology, Institute of Artificial Intelligence, Ulm University, Ulm, Germany

Michael Glodek Institute of Neural Information Processing, Ulm University, Ulm, Germany

Tatiana Gossen Data and Knowledge Engineering Group, Faculty of Computer Science, Otto-von-Guericke-Universität Magdeburg, Magdeburg, Germany

Jan Gugenheimer Faculty of Engineering, Computer Science and Psychology, Institute of Media Informatics, Ulm University, Ulm, Germany

Stephan Günther Institut für Wissens- und Sprachverarbeitung (IWS), Otto-von-Guericke Universität Magdeburg, Magdeburg, Germany

Matthias Haase Universitätsklinik für Psychosomatische Medizin und Psychotherapie, Otto-von-Guericke-Universität Magdeburg, Magdeburg, Germany

Sebastian Handrich Institute for Information Technology and Communications (IIKT), Otto-von-Guericke Universität Magdeburg, Magdeburg, Germany

Kim Hartmann Cognitive Systems Group, Institute for Information Technology and Communications (IIKT), Otto von Guericke Universität Magdeburg, Magdeburg, Germany

Ralph Heinemann Institute for Information Technology and Communications (IIKT), Otto-von-Guericke Universität Magdeburg, Magdeburg, Germany

Holger Hoffmann Medical Psychology, University of Ulm, Ulm, Germany

Daniel Höller Faculty of Engineering, Computer Science and Psychology, Institute of Artificial Intelligence, Ulm University, Ulm, Germany

Frank Honold Faculty of Engineering, Computer Science and Psychology, Institute of Media Informatics, Ulm University, Ulm, Germany

Thilo Hörnle Faculty of Engineering, Computer Science and Psychology, Institute of Artificial Intelligence, Ulm University, Ulm, Germany

Anke Huckauf General Psychology, Institute of Psychology and Education, Ulm University, Ulm, Germany

Markus Kächele Institute of Neural Information Processing, Ulm University, Ulm, Germany

Henrik Kessler University Clinic for Psychosomatic Medicine and Psychotherapy, Bochum, Germany

Anne Köpsel Institute of Psychology and Education, Ulm University, Ulm, Germany

Michael Kotzyba Data and Knowledge Engineering Group, Faculty of Computer Science, Otto-von-Guericke-Universität Magdeburg, Magdeburg, Germany

Gerald Krell Technical Computer Science Group, Institute for Information Technology and Communications (IIKT), Otto-von-Guericke Universität Magdeburg, Magdeburg, Germany

Julia Krüger Universitätsklinik für Psychosomatische Medizin und Psychotherapie, Otto-von-Guericke-Universität Magdeburg, Magdeburg, Germany

Georg Layher Institute of Neural Information Processing, Ulm University, Ulm, Germany

Rüdiger Mecke Fraunhofer-Institut für Fabrikbetrieb und-automatisierung (IFF), Magdeburg, Germany

Sascha Meudt Institute of Neural Information Processing, Ulm University, Ulm, Germany

Wolfgang Minker Faculty of Engineering, Computer Science and Psychology, Institute of Communications Engineering, Ulm University, Ulm, Germany

Heiko Neumann Institute of Neural Information Processing, Ulm University, Ulm, Germany

Robert Niese Institute for Information Technology and Communications (IIKT), Otto-von-Guericke Universität Magdeburg, Magdeburg, Germany

Florian Nielsen Faculty of Engineering, Computer Science and Psychology, Institute of Communications Engineering, Ulm University, Ulm, Germany

Andreas Nürnberger Data and Knowledge Engineering Group, Faculty of Computer Science, Otto-von-Guericke-Universität Magdeburg, Magdeburg, Germany

Frank W. Ohl Leibniz Institute for Neurobiology, Magdeburg, Germany

Center for Behavioral Brain Sciences (CBBS), Otto-von-Guericke-Universität Magdeburg, Magdeburg, Germany

Günther Palm Institute of Neural Information Processing, Ulm University, Ulm, Germany

Denis Ponomaryov A.P. Ershov Institute of Informatics Systems, Novosibirsk, Russia

Omer Rashid Institute for Information Technology and Communications (IIKT), Otto-von-Guericke Universität Magdeburg, Magdeburg, Germany

Stephan Reuter Faculty of Engineering, Computer Science and Psychology, Institute of Measurement, Control, and Microtechnology, Ulm University, Ulm, Germany

Felix Richter Faculty of Engineering, Computer Science and Psychology, Institute of Artificial Intelligence, Ulm University, Ulm, Germany

Dietmar Rösner Institut für Wissens- und Sprachverarbeitung (IWS), Otto-von-Guericke Universität Magdeburg, Magdeburg, Germany

Enrico Rukzio Faculty of Engineering, Computer Science and Psychology, Institute of Media Informatics, Ulm University, Ulm, Germany

Frerk Saxen Institute for Information Technology and Communications (IIKT), Otto-von-Guericke Universität Magdeburg, Magdeburg, Germany

Andreas Scheck Medical Psychology, University of Ulm, Ulm, Germany

Alexander Scheel Faculty of Engineering, Computer Science and Psychology, Institute of Measurement, Control, and Microtechnology, Ulm University, Ulm, Germany

Martin Schels Institute of Neural Information Processing, Ulm University, Ulm, Germany

Marvin Schiller Faculty of Engineering, Computer Science and Psychology, Institute of Communications Engineering and Institute of Artificial Intelligence, Ulm University, Ulm, Germany

Miriam Schmidt/Schmidt-Wack Institute of Neural Information Processing, Ulm University, Ulm, Germany

Andreas L. Schulz Leibniz Institute for Neurobiology, Magdeburg, Germany

Felix Schüssel Faculty of Engineering, Computer Science and Psychology, Institute of Media Informatics, Ulm University, Ulm, Germany

Reinhard Schwegler Faculty of Engineering, Computer Science and Psychology, Institute of Artificial Intelligence, Ulm University, Ulm, Germany

Friedhelm Schwenker Institute of Neural Information Processing, Ulm University, Ulm, Germany

Ingo Siegert Cognitive Systems Group, Institute for Information Technology and Communications (IIKT), Otto von Guericke Universität Magdeburg, Magdeburg, Germany

Sebastian Stober Machine Learning in Cognitive Science Lab, Research Focus Cognitive Sciences, University of Potsdam, Potsdam, Germany

Patrick Thiam Institute of Neural Information Processing, Ulm University, Ulm, Germany

Michael Tornow Institute for Information Technology and Communications (IIKT), Otto-von-Guericke Universität Magdeburg, Magdeburg, Germany

Harald C. Traue Medical Psychology, University of Ulm, Ulm, Germany

Johannes Tümler Volkswagen AG, Wolfsburg, Germany

Bogdan Vlasenko Cognitive Systems Group, Institute for Information Technology and Communications (IIKT), Otto von Guericke Universität Magdeburg, Magdeburg, Germany

Michael Weber Faculty of Engineering, Computer Science and Psychology, Institute of Media Informatics, Ulm University, Ulm, Germany

Andreas Wendemuth Cognitive Systems Group, Institute for Information Technology and Communications (IIKT) and Center for Behavioral Brain Sciences (CBBS), Otto-von-Guericke Universität Magdeburg, Magdeburg, Germany

Christian Winkler Faculty of Engineering, Computer Science and Psychology, Institute of Media Informatics, Ulm University, Ulm, Germany

Marie L. Woldeit Leibniz Institute for Neurobiology, Magdeburg, Germany

Dennis Wolf Faculty of Engineering, Computer Science and Psychology, Institute of Media Informatics, Ulm University, Ulm, Germany

Chapter 1
An Introduction to *Companion*-Technology

Susanne Biundo and Andreas Wendemuth

Abstract *Companion*-technology enables a new generation of intelligent systems. These *Companion*-systems smartly adapt their functionality to a user's individual requirements. They comply with his or her abilities, preferences, and current needs and adjust their behavior as soon as critical changes of the environment or changes of the user's emotional state or disposition are observed. *Companion*-systems are distinguished by characteristics such as competence, individuality, adaptability, availability, cooperativeness, and trustworthiness. These characteristics are realized by integrating the technical functionality of systems with a combination of cognitive processes. *Companion*-systems are able to perceive the user and the environment; they reason about the current situation, exploit background knowledge, and provide and pursue appropriate plans of action; and they enter into a dialog with the user where they select the most suitable modes of interaction in terms of media, modalities and dialog strategies. This chapter introduces the essence of *Companion*-technology and sheds light on the huge range of its prospective applications.

1.1 Motivation and Overview

When looking at the advanced technical systems we constantly use in our everyday lives, we make a striking observation: Although these systems provide increasingly complex and "intelligent" functionality, like modern household appliances, smart phones, cars, machines, and countless numbers of electronic services do, there is often a considerable lack of comfort and convenience in use. Extensive (or tenuous) operating instructions have to be downloaded from the Internet; lengthy menu promptings have to be passed; and in many cases the user is even left with no option but to explore the system's functionalities by him- or herself. Depending on the

S. Biundo (✉)
Institute of Artificial Intelligence, Ulm University, Ulm, Germany
e-mail: susanne.biundo@uni-ulm.de

A. Wendemuth
Institute for Information and Communications Engineering, Otto-von-Guericke University Magdeburg, Magdeburg, Germany
e-mail: andreas.wendemuth@ovgu.de

S. Biundo, A. Wendemuth (eds.), *Companion Technology*, Cognitive Technologies,
DOI 10.1007/978-3-319-43665-4_1

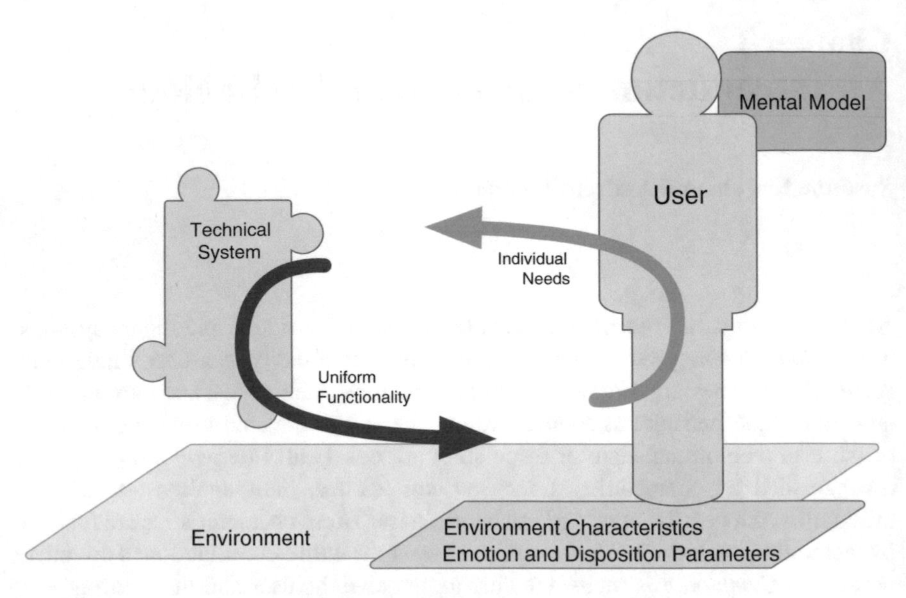

Fig. 1.1 Present-day human-technology interaction

particular user, the situation, and the system at hand, these obstacles may not only impede an exhaustive use of these innovative products and services, but may cause frustration and a reluctant attitude, and consequently the user may even lose interest in employing the system any further.

In other words, there is a wide gap between the growing functional intelligence of technical systems on the one hand and the lacking intelligence in providing this functionality to the user on the other hand. One reason, illustrated in Fig. 1.1, lies in the fact that technical systems offer their functionality in a strictly uniform way. They make no distinction between user types or even individual users, whether they are experienced with the system or not, request just a specific function, or have needs that demand some smart explanation of particular aspects of the system's functionality.

Companion-technology aims to bridge this gap by complementing the expanding functional intelligence of technical systems with an equivalent intelligence in interacting with the user and to integrate the two. It does so by enabling the realization of arbitrary technical systems as *Companion*-systems—cognitive technical systems that smartly adapt their functionality to the individual user's requirements, abilities, preferences, and current needs. They take into account the user's personal situation, emotional state, and disposition. They are always available, cooperative, and reliable and present themselves as competent and trustworthy partners to their users. *Companion*-systems are technical systems that exhibit so-called *Companion-characteristics*, namely *competence*, *individuality*, *adaptability*, *availability*, *cooperativeness*, and *trustworthiness*. These characteristics are implemented through the well-orchestrated interplay of cognitive processes based on advanced perception, planning, reasoning, and interaction capabilities.

In this chapter, we give an introduction to *Companion*-technology. We present the underlying theory and discuss its conceptual constituents. They include the acquisition, management, and use of comprehensive knowledge; the abilities to reason, to decide, and to recognize a user's context, emotion, and disposition; and the capacity for dialogue with individual users.

Up to now, the notion of a technical or artificial *Companion* has appeared in the literature only in a few contexts. The most prominent work is reported by Wilks [22]. Here, *Companions* are supposed to be conversational software agents, which accompany their owners over a (life-)long period. Rather than "just" providing assistance they are intended to give companionship by offering aspects of real personalization. In recent years, the paradigm of *Robot Companions* emerged in the field of cognitive robotics [1, 6, 13]. Those *Companions* are autonomous embodied systems, which accompany and assist humans in their daily life. Here, the main focus of research lies in the development of advanced training and learning processes to enable the robots to continuously improve their capabilities by acquiring new knowledge and skills.

In contrast, *Companion*-technology builds upon wide-ranging cognitive abilities of technical systems. Their realization and synergy have been, for roughly one decade, investigated under the research theme of *cognitive systems* or *cognitive technical systems*. The theme is focused on capabilities such as environment perception, emotion recognition, planning, reasoning, and learning, and their combination with advanced human-computer interaction. An overview on cognitive technical systems was initially published by Vernon et al. [19], while Putze and Schultz give a more recent introduction [15]. A comprehensive survey on the current state of the art in research and development towards *Companion*-technology is presented by Biundo et al. [5] in the special issue on *Companion Technology* of the KI journal [4].

However, up to now a systemized definition of the essence of *Companion*-technology or companionable systems has been lacking. The first attempt to come up with such a definition was made when establishing the interdisciplinary Transregional Collaborative Research Centre "*Companion*-Technology for Cognitive Technical Systems" [2, 3, 21]. In this chapter, we elaborate on this definition and draw the big picture of a novel technology.

1.2 The Big Picture

As the illustration in Fig. 1.1 shows, it is obvious that two important prerequisites for an individualized and context-sensitive functionality of technical systems are already given. First, both the system and the user are embedded in the environment. Provided with suitable perception capabilities the system would thus be able—in a way similar to human users—to perceive and recognize those context parameters that are relevant for the system's correct functioning and its interaction with the user. The system would also be able to observe its user and sense parameters that give an indication of the user's contentment and his or her emotional state and disposition.

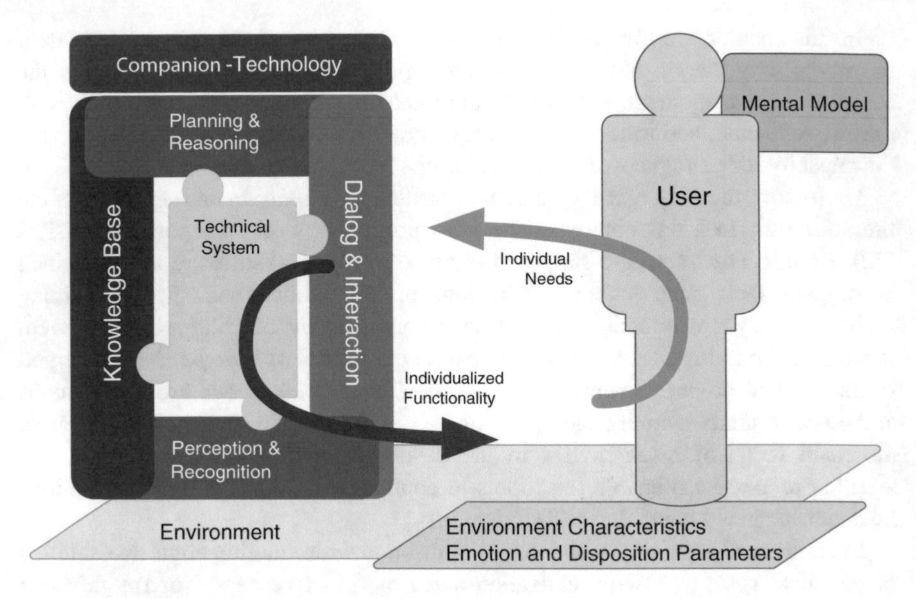

Fig. 1.2 Future human-technology interaction

Second, the user has a mental model of the technical system he or she is using and, in particular, has individual expectations concerning the system's behavior, including the way it should present its functionality and the interaction modes it should use to do so.

These prerequisites are utilized to provide input for the cognitive processes that establish a *Companion*-technology. Figure 1.2 shows the components that are required. Corresponding to the user's mental model of a technical system, a *Companion*-system will be equipped with a comprehensive knowledge base. It holds knowledge about the system itself such as declarative descriptions of its technical functionality and operation conditions as well as knowledge about the individual user, his or her abilities, preferences, and requirements. Based on this knowledge, advanced planning and reasoning facilities implement the technical functionality. Plans of action are automatically generated according to the user's profile. Depending on the application at hand, these plans either serve to directly control the system, or are passed on to the user as recommendation for action.

The situational context is perceived through various sensors, as are the user and his or her behavior. The emotional state and disposition are recognized by analyzing multiple modalities such as speech, facial expressions, hand and body gestures, and physiological measurements. With that, it is feasible to dynamically adapt the system's technical functionality according to sudden unexpected changes of the world and the user state.

A system's knowledge base does not only support the generation and adaptation of technical functionality, but also determines how the system and the user interact. Just as humans interact with their environment by employing various cognitive and motoric skills, *Companion*-technology enables systems to select appropriate communication devices and modalities according to both the current situational context and the user's tasks, preferences, emotional state, and disposition.

Companion-technology gives *Companion*-characteristics to technical systems. Competence, individuality, adaptability, availability, cooperativeness, and trustworthiness are realized through a three-stage approach. The first stage is advanced cognitive functions including perception, knowledge-based planning and reasoning, dialog management, and multi-modal interaction. By means of these cognitive functions, the second stage implements a number of cognitive competences. They include a robust recognition of the environmental situation and the user's emotional state; an individualized technical functionality and user-system interaction by continuously taking individual user characteristics and preferences into account; a consistent consideration of location, time, and behavioral context; and a robust activity recognition and plan execution monitoring.

In a third stage, a variety of meta-functions build upon the above-mentioned capabilities, thereby manifesting the *Companion*-characteristics. These meta-functions include:

- supporting the user with motivating comments and confirmation;
- sustaining the dialogue with the user and conducting meta-dialogs;
- recognizing a user's intentions;
- explaining the system's behavior and the system's recommendations;
- detecting erroneous situations and reacting appropriately;
- convincing the user of overarching goals;
- generating, presenting, and explaining possible alternatives for action;
- recognizing and accounting for changes in users' behavioral strategies;
- clarifying ambiguous user reactions through appropriate system intervention.

Companion-technology aims to lend *Companion*-characteristics to technical systems of all kinds: technical devices such as ticket vending machines, digital cameras, espresso machines, dishwashers, cars, and autonomous robots; electronic support systems such as navigation systems or fitness-apps; and complex application systems or electronic services such as travel and booking agents or planning assistants which help users in the accomplishment of a range of everyday tasks.

To give an impression, Fig. 1.3 shows the so-called *Companion*-space—a systematic view on application perspectives of *Companion*-technology. It indicates classes of prospective *Companion*-systems using the three dimensions of technical realization, *Companion*-task and application domain. Although not every point in this space describes a meaningful *Companion*-system, it does nonetheless demonstrate the great breadth of variation possible in such systems. For each application there exist various technical realizations and various tasks for which different cognitive functions, competences, and meta-functions are relevant. When analyzing user and situation parameters for the application "navigation", for example, it may be essential—depending on the actual device used—to first determine whether the user is traveling by car, by bicycle or on foot. Furthermore, the *Companion*-task in this context could be to give instructions on how to configure

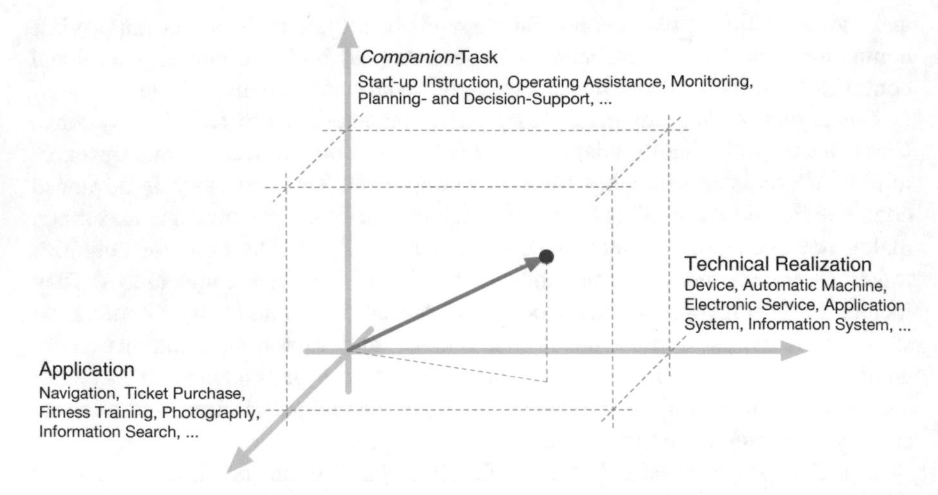

Fig. 1.3 Application perspectives of *Companion*-technology: the *Companion*-space

the navigation system or on how to operate the system in order to find a particular way to a certain destination. In a similar fashion, the implementation of the complex *Companion*-task of "monitoring" would require a whole range of cognitive functions and competences to be realized across a variety of devices.

By providing a novel paradigm for the operation of and interaction with technical systems of any kind, *Companion*-technology addresses important societal concerns. As increasingly complex technical systems continue to find their way into ever more areas of our lives, the requirements placed on individual users when using these systems also increases. At the same time, developments in technology continue to open up new and unforeseen opportunities for technical support and digital assistance. In this field of tension—especially as concerns the future of our aging society—*Companion*-technology is poised to make further important contributions. The areas of potential applications range from new types of individualized user assistance in operating technical devices over new generations of versatile organizational assistants and electronic service providers, to innovative support systems, for instance, for persons with limited cognitive abilities.

1.3 The Role of Knowledge

In order to function in a companionable manner, *Companion*-systems need to rely on comprehensive and multifaceted knowledge. Depending on the application domain at hand and the tasks to be performed, various kinds of knowledge are relevant. In order to provide this knowledge in a systematic way, knowledge bases of *Companion*-systems are structured along two lines. The static *Knowledge Model* holds a number of declarative models to supply the high-level cognitive functions

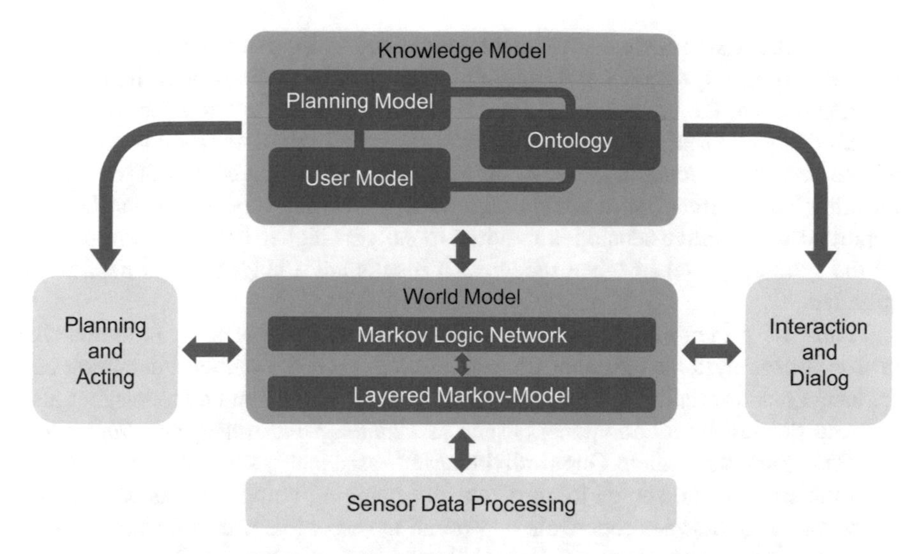

Fig. 1.4 Knowledge architecture for *Companion*-systems

of planning, reasoning, and decision-making. The dynamic *World Model* represents the current states of both the environment and the user and updates these states over time. Figure 1.4 shows the structure of these models.

The knowledge model has three components. The ontology represents static knowledge in terms of hierarchies of concepts and relations that characterize the application domain. The planning model holds the action portfolio. Single actions and entire courses of action describe the various possible ways of acting in the domain in order to achieve certain goals or accomplish certain tasks.

The user model basically enables the individualized and user-adaptive functionality of *Companion*-systems. It includes profiles indicating a user's technical knowledge level, his or her expertise w.r.t. the system's functionality, and preferences regarding ways of acting and interacting. Furthermore, information on the user's personality, abilities, general disposition, and motivational background is provided as are individual emotion patterns that help to assess the user's current emotional state when certain emotion parameters were sensed.

In order to ensure an effective use of the static knowledge as a whole, the various models are properly synchronized. This includes the coherent naming, the implied semantics, and the use of concepts and relations, which have to be established by a co-ordinated, tool-supported construction and maintenance of these models.

The dynamic world model reflects the current states of the application and the environment, the user's emotional, dispositional and motivational situation, and their development over time. Beyond that, the world model embodies the connection between the sub-symbolic processing of signal streams from various sensors, which collect audio, visual, and physiological data, and the inference-based information processing on the symbolic level. It consists of a Markov Logic Network and a multi-layered Markov-Model. The network encodes rules that are derived from the symbolic knowledge model and represent relevant information about the user, the application, and the environment. Its role is twofold. On the one hand, it enables the multi-layered Markov-Model that analyzes and interprets sensor data to put the recognized data in context, thereby improving the quality of recognition results. On the other hand, perception can be initiated and guided this way. This is particularly important when active sensing is required to support higher-level decision making on the symbolic level or when recognition results are ambiguous and need to be specified.

Figure 1.4 presents the knowledge architecture and the processing of knowledge in *Companion*-systems (see also Chap. 2). There is a close mutual interaction and exchange of information not only between the static and dynamic models, but also between the functional components such as *Planning and Acting* and *Interaction and Dialog* and the models. Once individualized assistance is requested, information from the user model serves to configure the planning component as well as the interaction and dialog components by providing respective user information. This way, it is guaranteed that the functional behavior of the system, its dialog strategies, the modalities, and media for interaction are geared to the needs of the particular user.

The sensor data processing modules recognize current parameters of the user and the environment over time. This information is further processed and combined with input from the plan execution, dialog, and interaction components. It leads to declarative descriptions of the environmental state and the user situation, which are stored and continually updated within the world model, thereby enabling the system to immediately react to changes of the state and the user situation.

Initially, the entire knowledge model is set up by a modeler. The knowledge stored in this model is not genuinely static, however. It may change in the long run and therefore needs to be updated from time to time. As far as the user model is concerned, this process is supported by the system itself. The world model stores a history of the user's reactions to the system's actions as well as to its interaction behavior. If for quite a number of episodes it is observed that the user behavior deviates from what is known about his or her expertise or preferences, for example, the user model is updated accordingly.

1.4 Planning and Decision Making

Companion-systems provide their technical functionality in a way such that each user is served individually, according to his or her specific needs, requirements, abilities, expertise, and current situation. This demands flexibility from the system's functional behavior, its responsiveness, and its ability to reason and reflect on its own behavior as well as on the user's reaction. *Companion*-systems meet these requirements by being provided with the high-level cognitive abilities of planning, reasoning, and decision making. Here, Artificial Intelligence planning technology [7] plays an essential role. Based on declarative descriptions of states, actions, and tasks, it allows for the construction of plans—courses of action—that are appropriate to reach a specified goal or accomplish a certain task.

Plans serve different purposes, depending on the particular application and the current *Companion*-task. They may be executed automatically to control a technical system directly; they may be used to instruct a human user on how to operate a technical system; or they may function as a guide when assisting a user in the accomplishment of a complex task (cf. Chaps. 5 and 6). Furthermore, the ability to plan enables *Companion*-systems to generate plans of action in close co-operation with a user by following a mixed-initiative strategy (see Chap. 7). Figure 1.5 shows an application example, where a *Companion*-system and a user co-operatively develop a workout plan.

Basically, plans are generated through causal reasoning. Actions are described by pre- and postconditions. The preconditions indicate in which states the action is applicable; the postconditions specify the effects of the action, i.e. the state changes it raises. Starting from a given goal and a description of the initial state, the planning component of the system selects appropriate actions from the action portfolio of the planning model. Appropriate actions are those whose effects coincide with the goal or with subgoals. Subgoals are preconditions of actions that in turn enable the execution of actions relevant to achieve the goal. By (partially) ordering the actions according to their causal dependencies, a plan is automatically generated. Executing this plan in the initial state finally leads to the goal. Action selection and the ordering of actions are determined by planning strategies and heuristics. They account for the system's ability to show a functional behavior that is customized to the individual user, his or her personal situation, and the current environmental situation.

Fig. 1.5 A *Companion*-system and a user co-operatively generate a workout plan

An Artificial Intelligence planning approach particularly well suited for *Companion*-systems is *Hybrid Planning*, which combines causal reasoning with reasoning about hierarchical dependencies between actions (cf. Chap. 5). Here, the planning model distinguishes between abstract and primitive tasks. Primitive tasks are actions that can be executed immediately, while abstract ones have to be refined over a cascade of hierarchy levels. For each abstract task the model provides one or more methods for refinement. A method represents a course of abstract and/or primitive tasks suitable to accomplish the respective abstract task. This way, predefined standard or individual solutions for problems and tasks can be specified in the planning model. This provides even more flexibility for the planning component of a *Companion*-system. It can decide to just use a predefined standard plan and thus speed up its response time, for example; it can modify a standard plan to meet specific user requests; or it can build a completely new plan from scratch.

Based on the plans of action a *Companion*-system creates and uses for support, feedback on the appropriateness of the system's functional behavior can find its way back into the underlying model. If it turns out, for example, that users regularly change their strategy of action or deviate from the procedures the system proposes, a careful analysis of this behavior may induce a modification of the planning model or the underlying user models, respectively. Chapter 8 discusses the issues of strategy change from a neuro-biological perspective.

One of the most prominent proficiencies that distinguishes *Companion*-systems from conventional technical systems as well as from today's cognitive systems is the ability to explain their own behavior. This ability is essential for implementing the *Companion*-characteristics of competence and trustworthiness. Explanations of the system's operations or the instructions for action it presents to the user are automatically generated by deriving and verbalizing information about causal and hierarchical relationships between actions. This information is obtained by analyzing the underlying plan of action and its generation process. The plan explanation technique is introduced in Chap. 5, whereas Chap. 7 presents a most useful combination of plan and ontology explanations.

Another essential functionality of *Companion*-systems is to adequately react if the execution of a plan fails. The reasons for an execution failure can be manifold and need to be ascertained carefully. To this end, information from various sources is used. It includes sensed data provided via the dynamic world model and information obtained through a multi-modal dialog with the user. Depending on the reason for failure the user is accordingly instructed and the plan is automatically repaired so as to provide a way out of the failed situation and to finally reach the original goal.

Chapter 24 describes a prototypical *Companion*-system where the functionalities of plan generation, plan explanation, and plan repair are integrated with components for multi-modal user interaction and dialog. This system provides advanced assistance to users in the task of setting up a complex home theater.

1.5 Interaction and Dialog

A main asset of *Companion*-systems is their dialogic nature. This characteristic reaches far beyond simple slot-filling interaction, but entails sustaining the dialogue with the user and conducting meta-dialogs.

Researchers therefore investigate the cognitive abilities that determine the design of the *interaction and dialog* between a human user and a technical system. Humans interact with their environment in multiple ways and, in doing so, they may use almost all of the senses, cognitive abilities, and motor skills available. Consequently, a *Companion*-system, as a peer communication and interaction partner to the human, is able to interact with its users through different modalities and a variety of input and output devices [8, 9], cf. Chap. 10. Modalities and media are determined according to the current situation and the individual user model that indicates the user's interaction preferences, cf. Chap. 11. This addresses the *Companion*-characteristics of individuality and adaptability. A prominent example is information seeking behavior, cf. Chap. 3.

Small latency in interaction is vital to ensure availability and cooperativeness of the perceived interaction. In a functional imaging study, it was observed that an unexpected delay of feedback by only 500 ms has an equally strong effect on brain activation as a complete omission of feedback [11]. Hence additional neural resources are needed in such potentially irritating situations, which also leads to further cognitive load and therefore should be avoided.

Understanding the interaction between a user, or multiple users, and a *Companion*-system as an adaptive *dialogue* is the natural choice, as it is made up of a sequence of consecutive interaction steps, including meta-dialogues when the train of mutual understanding is interrupted (cf. Chap. 9). An example of such a scenario is shown in Fig. 1.6. This cumulative interaction structure forms the basis for the determination of user intentions by the *Companion*-system. Under laboratory conditions, this calls for the development of an experimental paradigm involving the interaction history and presenting dedicated and reproducible stages of interaction, as presented in Chaps. 12 and 13. In a very practical industrial setting, *Companion*-systems have been used and evaluated as machine–operator assistance systems (Chap. 23).

For an effective and constructive dialog, the system is not only able to recognize the current dialog situation and user's disposition, but can choose among various strategies to keep the dialog going. Therefore, *Companion*-systems change the interaction strategy in the course of action, leading to *evolving search user interfaces* (cf. Chap. 4). Neurobiological fundamentals of strategy change are a basis for understanding and designing the dialog accordingly, see Chap. 8.

In this context, the *Companion*-characteristic of trustworthiness is of particular importance. One means to show trustworthiness is the ability to conduct explanation dialogs [14], i.e., a *Companion*-system is able to explain its own behavior and the situational circumstances that can be considered as a cause (cf. Chap. 7). The nature

Fig. 1.6 Users interacting with each other and with a *Companion*-system

and effect of such explanatory interventions can be measured when comparing to non-intervening situations, which was the subject of large field studies [12], cf. Chap. 13. Here, the focus was laid on identifying strategies which avoid mistrust and resistance. A main aspect was to investigate which intentional stance of the *Companion*-system is insinuated by the user.

1.6 Recognizing Users' Situation and Disposition

To ensure that the functionality of *Companion*-systems is customized to the individual user, adapting to his or her emotional state and current behavioral disposition, a pivotal facet consists of the cognitive abilities of *perception and recognition of the users' situation and disposition*. The technology must be able to recognize and appropriately interpret any relevant changes in the environmental conditions as well as the user's state on a continuous basis.

Changes in behavioral disposition and emotion occur in various ways, hence a wide range of parameters are used to detect them. They include prosodic and linguistic characteristics (Chap. 20), articulated motion (Chap. 17), head and body positioning and gestures (Chap. 16), facial expressions (Chap. 18), as well as psychobiological data. In total, a fully multimodal interpretation [17, 18] of the situation is required, see Chaps. 10 and 19. The dynamic evolution and prediction of emotions, dispositions, and moods is best captured under modeling hypotheses as detailed in Chaps. 4 and 21.

It is vital that the multi-modal recognition processes include location and time components, take into account the operational context and consider background information. The latter includes, among other things, typical behaviors and emotional patterns of the individual users and their environmental and situative disposition. Interactions between users and objects are modeled on the basis of the knowledge base of the *Companion*-system (Chap. 15) and serve as an environmental perception system. The environmental conditions of the user and the specific user parameters are then captured reliably and dynamically, interpreted and subsequently transformed into a total state description in a cascade of recognition and fusion processes (Chap. 19). In a dedicated demonstration scenario of a ticket vending task, the interplay of the various modalities and the subsequent information fusion aspects have been carefully studied (Chap. 25). It was revealed how stepwise dialogs are sensitive and adaptable within processing time to signals and background data, resulting in a user-adaptive and very efficient *Companion*-system.

Realization of *Companion*-systems must be based on real-world situational aspects and emotional processes in interactions between humans and computers, and it must make available system elements for realization of these effects. This is achieved through investigation and provision of decision-relevant and actionable *corpora*. The experimental settings must include non-linguistic, human behaviors, which are induced by a natural language dialog with delay of the commands, non-execution of the command, incorrect speech recognition, offer of technical assistance, lack of technical assistance, and request for termination and positive feedback [20]. Data acquisition is designed in a way such that many aspects of User-*Companion* interaction that are relevant in mundane situations of planning, re-planning, and strategy change (e.g. conflicting goals, time pressure, ...) are experienced by the subjects, with huge numbers and ranges in quality of recorded channels, additional data from psychological questionnaires, and semi-structured interviews [16] (Chap. 13). Established Wizard-of-Oz techniques as well as fully or semi-automated interactions have been employed, leading to general insights in the design and annotation of emotional corpora for real-world human-computer-interaction [10]. As multi-modal annotation is a novel and demanding task, software support systems such as ATLAS and *ikannotate* have been developed and tested (Chap. 19).

Eminently, corpora are a rich source of studying general feedback, planning and interaction activities in multiple modalities in real-world Human-Machine Interaction (HMI), see Chap. 14. Main assets of data for designing *Companion*-systems are elaborated hardware synchronicity over many modalities recorded in multiple sensory channels, and a setup with dedicated and standardized phases of subject-dispositional reactions (interest, cognitive underload and cognitive overload) as well as standardized HMI-related emotional reactions (such as fear, frustration, joy). Figure 1.7 shows such a multi-sensorial setup of a data recording during natural interaction. Corpora entailing these standards with up to ten modalities have been realized [18]. A careful system design serves as a model architecture for future *Companion*-systems, as detailed in Chap. 22.

Fig. 1.7 Multi-sensorial setup of a data recording where a user interacts with a *Companion*-system

Acknowledgements This work originates from the Transregional Collaborative Research Centre SFB/TRR 62 "*Companion*-Technology for Cognitive Technical Systems" (www.sfb-trr-62.de) funded by the German Research Foundation (DFG).

References

1. Albrecht, S., Ramirez-Amaro, K., Ruiz-Ugalde, F., Weikersdorfer, D., Leibold, M., Ulbrich, M., Beetz, M.: Imitating human reaching motions using physically inspired optimization principles. In: Proceedings of 11th IEEE-RAS International Conference on Humanoid Robots (2011)
2. Biundo, S., Wendemuth, A.: Von kognitiven technischen Systemen zu *Companion*-Systemen. Künstliche Intelligenz **24**(4), 335–339 (2010). doi:10.1007/s13218-010-0056-9
3. Biundo, S., Wendemuth, A.: *Companion*-technology for cognitive technical systems. Künstliche Intelligenz **30**(1), 71–75 (2016). doi:10.1007/s13218-015-0414-8
4. Biundo, S., Höller, D., Bercher, P.: Special issue on companion technologies. Künstliche Intelligenz **30**(1), 5–9 (2016). doi:10.1007/s13218-015-0421-9
5. Biundo, S., Höller, D., Schattenberg, B., Bercher, P.: Companion-technology: an overview. Künstliche Intelligenz **30**(1), 11–20 (2016). doi:10.1007/s13218-015-0419-3
6. Dautenhahn, K., Woods, S., Kaouri, C., Walters, M.L., Koay, K.L., Werry, I.: What is a robot companion - friend, assistant or butler. In: Proceedings of IEEE IROS, pp. 1488–1493 (2005)
7. Ghallab, M., Nau, D.S., Traverso, P.: Automated Planning: Theory & Practice. Morgan Kaufmann, Burlington (2004)
8. Gugenheimer, J., Knierim, P., Seifert, J., Rukzio, E.: Ubibeam: an interactive projector-camera system for domestic deployment. In: Proceedings of the 9th ACM International Conference on Interactive Tabletops and Surfaces, ITS'14, pp. 305–310. ACM, New York (2014)
9. Honold, F., Schüssel, F., Weber, M.: The automated interplay of multimodal fission and fusion in adaptive HCI. In: 10th International Conference on Intelligent Environments (IE), pp. 170–177 (2014)

10. Kächele, M., Rukavina, S., Palm, G., Schwenker, F., Schels, M.: Paradigms for the construction and annotation of emotional corpora for real-world human-computer-interaction. In: ICPRAM 2015 - Proceedings of the International Conference on Pattern Recognition Applications and Methods, Volume 1, Lisbon, 10–12 January 2015, pp. 367–373 (2015)
11. Kohrs, C., Angenstein, N., Scheich, H., Brechmann, A.: The temporal contingency of feedback: effects on brain activity. In: Proceedings of International Conference on Aging and Cognition (2010)
12. Lange, J., Frommer, J.: Subjektives Erleben und intentionale Einstellung in Interviews zur Nutzer-Companion-Interaktion. In: Lecture Notes in Informatics, Band P192: 41, Jahrestagung der Gesellschaft für Informatik, pp. 240–254. Gesellschaft für Informatik, Bonn (2011)
13. Matthews, J.T., Engberg, S.J., Glover, J., Pollack, M.E., Thrun, S.: Robotic assistants for the elderly: designing and conducting field studies. In: Proceedings of 10th IASTED International Conference on Robotics and Applications (2004)
14. Nothdurft, F., Heinroth, T., Minker, W.: The impact of explanation dialogues on human-computer trust. In: Human-Computer Interaction. Users and Contexts of Use, pp. 59–67. Springer, Berlin (2013)
15. Putze, F., Schultz, T.: Adaptive cognitive technical systems. J. Neurosci. Methods **234**, 108–115 (2014). doi:10.1016/j.jneumeth.2014.06.029
16. Rösner, D., Haase, M., Bauer, T., Günther, S., Krüger, J., Frommer, J.: Desiderata for the design of companion systems. Künstliche Intelligenz **30**(1), 53–61 (2016). doi:10.1007/s13218-015-0410-z
17. Schwenker, F., Scherer, S., Morency, L. (eds.): Multimodal pattern recognition of social signals in human-computer-interaction. Lecture Notes in Computer Science, vol. 8869. Springer, Berlin (2015)
18. Tornow, M., Krippl, M., Bade, S., Thiers, A., Siegert, I., Handrich, S., Krüger, J., Schega, L., Wendemuth, A.: Integrated health and fitness (iGF)-corpus - ten-modal highly synchronized subject-dispositional and emotional human machine interactions. In: Multimodal Corpora: Computer Vision and Language Processing (MMC) at LREC 2016, pp. 21–24 (2016)
19. Vernon, D., Metta, G., Sandini, G.: A survey of artificial cognitive systems: implications for the autonomous development of mental capabilities in computational agents. Trans. Evol. Comput. **11**(2), 151–180 (2007). doi:10.1109/TEVC.2006.890274
20. Walter, S., Scherer, S., Schels, M., Glodek, M., Hrabal, D., Schmidt, M., Böck, R., Limbrecht, K., Traue, H.C., Schwenker, F.: Multimodal emotion classification in naturalistic user behavior. In: 14th International Conference on Human-Computer Interaction (HCI), Orlando, FL, 9–14 July 2011, Proceedings, Part III, pp. 603–611 (2011)
21. Wendemuth, A., Biundo, S.: A companion technology for cognitive technical systems. In: Cognitive Behavioural Systems. Lecture Notes in Computer Science, pp. 89–103. Springer, Cham (2012)
22. Wilks, Y.: Close Engagements with Artificial Companions: Key Social, Psychological, Ethical and Design Issues. Natural Language Processing, vol. 8. John Benjamins Publishing, Amsterdam (2010)

Chapter 2
Multi-level Knowledge Processing in Cognitive Technical Systems

Thomas Geier and Susanne Biundo

Abstract Companion-Systems are composed of different modules that have to share a single, sound estimate of the current situation. While the long-term decision-making of automated planning requires knowledge about the user's goals, short-term decisions, like choosing among modes of user-interaction, depend on properties such as lighting conditions. In addition to the diverse scopes of the involved models, a large portion of the information required within such a system cannot be directly observed, but has to be inferred from background knowledge and sensory data—sometimes via a cascade of abstraction layers, and often resulting in uncertain predictions. In this contribution, we interpret an existing cognitive technical system under the assumption that it solves a factored, partially observable Markov decision process. Our interpretation heavily draws from the concepts of probabilistic graphical models and hierarchical reinforcement learning, and fosters a view that cleanly separates between inference and decision making. The results are discussed and compared to those of existing approaches from other application domains.

2.1 Introduction

Early computers were separated from the physical world they resided in by nearly impervious barriers—punch cards, light bulbs, and, later, monochrome low-resolution screens and keyboards. Experts were required to communicate with those machines using now-obsolete (and possibly obscure) machine languages.

But technology evolves, and today's technical systems have become complex; very distinguishable from their ancestors. A contemporary smart phone, easily fitting into a pocket, is connected to its environment through a multitude of sensors and output devices—high-resolution touch screens, video cameras, gyroscopes, heartbeat detectors, permanent connection to the internet, and much more. The technology that is used to process this plethora of information has also advanced

T. Geier (✉) • S. Biundo
Institute of Artificial Intelligence, Ulm University, Ulm, Germany
e-mail: thomas.geier@alumni.uni-ulm.de; susanne.biundo@uni-ulm.de

S. Biundo, A. Wendemuth (eds.), *Companion Technology*, Cognitive Technologies,
DOI 10.1007/978-3-319-43665-4_2

significantly. Automatic speech recognition has become viable, and development of touch screen text input has reached a point where organic swipes yield the intended words with high accuracy.

Current technical systems are thus embedded much deeper in their physical environment, and using these devices has become possible for laypersons; they do not require any exceptional skills anymore. Still, there appears to be a gap between the complexity the average user is able to handle, and the functionality that could be provided by a modern technical device.

To bridge this gap modern technical systems should be cognitive. This means that they should possess abilities such as attention, memory, reasoning, decision making and comprehension that let them exploit their deep embeddedness in the physical world to offer their functionality in a more accessible way to human users.

In this chapter we attempt to shed some light on a subset of the processes that appear to be required to realize these cognitive abilities. Starting from an abstract definition of what resembles a cognitive technical system (CTS), and what task it must accomplish, we analyze an exemplary implementation and try to identify a modularized architecture that encompasses abstraction and separation between inference and decision making. We draw parallels between our interpretation of a general CTS and the architectures that have been used in already matured fields, such as automatic speech recognition, robotics and dialogue systems. We discuss implications, and possible challenges.

2.2 The Home Theater Setup Task

Throughout this chapter we will use the example application of an intelligent system whose function is to assist a human user with the setup of the devices of a home theater system. We have been involved in the implementation of such a system [4, 5], and its technical aspects are described in Chap. 24. We begin by describing the task the user tries to solve, and go on to describe the required properties of a technical system that provides assistance.

In the home cinema setup task (HCST), a person is faced with the problem of correctly connecting cables between various audio/video devices, such as TV, AV-amplifier, satellite receiver or DVD-player, using a multitude of available cables (Fig. 2.1). The goal is to enable certain functionality, such as to watch DVDs on the TV and to watch satellite TV. This goal can be achieved by correctly connecting the cables.

The HCST contains various difficulties for the involved human:

- It is necessary to have a mental model of the devices to predict what a specific cable configuration achieves.
- There exists a combinatorial problem when the number of available cables is limited, or when the number of connectors of the devices is low.

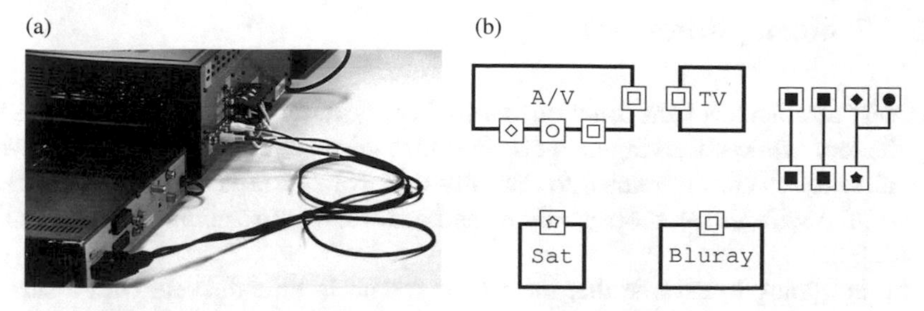

Fig. 2.1 A picture of two typical devices, and a symbolic representation of an instance of the Home Theater Setup Task. (**a**) Picture of two real devices. (**b**) Instance with four devices and three cables

- Different solutions can potentially result in different trade-offs, like being able to turn off the DVD-player while watching satellite TV in one configuration, but not in another one.
- If the final setup is not working, the source of the error has to be diagnosed.

By providing aide to a human user during the HCST, some of these problems can be mitigated. We assume that an assistive CTS provides its functionality merely by communicating with the user. In order to solve the HCST efficiently and satisfactorily, the system requires at least the following:

- knowledge about the problem domain, ports and functionality of the devices, connectors of cables, etc.
- a means to solve the combinatorial problem of identifying a connection scheme
- the ability to communicate with the user to provide directions and acquire feedback about the current state of the devices and cables

In addition, the following can provide an enhanced experience to the user:

- knowledge about the user: Which commands can he be trusted with executing correctly? How to shape the communication, and how to choose a good granularity for the provided instructions.
- sensors to detect the current state of the environment automatically: this comprises the available devices, cables, and the desired goal

The implemented prototype was also able to handle a scenario with multiple persons. This included identifying the dedicated user who has to be communicated the instructions, and who is supposed to connect the devices. In the examined scenario, the system was able to localize persons using a laser range finder. Since output of the system can be routed through multiple display devices, the system must try to leverage output close to the dedicated user to minimize cost of communication. The same also holds for input devices in the case of touch screens.

2.3 Problem Statement

At a very abstract level, the function of a CTS can be reduced to the operation of a set of effectors, while observing data obtained through a set of sensors. The following formalization is closely related to Partially Observable Markov Decision Processs (POMDPs) with factored observations and actions. That relation is formalized in Sect. 2.3.2.

We are going to assume that the whole system is time-discrete with a shared, synchronized heartbeat. When ignoring the computational problem of acting intelligently, the possibilities of a CTS are limited by its hardware realization: a set of sensors $\mathscr{S}$ and a set of effectors $\mathscr{E}$. From this point of view, the system is supposed to interact with its environment by triggering the effectors depending on the information that it acquires through its sensors (Fig. 2.2).

Each *sensor* $S \in \mathscr{S}$ is associated with a (not necessarily finite) set of possible observations Z^S. It produces a time-indexed sequence of observations z_t^S. The observation-sequence is not determined in advance, but depends on the stochastic evolution of, and the interaction with, the system's environment. Classical sensors of computer systems are keyboard and mouse. Modern smart phones have access to a much larger array of sensors, including but not limited to touch displays, accelerometers, cameras, and GPS. Access to the internet can also be treated as a sensor of some kind, as a computer is able to query various online information sources such as e-mail servers, postings of friends at social networks, or the weather forecast for next week.

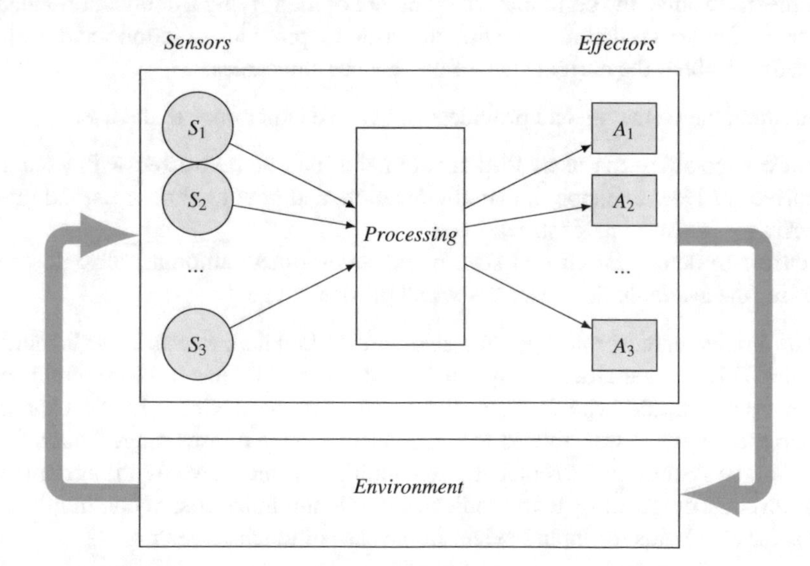

Fig. 2.2 A very abstract view of a CTS. The environment is perceived through a set of sensors and can be influenced via a set of effectors. This view leads to the interpretation as a POMDP with factored observation and action spaces

An effector $A \in \mathscr{E}$ is associated with a (not necessarily finite) set of possible actions O^A. It has to be supplied with an action o_t^A for every time step t. Examples of possible effectors of an assistive CTS can be displays or speakers, as can be found with ordinary computer systems and smart phones. The role of these effectors is usually communication with the user. But the triggering of events external to the technical system can also be modelled as effectors. These can include posting an e-mail, issuing a job to a printer, or even initiating a purchase at some online market platform.

The task of a CTS at time t is to find actions o_t^A for every effector $A \in \mathscr{E}$, based on the past observations $z_{1;t-1}^S$ it has obtained through its sensors. The effectors shall be operated in an intelligent manner, maximizing the system's positive properties. In the context of *Companion* technology, these properties could be the *Companion* properties of individuality, adaptability, availability, co-operativeness and trustworthiness (cf. Chap. 1, [6]). Finding an operationalization for the "positive properties" of an assistive CTS is a difficult problem that is not to be discussed in this chapter. Independently of the concrete nature of these desirable properties, we can demand that the system acts in a way to optimize the expected value of a single given utility (or reward) function R that depends on the inaccessible state of the system's environment and the action taken.

2.3.1 Markov Decision Processes

A Markov Decision Process (MDP) is a model for time-discrete, sequential decision problems within a fully observable, stochastic environment [3]. This means that actions can have an uncertain, probabilistic outcome. An agent that acts according to an MDP takes into account the possible failure of actions, and acts in anticipation, possibly avoiding such actions on purpose. As such, solutions to MDPs deal with risk in a way that can be considered rational, by maximizing the attained expected utility [48].

For simplicity we restrict our attention to finite state and action spaces. Then an MDP is a tuple $\langle S, A, T, R \rangle$, where S is a finite set of states, A is a finite set of actions, and $T: S \times A \rightarrow \mathscr{P}(S)$ maps state-action pairs to distributions over the following state.[1] In case the state space is continuous instead of finite, we talk of continuous MDPs, and the transition function is a probability density function. The reward $R: S \times A \rightarrow \mathbb{R}$ defines the immediate reward obtained from applying an action in some state.

A candidate solution for an MDP is called a *policy* $\pi: S \rightarrow A$ and maps states to actions. There exist different ways of defining the value of a policy. We present the traditional and most simple one—the discounted reward over an infinite horizon with discount factor γ. In this case, the expected discounted reward obtained under

[1] $\mathscr{P}(X)$ is the set of all probability mass functions (or density functions) over the set X.

policy π when starting in state s is given by

$$J_{\pi}(s) = \mathbb{E}\left[R(s, \pi(s), s') + \gamma \cdot J_{\pi}(s')\right], \tag{2.1}$$

where the expectation is over the successor state s' according to the transition distribution $T(s|s', \pi(s))$. Alternatives are the average reward over an infinite horizon [17] or the accumulated reward over a finite horizon [20]. The goal is to find a policy π that maximizes $J_{\pi}(s)$ for some initial state $s \in S$.

2.3.2 Partially Observable Markov Decision Processes

The model of POMDPs [20] further extends the MDP model by concealing the state of the environment from the acting agent. On each step, the agent is only provided with a limited observation that depends stochastically on the current state, but not with the state itself. Acting successfully within a POMDP can require performing actions whose sole purpose is the acquisition of more information about the true state of the world.

Formally, a POMDP is a tuple $\langle S, A, T, O, Z, R\rangle$, where $\langle S, A, T, R\rangle$ is an MDP. Additionally, O is a finite set of observations, where the acting agent is given an observation $o \in O$ with probability $Z(o|s, a)$ after executing action a in (concealed) state s.

Since the agent does not have access to the true state, it has to base its decision on knowledge about past observations and applied actions. Thus a policy π for a POMDP is a mapping from a history of actions and observations $a_0, o_0, a_1, o_1, \cdots, a_{t-1}, a_{t-1}$ to the next action a_t. It can be shown that policies for POMDPs can also be defined by mapping a probability distribution $b \in \mathscr{P}(S)$ over the state to actions [20]. Such a distribution is then called a *belief state*. A belief state can be updated to reflect the information gained after acting and obtaining a certain observation. A *belief update* on b when observing o after applying action a is defined by the following formula:

$$b'(s) \propto Z(o|s, a) \sum_{s'} b(s') \cdot T(s|s', a) \tag{2.2}$$

Using this update, one can define a *belief MDP* that is equivalent to a given POMDP [20]. The belief MDP has the space of distributions over states $\mathscr{P}(S)$ as its state space; it is thus a continuous state MDP. Thus, a policy for the belief MDP maps belief states to actions, and the value can be defined analogously to Eq. (2.1).

A popular example of a POMDP is the tiger domain, where the agent faces two doors, behind one of which a dangerous tiger is waiting to devour the unwary (or unaware) agent. Solving the problem successfully involves listening closely to one of the doors, an action that does not influence the state, but yields an observation possibly revealing the true location of the beast. With full access to the world state, the listening action is pointless as the agent always knows which door is safe to enter.

While MDPs allow a planning agent to respond to stochastic/unexpected changes of the environment by using information gained from (perfectly) observing its state, solving a POMDP requires a *two-way* communication between the agent and the environment, which includes taking actions that elicit useful observations. It is thus not very surprising that dialogue systems [49, 50] and human-computer interaction [30] are one of the applications of POMDPs.

When looking at our initial formalization of CTSs by a set of senors $\mathcal{S}$ and a set of effectors $\mathcal{E}$, we can identify the sensors as the means to obtaining observations within the POMDP framework. Joint assignments to the effectors form the space of possible actions. The POMDP framework fills the remaining gaps of our initial formalization—defining the system dynamics, the observation dynamics and the objective via a reward function.

The main difference lies in the fact that both observations and actions appear factored in our CTS formalization, e.g. $O = Z^{S_1} \times Z^{S_2} \times \cdots \times Z^{S_k}$ with $S_i \in \mathcal{S}$ and $A = O^{A_1} \times O^{A_2} \times \cdots \times O^{A_l}$ $A_i \in \mathcal{E}$. Factoring observation, state and action spaces is common for complex MDP and POMDP models [8, 41, 49].

2.4 A System Architecture: Divide and Conquer

When designing complex things like cars, robots or large software artifacts, we strive to conquer complexity by dividing the task into smaller, more manageable, parts. By defining what happens on the boundaries of those parts we create conceptual independence and enable working on one local problem at a time. By fixing the boundary conditions between modules we limit ourselves to those solutions of the problem satisfying the boundary conditions. Usually the constraints will rule out the optimal solution, but they reduce the cognitive cost of designing the system and enable us to construct the system [9]. This form of modularization is one ingredient of this section.

Another one is constituted by the observation that *every way* of solving a problem—no matter how pragmatic it is—also solves an instance of the idealized problem. For us the idealized problem will be a complex POMDP, or a CTS problem, exemplarily capturing the HCST. We have "solved" this problem by building a working prototype [5] (see also Chap. 24). The prototype solves the combinatorial aspect of the HCST using a deterministic hybrid planning approach (Chap. 5). Actions decided on by the planner are passed on to a dialogue manager that implements them by means of issuing instructions to the user. The dialogue manager passes on communication directives to a fission component that distributes them over available output channels.

While our system solves a task whose complete formalization requires the expressiveness of POMDPs, we have never formulated a joint problem, as this would be difficult. Somehow we shy away from modelling the environment,

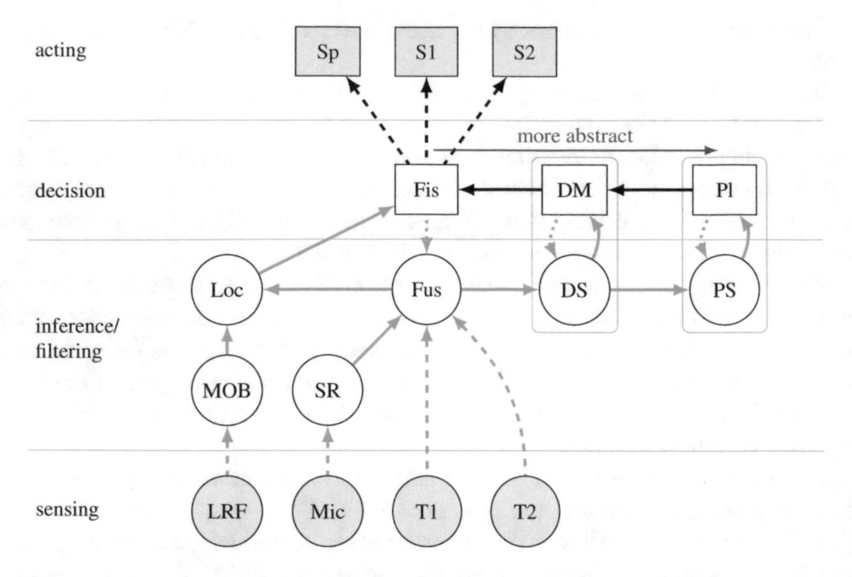

Fig. 2.3 Data-flow within a prototype implementation of a cognitive technical system

including a human user, as a single, hand-crafted Markov process, because of the many unknowns. But we have implicitly encoded a lot of assumptions about human communication behaviour within a dialogue manager [29] (and Chap. 9) and a multi-modal fusion and fission approach (Chap. 10). We formalized the technical aspects of the HCST as a planning problem (see Chaps. 5 and 6). So we have solved many subproblems, but the solution to the idealized joint problem lies hidden between pragmatic decisions and implementations.

Figure 2.3 gives a model-based view on the internal workings of the prototype implementation (Chap. 24). On the lowest layer we have the sensors (from left to right: laser range finder (LRF), microphone (Mic), touch screens (T1, T2)). On the next layer, components are shown that have the purpose of tracking the current state of the environment. The data from the LRF is processed by a multi-object Bayes filter (MOB). Object localization data together with information from the input fusion (Fus) is used to locate the user (Loc). Audio is processed by a stock speech recognizer (SR), whose output is then processed by the multi-modal input fusion. The current dialogue state (DS) uses the output of the input fusion, and provides information required to track the state of plan execution (PS). The belief state produced by the filtering stage is used to achieve decisions on how to control the effectors. The dedicated planner (Pl), the dialogue manager (DM), and the multi-modal output fission stage (Fis) are decision modules deciding upon actions. The decisions of the planner and the dialogue manager are only used to control the policy of other decision components, with only the fission being in direct control of effectors (speakers (Sp) and screens (S1, S2)). The type of data that flows between parts/components of the model is completely determined by the region boundaries crossed by the arrows. Observations flow from sensors to the filtering stage (◄- -).

Nodes within the filtering stage produce marginal distributions over the complete belief state (←). The actions taken by the decision components get fed into the filtering stage as observations for the next time step (←··). Outputs of decision components are either abstract, internal actions when used by another decision component (←), or primitive actions that control effectors (← -).

The rest of this chapter is dedicated to the interpretation and generalization of the described prototype implementation.

2.5 Inference/Filtering

The observations that are produced by the sensors are consumed by model parts that incorporate this data into their probabilistic prediction of the current world state. This process of estimating the current world state based on past observation is sometimes called filtering, and algorithms implementing this process are called filters. This process is equivalent to the belief update for belief MDPs given in Eq. (2.2). The nature of the components found within the filtering stage of Fig. 2.3 requires some explanation. First, we can identify components that are true probabilistic filters. The laser-range-finder data is processed by a multi-object Bayes filter (see Chap. 15, [39]) to track objects near the system. Input from the speakers is processed by a stock automatic speech recognizer (SR). Speech recognition can be performed using hidden Markov models, which are themselves temporal probabilistic models with latent variables [12]. Fusion of input events (Fus) is performed using an approach based on the transferable belief model [43] (Chap. 10, [42]), and is thus able to treat uncertainty. Information about user interaction is fused with the tracking results obtained from (MOB) to identify a possible user among tracked objects (Loc) [15]. This is achieved using a high-level temporal, probabilistic model formulated in Markov logic [14, 40]. All the model parts of the filtering stage mentioned so far are either truly probabilistic filters, or at least use a different form of uncertainty management, as in the case of input fusion.

The planner (Pl) and the dialogue manager (DM) work with models that assume full observability, equivalent to solving MDP problems. The planner even follows a deterministic modelling approach without stochastic change of the environment, although it is able to handle unforeseen events by performing replanning [4]. Both modules have in common that they can be thought of tracking the current state using a deterministic finite automaton (DFA) (represented by DS, PS in Fig. 2.3). In the prototype, changes to the planning model, which captures which cables have been connected to which devices, are tracked by observing the acknowledgement by the user of having executed a given instruction. Since a DFA can be considered a special case of a probabilistic temporal filter (one for a totally deterministic model with no uncertainty about the initial state), tracking of dialogue and planner states can be realized via Eq. (2.2), too, and thus fits our interpretation.

The product of a filter component consists of a probability distribution for the current state. Models of different components cover different variables, and the overlap between those variables is represented by arrows in Fig. 2.3. This probabilistic knowledge is passed along the black arrows, which can be thought of as marginalizations of the distributions represented by their source. These marginal distributions are relevant either for other filters or for the decision modules within the next layer.

2.5.1 *Factorizing the Belief State Using Graphical Models*

As we have argued, the task that has to be performed jointly by the filtering modules in Fig. 2.3 is to calculate the current belief. If we expand Eq. (2.2) recursively to compute the belief b_t at time t, and define the initial belief state as b_0, we obtain

$$b_t(s_t|o_1,...,o_t) = \sum_{s_0,\ldots,s_{t-1}} b_0(s_0) \prod_{i=1}^{t} T(s_i|s_{i-1},a_{i-1})Z(o_i|s_i,a_{i-1}). \tag{2.3}$$

We can observe that the filtering stage needs to perform a marginalization over the past states $s_0,\ldots,s_{t-1}$ for a probability distribution that is given in factored form by the conditional probabilities T and Z (see Fig. 2.4). This factorization corresponds to a hidden Markov model, or more generally a dynamic Bayesian network [21, 28].

Multivariate probability distributions that are defined in a factored form are commonly known under the name of *probabilistic graphical models (PGMs)* [21]. The factorization of the distribution implies a set of stochastic (conditional) independencies that both reduce the required number of parameters to specify a distribution, and simplify probabilistic inference. PGMs come in two major flavors, namely Bayesian networks (BNs) and Markov random fields (MRFs). They differ in the type of used graphical representation (directed vs. undirected) and the type of factors they consist of. BNs consist of conditional probability tables while MRFs can consist of arbitrary non-negative multivariate functions.

BNs were proposed for use in the area of Artificial Intelligence by Judea Pearl in the 1980s [34], although their roots range further back. A distribution P over variables $\mathbf{X} = \{X_1,...,X_n\}$ represented by a BN is defined as the product

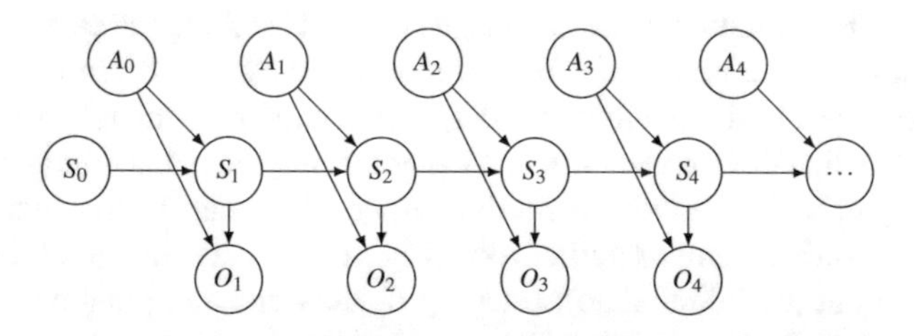

Fig. 2.4 Belief update for a POMDP interpreted as a dynamic Bayesian network [8], or a hidden Markov model. The prior distribution over the initial state S_0 is given by the initial belief state b_0

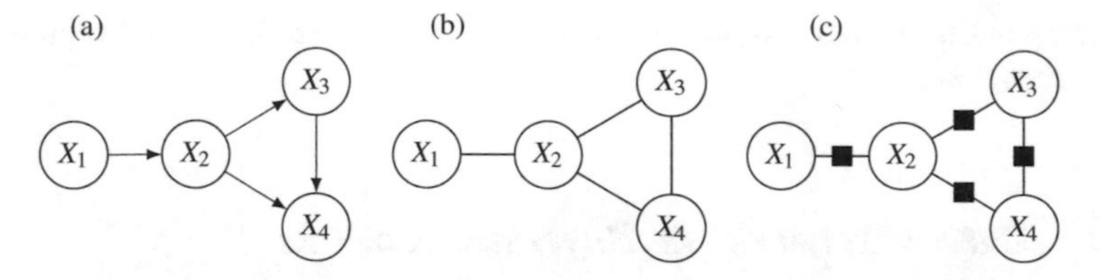

Fig. 2.5 Different flavors of graphical models. The Bayesian network shows the directed graph for distributions expressible through the factorization $P(X_1,...,X_4) = P(X_1) \cdot P(X_2|X_1) \cdot P(X_3|X_2) \cdot P(X_4|X_2,X_3)$. The Markov network and the factor graph are two different undirected graphical representations of the distribution defined by the factorization $P(X_1,...,X_4) \propto \phi_1(X_1,X_2) \cdot \phi_2(X_2,X_3) \cdot \phi_3(x_3,x_4) \cdot \phi_4(x_4,x_2)$. (**a**) Bayesian network. (**b**) Markov network. (**c**) Factor graph

of conditional probabilities (called conditional probability tables). A conditional probability table $P(X_i|\text{Par}(X_i))$ encodes the distribution over variable X_i given the ancestor variables $\text{Par}(X_i)$ in a directed acyclic graph that defines the dependency structure of the BN (see Fig. 2.5a). The joint distribution is given by

$$P(X_1,...,X_n) = \prod_i P(X_i|\text{Par}(X_i)). \tag{2.4}$$

The undirected version of PGMs is called MRF or Markov network (MN) [21]. MRFs can be seen as a generalization of BNs where the conditional probabilities are replaced by arbitrary functions. The dependency structure of MRFs can be represented graphically either by an undirected graph (Markov network) or a factor graph ([22]; Fig. 2.5). A MRF over a set of variables $\mathbf{X} = \{X_1,...,X_n\}$ is defined by a set of factors ϕ_a with index set $A \ni a$. A factor maps an assignment to a subset of variables $\mathbf{X}_a \subset \mathbf{X}$ to the positive reals, and the distribution is defined by the product over all factors:

$$P(X_1,...,X_n) = \frac{1}{Z} \prod_{a \in A} \phi_a(\mathbf{X}_a) \tag{2.5}$$

As MRFs are not normalized, turning them into a proper distribution requires scaling by a normalizing constant Z, called partition function, with

$$Z = \sum_{X_1,...,X_n} \prod_{a \in A} \phi_a(\mathbf{X}_a). \tag{2.6}$$

We have already observed that the joint filtering task consists of a marginalization in a BN whose conditional probability tables are given by the transition and observation probabilities (cf. Eq. (2.3)). These CPTs can be further broken down by assuming that the state S is factorized, just like observations and actions are factorized for the CTS task (cf. Sect. 2.3). This *inner* factorization enables the distributed filtering as can be seen in Fig. 2.3. The factorization can be either of a directed or undirected nature, and we shall discuss the trade-offs between the two approaches now.

2.5.2 Markov Networks vs. Bayesian Networks

The conditional independence structures that can be represented by BNs and MNs are slightly different [34, pp. 126], although both models are able to represent any probability distribution over a finite set of discrete-valued random variables.

Causal models are more faithfully represented by BNs, as the independence structure expressible by directed edges allows for a causal interpretation [35]. Under the assumption that every dependence between random variables is due to a causal mechanism [37, p. 156], it is reasonable to assume that a POMDP (in particular T and Z) can be expressed in a factorized form using only directed edges. But since the marginal abstraction over variables in a BN can lead to undirected dependencies, i.e., spurious correlations that are caused by latent confounding variables, it is reasonable to assume that high-level models abstract over enough variables such that some dependencies cannot be considered causal anymore.

Since dependencies along the progression of time are usually of a causal nature, it is reasonable to represent transition probabilities T as directed dependencies. Indeed, it has been argued that undirected graphical models are not suited to describe temporal data [18, 33], although there are successful applications of temporal MRFs for labeling and segmentation of time series data [46].

An advantage of MRFs is their relatively unproblematic composability. Two MRFs over the same variables can easily be combined by multiplying them. The combination of BNs is not as painless, because one has to make sure that the resulting directed graph must be free of cycles. In addition MRFs can be regarded as a generalization of constraint satisfaction problems and Boolean formulas in conjunctive normal form. This makes the use of existing deterministic formalizations trivial.

There exist numerous flavours of graphical models for the representation of time series data, such as maximum entropy Markov models [26], conditional random fields [23], and slice-normaliced dynamic Markov logic networks [33]. All models address various shortcomings of the others. Chain graphs [21, 24] are a hybrid between BNs and MRFs, as they pose as a directed model on the high level, but allow conditional probability tables to be factorized in an undirected fashion. This allows us to model the temporal progression between time steps as causal, while the interaction between variables of the same time step can be undirected. Thus, chain graphs pose as a potentially useful modelling paradigm in the context of CTSs.

2.6 Decision

The filtered estimates of the current state are used by the decision components to determine the behavior of the system, namely control of the effectors. In our interpretation of the prototype, there exist three distinguishable decision components.

At the lowest level we have the multi-modal fission (Fis). Fis is given abstract descriptions of dialogues and turns these into a concrete visual or audio output. For each information item that is to be conveyed to the user, Fis has to determine a modality (text, speech, image, video, etc.) and concrete output devices. For selection of a most suited device, Fis uses probabilistic information about the location of the target user (Loc) with the intent of choosing devices close to the user. Fis makes its decision according to the principle of maximizing expected utility [48]. The chosen output realization must be published to the multi-modal fusion (Fus), as e.g., pointing gestures can only be interpreted with knowledge about the on-screen location of involved entities. This is in accordance with our observation that the past action is necessary to update the current belief (Eq. (2.2)). Fis receives its control input (abstract dialogue descriptions) from the dialogue manager (DM).

The DM realizes communicative acts that may result in several dialogue turns each. An exemplary communicative act could be about issuing instructions to the user on connecting two devices with an available video-capable cable. The communicative act could start with querying the user about the availability of some suitable cables, and then instructing the user to plug in a user-chosen cable correctly. The implementation of the DM can be thought of as a POMDP policy that is predefined by an expert, instead of being obtained by planning/optimization. It follows that the DM component consumes user input to update its internal state, and it issues dialogue actions. For Fig. 2.3 we have divided the dialogue manager into the policy (DM) and the state tracker (DS), though this separation was not manifested in the real implementation.

At the highest decision instance we employed a deterministic planner (Pl, see Chap. 5). The deterministic planning component uses replanning to handle the case when the observed state deviates from the expected state trajectory. This construction turns the planner into an MDP planner/policy, although it cannot anticipate risky situations. To close the gap between MDP and POMDP expressiveness, we used the most probable state for detecting plan failures—a construction which only works in settings where the relevant transitions are nearly deterministic. This approach corresponds to the *belief replanning* paradigm in [11]. As for the DM, we have divided this replanning capable planner into a state tracker, PS, and the policy, Pl. The state tracker has to maintain merely a most probable current world state, instead of a more general probabilistic belief state.

Following our architectural interpretation—where we have separated the tracking of the current state (filtering) into the inference stage—decision modules base their judgement solely on this belief state. This separation follows from the idea of the belief MDP for solving POMDP problems. And the equivalence between POMDP and belief MDP implies that separating filtering from decision making allows us to still obtain the optimal solution. To calculate the correct belief update, the inference stage requires knowledge of past actions and past sensory observations (cf. Eq. (2.2)). Notably, the separation results in purely functional (as in functional programming) decision components that do not rely on an internal state that changes over time.

2.7 Abstraction

Within the prototype, the only decision module that was interfacing with the effectors was the multi-modal fission (Chap. 10). It was given abstract decisions from the dialogue manager, which in turn was given abstract decisions from the planner. While it is perceivable that DM and/or Pl are also issuing commands to effectors directly, the configuration we find in the prototype appears to be common in other CTSs, as we shall see in Sect. 2.8. We are now going to discuss the types of abstractions we can identify within the prototype.

The abstraction in the prototype between Fis, DM and Pl is of two kinds. We can identify *temporal* abstraction in the sense that decisions made on higher levels have a longer duration and occur less frequently. In addition, an *arbitrational* abstraction reduces the size/dimensionality of the decision space towards the Pl end. In an extreme sense, the lowest layer has to assign a color to every pixel of a screen, while on the higher level, between DM and Pl, the action is merely to "instruct the user to connect cable X with device Y".

2.7.1 Temporal Abstraction

Temporal abstraction is a well-researched topic in reinforcement learning [7, 19, 27, 36, 47], a field that deals with solving POMDP problems through interaction with the environment.[2] Within the MDP setting, options [45] formalize time-extended actions that are equipped with specialized sub-policies. The abstract decision propagation between decision components in the prototype bears much resemblance to options, as, e.g., the actions decided upon by the planner are implemented by the dialogue manager by a sequence of dialogues. A peculiarity of options is that the abstract policy cannot look inside options, and it is not able to interrupt them. This property also holds in both abstraction stages within the prototype. Because they have to choose between different actions less frequently, the higher-level decision components are also able to use more complex algorithms that show a slower

[2]This is also called model-free reinforcement learning, as the model has to be learned together with a policy.

response. Decisions on the lower levels have to be found quickly, and thus it is not possible to take into account much context—they are often of a reactive character.

When examining Fig. 2.3, we can observe a hierarchy of filtering models, mirroring the hierarchical arrangement of the decision components with the pairings Fus-Fis, DS-DM, and PS-Pl. Apparently it is natural to have abstraction within the state space of the joint POMDP, too. In contrast to hierarchical abstraction during decision making, the literature on temporal abstraction of probabilistic models appears to be much sparser. Brendan Burns and others describe an approach to temporal abstraction in Dynamic Bayesian networks (DBNs) [10]. Several works by Kalia Orphanou [31, 32] address temporal abstraction in DBNs for the evaluation of experimental data. Another very well-researched field that strongly relies on temporal abstraction is automatic speech recognition, where approaches already span the range from the waveform over phonemes and words to the grammar level [12].

2.7.2 *Arbitrational Abstraction*

We have introduced the term *arbitrational* abstraction to describe the constellation where the action space or state is coarsened without changing the temporal extent. This concept on its own appears to be much less researched than temporal abstraction for action abstraction. This is potentially due to the fact that both types often occur together and thus are not recognized as separate phenomena, at least in the setting of planning and decision making. In [41] an approach to reinforcement learning with factored state and action spaces is described, where planning occurs by inference within a restricted Boltzmann machine. In the analysis, Sallans and Hinton identify some variables that act as "macro actions". Their activation defines the joint behavior of many primitive decision variables. They also observe that the variables acting as macro actions are activated over longer periods of time, and thus arbitrational abstraction coincides with temporal abstraction again.

For probabilistic inference (not necessarily the temporal kind), arbitrational abstraction is omnipresent. The task of classification consists in mapping a high-dimensional input, for example an image, to a single output variable that represents an abstract concept like the proposition "contains a face". Only recently, larger progress has been made in the field of *deep learning* [1], where classification is done over a sequence of layers. This layering of classificators has proven to yield a good boost in classification performance to earlier "shallow" architectures, and the used layer model can also be a probabilistic graphical model. As such, the crafted abstraction hierarchy on the inference side of Fig. 2.3 bears some resemblance to "learned" abstraction hierarchies found by deep learning approaches.

2.8 Related Work

There are many examples of existing complex systems that can be thought of as CTSs. There exist problem areas where the complete issue of acting based on observations has to be (and to some extent *is*) solved; for example, robotics and dialogue management.

2.8.1 *Control Theory*

Control theory is concerned with the control of systems with simple dynamics. It has a long tradition and rests on solid mathematical foundations [2]. At the heart of the field lies the idea of exploiting feedback from measurements of the controlled system (see Fig. 2.6). The simple closed-loop controller can be interpreted within our framework as having a degenerate filtering stage without latent variables. On the more complex end of control theory, the concept of multiple-input/multiple-output controllers [16, pp. 591] approaches the complexity of POMDPs with factored actions and observations.

2.8.2 *Robot Architectures*

It has been observed that many independently designed robot architectures have a similar structure. In robotics, three-layer architectures are common [13]. They consist of a *controller*, a *sequencer* and a *liberator*. The controller consists of computations that realize a tight coupling between sensors and effectors, and it is often using approaches from control theory [16, 44]. The sequencer determines the behavior the controller is to follow, and possibly supplies additional parameters. According to [13], the sequencer resembles an MDP policy. On the highest level, the liberator implements computations that may require an arbitrary amount of

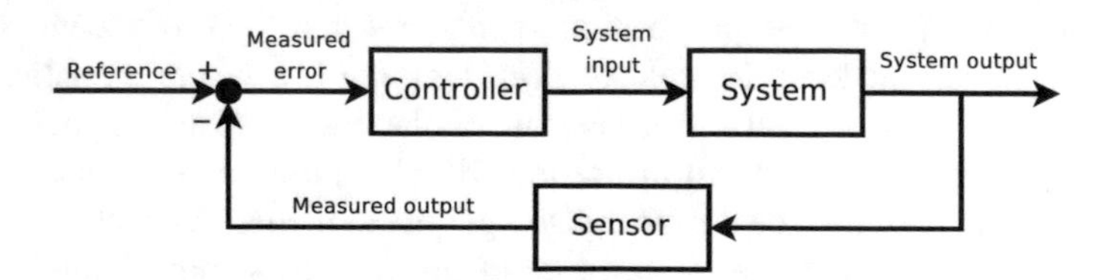

Fig. 2.6 The difference between a control input and a measurement of the target are used to control the dynamic system in a closed-loop architecture

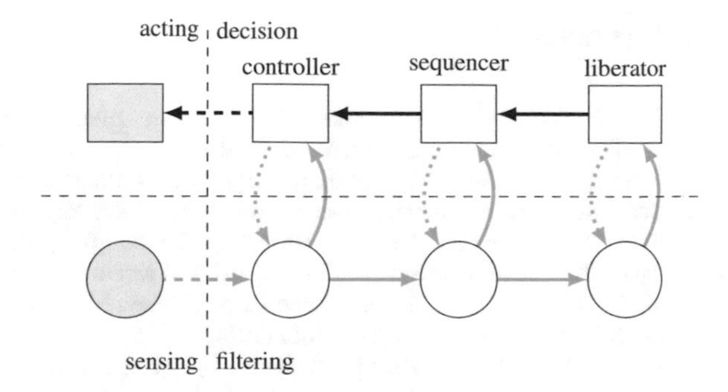

Fig. 2.7 Interpretation of the general three-layer architecture

processing time and cannot implement any real-time requirements. Our rendition of the general three-layer architecture is presented in Fig. 2.7.

2.8.3 Dialogue Systems

Multi-layer architectures have been used for dialogue management. For example, Lemon et al. [25] proposes a two-layer architecture. The task of the lower layer is to react to phenomena related to maintenance of the communication channel, such as turn taking, back-channel feedback (from the listener to the speaker), or interruption. The higher layer is concerned with structuring and planning of the conversation. There are also attempts at tracking the dialogue state using graphical models [38].

2.9 Conclusion

We have analyzed a prototypical implementation of an assistive CTS. We have provided a descriptive problem specification by arguing that the joint behavior of the system can be considered an attempt to solve a POMDP problem. Based on the POMDP formalization, we have interpreted implemented software components as contributions of solving the POMDP using the belief MDP approach, thus partitioning functionality into filtering and decision making. We have identified temporal and arbitrational abstractions as major components of the architecture. We have also discussed approaches to the modularization of filtering and decision making, and given references to related work.

Acknowledgements This work was done within the Transregional Collaborative Research Centre SFB/TRR 62 "*Companion*-Technology for Cognitive Technical Systems" funded by the German Research Foundation (DFG).

References

1. Arel, I., Rose, D.C., Karnowski, T.P.: Deep machine learning-a new frontier in artificial intelligence research [research frontier]. IEEE Comput. Intell. Mag. **5**(4), 13–18 (2010)
2. Åström, K.J., Kumar, P.: Control: a perspective. Automatica **50**(1), 3–43 (2014)
3. Bellman, R.: A markovian decision process. Technical Report, DTIC Document (1957)
4. Bercher, P., Biundo, S., Geier, T., Hoernle, T., Nothdurft, F., Richter, F., Schattenberg, B.: Plan, repair, execute, explain - how planning helps to assemble your home theater. In: Proceedings of the 24th International Conference on Automated Planning and Scheduling (ICAPS 2014), pp. 386–394. AAAI Press, Palo Alto (2014)
5. Bercher, P., Richter, F., Hörnle, T., Geier, T., Höller, D., Behnke, G., Nothdurft, F., Honold, F., Minker, W., Weber, M., Biundo, S.: A planning-based assistance system for setting up a home theater. In: Proceedings of the 29th National Conference on Artificial Intelligence (AAAI 2015). AAAI Press, Palo Alto (2015)
6. Biundo, S., Wendemuth, A.: *Companion*-technology for cognitive technical systems. Künstliche Intelligenz **30**(1), 71–75 (2016). doi:10.1007/s13218-015-0414-8
7. Botvinick, M.M.: Hierarchical reinforcement learning and decision making. Curr. Opin. Neurobiol. **22**(6), 956–962 (2012)
8. Boutilier, C., Dean, T.L., Hanks, S.: Decision-theoretic planning: structural assumptions and computational leverage. J. Artif. Intell. Res. (JAIR) **11**, 1–94 (1999). doi:10.1613/jair.575
9. Brusoni, S., Marengo, L., Prencipe, A., Valente, M.: The value and costs of modularity: a cognitive perspective. SPRU Electronic Working Paper Series. SPRU, Brighton (2004)
10. Burns, B., Morrison, C.T.: Temporal abstraction in Bayesian networks. In: AAAI Spring Symposium. Defense Technical Information Center (2003)
11. Cassandra, A.R., Kaelbling, L.P., Kurien, J.: Acting under uncertainty: discrete Bayesian models for mobile-robot navigation. In: Proceedings of IEEE/RSJ International Conference on Intelligent Robots and Systems. IROS 1996, November 4–8, 1996, Osaka, pp. 963–972 (1996). doi:10.1109/IROS.1996.571080
12. Gales, M., Young, S.: The application of hidden Markov models in speech recognition. Found. Trends Signal Process. **1**(3), 195–304 (2008)
13. Gat, E.: Three-layer architectures. In: Kortenkamp, D., Peter Bonasso, R., Murphy, R.R. (eds.) Artificial Intelligence and Mobile Robots, pp. 195–210. AAAI Press (1998)
14. Geier, T., Biundo, S.: Approximate online inference for dynamic Markov logic networks. In: International IEEE Conference on Tools with Artificial Intelligence, pp. 764 –768 (2011)
15. Geier, T., Reuter, S., Dietmayer, K., Biundo, S.: Goal-based person tracking using a first-order probabilistic model. In: Proceedings of the Nineth UAI Bayesian Modeling Applications Workshop (2012)
16. Goodwin, G.C., Graebe, S.F., Salgado, M.E.: Control System Design. Prentice Hall, Upper Saddle River (2001)
17. Gosavi, A.: Reinforcement learning: a tutorial survey and recent advances. INFORMS J. Comput. **21**(2), 178–192 (2009)
18. Jain, D., Barthels, A., Beetz, M.: Adaptive Markov logic networks: learning statistical relational models with dynamic parameters. In: ECAI, pp. 937–942 (2010)
19. Jong, N.K., Hester, T., Stone, P.: The utility of temporal abstraction in reinforcement learning. In: Proceedings of the 7th International Joint Conference on Autonomous Agents and Multiagent Systems - Volume 1, AAMAS'08, pp. 299–306. International Foundation for Autonomous Agents and Multiagent Systems, Richland, SC (2008)
20. Kaelbling, L.P., Littman, M.L., Cassandra, A.R.: Planning and acting in partially observable stochastic domains. Artif. Intell. **101**(1), 99–134 (1998)
21. Koller, D., Friedman, N.: Probabilistic Graphical Models: Principles and Techniques. MIT Press, Cambridge (2009)
22. Kschischang, F., Frey, B., Loeliger, H.A.: Factor graphs and the sum-product algorithm. IEEE Trans. Inf. Theory **47**(2), 498–519 (2001). doi:10.1109/18.910572

23. Lafferty, J., McCallum, A., Pereira, F.: Conditional random fields: probabilistic models for segmenting and labeling sequence data. In: Proceedings of the 18th International Conference on Machine Learning (2001)
24. Lauritzen, S.L., Richardson, T.S.: Chain graph models and their causal interpretations. J. R. Stat. Soc. Ser. B Stat. Methodol. **64**(3), 321–348 (2002)
25. Lemon, O., Cavedon, L., Kelly, B.: Managing dialogue interaction: a multi-layered approach. In: Proceedings of the 4th SIGdial Workshop on Discourse and Dialogue, pp. 168–177 (2003)
26. McCallum, A., Freitag, D., Pereira, F.C.N.: Maximum entropy Markov models for information extraction and segmentation. In: Proceedings of the Seventeenth International Conference on Machine Learning (ICML 2000), Stanford University, Stanford, CA, June 29–July 2, 2000, pp. 591–598 (2000)
27. Montani, S., Bottrighi, A., Leonardi, G., Portinale, L.: A CBR-based, closed-loop architecture for temporal abstractions configuration. Comput. Intell. **25**(3), 235–249 (2009). doi:10.1111/j.1467-8640.2009.00340.x
28. Murphy, K.: Dynamic Bayesian networks: representation, inference and learning. Ph.D. Thesis, University of California (2002)
29. Nothdurft, F., Honold, F., Zablotskaya, K., Diab, A., Minker, W.: Application of verbal intelligence in dialog systems for multimodal interaction. In: 2014 International Conference on Intelligent Environments (IE), pp. 361–364. IEEE, New York (2014)
30. Nothdurft, F., Richter, F., Minker, W.: Probabilistic human-computer trust handling. In: 15th Annual Meeting of the Special Interest Group on Discourse and Dialogue, p. 51 (2014)
31. Orphanou, K., Keravnou, E., Moutiris, J.: Integration of temporal abstraction and dynamic Bayesian networks in clinical systems. A preliminary approach. In: Jones, A.V. (ed.) 2012 Imperial College Computing Student Workshop, OpenAccess Series in Informatics (OASIcs), vol. 28, pp. 102–108. Schloss Dagstuhl–Leibniz-Zentrum fuer Informatik, Dagstuhl (2012). doi:http://dx.doi.org/10.4230/OASIcs.ICCSW.2012.102
32. Orphanou, K., Stassopoulou, A., Keravnou, E.: Temporal abstraction and temporal Bayesian networks in clinical domains: a survey. Artif. Intell. Med. **60**(3), 133–149 (2014). doi :http://dx.doi.org/10.1016/j.artmed.2013.12.007
33. Papai, T., Kautz, H., Stefankovic, D.: Slice normalized dynamic Markov logic networks. In: Pereira, F., Burges, C., Bottou, L., Weinberger, K. (eds.) Advances in Neural Information Processing Systems, vol. 25, pp. 1907–1915. Curran Associates, Red Hook (2012)
34. Pearl, J.: Probabilistic Reasoning in Intelligent Systems: Networks of Plausible Inference. Morgan Kaufmann, San Francisco (1988)
35. Pearl, J.: Causality: Models, Reasoning and Inference, vol. 29. Cambridge University Press, Cambridge (2000)
36. Rafols, E., Koop, A., Sutton, R.S.: Temporal abstraction in temporal-difference networks. In: Weiss, Y., Schölkopf, B., Platt, J. (eds.) Advances in Neural Information Processing Systems, vol. 18, pp. 1313–1320. MIT Press, Cambridge (2006)
37. Reichenbach, H., Reichenbach, M.: The Direction of Time. Philosophy (University of California (Los Ángeles)). University of California Press, Berkeley (1991)
38. Ren, H., Xu, W., Zhang, Y., Yan, Y.: Dialog state tracking using conditional random fields. In: Proceedings of the SIGDIAL 2013 Conference, pp. 457–461. Association for Computational Linguistics, Metz (2013)
39. Reuter, S., Dietmayer, K.: Pedestrian tracking using random finite sets. In: Proceedings of the 14th International Conference on Information Fusion, pp. 1–8 (2011)
40. Richardson, M., Domingos, P.: Markov logic networks. Mach. Learn. **62**(1–2), 107–136 (2006)
41. Sallans, B., Hinton, G.E.: Reinforcement learning with factored states and actions. J. Mach. Learn. Res. **5**, 1063–1088 (2004)
42. Schüssel, F., Honold, F., Weber, M.: Using the transferable belief model for multimodal input fusion in companion systems. In: Multimodal Pattern Recognition of Social Signals in Human-Computer-Interaction, pp. 100–115. Springer, Berlin (2013)
43. Smets, P., Kennes, R.: The transferable belief model. Artif. Intell. **66**(2), 191–234 (1994)

44. Sontag, E.D.: Mathematical Control Theory: Deterministic Finite Dimensional Systems, vol. 6. Springer, New York (1998)
45. Sutton, R.S., Precup, D., Singh, S.: Between MDPs and semi-MDPs: a framework for temporal abstraction in reinforcement learning. Artif. Intell. **112**(1), 181–211 (1999)
46. Sutton, C., McCallum, A., Rohanimanesh, K.: Dynamic conditional random fields: factorized probabilistic models for labeling and segmenting sequence data. J. Mach. Learn. Res. **8**, 693–723 (2007)
47. Theocharous, G., Kaelbling, L.P.: Approximate planning in POMDPs with macro-actions. In: Thrun, S., Saul, L., Schölkopf, B. (eds.) Advances in Neural Information Processing Systems, vol. 16, pp. 775–782. MIT Press, Cambridge (2004)
48. Von Neumann, J., Morgenstern, O.: Theory of Games and Economic Behavior. Princeton University Press, Princeton (1944)
49. Williams, J.D., Poupart, P., Young, S.: Factored partially observable Markov decision processes for dialogue management. In: 4th Workshop on Knowledge and Reasoning in Practical Dialog Systems, International Joint Conference on Artificial Intelligence (IJCAI), pp. 76–82 (2005)
50. Young, S., Gasic, M., Thomson, B., Williams, J.D.: POMDP-based statistical spoken dialog systems: a review. Proc. IEEE **101**(5), 1160–1179 (2013)

Chapter 3
Model-Based Frameworks for User Adapted Information Exploration: An Overview

Michael Kotzyba, Tatiana Gossen, Sebastian Stober, and Andreas Nürnberger

Abstract The target group of search engine users in the Internet is very wide and heterogeneous. The users differ in background, knowledge, experience, etc. That is why, in order to find relevant information, such search systems not only have to retrieve web documents related to the search query but also have to consider and adapt to the user's interests, skills, preferences and context. In addition, numerous user studies have revealed that the search process itself can be very complex, in particular if the user is not providing well-defined queries to find a specific piece of information, but is exploring the information space. This is very often the case if the user is not completely familiar with the search topic and is trying to get an overview of or learn about the topic at hand. Especially in this scenario, user- and task-specific adaptations might lead to a significant increase in retrieval performance and user experience. In order to analyze and characterize the complexity of the search process, different models for information(-seeking) behavior and information activities have been developed. In this chapter, we discuss selected models, with a focus on models that have been designed to cover the needs of individual users. Furthermore, an aggregated framework is proposed to address different levels of information(-seeking) behavior and to motivate approaches for adaptive search systems. To enable Companion-Systems to support users during information exploration, the proposed models provide solid and suitable frameworks to allow cooperative and competent assistance.

M. Kotzyba (✉) • T. Gossen • A. Nürnberger
Data and Knowledge Engineering Group, Faculty of Computer Science, Otto von Guericke University Magdeburg, Magdeburg, Germany
e-mail: michael.kotzyba@ovgu.de; tatiana.gossen@ovgu.de; andreas.nuernberger@ovgu.de

S. Stober
Machine Learning in Cognitive Science Lab, Research Focus Cognitive Sciences, University of Potsdam, Potsdam, Germany
e-mail: sstober@uni-potsdam.de

S. Biundo, A. Wendemuth (eds.), *Companion Technology*, Cognitive Technologies,
DOI 10.1007/978-3-319-43665-4_3

3.1 Introduction

In the last few decades the amount of digitally stored data and information has increased enormously. Especially the Internet-based platforms such as forums, encyclopedias, home pages, social networks and multimedia portals have acquired more and more information due to their popularity; and vice versa, i.e. due to their popularity, the platforms have acquired continuously more information. The Internet (and its traffic) continuously grows with an increasing rate[1] [8, 31] and a corresponding number of users. Rainie and Shermak [38] listed the top Internet activities of Americans in 2005. After email activities, with 77%, the usage of search engines, with 63%, was the one second most frequently carried out Internet activity. In 2012, the study of Jean et al. [21] showed that the most stated information behavior of people who frequently use the Internet is reading and searching, with 69.1%. Furthermore, the most stated intentions to use the Internet were to keep up-to-date (40.4%) and to gather data (35%).

In addition to the growing availability and consumption of information, the target group of Internet users is very wide and heterogeneous. The users differ in cultural background, knowledge, experience, etc. This leads to high requirements for the systems and tools that provide information access, such as search engines. In order to find relevant information, search systems not only have to retrieve web documents related to the search query in general but also have to consider and adapt to the user's interests, skills, preferences and context. Current web search engines only partially provide such features, e.g. by considering the location, previously used search queries or already visited result pages, to adapt query suggestions or the search result set. But a holistic and trustworthy model for user-adapted information exploration that also adapts to the user's skills and utilizes further context information is still missing. In addition, search engines should go beyond the simple fact-finding paradigm and should support the user during the search process itself.

Numerous user studies revealed that the search process can be very complex [12–14, 21, 22, 25, 42], in particular if the user has difficulties specifying his or her information needs. This is often the case if a user has only little knowledge about the search domain or the search tool that is utilized. Here the user tries to get an overview and learn about the topic or tools. If the search domain is not well known, the user's search behavior during the information acquisition tends to be more exploratory than directly searching and navigating for already known sources. To support exploratory search behavior, a holistic user description in the background is necessary. There are different abstract models that reflect the user's search process and the user's information behavior in general [1–3, 12, 23, 25, 45]. To provide adequate search assistance during the information exploration, search systems can benefit from such abstract user models. An appropriate user model in the background allows a system to adapt to the user's interests, preferences, and

[1]According to http://www.internetlivestats.com/total-number-of-websites/, accessed: 29 July 2015.

needs, and hence might lead to a significant increase in retrieval performance and user experience.

For *Companion*-Systems as addressed in this book, an essential property is the ability to support the user during the current task(s) with consideration of the context and the user's individual characteristics. With the growing availability of information the associated task to find desired information also emerges more frequently. Simultaneously the task gets more complex. Especially if the user needs to perform an information exploration, e.g. because of the inability to specify the information need, adequate assistance is necessary. A holistic and adaptive information behavior model can provide a solid and suitable framework for *Companion*-Systems and support users during information exploration.

This chapter presents a theoretical perspective on the user's information exploration process and provides suggestions for adaptive search systems. It is structured as follows: The next section describes the need of an individual to discover his or her own environment as an elementary process to enable learning and to perform actions to satisfy an emerging information need. The information need often serves as a source for different information(-seeking) behavior that is discussed in Sect. 3.3. Furthermore, the section addresses information activities like searching, browsing as well as information exploration. Information behavior models describe the user's search process on a macroscopic level; information-seeking models and information activities describe a meso- and microscopic perspective. Section 3.4 aggregates the different levels and describes different suggestions for adaptive search systems to provide adequate support and assistance for the search process as an essential part of a *Companion*-System. The last section concludes this chapter and provides an outlook for prospective adaptive search systems.

3.2 The Need for Information

Any (biological) creature is inherently driven by the elementary and essential need to discover the surrounding environment (cf. [4], Sect. *Summary and Conclusion*). Originally the process is triggered by an instinct to guarantee the very first (unconscious) learning processes. During the discovery, an individual accumulates information of the environment by observing and interacting. The gathered information is stored, processed and abstracted to build patterns that manipulate/influence the individual's behavior (i.e. to learn) for the following exploration steps and situations. This iterative process enables the individual to understand the environment (which the individual itself belongs to), to derive meaning/associations and knowledge and to execute (and plan for) following actions. With a sufficient amount of experience the individual can perform more and more conscious actions, e.g. to satisfy particular needs.

In Maslow's classic hierarchy of needs [30], the physiological and safety needs and implicitly all actions to satisfy these needs build the fundamental layer for an individual. Once these basic two layers are satisfied, individuals turn towards the

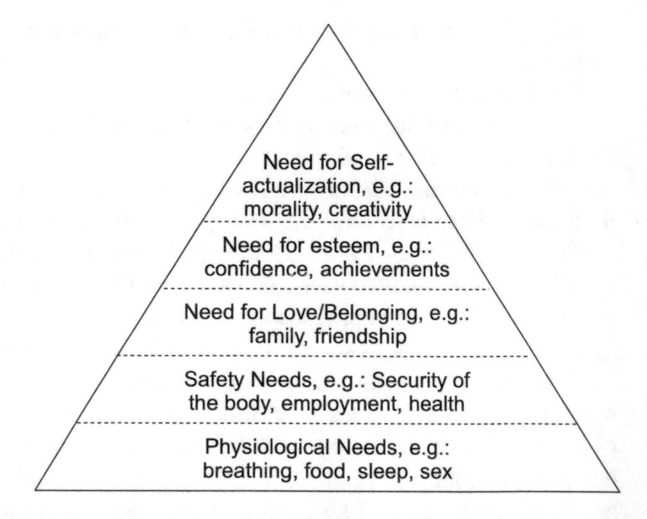

Fig. 3.1 Illustration of Maslow's hierarchy of needs according to [30]

next higher ones, the social and psychological needs; cf. Fig. 3.1. In a developed society these high-level needs play a more and more important role. With the degree of integration of data- and knowledge bases and social components (e.g. by social networks) in particular, the need for and the value of information (e.g. to satisfy the need for esteem or love/belonging) increases. Also, the majority of models describing the information behavior of a user in general are based on the assumption that the user is driven by the demand to satisfy an information need. In the following section different information(-seeking) behavior models are discussed, to provide an overview and to lay out the basic principles for the adaptive framework proposed in Sect. 3.4.

3.3 Models for Information Behavior and Activities

Information behavior models are the most general approach to describing a user during information acquisition and exploration. There is a huge variety of models that addresses different levels and aspects of information behavior. Wilson [45] summarized several models, possible extensions and (first) approaches of aggregation. An alternative overview of different models can be found in Knight and Spink [23]. Models for information(-seeking) behavior describe the user's search process on a high level of abstraction. Therefore, all systems that serve as a source for information (such as search engines) are in the following also called information systems. Furthermore, the addressed models assume that the emerging information need is always internally motivated by the user.

3.3.1 Hierarchy of Models

The models of information behavior are complementary rather than contradictory and build a hierarchy of concepts; cf. Fig. 3.2. *Information behavior* models in general build the framework to categorize a user attempting to satisfy an information need, including context information about the user, possible dialog partners and the system. All methods describing a user who is conducting a search to discover and gain access to information sources, i.e. all used strategies and tactics, are covered by *information-seeking behavior* models. The models of information-seeking behavior are included in the framework of information behavior and hence can be considered as a subset. If a model is rather an instance or a class of information-seeking behavior (e.g. exploratory search), then it is often categorized as *information activity*. The last and most concrete subset of the hierarchy builds the *information search behavior* (sometimes also called *information searching behavior*). Here, all directly conducted interactions with the information system (e.g. mouse, keyboard or touchscreen actions, eye movements or speech controls) and its components such as query input and result visualization or comparison are addressed. This hierarchy allows us to categorize the following information behavior models and their different levels.

3.3.2 Information Behavior

According to Wilson [45], information behavior is defined as follows:

> "By information behaviour is meant those activities a person may engage in when identifying his or her own needs for information, searching for such information in any way, and using or transferring that information." (p. 249)

This description covers all actions to gather information and to satisfy the information need of a user. Wilson published his first model back in 1981 [43]; cf. Fig. 3.3. In this model, the user recognizes an information need and starts with information-seeking behavior on different formal or informal information sources. Alternatively the user can seek information exchange with other people. If successful, the user may use the gained information to further refine his or her information-seeking behavior or to transfer it to other people. Furthermore, successfully gained information may be used to evaluate the current state of satisfaction and to (re-)formulate the (new) information need.

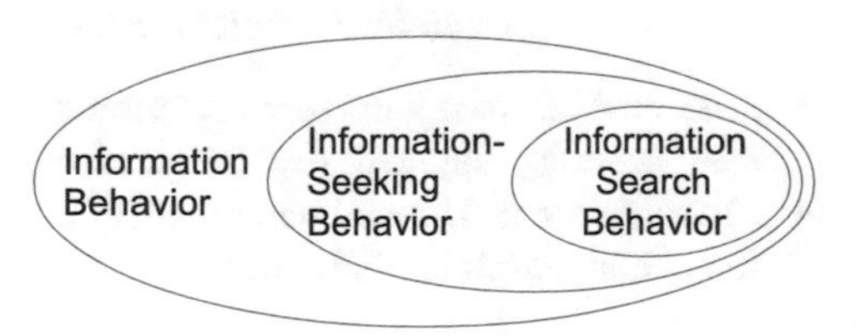

Fig. 3.2 Wilson's nested model of information behavior areas [45]

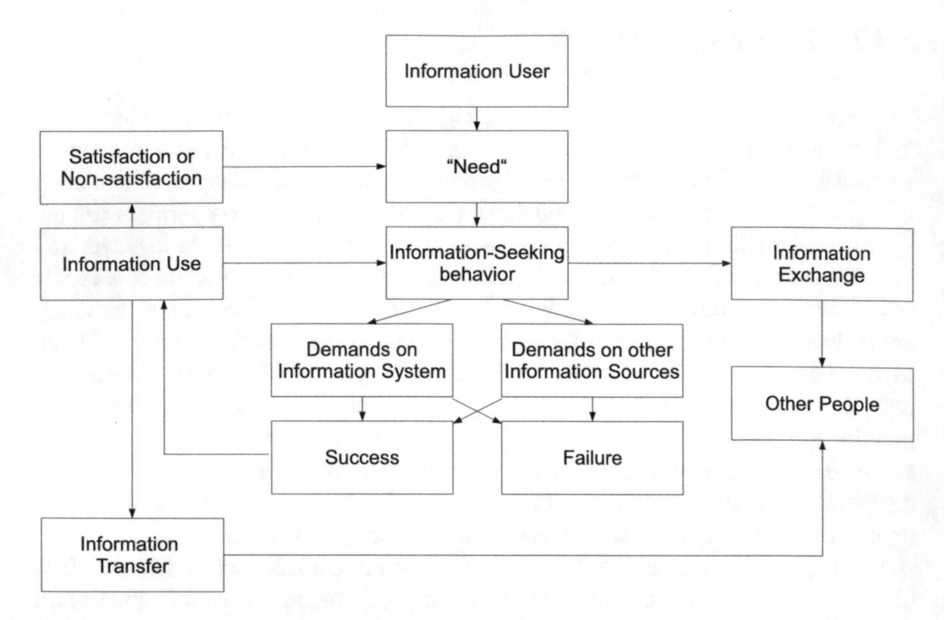

Fig. 3.3 Wilson's (first) model of information behavior from 1981

The diagram in Fig. 3.3 gives first insights into the process of a user's information behavior but it is rather a representation of involved fields and activities than a model to derive the user's states or to illustrate the relationship in between. That is why in 1996 an improved version was published [44, 46]; cf. Fig. 3.4. This revision gives a more detailed view on stages around the information-seeking behavior and hence information behavior in general. Especially the intervening variables, influencing the information-seeking behavior, play an important role for the search process and should be part of the consideration of adaptive information systems to provide an appropriate guidance.

Psychological and demographic variables represent the user's personal characteristics, preferences and skills. Role-related variables become crucial if the information system has to switch from a single- to a multi-user scenario. Environmental variables help to derive the physical situation of the user and his or her context. Source characteristics are addressed if different information sources are used. A more detailed description of the variables, their influence on the search process and the importance for adaptive information systems as part of a *Companion*-System is given in Sect. 3.4.

With respect to information acquisition, the model in Fig. 3.4 furthermore distinguishes between four different modes:

- *Passive Attention*: Passively consuming/acquiring information without any conscious, strong cognitive resonance. Here no information seeking is intended.
- *Passive Search*: Here all parenthetical but relevant information acquired, besides the actual search topic, is addressed.

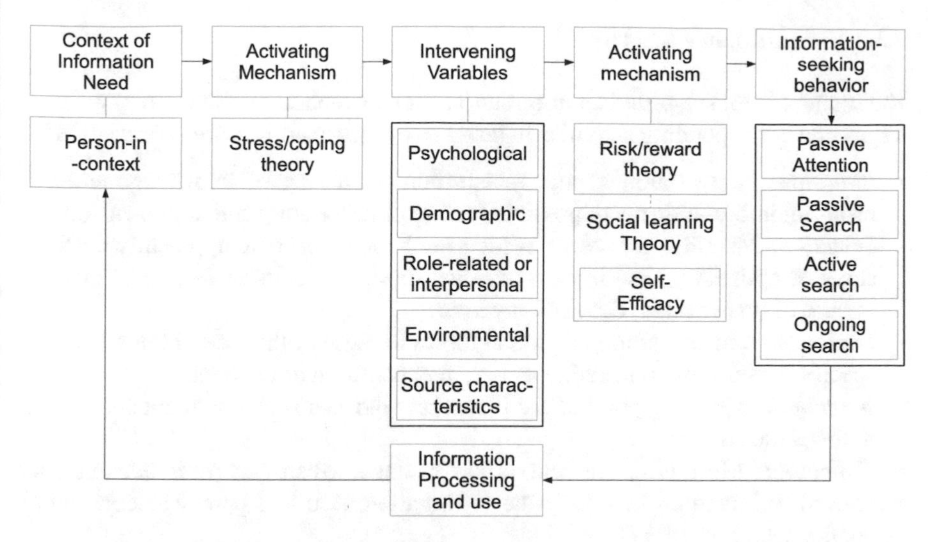

Fig. 3.4 Wilson's extended version of information behavior from 1996

- *Active Search*: Active, conscious search and seeking behavior to satisfy the present information need.
- *Ongoing Search*: This mode is related to all motives to build knowledge for future information behavior, e.g. in the context of consumer research, the knowledge for future purchases [6].

While the active search mode is manifested in observable interactions between the user and the information system (e.g. mouse clicks or eye movements), the other modes have to be estimated by more advanced methods related to implicit attention.

3.3.3 Information-Seeking Behavior

As illustrated in Sect. 3.3.1, information-seeking behavior is integrated within the general framework of information behavior as a (major) component and hence builds the first step from a macro- to a mesoscopic view on the user's search process. Here cognitive mechanisms to bring together the requirements of the information need and specific performed (inter-)actions with the information system are addressed. Nevertheless, the models can focus on different aspects. In the following, two strongly empirically supported information-seeking models are presented that are complementary.

3.3.3.1 Kuhlthau's Model

Kuhlthau [25, 26] suggested an information-seeking model (or rather a list) with six stages and corresponding activities in terms of an information search process (ISP):

- *Initiation*: First awareness and recognition for a lack of knowledge and the resulting information need; possible feeling of uncertainty and apprehension.
- *Selection*: Identification of relevant search domains that apparently lead to success; optimistic view in case of quick, positive results or feeling of anxiety in case of (unexpected) delay of any kind.
- *Exploration*: Investigation of general topics because of the lack of knowledge to specify the information need; user may feel confused or uncertain.
- *Formulation*: Turning point of the ISP, where the user feels confident and is able to focus the search.
- *Collection*: Most effective and efficient stage, where the user accumulates relevant information to satisfy the information need and perceives continuing confidence.
- *Presentation*: Retrospective evaluation of the search process to estimate the satisfaction; if satisfied, the feeling of relief occurs.

This model describes the information acquisition from the point of view of the user's feelings, thoughts and actions and hence additionally has a phenomenological perspective. The six stages as a whole suggest an internal order; however, the order between the selection and exploration stages is not binding. If the user has a promising seed of knowledge, he or she can directly perform the selection and afterwards explore the results of the information system. Otherwise, first investigations are necessary by exploring different general topics and information resources to enable the selection stage afterwards.

3.3.3.2 Ellis' Model

Ellis et al. [12–14] discussed a model of different behavior involved in information-seeking that is empirically supported by studies with scientists, e.g. physicists, chemists or social scientists. The eight categories are termed *features* instead of *stages* as in Kuhlthau's model:

- *Starting*: all activities that initiate an information acquisition, such as asking other people or typing a query into a search engine and identifying a first key document.
- *Chaining*: building chains of (relevant) documents by following references such as hyperlinks (on result pages of search engines), citations, footnotes or other index structures.
- *Browsing*: performing a semi-directed search or exploration in a promising domain.
- *Differentiating*: differentiation between several information sources to exploit their particular/specific characteristics as a filter.

- *Monitoring*: keeping awareness of developments in a (search) domain.
- *Extracting*: identifying specific pieces of information that apparently lead to success.
- *Verifying*: checking the retrieved information.
- *Ending*: all activities that complete an information acquisition, such as validating or using the new gained information.

Ellis' model does not explain the connections between the features. However, Wilson [45] suggested a relation between the features and in addition enriched the model with the stages and activities of Kuhlthau to complement both models; cf. Sect. 3.4.

3.3.4 Search, Explore and Browse for Information

The models illustrated in Sects. 3.3.2 and 3.3.3 embed the user in the context of information acquisition initiated by a recognized information need. Nevertheless, the models do not specify the actions a user is performing by interacting with an information system, especially during Kuhlthau's selection and exploration stages. This section addresses different information activities such as searching and browsing and connects them to the paradigm of information-seeking.

3.3.4.1 Exploratory Search

The way a user interacts with the system during the search process strongly depends on the user's experience with technical systems and on the user's knowledge about the present search domain. An expert for a specific search domain is able to define his or her information need precisely. The more experience the user has with the system, in particular with the user interfaces, the more precisely the information need can be formulated as input query. In contrast to that, if the user has little knowledge about the search domain at first, the user can only formulate vague input queries and has to explore the domain.

Section 3.2 described the discovery and exploration of the individual's environment as a very fundamental but also general process to learn and hence to satisfy (basic) needs. In the research area of Information Retrieval the term *exploration* or *exploratory search* often is embedded into the context of a search process with a vague information need and is considered as a class of information-seeking. A discussion about different meanings of exploration can be found in [17]. Furthermore, exploration is described as a stage of Kuhlthau's model (c.f. Sect. 3.3.3).

Marchinonini's Search Activities

Marchionini addresses the term *exploratory search* [29] and integrates it into a framework of different search activities; cf. Fig. 3.5. Here, exploratory search extends a standard lookup(-search) with the activities learning (cf. Sect. 3.2) and investigation. Lookup can be understood as a standard fact-finding search with a specified query request that leads to Kuhlthau's selection stage. Thus search is an elementary, conscious and purposeful action to satisfy a need for (a specific piece of) information. But to execute or plan that kind of action, it is necessary to have knowledge about the search domain the individual is acting in, and knowledge about the available means the individual can utilize. Learning and investigation are both iterative processes that involve different search strategies.

On the one hand, Machinonini's description is abstract enough to bring exploratory search and information-seeking behavior in line. On the other hand, it is specific enough to address information activities. Hence exploratory search merges querying and browsing strategies.

Three Dimensions of Exploratory Search

During the exploration process it is important to provide an appropriate support for the user and to assist him or her during the discovery process in new domains. If the user searches for topics that differ from his or her common interests, the system at least can consider this indication as a possible demand on exploration. For example, to get an overview, some types of information sources like encyclopedias

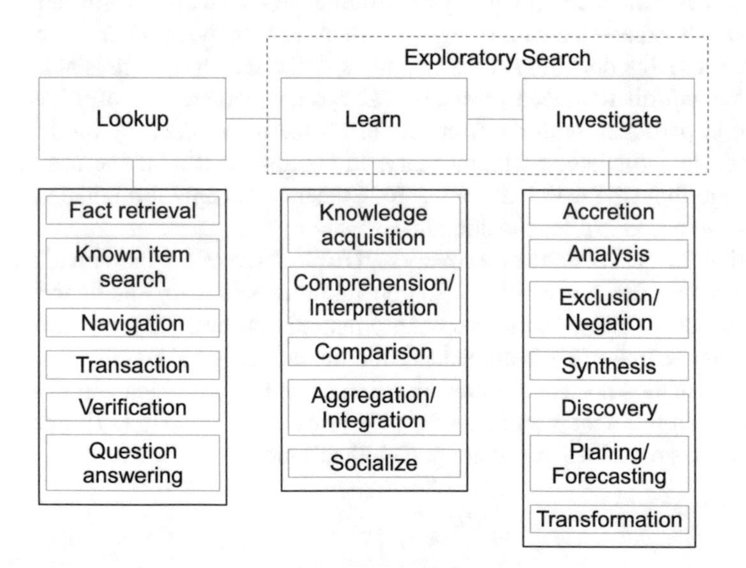

Fig. 3.5 Marchinonini's exploratory search embedded in search activities

can be more helpful than others, such as specific topics discussed in forums, and hence the information system could ask the user whether the result ranking should be influenced. To support the user during the exploration, Noël proposes a model that describes the characteristics of exploratory search by a three-dimensional model [34, 35]:

- *Vertical Axis*: The user can change the level of focus for the relevant information (sub-)space.
- *Horizontal Axis*: The user can differentiate (cf. Ellis' features in Sect. 3.3.3) between the retrieved results to identify those that best match the current information need. On the horizontal axis the user furthermore derives information about the focus level, cf. vertical axis.
- *Transversal Axis*: The user can change the perspective on the retrieved information pieces to identify the relation(s) in between.

The three axes state that information exploration comes along with three tasks, namely to adjust the focus, to differentiate between the focus levels and to change the perspective. Furthermore, the three axes can also be used for the design of search user interfaces [33]. An illustration of the three-dimensional character of exploratory search is given in Fig. 3.6. The user can explore the information space related to domain X, can change the focus levels on the vertical axis (i.e. the level of detail), or can differentiate between the sub-domains of domain X on the horizontal axis. On the transversal axis the user can switch between different domains (e.g. domain X and Y) to change the perspective to the same or similar piece of information. For example, methods of clustering are often used in the domain of scientific data analysis. The complexity of the methods or the details of discussions about the methods are related to the vertical axis. Clustering methods in the domain of data analysis also can be divided into sub-domains (cf. horizontal axis), which are represented by different web results. Furthermore, methods of clustering are also

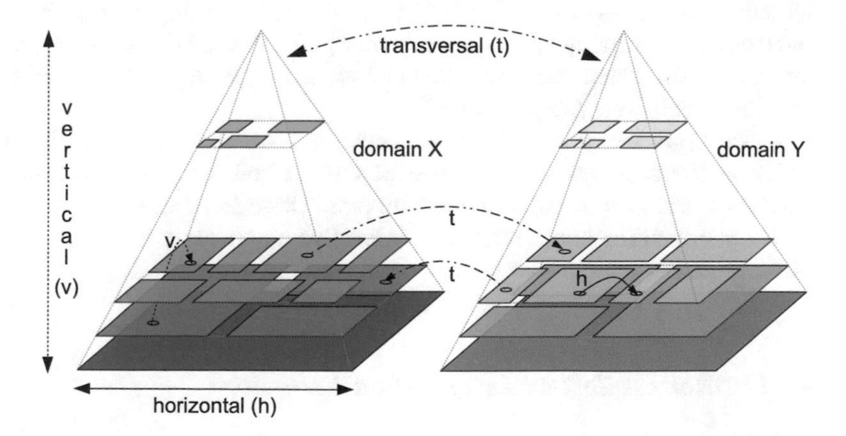

Fig. 3.6 Illustration of the three-dimensional character of exploratory search according to [34, 35]

used to structure retrieved (web) results in the domain of interactive information systems, e.g. to provide an overview for the user. By switching between the two domains the transversal axis is addressed. The same or similar clustering method can be used in both domains for different application areas with different purposes.

3.3.4.2 Browse and Pick Information

In information science the term *browsing* has been discussed long and exhaustively [18, 27, 39]. However, there is still no clear definition of what browsing means. Bates [4] accumulates different approaches and proposes browsing as an episode of four steps, whereby each episode does not necessarily need to contain all steps:

- *Glimpsing*: Browsing consists (at least) of glimpsing at the field of vision, abandoning it and glimpsing again.
- *Selecting or Sampling*: Selecting a physical or informational/representational object within the field of vision.
- *Examining*: Examining the object.
- *Acquiring the Object*: Physically or conceptually acquiring the object or abandoning it.

Furthermore, browsing can be seen as an instance or a behavioral expression of exploratory behavior [4]. Ellis also emphasized browsing as an important part of information seeking [12]; cf. Sect. 3.3.3.

During information exploration, the user retrieves new pieces of information, bit by bit. On the one hand, the new information influences the degree of satisfaction of the current information need. On the other hand, the information need can change, e.g. be specified, or be moved to a different topic. That is, the search is not static and can evolve over time. This process is described by Bates [3] and called *Berry Picking*; cf. Fig. 3.7. The Information Space is usually high dimensional but in Fig. 3.7 only two dimensions are illustrated for simplicity. During the search, it is possible that a former query variation was closer to the desired document or piece of information than a later query variation and hence a loop way is performed (cf. retrieved document(s) via $Q(t_3)$ and $Q(t_5)$).

Bates also emphasized the difference between browsing and berry-picking. Berry-picking characterizes information search as a whole with evolving queries and provides a wide variety of search techniques. These techniques can be more related to a standard (lockup-)search, or can involve a number of browsing steps. That is, berry-picking can include (steps of) the browsing process.

3.3.5 Further Models on Information Behavior

This section gave a rough overview about selected information(-seeking) behavior models good frameworks for adaptive information systems. However, the list of

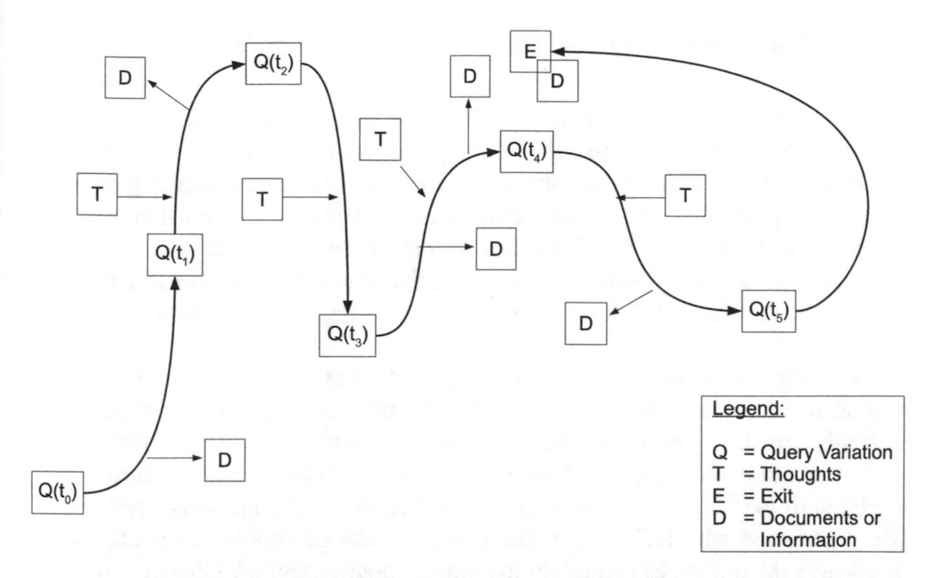

Fig. 3.7 Bates' Berry Picking Model [3] to illustrate a sequence of search behavior. The *arrows* describe the user's search path and are time-correlated

presented models is not complete. More models and theories which address the topic from different perspectives, e.g. more user- or search process-oriented, can be found: Dervin's Sense-Making theory [11], Ingwersen's cognitive model [20] and Belkin's model of information-seeking episodes [5]. Besides the illustration of exploratory search [29], Marchionini also investigated a model-based perspective on information seeking in electronic environments [28]. In addition, related work on information search activities and the information retrieval process can be found in Saracevis's [40] and Spink's [41] models.

3.4 Aggregation of Information-Seeking Models and Implications for Adaptive Information Exploration

As already mentioned, the models described in Sect. 3.3 are rather complementary than contradictory and provide a helpful framework for a holistic user description in the background of adaptive information systems. In the following, a combination of two information-seeking behavior models is illustrated. Afterwards, a framework that aggregates important components of information(-seeking) behavior models is proposed to describe crucial aspects for adaptive information systems. The framework links the theoretical aspects of information behavior with the interactions as defined by technical variables and states and thus closes the semantic gap between the two worlds. This allows us to develop *Companion*-Systems that are responsive to the user's search process, adapt to the user's personal characteristics and needs and hence improve the user's search experience.

3.4.1 Combination of Kuhlthau's and Ellis' Models

According to Wilson's argumentation [45], the stages of Kuhlthau's model [25, 26] support Ellis' features [12–14] and the suggested relation in between the features; cf. Sect. 3.3.3. Furthermore, Kuhlthau's model can be considered as slightly more general and phenomenological than Ellis' features. The combination of both models allows us to look at information-seeking from different perspectives. In addition the fusion provides a helpful infrastructure for approaches related to adaptation and affirms the implied development over time. In Fig. 3.8 the combined model is illustrated.

The initiation of information-seeking behavior starts with recognizing an information need. During this very first step of information-seeking, a system that carefully considers environmental variables can help to introduce the user to the information acquisition. Afterwards, the user continues with the next stages, selection or exploration. These stages depend on the user's age and competencies, like experience with technical systems and knowledge about the search domain and hence the ability to formulate the search input query; cf. Gossen and Nürnberger [15]. The user performs a mixture of browsing through and differentiating between different information sources as well as following and monitoring promising results. After the formulation and collection stage, the search process goes into a semi-structured and confident phase. During this phase, the user extracts and gathers promising information sources for verification. An information system that considers the source characteristics may contribute to the search performance because it makes it easier for the user to estimate the quality of the received results. In the last stage of information-seeking, the user is completing the search or the exploration to validate whether the information need is satisfied by the found sources and, if necessary, to proceed.

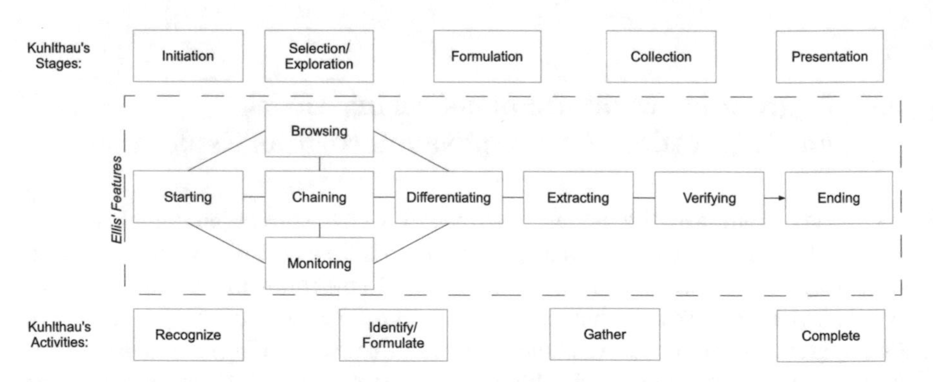

Fig. 3.8 Wilson's aggregation of Kuhlthau's and Ellis' models [45]

3.4.2 *An Aggregated Framework for Adaptive Information Systems*

In the following, an aggregated framework is proposed that describes the intervening variables (cf. Sect. 3.3.2, Fig. 3.4) and different aspects of information-seeking behavior. Furthermore, an adaptive system core utilizes different user and context data to provide an individualized information exploration.

The framework is illustrated in Fig. 3.9. On the left-hand side the user and his or her personal characteristics in a social and physical contexts are depicted. To satisfy the information need, the user interacts with the adaptive information system, in particular with the user interface. Here the user performs information-seeking behavior on different levels. This information acquisition contains explicit interactions such as typing in a query or clicking on results, but also implicit interactions such as differentiating or verifying information sources (cf. Sect. 3.4.1). Furthermore, system commands or controls, like user-adjusted system settings or feedback, can be considered as explicit interaction that represents a portion of the user's preferences. A system also considering the physical environment of the user and the physiological constitution of the user him- or herself is able to derive more implicit feedback. All interactions, user characteristics and sensory input, e.g. mimic (cf. Chap. 18), voice (cf. Chap. 20) or eye movements are accumulated in the system core and processed by the adaptation logic to provide appropriate result ranking, visualization, etc. After the adaptation and the system's output, the user continues with the search and/or evaluates the degree of satisfaction for his or her information need. In the following sections crucial aspects and intervening variables for the framework are described. While several variables such as environment or source

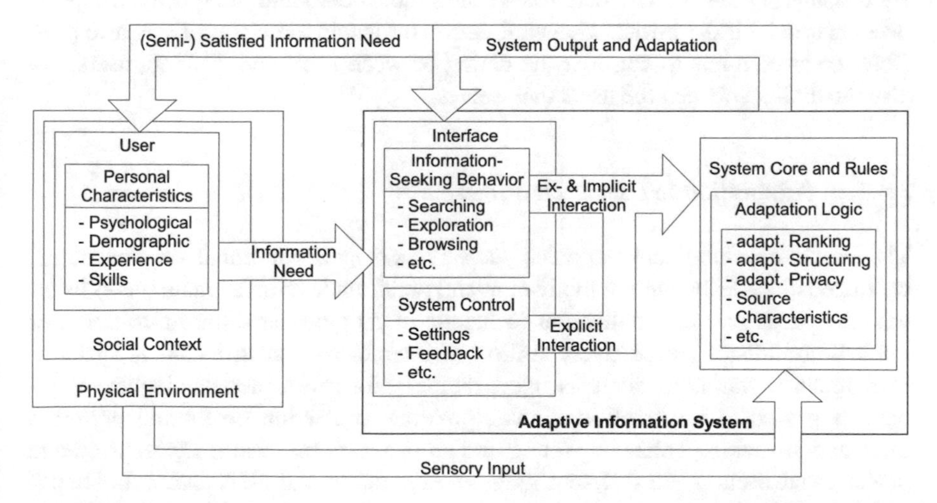

Fig. 3.9 Aggregated information model-based framework to illustrate crucial aspects for adaptive information systems

characteristics can be immediately recorded by technical sources, psychological or demographic variables have to be queried from other sources by asking the user or have to be derived over several search steps or even sessions.

3.4.2.1 Consideration of Context

Environmental variables are related to the physical context of the user. As mentioned in the previous subsection in Wilson's aggregation of Kuhlthau's and Ellis' models (Sect. 3.4.1) for the initiation of information-seeking behavior, a system can help to introduce the user to the information acquisition. Location, time and the surrounding circumstances allow us to adapt the results, e.g. the type of information source or content, to maximize the probability of the user's satisfaction. For example, a user at a private location (e.g. at home), may prefer information more related to hobbies or personal interests. The same user on vacation or while traveling may rate information about traffic options, the current city or hotels higher, and during shopping it becomes more relevant to get additional information about specific products or to compare prices in online offerings. Furthermore, the location allows us to adapt the degree of privacy such as by showing or hiding personal related information on the user interface that may be seen by other people. Of course the user him- or herself also belongs to the environmental context. For the initiation step the system can consider the user's appearance such as mimic or gesture. If the user appears to be stressed, e.g. because he or she wants to catch the right train, the system should try to restrict the given information to the most important. In addition, the reading style of the user [37], recorded via eye-tracking, can provide hints to the user's time restrictions or the search strategy and thus allow for adaptation. By considering the context also, role-related (i.e. social) and interpersonal aspects become crucial if the information exploration turns into a multi-user scenario [36]. Here the system has to consider the dialog between itself and different users, but also the dialog between the users themselves.

3.4.2.2 Adaptation to User's Characteristics

The user's psychological variables such as personal, emotional or educational characteristics, designate the motivation and persistence to investigate (new) information spaces and the motivation to handle stereotype-matching information or the self-confidence (respectively ability) to interact with such technical systems. Demographic variables such as age provide information about adequate numbers of presented information chunks, possible interaction errors and preferred information-seeking behavior. Young and adult users, but also inexperienced and professional users, need different kinds of adaptation and assistance (cf. Chap. 4 and [15, 16]). For example, young users have difficulties with typing [7] and tend to formulate informational queries while adults' most frequent queries are navigational [16]. Furthermore, young users have less expert knowledge [19] and the emotional/affective state often plays a large role in children's information-seeking behavior [9, 10, 24, 32]. To provide facilitated interaction for the selection or exploration stage (cf. Fig. 3.8), especially psychological and demographic variables and the consideration of different system components become more relevant. For example, by typing the input query, the information system can provide query suggestions related to the user's characteristics and current context to state the information need more precisely. In addition, the result ranking can be influenced by considering the result properties such as text complexity, topics, number of images, website type (e.g. forum, encyclopedia, social network, etc.) or age-based content. If the user's search is information-oriented the results should focus more on the information itself. In contrast to that, a user who is interested in a navigation-oriented search may prefer more resource-focused results to obtain the desired specific website or text document immediately. Furthermore, an adaptive system can parametrize the preferred result presentation form, such as the number of results per search engine result page, or the information about the found sources.

3.4.2.3 Consideration of Source Characteristics

An information system that considers the source characteristics (as one of the intervening variables) may contribute to the search performance because it makes it more easy for the user to estimate the quality of the received results. For example, if the user knows that the retrieved information is from a more reliable, trusted or objective source such as Wikipedia, a double-check is less necessary. If it is important for the user to retrieve the most recent information from a domain (e.g. state of the art) or an event (e.g. news), the source characteristic update time is crucial. For a young user or a user who is interested in understanding a (new) domain in detail, the readability is essential. Furthermore, a system with a reliable evaluation on the source characteristics can increase the confidence of the user in the system.

3.5 Conclusion

To satisfy the user's need for information, advanced methods and systems are necessary. On the one hand, information systems have to access a huge amount of unstructured data. On the other hand, they need to consider a huge variety of users. Users differ in background, knowledge, experience, interests, skills, preferences and context. Furthermore, the search process itself can be very complex, in particular if the user has only little knowledge about the search domain. This chapter presented the description of the user's information exploration process and presented different information(-seeking) behavior models as potential frameworks for user models in the background. In the last section an aggregated framework was proposed to address different levels of information(-seeking) behavior and to describe suggestions for adaptive search systems to provide adequate support and assistance for current tasks during the search process. Here, in particular the adaption to the user's personal characteristics, and the consideration of physical context and source characteristics, turned out to be crucial aspects for such adaptive information systems.

Acknowledgements This work was done within the Transregional Collaborative Research Centre SFB/TRR 62 "*Companion*-Technology for Cognitive Technical Systems" funded by the German Research Foundation (DFG).

References

1. Al-Suqri, M.N., Al-Aufi, A.S.: Information Seeking Behavior and Technology Adoption: Theories and Trends. IGI Global, Hershey (2015)
2. Bates, M.J.: Information search tactics. J. Am. Soc. Inf. Sci. **30**(4), 205–214 (1979)
3. Bates, M.J.: The design of browsing and berrypicking techniques for the online search interface. Online Inf. Rev. **13**(5), 407–424 (1989)
4. Bates, M.J.: What is browsing-really? a model drawing from behavioural science research. Inf. Res. **12**(4) paper 330, 1–22 (2007)
5. Belkin, N.J., Cool, C., Stein, A., Thiel, U.: Cases, scripts, and information-seeking strategies: on the design of interactive information retrieval systems. Expert Syst. Appl. **9**(3), 379–395 (1995)
6. Bloch, P.H., Sherrell, D.L., Ridgway, N.M.: Consumer search: an extended framework. J. Consum. Res. **13**, 119–126 (1986)
7. Budiu, R., Nielsen, J.: Usability of websites for children: Design guidelines for targeting users aged 3–12 years, 2nd edn. Nielsen Norman Group Report. (2010)
8. Coffman, K.G., Odlyzko, A.M.: Growth of the internet. In: Kaminow, I.P., Li, T. (eds.) Optical Fiber Telecommunications IV B: Systems and Impairments, pp. 17–56. Academic Press, San Diego (2002)
9. Cooper, L.Z.: A case study of information-seeking behavior in 7-year-old children in a semistructured situation. J. Am. Soc. Inf. Sci. Technol. **53**(11), 904–922 (2002)
10. Cooper, L.Z.: Methodology for a project examining cognitive categories for library information in young children. J. Am. Soc. Inf. Sci. Technol. **53**(14), 1223–1231 (2002)
11. Dervin, B.: An overview of sense-making research: concepts, methods and results to date. In: International Communications Association Annual Meeting (1983)
12. Ellis, D.: A behavioral approach to information retrieval system design. J. Doc. **45**(3), 171–212 (1989)
13. Ellis, D., Haugan, M.: Modelling the information seeking patterns of engineers and research scientists in an industrial environment. J. Doc. **53**(4), 384–403 (1997)
14. Ellis, D., Cox, D., Hall, K.: A comparison of the information seeking patterns of researchers in the physical and social sciences. J. Doc. **49**(4), 356–369 (1993)
15. Gossen, T., Nürnberger, A.: Specifics of information retrieval for young users: a survey. Inf. Process. Manag. **49**(4), 739–756 (2013)
16. Gossen, T., Low, T., Nürnberger, A.: What are the real differences of children's and adults' web search? In: Proceedings of the 34th International ACM SIGIR Conference on Research and Development in Information, pp. 1115–1116. ACM, New York (2011)
17. Gossen, T., Nitsche, M., Haun, S., Nürnberger, A.: Data exploration for bisociative knowledge discovery: a brief overview of tools and evaluation methods. In: Bisociative Knowledge Discovery. Lecture Notes in Computer Science, pp. 287–300. Springer, Heidelberg (2012)
18. Herner, S.: Browsing. Encyclopedia of Library and Information Science, vol. 3, pp. 408–415. Marcel Dekker, New York (1970)

19. Hutchinson, H., Druin, A., Bederson, B.B., Reuter, K., Rose, A., Weeks, A.C.: How do I find blue books about dogs? The errors and frustrations of young digital library users. In: Proceedings of HCII 2005, pp. 22–27 (2005)
20. Ingwersen, P.: Cognitive perspectives of information retrieval interaction: elements of a cognitive ir theory. J. Doc. **52**(1), 3–50 (1996)
21. Jean, B.S., Rieh, S.Y., Kim, Y., Yang, J.Y.: An analysis of the information behaviors, goals, and intentions of frequent internet users: findings from online activity diaries. First Monday **17** (2012). https://doi.org/10.5210/fm.v17i2.3870, ISSN: 1396-0466
22. Johnson, J.D., Meischke, H.: A comprehensive model of cancer-related information seeking applied to magazines. Hum. Commun. Res. **19**(3), 343–367 (1993)
23. Knight, S.A., Spink, A.: Toward a web search information behavior model. In: Spink, A., Zimmer, M. (eds.) Web Search. Information Science and Knowledge Management, vol. 14, pp. 209–234. Springer, Berlin (2008)
24. Kuhlthau, C.C.: Meeting the information needs of children and young adults: basing library media programs on developmental states. J. Youth Serv. Libr. **2**(1), 51–57 (1988)
25. Kuhlthau, C.C.: Inside the search process: Information seeking from the user's perspective. J. Am. Soc. Inf. Sci. **42**(5), 361–371 (1991)
26. Kuhlthau, C.C.: Seeking Meaning: A Process Approach to Library and Information Services. Ablex Publishing, Norwood, NJ (1994)
27. Kwasnik, B.H.: A descriptive study of the functional components of browsing. In: Proceedings of the IFIP TC2/WG2. 7 Working Conference on Engineering for Human Computer Interaction, p. 191 (1992)
28. Marchionini, G.: Information Seeking in Electronic Environments. Cambridge University Press, Cambridge (1997)
29. Marchionini, G.: Exploratory search: from finding to understanding. Commun. ACM **49**(4), 41–46 (2006)
30. Maslow, A.H.: A theory of human motivation. Psychol. Rev. **50**(4), 370 (1943)
31. Murray, B.H., Moore, A.: Sizing the internet. White Paper, Cyveillance, p. 3 (2000)
32. Nesset, V.: Two representations of the research process: the preparing, searching, and using (PSU) and the beginning, acting and telling (BAT) models. Libr. Inf. Sci. Res. **35**(2), 97–106 (2013)
33. Nitsche, M., Nürnberger, A.: Trailblazing information: an exploratory search user interface. In: Yamamoto, S. (ed.) Human Interface and the Management of Information. Information and Interaction Design. Lecture Notes in Computer Science, vol. 8016, pp. 230–239. Springer, Berlin (2013)
34. Noël, L.: From semantic web data to inform-action: a means to an end. In: Workshop: Semantic Web User Interaction, CHI 2008, 5–10 April, Florence, pp. 1–7 (2008)
35. Noël, L., Carloni, O., Moreau, N., Weiser, S.: Designing a knowledge-based tourism information system. Int. J. Digit. Cult. Electron. Tour. **1**(1), 1–17 (2008)
36. Nürnberger, A., Stange, D., Kotzyba, M.: Professional collaborative information seeking: on traceability and creative sensemaking. In: Cardoso, J., Guerra, F., Houben, G.J., Pinto, A., Velegrakis, Y. (eds.) Semantic Keyword-based Search on Structured Data Sources. First COST Action IC1302 International KEYSTONE Conference, IKC 2015, Coimbra, September 8–9, 2015, Revised Selected Papers. Lecture Notes in Computer Science, vol. 9398. Springer, Berlin (2015)
37. O'Regan, J.K.: Optimal viewing position in words and the strategy-tactics theory of eye movements in reading. In: Eye Movements and Visual Cognition, pp. 333–354. Springer, New York (1992)
38. Rainie, L., Shermak, J.: Search engine use November 2005. Technical Report, PEW Internet & American Life Project (2005)
39. Rice, R.E., McCreadie, M., Chang, S.L.: Accessing and Browsing Information and Communication. MIT Press, Cambridge (2001)
40. Saracevic, T.: Modeling interaction in information retrieval (IR): a review and proposal. In: Proceedings of the ASIS Annual Meeting, vol. 33, pp. 3–9. ERIC (1996)

41. Spink, A.: Study of interactive feedback during mediated information retrieval. J. Am. Soc. Inf. Sci. **48**(5), 382–394 (1997)
42. Weiler, A.: Information-seeking behavior in generation Y students: motivation, critical thinking, and learning theory. J. Acad. Librariansh. **31**(1), 46–53 (2005)
43. Wilson, T.D.: On user studies and information needs. J. Doc. **37**(1), 3–15 (1981)
44. Wilson, T.D.: Information behaviour: an interdisciplinary perspective. Inf. Process. Manag. **33**(4), 551–572 (1997)
45. Wilson, T.D.: Models in information behaviour research. J. Doc. **55**(3), 249–270 (1999)
46. Wilson, T.D., Walsh, C.: Information Behaviour: An Interdisciplinary Perspective. The British Library, London (1996)

Chapter 4
Modeling Aspects in Human-Computer Interaction: Adaptivity, User Characteristics and Evaluation

Tatiana Gossen, Ingo Siegert, Andreas Nürnberger, Kim Hartmann, Michael Kotzyba, and Andreas Wendemuth

Abstract During system interaction, the user's emotions and intentions shall be adequately determined and predicted to recognize tendencies in his or her interests and dispositions. This allows for the design of an *evolving search user interface* (ESUI) which adapts to changes in the user's emotional reaction and the users' needs and claims.

Here, we concentrate on the front end of the search engine and present two prototypes, one which can be customised to the user's needs and one that takes the user's age as a parameter to roughly approximate the user's skill space and for subsequent system adaptation. Further, backend algorithms to detect the user's abilities are required in order to have an adaptive system.

To develop an ESUI, user studies with users of gradually different skills have been conducted with groups of young users. In order to adapt the interaction dialog, we propose monitoring the *user's emotional state*. This enables monitoring early detection of the user's problems in interacting with the system, and allows us to adapt the dialog to get the user on the right path. Therefore, we investigate methods to detect changes in the user's emotional state.

We furthermore propose a *user mood modeling* from a technical perspective based on a mechanical spring model in PAD-space, which is able to incorporate

---

T. Gossen (✉) • A. Nürnberger • M. Kotzyba
Data & Knowledge Engineering Group, Otto von Guericke University, 39016 Magdeburg, Germany
e-mail: tatiana.gossen@ovgu.de; andreas.nuernberger@ovgu.de; michael.kotzyba@ovgu.de

I. Siegert • K. Hartmann
Cognitive Systems Group, Otto von Guericke University, 39016 Magdeburg, Germany
e-mail: ingo.siegert@ovgu.de; kim.hartmann@ovgu.de

A. Wendemuth
Cognitive Systems Group, Otto von Guericke University, 39016 Magdeburg, Germany

Center for Behavioral Brain Sciences, 39118 Magdeburg, Germany
e-mail: andreas.wendemuth@ovgu.de

S. Biundo, A. Wendemuth (eds.), *Companion Technology*, Cognitive Technologies,
DOI 10.1007/978-3-319-43665-4_4

several psychological observations. This implementation has the advantage of only three internal parameters and one user-specific parameter-pair.

We present a technical implementation of that model in our system and evaluate the principal function of the proposed model on two different databases. Especially on the EmoRecWoz corpus, we were able to show that the generated mood course matched the experimental setting.

By utilizing the user-specific parameter-pair the *personality trait extraversion* was modeled. This trait is supposed to regulate the individual emotional experiences.

Technically, we present an implementable feature-based, dimensional model for emotion analysis which is able to *track and predict the temporal development of emotional reactions* in an evolving search user interface, and which is adjustable based on mood and personality traits.

4.1 Introduction

In this chapter, we are going to discuss modeling aspects in human-computer interaction. To be more precise, we focus on a specific scenario, namely, information search. The possibility to acquire new information is an important functionality of *Companion*-Systems. If the potential user requires explanation of some previously unknown facts and this information is not available in the knowledge base of a *Companion*-System directly, then the *Companion*-System can gather it from various sources like the Web. Through a dialog with a user the *Companion*-System clarifies the user's intention, eliminates any ambiguities and guides the user towards successful information gain.

Search systems are an integral part of our lives. Most common known search systems come in the form of web search engines with an audience of hundreds of millions of people all over the world. This is a very wide and heterogeneous target group with different backgrounds, knowledge, experiences, etc. Therefore, researchers suggest providing a customized solution to cover the needs of individual users, e.g. [15]. Nowadays, solutions in personalization and adaptation of backend algorithms have been proposed in order to support the search of an individual user. These solutions include query adaptation, adaptive retrieval, adaptive result composition and presentation [38, 39]. But the front end, i.e. the search user interface (SUI), is usually designed and optimized for a certain user group and does not support personalization.

Common search engines allow the personalization of a SUI in a limited way: Users can choose a color scheme or change the settings of the browser to influence some parameters like font size. Some search engines also detect the type of device the user is currently using—e.g. a desktop computer or a mobile phone—and present an adequate user interface (UI). Current research concentrates on designing SUIs for specific user groups, e.g. for children [11, 15, 22] or elderly people [3, 4]. These SUIs are optimized and adapted to general user group characteristics. However, especially young and elderly users undergo fast changes in cognitive, fine motor and other abilities. Thus, design requirements change rapidly as well and a flexible modification of the SUI is needed.

Another important, currently under-emphasized aspect is the adaptation of the interaction dialog with the user. Different users require different dialog characteristics, e.g. a simple language should be used in dialogs with children. Moreover, the same user can have different cognitive and emotional states during the interaction with the system. Thus, modeling the user's emotional development during an interaction with a system could be a first step towards a representation of the user's inner mental state (cf. [19]) and therefore a step towards an adaptive interaction. This provides the possibility to evolve a user interaction model and gives the opportunity to predict the continuous development of the interaction from the system's perspective. As stated in [27], moods reflect medium-term affects, generally not related to a concrete event. They last longer and are more stable than emotions and influence the user's cognitive functions directly. Furthermore, the user's mood can be influenced by emotional experiences. These affective reactions can be measured by a technical system. In this case, the mood can be technically seen as a long-time integration over the occurring emotional events, to dampen their strength. User emotions and dispositions were considered in information retrieval before, however, mostly only for relevance assessment [2] or as additional features, for instance, in collaborative filtering [28, 29].

To have a customized solution to cover the needs of individual users, we suggest an users with an evolving user interface that adapts to providing individual user's characteristics and not only allows for changes in properties of UI elements but also influences the UI elements themselves and the human computer interaction dialog.

4.2 Model of an Evolving User Interface

We exploit a generic model of an adaptive system based on [39] and propose the model of an evolving user interface (EUI) as follows (see Fig. 4.1): In general, we suggest designing a mapping function and adapting the UI using it, instead of

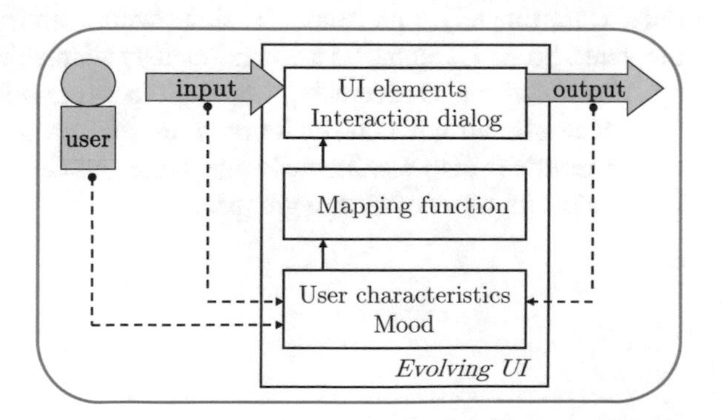

Fig. 4.1 Model of an evolving user interface based on [39]. Users' mood and characteristics are determined based on the information about the user as well as their interactions with search input and result output (see *dashed lines*)

building a UI for a specific user group. We have a set of user characteristics or skills and emotional states on one side. We suggest considering cognitive skills, information processing rates, fine motor skills, different kinds of perception, the knowledge base, emotional state, and reading and writing skills. In the ideal case, the system detects the characteristics automatically, e.g. based on the user's interaction with the system. In information retrieval we can, for example, use the user's queries and selected results for detection. The user's age can also be utilized as a fuzzy indicator of his or her skills. On the other side, there is a set of options to adapt the UI and the interaction dialog. In information retrieval, we can adapt the UI using different UI elements for querying or visualization of results. In between, an adaptation component contains a set of logic rules to map the user characteristics to the specific UI elements of the evolving user interface.

When designing an EUI, we first have to define the components of a UI that should be adapted. As different systems may have different UI designs, we will further discuss the user interface of a search engine (SUI). An overview of different SUI elements is given in [20]. Here, we consider three main components. The first component is the search *input*, i.e. UI elements which allow a user to transform his information need into a textual format. This component is traditionally represented by an input field and a search button. Other variants are a menu with different categories or voice input. The second component is the result *output* of an information retrieval (IR) system. The output consists of UI elements that provide an overview of retrieved search results. This component is traditionally represented by a vertical list of results. The third is the *management* component. Management covers UI elements that support users in information processing and retaining. Examples of management UI elements are bookmark management components or history mechanisms like breadcrumb trail.[1] Historically, UI elements for management are not part of a SUI. But recent research [15] shows that young users are highly motivated to use elements such as storage for favorite results. Besides these main components, there also exist general properties of UI elements which might affect all the three categories, e.g. font size or color.

Ideally, an evolving search user interface (ESUI) should be continuously adaptable. Unfortunately, this cannot be done with a high level of granularity for all elements. Some UI elements are continuously adaptable (e.g. font size, button size, space required for UI elements), whereas others are only discretely adaptable (e.g. type of results visualization). Not only are SUI properties, but the complexity of search results is also continuously adaptable and can be used as a personalization mechanism for users of all age groups.

[1]Breadcrumb is a navigation aid that shows a user's location in a website or Web application [20].

4.3 From Customization to Adaptation: Experimental Design

In order to demonstrate the idea of an ESUI, we developed a search engine for children that is age-adaptable. The search engine was developed in two steps.

4.3.1 Step 1: Customization

During the first stage we designed a search user interface, *Knowledge Journey*, that can be customized to the user's needs [15–17]. To achieve a coherent design between different variations of the ESUI, we suggest fixing the positions of different SUI parts. Figure 4.2 depicts the general structure of the developed ESUI. It consists of five groups of elements: a help section, a storage for bookmarked results, visualization of results, elements for keyword search and a menu for navigation. The search input consists of elements for text search and a menu for navigation to provide different possibilities for children to formulate their information need. A menu supports children who have problems with query formulation [15]. We implement parameters having some advantages for children, for example, having a browsing menu, coverflow result visualization, graphical separation of search results, etc. The summary of adaptable elements, their properties and the options that we implemented are given in Table 4.1. In our implementation, we also offer some parameters that, based on the state of the art, are considered to be unsuitable for kids, e.g. having no categories [21]. This makes the changes stand out by switching between different SUIs. Furthermore, the set of options that would be the best for a particular age group is unknown due to the lack of comparative studies. In order to customize the SUI, we implemented a configuration unit that allows users to manipulate the SUI directly. The adaptable elements were ordered according to their decreasing influence, which we have determined, on the entire SUI from theme to

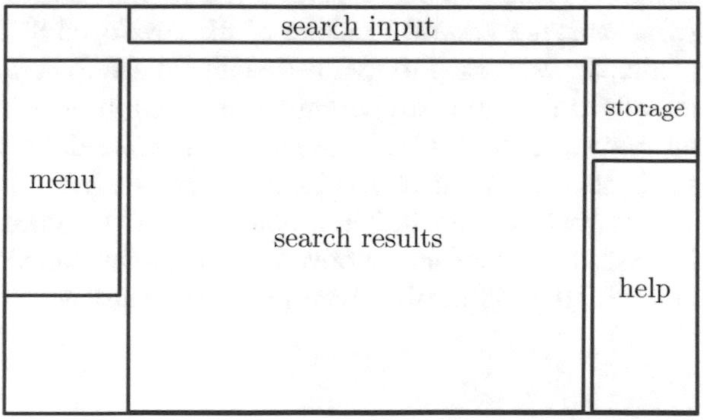

Fig. 4.2 General structure of the developed SUI with search elements, menu, help, storage for bookmarked pages and result visualization (modified from [17])

Table 4.1 Adaptable elements of the implemented ESUI, their parameters and their options [17]

Element	Parameter	Options
Menu	Type	No menu, classic, pie-menu
	Categories	For children, for adults
	Structure	Number of categories, hierarchy depth
Results	Visualization	List, tiles, coverflow
	Number	On a page for coverflow
	Separation	No separation, lines, boxes
	Page view	Preview on/off, same window, same tab, new window
Surrogate	Website picture	On/off
	Thumbnail size	Thumbnail vs. snippet size
	URL	On/off
	Keyword highlighting	Different color, on/off
Font	Size	From 10 pt to 18 pt
	Type	Comic Sans MS, Arial, Times New Roman, Impact
Theme	Type	No theme, different themes
Avatar	Type	No avatar, different avatars
Audio	Active	On/off
	Voice gender	Male, female, girl, boy
	Number of repetitions	Only the first time, twice, always

audio. As the backend, the *Bing Search API*[2] with the safe search option turned on is used.

We conducted a user study in order to derive the mapping function between users with different abilities and the UI elements of a search engine. Adults were chosen as a reference group as their abilities are different [14]. In particular, our hypothesis was that users from different age groups would prefer other UI elements and different general UI properties. We will apply our findings to offer default SUI settings.

The general evaluation procedure was as follows. A pre-interview was conducted to obtain the participants' demographic data and their Internet experience. Afterwards the general structure of the developed SUI was explained. Then each participant was asked to try the system out and to perform a free search. However, initially child-unfriendly settings were used in the SUI. There was no menu and no theme, impact font of size 10 pt was selected, no picture was provided in the surrogate, etc. We also visualized results as a list with no separation of items. A search result was opened in a new window. We chose these settings in order to increase the participants' motivation to configure the SUI and also for changes to be more striking. During this stage, participants also got more familiar with the system.

[2]https://datamarket.azure.com/dataset/5BA839F1-12CE-4CCE-BF57-A49D98D29A44.

Table 4.2 Default children and adults settings for an ESUI found in the user study[17]

	Children	Adults
Result visualization	Coverflow	Tiles
Website preview	Add	Add
URL	Add	Add
Result page view	User choice	New tab
Font size	14 pt	12 pt
Font type	Comic Sans MS	Arial
Menu type	User choice	Classic menu
Menu categories	For children	For adults
Number of categories	As many as possible	As many as possible
Audio support	On	Off

In the next step of the evaluation, all the configuration options were introduced. To provide a better overview for some options, like font type or result visualization, we prepared a printed sheet where all the options for each UI element or property could be seen at once. This made it easier for participants to be aware of all options. Using the configuration unit, each participant went through all adaptable elements, starting from those that had the strongest influence on the whole SUI, like theme, then selecting the result set visualization and customizing a surrogate,[3] etc. At the end, the participant was able to select whether to turn on the voice and, if so, to customize voice gender and the number of times to repeat the voice explanation.

After a participant customized his own SUI to his preferences, a search task was given. This step was designed for subjects actually to use their own created SUI. Afterwards, they were given the possibility to change SUI settings. Our search task was gender-independent, and it could be solved in a reasonable amount of time using a menu or a keyword search. We asked the participants to find out how many moons the planet Uranus has. In the last step, a post-interview was conducted to gather the user's opinion about the proposed ESUI. Each test session lasted about 30 min.

In order to conduct a user study with children, we collaborated with a trilingual international primary school in Magdeburg, Germany. Our evaluation was done using 17" displays which were kindly provided by the school director. Adults were recruited from an academic context and tested the SUI in a lab. Forty-four subjects participated in the study, 27 children and 17 adults. The children were between 8 and 10 years old (8.9 on average), 19 girls and 8 boys from third (18 subjects) and fourth (9 subjects) grades. The adults were between 22 and 53 years old (29.2 on average), 5 women and 12 men.

To summarize, we found differences in preferences between adults and young users. Table 4.2 depicts the main finding of the user study. Pupils chose a coverflow result view, whereas adults preferred to use a tiles view. Both children (78%) and

[3] A document surrogate summarizes important information about the document for users to be able to judge its relevance without opening the document.

adults (100%) chose to add the URL to a document surrogate. Eighty-two percent of the adults and 41% of the children preferred to open the results in a new tab. Those children were already familiar with tab functionality. Thirty-seven percent of the children preferred to open the results in the same window and 22% in a new window. All the pupils and 82% of the adults wished to have a menu in addition to the text input. However, there was no clear choice regarding the menu type made by children. Ninety-three percent of the adult users chose adult topics for the menu. Ninety-two percent of the children chose topics meant for them. However, many adults wished to select the menu topics by themselves. Both children and adults wanted to have as many menu categories as possible. Sixty-seven percent of the children chose to have the audio option on, whereas only 6% of the adults found this option to be useful. There was a clear tendency regarding font and font size. Fifty-six percent of young participants chose Comic Sans MS and 76% of the adults preferred Arial. The majority of pupils selected a 12 pt (37%) or 14 pt (48%) font size. Adults preferred font size of 10 or 12 pt. They said that they could read text in 10 pt. However, they chose 12 pt as there was enough place within the UI for larger texts. Based on the findings of the user study, initial settings for the ESUI can be determined.

4.3.2 Step 2: Adaptation

During the second stage we developed the *Knowledge Journey Exhibit (KJE)* [18] that implements the *Knowledge Journey* as an information terminal device and has an age-adaptable SUI. The previous version had a configuration window where a user was able to customize the SUI. However, we considered this window to be too difficult for children to operate without supervision and in a public place. Therefore, a decision was made to replace the configuration window with a slider where each point on the slider corresponded to a SUI configuration for a specific age starting with configuration for young children and ending with a setting for young adults (cf. Fig. 4.3). We use the age parameter to adapt the SUI. At the beginning a user is asked to input his or her age. Then, the user is forwarded to the corresponding search user interface, where he can explore other settings of the SUI using the slider. The settings for the slider were derived based on the results from the user study described in Table 4.2. The settings for young children are a pirate theme, coverflow result visualization, and large font size in Comic Sans MS. The settings for young adults are no theme, tiles result visualization, and smaller font size in Arial. The search results for adults contain twice as much text in summaries and smaller thumbnails. Each point on the slider changes one of the setting parameters, e.g. the font.

The system provides spelling correction after the query is submitted and suggestions for the term the user is currently typing. In addition, users can bookmark the relevant search results using the storage functionality. We used a star symbol that was added to the result surrogate to indicate if the search result is already bookmarked. Users can click directly on the star symbol to bookmark

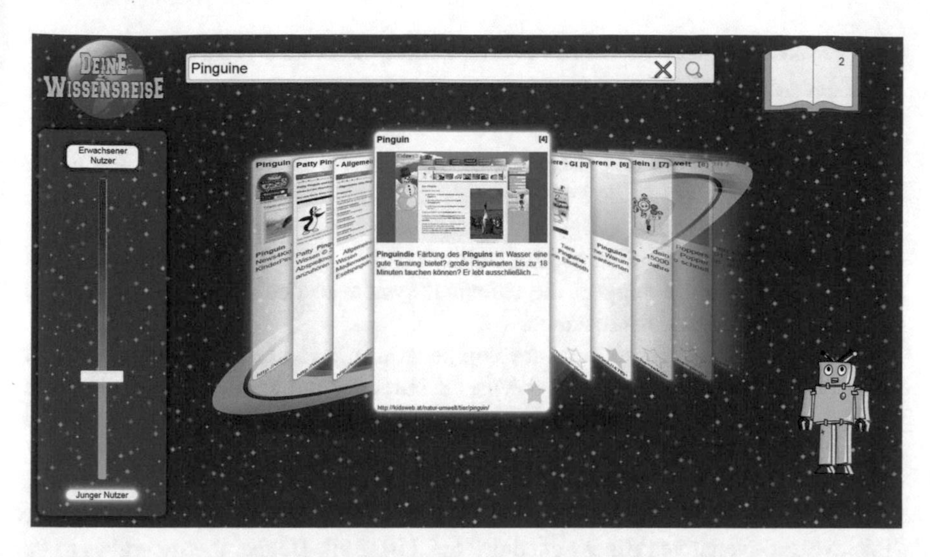

Fig. 4.3 User interface of Knowledge Journey Exhibit. An example query for Pinguine (German for penguins) and the corresponding results are shown. Two of the shown search results marked with a *yellow star* are bookmarked. The user can adapt the SUI using the slider on the left side

or unbookmark the result or they can place the search result in the storage using drag-and-drop. They can review the stored results, which are grouped by the issued query, in order to provide more context information.

Furthermore, we use information about the web page complexity that is calculated using the *Flesch-Reading-Ease* (FRE) readability index for German language [1]. We applied a traffic light metaphor and visualized each search result that is easy to understand in a green frame, while a search result that is hard to understand is visualized in a red frame, with varying levels of color in between. The traffic light metaphor is also applied to the slider.

So far, we have developed a search engine prototype with adaptable elements of the search user interface. Another aspect of the interface is the interaction dialog. It is also beneficial to have a means of adaptation for the interaction dialog between the system and the user. An important parameter for the dialog adaptation is the user's emotional state, because it makes it possible to predict the probability of the user's success during the interaction with the system. In the case of a low probability, the system can adapt the dialog with the user in order to get the user on the right track. To make this adaptation possible, approaches to continuously monitor the user's emotional state are required. Therefore, in the following we present our research about mechanisms to detect fluctuations caused by the user's emotional experiences.

4.4 Tracking Temporal Development of Users' Emotional Reactions

The mood specifies the actual feeling of the user and is influenced by the user's emotional experiences. As an important fact for human-computer-interaction (HCI), moods influence the user's cognitive functions, behavior and judgements, and the individual (creative) problem solving ability (cf. [27, 30]). Thus, knowledge about the user's mood could support the technical system to decide whether additional assistance is necessary, for instance.

Modeling the user's emotional development during an interaction could be a first step towards a representation of the user's inner mental state. This provides the possibility to evolve a user interaction model and gives the opportunity to predict the continuous development of the interaction from the system's perspective. However, it is not known how a user's mood can be deduced directly without utilizing labelling methods based on questionnaires, for instance "Self Assessment Manikins" or "Positive and Negative Affect Schedule" (cf. [10, 26]). Hence, the mood has to be modeled either based on observations or using computational models. The modeling described in this section is based of the following publications: [33, 35, 36].

To date, only limited research deals with the problem of mood modeling for human-computer-interaction. In [13], the Ortony, Clore and Collins's model of emotions is implemented (cf. [31]), outputting several co-existing emotions, where the computed emotions are afterwards mapped into the PAD-space. The mood is derived by using a mood change function in dependence of the computed emotion center of all active emotions and their averaged intensity. The direction of the mood change is defined by a vector pointing from the PAD-space origin to the computed emotion center. The strength of change is defined by the averaged intensity. Additionally, the authors utilize a time-span defining the amount of time the mood change function needs to move a current mood from one mood octant center to another. A mood simulation presented in [5] also relies on precomputed emotional objects located in the PAD-space. In contrast to [13], this model used the `valence` dimension to change the mood value. Thus, this model does not locate the mood within the PAD-space. Furthermore, in this model a computed emotional `valence` value results in a pulled mood adjusted by a factor indicating the "temperament" of an agent. A spring is then used to simulate the reset force to decrease steadily until neutrality is reached. Both mood models are used to equip virtual humans with realistic moods to produce a more human-like behavior.

4.4.1 Mood Model Implementation

Starting from the observation of the mood as a quite inert object within the pleasure-arousal-dominance (PAD)-space, a mood modeling algorithm is presented enabling technical systems to model mood by using emotional observations as input values. Therefore, the following behavior will be modeled:

- mood transitions in the PAD-space are caused by emotions
- single emotional observations do not directly change the mood's position
- repeated similar emotional observations facilitate a mood transition in the direction of the emotional observation
- repeated similar emotional observations hinder a mood transition in the opposite direction
- incorporation of the personality trait `extraversion` to adjust the emotional observation

For the approach presented in this chapter, the mood modeling is independent of the source of emotional assessment. It can be derived either from a computational model based on appraisals, as it is described here, or implicitly from observed emotional reactions of the user.

Both the observation of single short-term affects (the emotions) as well as the recognized mood will be located within the PAD-space in the range of $[-1, 1]$ for each dimension. This abstract definition of the modeled location by using the PAD-space allows the model to be independent of the chosen observed modality and to have the same representation for the emotional input values.

To illustrate the impact of recognized emotions on the mood, the observed emotion e_t at time t is modeled as the force F_t with the global weighting factor κ_0 (Eq. (4.1)). Furthermore, the emotions e_t are modeled for each dimension in the PAD-space separately. Thus, the calculation of the mood is conducted component-wise [35, 36]. The force F_t is used to update the mood M for that dimension by calculating a mood shift ΔL_M (Eq. (4.2)) utilizing the damping D_t, which is updated by using the current emotion force F_t and the previous damping D_{t-1}. This modeling technique is loosely based on a mechanical spring model: The emotional observation performs a force on the mood. This force is attenuated by a damping term, which is modified after each pulling.

$$F_t = \kappa_0 \cdot e_t \tag{4.1}$$

$$\Delta L_M = \frac{F_t}{D_t} \tag{4.2}$$

$$M_t = M_{t-1} + \Delta L_M \tag{4.3}$$

$$D_t = f(F_t, D_{t-1}, \mu_1, \mu_2) \tag{4.4}$$

The main aspect of this model is the modifiable damping D_t. It is calculated according to Eqs. (4.5) and (4.6). The damping is changed in each step by calculating ΔD_t, which is influenced by the observed emotion force F_t. The underlying function has the behavior of a tanh-function. It has two parameters. The parameter μ_1 changes the oscillation behavior of the function and the parameter μ_2 adjusts the

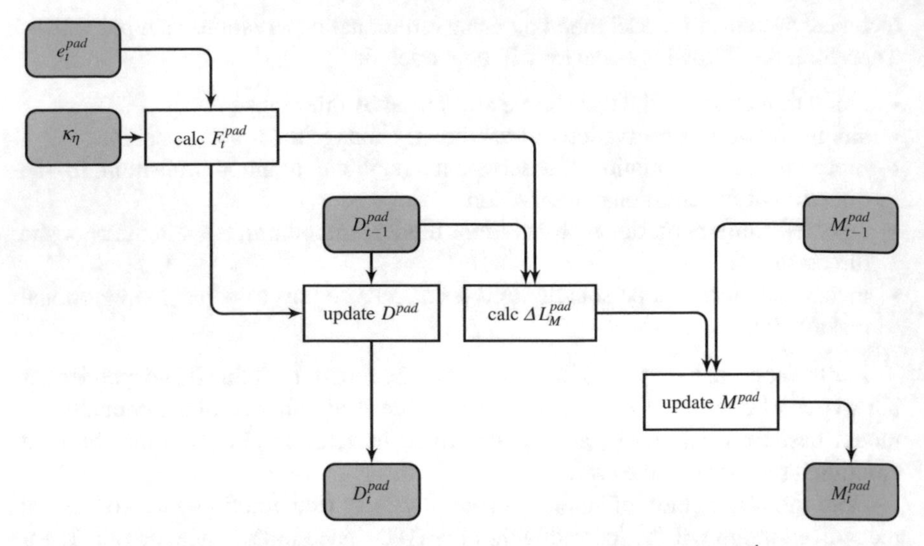

Fig. 4.4 Block scheme of the presented mood model. The *red rounded box* (e_t^{pad}) is an observed emotion, the *blue rounded box* (M_t^{pad}) represents the modeled mood, all other *grey rounded boxes* are inner model values, and *white boxes* are calculations. For simplification the combined components are used; internally the calculation is done for each dimension separately

range of values towards the maximum damping.

$$D_t = D_{t-1} - \Delta D_t \tag{4.5}$$

$$\Delta D_t = \mu_2 \cdot \tanh(F_t \cdot \mu_1) \tag{4.6}$$

The block scheme for the mood modeling is illustrated in Fig. 4.4.

The mood model consists of two calculation paths. The diagonal one calculates the actual mood M_t^{pad}. The vertical one updates the inner model parameter D_t^{pad}. As prerequisite, the emotional force F_t^{pad} is compiled from the observed emotion e_t^{pad}.

The mood is the result of the influence of the user's emotions over time with respect to the previous mood. The impact of the user's emotions on his mood depends also on the user's personality (cf. [33]). The observed (external) affect needs not be felt (internally) with the same strength. For this, the observed emotion have to be translated into an adequate emotional force with respect to the known differences in the external and internal representations. It is known from the literature (cf. [8, 24, 40]) that an external and an internal assessment of the emotional cause lead to different interpretations. Hence, this must be considered in the development of the mood model. For this, the presented model focusses on the emotional intensity as an adjustment factor to determine the difference between an external observation and an internal feeling of the user's emotion. For this case, the personality trait `extraversion` is used to adjust the emotional force (F_t^{pad}).

4.4.2 Generate Continuous Emotional Assessments

The presented modeling technique needs sequences of emotion values to allow a mood prediction. Since this type of data is hardly obtainable, we utilize a method described in detail in [23] to obtain emotional sequences. This method is based on the appraisal theory (cf. [32]) and implements a computational model enriched with fuzzy sets to obtain a continuous emotional evaluation.

In appraisal theory (cf. [32]), it is supposed that the subjective significance of an event is evaluated against a number of variables. The important aspect of the appraisal theory is that it takes into account individual variances of emotional reactions to the same event. This results in the following sequence: event, evaluation, emotional body reaction. The body reactions then result in specific emotions.

Appraisal theories attempt to specify the nature of criteria used in evaluation in terms of different appraisal variables or dimensions. Examples of these dimensions are "goal significance", representing the goals and needs that are high in priority at the moment (e.g. the goals of survival, maintaining social relationships or winning a game), or "urgency", representing the need for an action. In [12] it is suggested that the given appraisal variables allow us to deduce an emotion as the most probable emotional reaction to a certain event. For example, joy or happiness will occur if the appraisal values for the variable "goal significance" are high, while the value for "urgency" is low, etc. To generate a continuous emotional assessment the evaluation process maps from appraisal theory to a fuzzy model in order to derive the emotions of a user in a specific situation or event.

In our implementation ten different appraisal variables are used (cf. Table 29.2 in [12]). To operationalize the appraisal variables and their linguistic values, each appraisal variable is modeled as a fuzzy set (cf. [23]). The fuzzy sets consist of several fuzzy variables, depending on the number of postulated values of the appraisal variable. For example, the appraisal variable "urgency" has five different values ("very low", "low", "medium", "high", "very high") and thus is modeled as a fuzzy set with five corresponding fuzzy variables. Each appraisal variable is mapped to a uniform number in the range from 0 to 100; cf. Fig. 4.5. The corresponding fuzzy variables are distributed uniformly in this range. For simplicity we use a

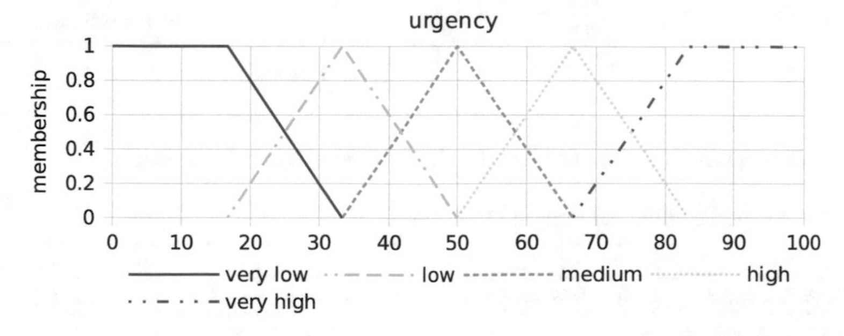

Fig. 4.5 Fuzzification of the appraisal variable "urgency" using a triangle function

triangle function to model a fuzzy variable. Using all ten appraisal variables, we are able to model both the emotional state of the user and a possible evaluation of an arising event in terms of appraisal variables.

In Fig. 4.6 a schematic overview of the framework is given.

According to appraisal theory, the emergence of an event triggers an evaluation process influencing the user's appraisal variables. To model this process, the values of the fuzzyfied appraisal variables of an event are estimated and used to adapt the corresponding variables in the user state. For example, an event with high "goal significance" should positively affect the user's condition for "goal significance". In which manner the events affect the user's state is not trivial because humans interpret events and their relevance differently. In the simplest way, the appraisal variables of an event affect the user's emotional state directly by overwriting any old value with the actual one (cf. left-hand side of Fig. 4.6). As stated above, the appraisal variables can be used to deduce an emotion as the most probable emotional reaction to a certain event. Each emotion is described by appraisal variables with specific values. From this specification a rule base is generated for each emotion (cf. right-hand side of Fig. 4.6). To derive an emotion, the postulated emotion profile and the user state are used to calculate the fulfillment of the emotion rule base. This framework can then be used to gain a continuous emotional assessment. Therefore, we rely on EmoRec-Woz I, a subset of the EmoRec corpus (cf. [41]). It was generated within the SFB/TRR 62 during a Wizard-of-Oz experiment containing audio, video, and bio-physiological data. The users had to play games of concentration (Memory) and each experiment was divided into two rounds with several experimental sequences (ESS); cf. Table 4.3.

The experiment was designed in such a way that different emotional states were induced through positive and negative feedback, wizard responses, and game difficulty levels. The ESS with their expected PAD octants are shown in Table 4.3.

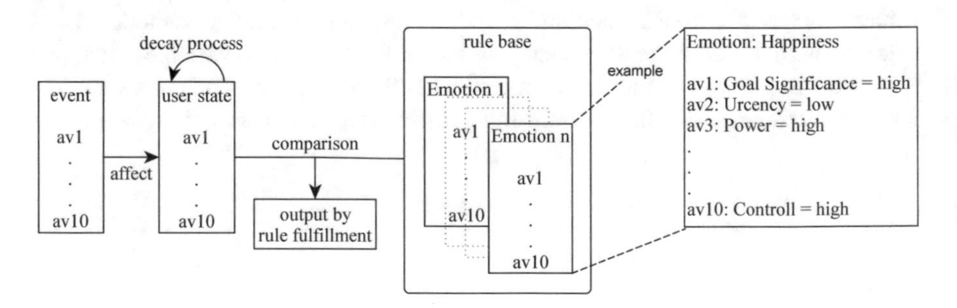

Fig. 4.6 Conceptual block scheme of our fuzzified appraisal-based framework

Table 4.3 Sequence of ES and expected PAD positions

ES	Intro	1	2	3	4	5	6
User's PAD location	All	+ + +	+ − +	+ − +	− + −	− + −	+ − +
Pleasure development	–	↗	↗	↗	→	↓	↑

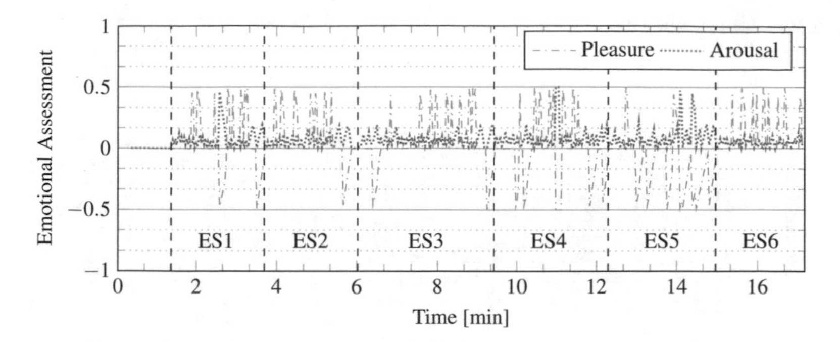

Fig. 4.7 Calculated arousal and valence values of our appraisal based computational model accumulating the output of all fuzzy sets

The octants are indicated by their extreme points. A session has an average length of about 18 min. In the following, the output of the model is exemplary, shown using the experiment of one participant. For more details concerning the data set (exemplarily cf. [41]). To apply the model, all possible events (like a hit or negative feedback) were extracted and all possible evaluations regarding the corresponding appraisal variables were designed using a rule-based approach. For example, turning over a pair of matching cards is considered as "goal significant" while receiving negative feedback is considered as unpleasant. For simplicity it is assumed that an event affects the emotional state of the user directly. Since the appraisal process has a short duration, a decay process can be used, e.g. via stepwise convergence to a "neutral" state (cf. "user state" in Fig. 4.6). At this point it is assumed that a "neutral" user state will be reached if all appraisal variables receive the value "medium".

In Fig. 4.7 the results for the ES1 up to ES6 of one test person are illustrated. In this experiment we only use the rule base to derive pleasure and arousal. The values in the diagram represent the fulfillment of the two emotion rules, calculated by the model, based on the events that occurred in the sessions. In comparison to ES5 the values in ES2 for pleasure are higher and arousal occurs only slightly. In ES5 the values for pleasure are reduced and arousal is increased, which corresponds to the expected pleasure development; cf. Table 4.3.

4.4.3 Experimental Model Evaluation

For the experimental model evaluation, we also rely on EmoRec-Woz I. To model the mood development, we limited the investigation to `pleasure`, as this dimension is already an object of investigation (cf. [6, 7, 41]), which simplifies the comparison of our mood modeling technique with other results. After performing some pre-tests to ensure that the mood remains within the limits of $[-1, 1]$ of the PAD-space, we defined the initial model parameters as given in Table 4.4.

Table 4.4 Initial values for mood model

M_0	D_0	μ_1	μ_2
5	5	0.1	0.1

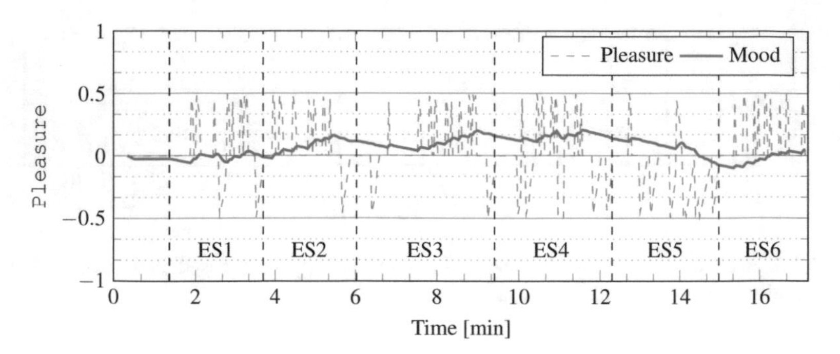

Fig. 4.8 Course of the mood model using the whole experimental session of one participant. The subject's pleasure value is additionally depicted

To gather the emotion data, we rely on the computation of the emotional course based on the appraisal modeling described in Sect. 4.4.2. The same session of one experiment is used for illustrations. This emotional computation serves as input value for the mood modeling. By doing so, it makes it possible to form a mood development over a whole experiment and compare the calculated model with the experimental descriptions, as ground truth, of the complete experiment (cf. Table 4.3). The whole mood development and the division into the single ES are shown in Fig. 4.8. We concentrated on the `pleasure`-dimension, as, for this, secured studies on the EmoRec corpus are available (cf. [6, 41]) showing that the experiment induces an "emotional feeling" that is measurable. Investigations with emotion recognizers using prosodic, facial, and bio-physiological features and the comparison to the experimental design could support the fact that the participants experienced ES2 as mostly positive and ES5 as mostly negative (cf. [41]). The underlying experimental design—the ground truth for the presented mood model—is described as follows: in ES1, ES2, ES3, and ES6 mostly positive emotions; in ES4 the emotional inducement goes back to a neutral degree; In ES5 mostly negative emotions were induced.

Using the computed emotional labels as input data for the modeling, we were able to demonstrate that the mood follows the prediction for the pleasure dimension of the given ES in the experiment (cf. Fig. 4.8). The advantage of this modeling is that the entire sequence can be represented in one course. Furthermore, the influence of a preceding ES on the actual ES is included in the mood modeling.

The resulting mood development is as follows: in the beginning of ES1 the mood rests in its neutral position and it takes some time, until the mood starts to shift towards the positive region. In ES2 and ES3 the mood continues to rise. During ES4 the mood reaches its highest value of 0.204 at 11:33 min. As in this ES the inducement of negative emotions has started, the mood is decreasing afterwards.

The previous induced positive emotions lead to a quite high damping in the direction of a negative mood; thus the mood falls at the end of ES4. In ES5, when more negative emotions are induced due to negative feedback and time pressure, the mood is decreasing quite fast. Shortly after the end of ES5 the mood reaches its lowest value with −0.100 at 15:21 min. Here, it should be noted that negative emotional reactions are observed already in the very beginning of ES5; otherwise the strong decreasing of the mood could not have been observed. The mood remains quite low in the beginning of ES6. During the course of ES6, where many positive emotions were induced, the mood rises again and reaches 0.04 at the end of the experiment (17:07 min).

When including personality traits into the mood model, the following must be given: (1) The personalities of the participants and (2) their subjective feelings. The first prerequisite is fulfilled by EmoRec-Woz I, as the Big Five personality traits for each participant were captured with the NEO-FFI questionnaire (cf. [9]). The personality trait `extraversion` is particularly useful to divide subjects into the groups of users "showing" emotions and users "hiding" emotions (cf. [25]). Additionally, users with `high extraversion` are more stable on positive affects. These considerations lead to a sign-dependent factor to distinguish between positive and negative values for emotional dimensions. Thus, we expand the adjustment factor κ_η (cf. Fig. 4.9).

For this case, κ_η can be modified by choosing different values for κ. We will depict results for values in the range of 0.05–0.4. These values reproduce the strength of how an observed emotion is experienced by the observed person itself. An example of the different values for κ_η is given in Fig. 4.9. For this experiment, emotional traces based on computed emotions for ES2 are used. It can be seen that values higher than 0.3 led to the mood rising too fast. This causes implausible moods, since the upper boundary of 1 for the PAD-space is violated. Hence, a $\kappa_\eta > 0.3$ should be avoided.

In contrast, for very small values ($\kappa_\eta < 0.1$) the mood becomes insensitive to emotional changes. Therefore, we suggest using values in the range from 0.1 to 0.3, as they seem to provide comprehensible mood courses. In Fig. 4.10 it is depicted how the difference between κ_{pos} and κ_{neg} influences the mood development.

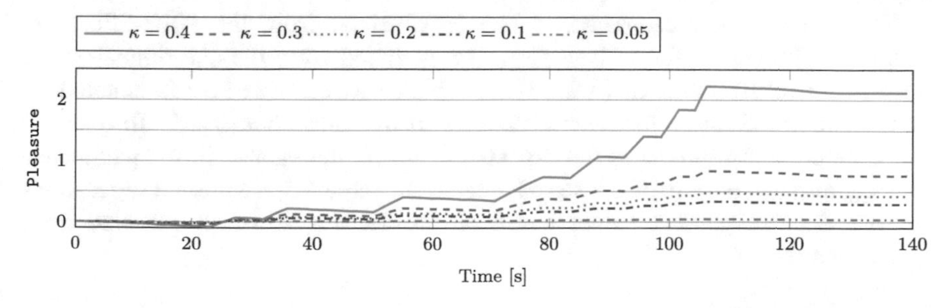

Fig. 4.9 Mood development for different settings of κ_η, but not differing between κ_{pos} and κ_{neg}

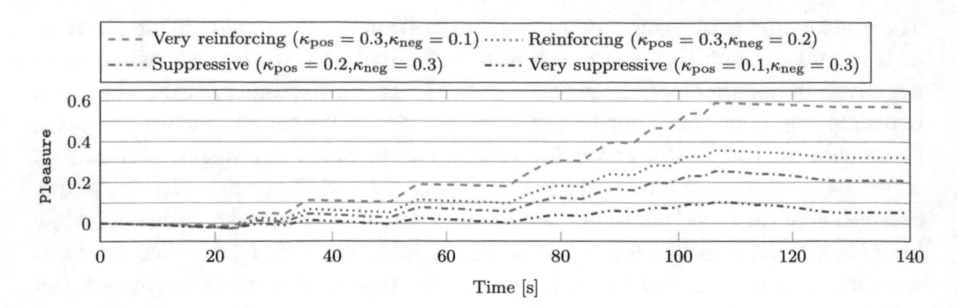

Fig. 4.10 Mood development for different settings of κ_{pos} and κ_{neg}

Table 4.5 Suggested κ_{pos} and κ_{neg} values based on the `extraversion` personality trait

`Extraversion`	κ_{pos}	κ_{neg}
>0.7	0.3	0.1
0.6–0.7	0.3	0.2
0.4–0.6	0.2	0.2
0.2–0.4	0.2	0.3
<0.2	0.1	0.3

According to the phenomenon described earlier, namely that persons with a `high extraversion` are more stable on positive affects, we tested different settings for the difference between κ_{pos} and κ_{neg} (cf. Fig. 4.10). Here, we basically distinguish between two different settings. First, a very reinforcing setting, where positive observations are emphasized and negative observations are suppressed. Secondly, a very suppressive setting where the mood development behaves the other way around, positive values are suppressed and negative ones emphasized. Again, the annotated emotional traces of ES2 are used and the previous considerations on the κ_η-values are included to choose only values between 0.1 and 0.3. More details on annotation can be found in [34, 37]. By varying these values, we could change the behavior of the model to match the different settings of emotional stability. Although the input data remains the same, the emotional influence of positive observations on the mood can be either very suppressive or very reinforcing, depending only on the adjustment factor κ_η, as seen in Fig. 4.10.

The subject's `extraversion` can be obtained from the NEO-FFI questionnaire. The values for `extraversion` gathered from the questionnaires are normalized in the range of $[0, 1]$. Thus, a high `extraversion` is denoted by values above 0.5 and a low `extraversion` by values below 0.5. To obtain a mood model that reproduces the expected behavior, the values for the parameter-pair κ_{pos} and κ_{neg} have to be chosen adequately. In Table 4.5, we present suggestions for plausible adjustment values based on the `extraversion` gathered from questionnaires.

4.5 Summary

In this chapter, we described our vision of an evolving search user interface. An evolving search user interface should adapt itself (both, user interface and interaction dialog) to the specific characteristics of an individual user. Here, we concentrated on the front end of the search engine and developed two prototypes, one which can be customized to the user's needs and one that takes the user's age as a parameter for adaptation. However, in order to have an adaptive system, backend algorithms to detect a user's abilities are required. Moreover, in this work, age was used in order to approximate the user's skill space. However, the age is only a fuzzy indicator of what the user's skills are. The design of a more fine-grained mapping function between the user's skill space and the options to adapt the UI elements of the SUI is a great challenge for future work. To develop an ESUI, user studies with users of gradually different skills should be conducted.

In order to adapt the interaction dialog, we propose to monitor the user's emotional state. This makes an early detection of users' problems with the system possible and allows us to adapt the dialog to get the user on the right path. Therefore, we investigated methods to detect changes in the user's emotional state.

We furthermore proposed a mood modeling from a technical perspective that is able to incorporate several psychological observations. After describing the desired mood development, we presented a technical mood implementation. This implementation has the advantage of only three internal parameters (D_0, μ_1 and μ_2) and one user-specific parameter-pair, κ_{pos} and κ_{neg}. The mood development is based on a mechanical spring model.

Using the EmoRec dataset, we were able to evaluate the principal function of the proposed model on two different databases. Especially on the EmoRec Woz corpus, we were able to show that the generated mood course matched the experimental setting. By utilizing the user-specific parameter-pair κ_{pos} and κ_{neg} the personality trait `extraversion` was integrated. This trait is supposed to regulate the individual emotional experiences.

The presented results are based on locating both emotions and calculated mood in the PAD-space according to their value. A problem that has to be addressed in future work is the need for smooth transitions of emotion values over time. To date, the emotional assessments are processed without regarding the gap between them, but this cannot always be guaranteed, especially when using automatically recognized emotion values. Here a further extension of the proposed model is needed by, for example, a temporal averaging or weighting technique.

In the future, we are going to use the mood model for the search scenario and incorporate it in the search engine prototype we developed. User emotions can be detected by analyzing the facial expression of a user. It is also possible to implement a voice-controlled search interface which allows a voice interaction with the search engine and to use the user's voice for emotion detection.

Acknowledgements This work was done within the Transregional Collaborative Research Centre SFB/TRR 62 "*Companion*-Technology for Cognitive Technical Systems" funded by the German Research Foundation (DFG).

References

1. Amstad, T.: Wie verständlich sind unsere Zeitungen? Ph.D. Thesis, University of Zurich (1978)
2. Arapakis, I., Athanasakos, K., Jose, J.M.: A comparison of general vs personalised affective models for the prediction of topical relevance. In: Proceedings of the 33rd ACM SIGIR'10, pp. 371–378 (2010)
3. Aula, A.: User study on older adults'use of the web and search engines. Univ. Access Inf. Soc. **4**(1), 67–81 (2005)
4. Aula, A., Käki, M.: Less is more in web search interfaces for older adults. First Monday **10**(7) (2005)
5. Becker-Asano, C.: WASABI: affect simulation for agents with believable interactivity. Ph.D. Thesis, University of Bielefeld (2008)
6. Böck, R., Limbrecht, K., Walter, S., Hrabal, D., Traue, H.C., Glüge, S., Wendemuth, A.: Intraindividual and interindividual multimodal emotion analyses in human-machine-interaction. In: Proceedings of the IEEE CogSIMA, New Orleans, pp. 59–64 (2012)
7. Böck, R., Limbrecht-Ecklundt, K., Siegert, I., Walter, S., Wendemuth, A.: Audio-based pre-classification for semi-automatic facial expression coding. In: Kurosu, M. (ed.) Human-Computer Interaction. Towards Intelligent and Implicit Interaction. Lecture Notes in Computer Science, vol. 8008, pp. 301–309. Springer, Cham (2013)
8. Carpenter, S.M., Peters, E., Västfjäll, D., Isen, A.M.: Positive feelings facilitate working memory and complex decision making among older adults. Cognit. Emot. **27**, 184–192 (2013)
9. Costa, P.T., McCrae, R.R.: The NEO Personality Inventory Manual. Psychological Assessment Resources, Odessa (1985)
10. Crawford, J.R., Henry, J.D.: The positive and negative affect schedule (PANAS): construct validity, measurement properties and normative data in a large non-clinical sample. Br. J. Clin. Psychol. **43**, 245–265 (2004)
11. Eickhoff, C., Azzopardi, L., Hiemstra, D., De Jong, F., de Vries, A., Dowie, D., Duarte, S., Glassey, R., Gyllstrom, K., Kruisinga, F., Marshall, K., Moens, S., Polajnar, T., van der Sluis, F.: Emse: initial evaluation of a child-friendly medical search system. In: Proceedings of the 4th ACM Information Interaction in Context Symposium, pp. 282–285 (2012)
12. Ellsworth, P., Scherer, K.: Appraisal Processes in Emotion, pp. 572–595. Oxford University Press, New York (2003)
13. Gebhard, P.: ALMA a layered model of affect. In: Proceedings of the 4th ACM AAMAS, Utrecht, pp. 29–36 (2005)
14. Gossen, T., Nürnberger, A.: Specifics of information retrieval for young users: a survey. Inf. Process. Manag. **49**(4), 739–756 (2013)
15. Gossen, T., Nitsche, M., Nürnberger, A.: Knowledge journey: a web search interface for young users. In: Proceedings of the Symposium on Human-Computer Interaction and Information Retrieval (HCIR'12), pp. 1:1–1:10. ACM, Cambridge (2012)
16. Gossen, T., Nitsche, M., Nürnberger, A.: Evolving search user interfaces. In: Proceedings of EuroHCIR 2013 Workshop. Dublin, Ireland (2013)
17. Gossen, T., Nitsche, M., Vos, J., Nürnberger, A.: Adaptation of a search user interface towards user needs - a prototype study with children & adults. In: Proceedings of the Symposium on Human-Computer Interaction and Information Retrieval (HCIR'13), Vancouver, BC (2013)

18. Gossen, T., Kotzyba, M., Nürnberger, A.: Knowledge journey exhibit: towards age-adaptive search user interfaces. In: Hanbury, A., Kazai, G., Rauber, A., Fuhr, N. (eds.) Advances in Information Retrieval. Lecture Notes in Computer Science, vol. 9022, pp. 781–784. Springer, Cham (2015)
19. Hartmann, K., Siegert, I., Glüge, S., Wendemuth, A., Kotzyba, M., Deml, B.: Describing human emotions through mathematical modelling. In: Proceedings of the 7th MATHMOD, Vienna, pp. 463–468 (2012)
20. Hearst, M.: Search User Interfaces. Cambridge University Press, Cambridge (2009)
21. Hutchinson, H., Druin, A., Bederson, B., Reuter, K., Rose, A., Weeks, A.: How do I find blue books about dogs? The errors and frustrations of young digital library users. In: Proceedings of the 11th International Conference on Human-Computer Interaction (2005)
22. Jansen, M., Bos, W., van der Vet, P., Huibers, T., Hiemstra, D.: TeddIR: tangible information retrieval for children. In: Proceedings of the 9th ACM International conference on Interaction Design and Children, Barcelona, 2010, pp. 282–285
23. Kotzyba, M., Deml, B., Neumann, H., Glüge, S., Hartmann, K., Siegert, I., Wendemuth, A., Traue, H.C., Walter, S.: Emotion detection by event evaluation using fuzzy sets as appraisal variables. In: Proceedings of the 11th ICCM, Berlin, pp. 123–124 (2012)
24. Larsen, R.J., Fredrickson, B.L.: Measurement Issues in Emotion Research, pp. 40–60. Russell Sage Foundation, New York (1999)
25. Larsen, R.J., Ketelaar, T.: Personality and susceptibility to positive and negative emotional states. J. Pers. Soc. Psychol. **61**, 132–140 (1991)
26. Morris, J.D.: SAM: the self-assessment manikin an efficient cross-cultural measurement of emotional response. J. Advert. Res. **35**, 63–68 (1995)
27. Morris, W.N.: Mood: The Frame of Mind. Springer, New York (1989)
28. Moshfeghi, Y.: Role of emotion in information retrieval. Ph.D. Thesis, University of Glasgow, School of Computing Science (2012)
29. Moshfeghi, Y., Piwowarski, B., Jose, J.: Handling data sparsity in collaborative filtering using emotion and semantic based features. In: Proceedings of the 34th ACM SIGIR'11, pp. 625–634 (2011)
30. Nolen-Hoeksema, S., Fredrickson, B., Loftus, G., Wagenaar, W.: Atkinson & Hilgard's Introduction to Psychology, 15 edn. Cengage Learning EMEA, Hampshire (2009)
31. Ortony, A., Clore, G.L., Collins, A.: The Cognitive Structure of Emotions. Cambridge University Press, Cambridge (1990)
32. Scherer, K.R.: Appraisal Considered as a Process of Multilevel Sequential Checking, pp. 92–120. Oxford University Press, Oxford (2001)
33. Siegert, I.: Emotional and user-specific cues for improved analysis of naturalistic interactions. Ph.D. Thesis, Otto von Guericke University Magdeburg (2015)
34. Siegert, I., Böck, R., Wendemuth, A.: The influence of context knowledge for multimodal annotation on natural material. In: Joint Proceedings of the IVA 2012 Workshops, Santa Cruz, pp. 25–32 (2012)
35. Siegert, I., Böck, R., Wendemuth, A.: Modeling users' mood state to improve human-machine-interaction. In: Esposito, A., Esposito, A.M., Vinciarelli, A., Hoffmann, R., Müller, V.C. (eds.) Cognitive Behavioural Systems. Lecture Notes in Computer Science, vol. 7403, pp. 273–279. Springer, Berlin (2012)
36. Siegert, I., Hartmann, K., Glüge, S., Wendemuth, A.: Modelling of emotional development within human-computer-interaction. Kognitive Systeme **1** s.p. (2013)
37. Siegert, I., Böck, R., Wendemuth, A.: Inter-rater reliability for emotion annotation in human-computer interaction – comparison and methodological improvements. J. Multimodal User Interfaces **8**, 17–28 (2014)
38. Steichen, B., Ashman, H., Wade, V.: A comparative survey of personalised information retrieval and adaptive hypermedia techniques. Inf. Process. Manag. **48**(4), 698–724 (2012)
39. Stober, S., Nürnberger, A.: Adaptive music retrieval–a state of the art. Multimed. Tools Appl. **65**, 467–494 (2013)

40. Truong, K.P., Neerincx, M.A., Leeuwen, D.A.V.: Assessing agreement of observer- and self-annotations in spontaneous multimodal emotion data. In: Proceedings of the INTERSPEECH-2008, Brisbane, pp. 318–321 (2008)
41. Walter, S., Scherer, S., Schels, M., Glodek, M., Hrabal, D., Schmidt, M., Böck, R., Limbrecht, K., Traue, H.C., Schwenker, F.: Multimodal emotion classification in naturalistic user behavior. In: Jacko, J. (ed.) Human-Computer Interaction. Towards Mobile and Intelligent Interaction Environments. Lecture Notes in Computer Science, vol. 6763, pp. 603–611. Springer, Cham (2011)

Chapter 5
User-Centered Planning

Pascal Bercher, Daniel Höller, Gregor Behnke, and Susanne Biundo

Abstract User-centered planning capabilities are core elements of *Companion*-Technology. They are used to implement the functional behavior of technical systems in a way that makes those systems *Companion*-able—able to serve users individually, to respect their actual requirements and needs, and to flexibly adapt to changes in their situation and environment. This chapter presents various techniques we have developed and integrated to realize user-centered planning. They are based on a hybrid planning approach that combines key principles also humans rely on when making plans: stepwise refining complex tasks into executable courses of action and considering causal relationships between actions. Since the generated plans impose only a partial order on actions, they allow for a highly flexible execution order as well. Planning for *Companion*-Systems may serve different purposes, depending on the application for which the system is created. Sometimes, plans are just like control programs and executed automatically in order to elicit the desired system behavior; but sometimes they are made for humans. In the latter case, plans have to be adequately presented and the definite execution order of actions has to coincide with the user's requirements and expectations. Furthermore, the system should be able to smoothly cope with execution errors. To this end, the plan generation capabilities are complemented by mechanisms for plan presentation, execution monitoring, and plan repair.

5.1 Introduction

Companion-Systems are able to serve users individually, to respect their actual requirements and needs, to flexibly adapt to changes in their situation and environment, and to explain their own behavior (cf. Chap. 1, the survey on *Companion*-Technology [21], or the work of the collaborative research centre SFB/TRR 62 [19]). A core element when realizing such systems is user-centered planning. This chapter

---

P. Bercher (✉) • D. Höller • G. Behnke • S. Biundo
Institute for Artificial Intelligence, Ulm University, Ulm, Germany
e-mail: pascal.bercher@uni-ulm.de; daniel.hoeller@uni-ulm.de; gregor.behnke@uni-ulm.de; susanne.biundo@uni-ulm.de

S. Biundo, A. Wendemuth (eds.), *Companion Technology*, Cognitive Technologies,
DOI 10.1007/978-3-319-43665-4_5

presents various techniques we have developed and integrated to realize user-centered planning [20]. They are based on a hybrid planning approach [18] that combines key principles also humans rely on when making plans by refining complex tasks stepwise into executable courses of action, assessing the various options for doing so, and considering causal relationships.

Planning for *Companion*-Systems may serve different purposes, depending on the application for which the system is created. Sometimes, plans are just like control programs and executed automatically in order to elicit the desired system behavior; but sometimes they are made for humans. In the latter case, plans have to be adequately presented to the user. Since the generated plans impose only a partial order on actions, they allow for a highly flexible execution order. A suitable total order must be selected for step-wise presentation: it should coincide with the user's requirements and expectations. We ensure this by a technique that allows finding *user-friendly* linearizations [32].

In particular when planning for humans, a plan execution component must monitor the current state of execution so that the system can detect failures, i.e., deviations from the expected execution outcome. In such a case, the hybrid plan repair mechanism finds a new plan that incorporates the execution error [12, 17].

Companion-Systems assist users in completing demanding tasks, so the user may not understand the steps that the system recommends doing. In particular after execution errors, question might arise due to the presentation of a new plan. To obtain transparency and to increase the user's trust in the system, it is essential that it is able to explain its behavior. Therefore, the purpose of any action within a plan may be automatically explained to the user in natural language [53].

These user-centered planning capabilities of plan generation, plan execution and linearization, plan repair, and plan explanation are essential for providing intelligent user assistance in a variety of real-world applications [20]. As an example, we integrated all those techniques in a running system that assists a user in the task of setting up a complex home theater [12, 14, 35]. The respective system and, in particular, the integration of the user-centered planning capabilities with a knowledge base and components for user interaction is described in Chap. 24. Here, we focus on the underlying planning capabilities and explain them in detail. We use the planning domain of that application scenario as a running example.

The hybrid planning framework is explained in Sect. 5.2. Section 5.3 is devoted to plan execution. Plan execution consists of various key capabilities when planning for or with humans: the monitoring of the executed plans to trigger plan repair in case of the execution errors (explained in Sect. 5.4), the linearization of plans to decide which plan step to execute next, and the actual execution of the next plan step, which includes the adequate presentation to the user. Section 5.5 introduces the plan explanation technique that allows us to generate justifications for any plan step questioned by the user. Finally, Sect. 5.6 concludes the chapter.

5.2 Hybrid Planning Framework

Hybrid planning [18, 36] fuses Hierarchical Task Network (HTN) planning [27] with concepts known from Partial-Order Causal-Link (POCL) planning [41, 50].

The smooth integration of hierarchical problem solving (inherited from HTN planning) with causal reasoning (inherited from POCL planning) provides us with many capabilities that are beneficial when planning for or with humans:

HTN Planning In HTN planning, problems are specified in terms of abstract activities one would like to have accomplished. For this, they have to be refined step-wise into more specific courses of action that can be executed by the user. This provides us with certain benefits:

- First of all, a domain expert has more freedom in modeling a domain. Often, expert knowledge is structured in a hierarchical way. Hence, it is often known to the expert what actions need to be taken in what order to accomplish some high-level goal. Such knowledge can easily be modeled by introducing a hierarchy among the available actions. Many real-world application scenarios are hence modeled using hierarchical planning approaches such as hybrid planning or the SHOP approach [20, 39, 44]. Further, a domain modeler can be assisted in the task of creating a hierarchical domain model by techniques that automatically infer abstractions for hierarchical planning [6].
- That action hierarchy may then be exploited for generating and improving explanations [53]. When the user wants to know about the purpose of a presented action during execution, the hierarchy can be used to come up with a justification.
- The hierarchy defined on the actions may also be exploited to come up with plausible linearizations of plans [32]. The actions in the plans are presented to a user one-by-one. Some linearization might be more plausible to a user than others. So, presenting those actions close to each other that "belong to each other" with respect to the action hierarchy might achieve reasonable results.
- The way in which humans solve tasks is closely related to the way hierarchical problems are solved by a planning system. That makes it more natural to a user to be integrated into the planning process [7], as it resembles his or her idea of problem solving. The integration of the user into this decision making process is called *mixed initiative planning*. It is presented in Chap. 7.

POCL Planning In POCL planning, problems are specified in terms of world properties that one would like to hold. The problem is solved via analyzing causal dependencies between actions to decide what action to take in order to fulfill a required goal. The way in which plans are found and how they are represented can be exploited in various ways:

- The causal dependencies between actions within a plan are explicitly represented using so-called causal links. Analogously and complementarily to the exploitation of the hierarchy, these causal relations can be analyzed and exploited to generate explanations about the purpose of any action within a plan [53].

- The causal structure of plans may be used to find plausible linearizations of the actions within a plan. For instance, given there are causal relationships between two actions, it seems more plausible to present them after each other before presenting another action that has no causal dependencies with either of them [32].
- Finally, the POCL planning approach also seems well suited for a mixed initiative planning approach, since humans do not only plan in a hierarchical manner, but also via reasoning about what action to take in order to fulfill a requirement that has to hold later on.

Hybrid planning combines HTN planning with POCL planning; it hence features all previously mentioned user-centered planning capabilities.

5.2.1 Problem Formalization

A hybrid planning problem is given as a pair consisting of a domain model $\mathscr{D}$ and the problem instance $\mathscr{I}$. The domain describes the available actions required for planning. The problem instance specifies the actual problem to solve, i.e., the available world objects, the current initial state, the desired goal state properties, and an initial plan containing the abstract tasks that need to be refined.

More specifically, a domain is a triple $\mathscr{D} = \langle T_p, T_a, M \rangle$. T_p and T_a describe the *primitive* and *abstract tasks*, respectively. The primitive tasks are also referred to as *actions*—those can be executed directly by, and hence communicated to, the user. Actions are triples $\langle a(\bar{\tau}), pre(\bar{\tau}), eff(\bar{\tau}) \rangle$ consisting of a name a that is parametrized with variables $\bar{\tau}$, a parametrized precondition *pre* and the effects *eff*. As an example, Eq. (5.1) depicts the name and parameters of an action of the home assembly task (see Chap. 24) for plugging the audio end of a SCART cable into the audio port of an audio/video receiver. In the depicted action, the parameters are bound to constants, which represent the available objects—in the example domain those are the available hi-fi devices, cables, and their ports.

$$plugIn(\text{SCART-CABLE}, \text{AUDIO-PORT}, \text{AV-RECEIVER}, \text{AUDIO-PORT}) \tag{5.1}$$

The precondition describes the circumstances under which the action can be executed, while the effects describe the changes that an execution has on the respective world state. Formally, preconditions and effects are conjunctions of literals that are defined over the variables $\bar{\tau}$. For instance, the (negative) literal $\neg used(\text{SCART-CABLE}, \text{AUDIO-PORT})$ is part of the precondition of the action depicted in Eq. (5.1). It describes that the audio port of the SCART cable may only be plugged into a port if it is currently not in use. The effects of that action mark the port as blocked. Abstract tasks syntactically look like primitive ones, but they are regarded to be not directly executable by the user. Instead, they are abstractions of one or more primitive tasks. That is, for any abstract task t, a so-called *(decomposition) method* $m = \langle t, P \rangle$ relates that task to a plan P that "implements" t [15, 18]. The set of all methods is given by M. The implementation (or legality) criteria ensure that t is a legal abstraction of the plan P, which can be verified by comparing the preconditions and effects of t with those of the tasks within P. One can also regard it the other way round: the implementation criteria ensure that only those plans may be used within a method that are actual implementations of the respective abstract task. That way, a human user (in that case the domain modeler) can be actively supported in the domain modeling process—independently of whether he or she uses a top-down or bottom-up modeling approach.

Plans are generalizations of action sequences in that they are only partially ordered. They are knowledge-rich structures, because causality is explicitly represented using so-called *causal links*. Formally, a plan P is a tuple $\langle PS, V, \prec, CL \rangle$ consisting of the following elements. The set PS is referred to as *plan steps*. Plan steps are uniquely labeled tasks. Thus, each plan step $ps \in PS$ is a tuple $l : t$, with l being a label symbol unique within P and t being a task taken from $T_p \cup T_a$. Unique labeling is required to differentiate identical tasks from each other that are all within the same plan. The set V contains the variable constraints that (non-)codesignate task parameters with each other or with constants. Codesignating a variable with a constant means assigning the respective constant to that variable. Codesignating two variables means that they have to be assigned to the same constant. Non-codesignating works analogously. The set $\prec$ is a strict partial order defined over $PS \times PS$. The causal links CL represent causal dependencies between tasks: each link $cl \in CL$ is a triple $\langle ps, \varphi, ps' \rangle$ representing that the literal φ is "produced" by (the task referenced by) ps and "consumed" by ps'. Due to that causal link, thc precondition literal φ of ps' is called *protected*, since the solution criteria ensure that no other task is allowed to invalidate that precondition anymore (see Solution Criterion 2c given below).

A problem instance $\mathcal{I}$ is a tuple $\langle C, s_{init}, P_{init}, g \rangle$ consisting of the following elements. The set C contains all available constants. The conjunction s_{init} of ground positive literals describes the initial state. We assume the so-called *closed world assumption*. That is, exactly the literals in s_{init} are assumed to hold in the initial state, while all others are regarded as false. The conjunction of (positive and negative) literals g describes the goal condition. All these goal state properties *must* hold after the execution of a solution plan. Hence, g implicitly represents a set of world states that satisfy g. In particular when planning for humans, not all goals are necessarily mandatory. Instead, some of them might only be *preferred* by the user. That is, while some goals might be declared as non-optional (those specified by g), a user might also want so specify so-called soft goals that he or she would like to see satisfied, but that are regarded as optional. A planner would then try to achieve those goals to increase plan quality, but in case a soft-goal cannot be satisfied, the planning process does not fail altogether. Some work has been done in incorporating such soft-goals into hierarchical planning in general [39, 55] and into hybrid planning in particular [8]. For the sake of simplicity, we focus on the non-optional goals in this chapter. Note that this is not a restriction, since any planning problem with soft goals can be translated into an equivalent problem without soft goals [23, 37]. The initial plan

P_{init} complements the desired goal state properties by the tasks that the user would like to have achieved. This plan may contain primitive tasks, abstract tasks, or both. In addition, it contains two special actions a_{init}, and a_{goal}, that encode the initial state and goal description, respectively. The respective encoding is done as usual in POCL planning: a_{init} is always the very first action in every refinement of P_{init}, while a_{goal} is always the very last. The action a_{init} has no precondition and uses s_{init} as effect,[1] while a_{goal} uses g as precondition and has no effect.

5.2.1.1 Solution Criteria

Informally, a solution is any plan that is executable in the initial state and satisfies the planning goals and tasks, i.e., after the execution of a solution plan P_{sol}, g holds, and P_{sol} is a refinement of P_{init}, thereby ensuring that the abstract activities the user should accomplish (specified in P_{init}) have actually been achieved. More formally, a plan P_{sol} is a solution if and only if two criteria hold:

1. P_{sol} is a refinement of P_{init}. That is, one must be able to obtain P_{sol} from P_{init} by means of the application of the following refinement operators:
   a. ***Decomposition.*** Given a plan $P = \langle PS, V, \prec, CL \rangle$ with an abstract plan step $l : t \in PS$, the decomposition of the abstract task t using a decomposition method $m = \langle t, P' \rangle$ results in a new plan P'', in which $l : t$ is removed and replaced by P'. Ordering and variable constraints, as well as causal links pointing to or from $l : t$, are inherited by the tasks within P' [15]. This is a generalization of Definition 3 by Geier and Bercher [29] for standard HTN planning without causal links. This decomposition criterion ensures that the abstract tasks specified in P_{init} are accomplished by any solution. That criterion is the reason why HTN or hybrid planning is undecidable in the general case [3, 15, 27, 29]. It also makes the verification of plans (i.e., answering "is the given plan a valid solution to the given problem?") hard (NP-complete) even under severe restrictions [5, 15].
   b. ***Task Insertion.*** In hybrid planning, both primitive and abstract tasks may be inserted into a plan. Note that this feature is optional. Allowing or disallowing task insertion might influence the complexity of both solving the planning problem [4, 29] and of the solutions themselves [31, 33]. Allowing task insertion allows for more flexibility for the domain modeler, as it allows us to define partial hierarchical models [4, 36]. That is, the domain modeler does not need to specify decomposition methods that ensure that any decomposition is an executable solution, as the planner might insert tasks to ensure executability.

[1] More technically, it uses not just s_{init} as effect, but—because s_{init} consists only of positive literals due to the closed world assumption—also all negative ground literals that unify with any negative precondition that are not contradicting s_{init}. Otherwise, there might be a negative task precondition literal that could not be protected by a causal link rooting in the initial state.

Thus, allowing task insertion moves some of the planning complexity from the modeling process (which is done by a user/domain expert) to the planning process (which is done automatically).

c. ***Causal Link Insertion* and *Ordering Insertion*.** Given two plan steps, *ps* and ps', within a plan, a causal link can be inserted from any literal in *ps*'s effect to any (identical) literal in the precondition of ps'. The parameters of the two literals become pairwise codesignated. Also, an ordering constraint may be inserted between *ps* and ps'. Both these refinement options are inherited from standard POCL planning. They are a means to ensure the executability of plans [41, 50].

2. P_{sol} is executable in the initial state s_{init} and, after execution of that plan, the goal condition g is satisfied. Since s_{init} and g are encoded within P_{init} by means of the two special actions a_{init} and a_{goal}, respectively, and because P_{sol} is a refinement of P_{init} due to Solution Criterion 1, both planning goals can be achieved by using standard POCL solution criteria. Thus, $P_{sol} = \langle PS_{sol}, V_{sol}, \prec_{sol}, CL_{sol}\rangle$ is executable in s_{init} and satisfies g if and only if:

a. ***All tasks are primitive and ground.*** Only primitive actions are regarded as executable. Grounding is required to ensure unique preconditions and effects.

b. ***There are no open preconditions.*** That is, for each precondition literal φ of any plan step $ps \in PS_{sol}$ there is a causal link $\langle ps', \varphi, ps\rangle \in CL_{sol}$ with $ps' \in PS_{sol}$, protecting φ.

c. ***There are no causal threats.*** We need to ensure that the literals used by the causal links are actually "protected". This is the case if there are no so-called *causal threats*. Within a primitive ground plan $P = \langle PS, V, \prec, CL\rangle$, a plan step *ps* is threatening a causal link $\langle ps', \varphi, ps''\rangle \in CL$ if and only if the set of ordering constraints allows *ps* to be ordered between ps' and ps'' (that is, $\prec \cup \{(ps', ps), (ps, ps'')\}$ is a strict partial order) and *ps* has an effect $\neg\varphi$.

5.2.2 *Finding a Solution*

Hierarchical planning problems may be solved in many different ways [1]; hence various hierarchical planning systems and techniques exist, such as SHOP/SHOP2 [43], UMCP [26], or HD-POP [52, edition 1, p. 374–375], to name just a few.

We follow the approach of the HD-POP technique. The resulting planning system, PANDA [13, Algorithm 1], performs heuristic search in the space of plans via refining the initial plan P_{init} until a primitive executable plan has been obtained. The algorithm basically mimics the allowed refinement options: it decomposes abstract tasks (thereby introducing new ones into the successor plan), inserts new tasks from the domain (if allowed; cf. Solution Criterion 1b), and inserts ordering constraints and causal links to ensure executability. Hierarchical planning is quite difficult. In the general case, it is undecidable, but even for some quite restricted special cases, it is still at least PSPACE-hard [3, 4, 15, 27, 29]. During search, that hardness corresponds to the choice of which task to insert and which decomposition method to pick when decomposing an abstract task. In the approach taken by PANDA, these questions are answered by heuristics: each candidate plan is estimated in terms of the number of required modifications to refine it into a solution, or by means of the number of actions that need to be inserted for the same purpose [13].

For standard POCL planning, i.e., in case the initial plan P_{init} does not contain abstract tasks, there are basically two different kinds of heuristics. The first kind is based on delete-relaxation,[2] as this reduces the complexity of deciding the plan existence problem from PSPACE to P or NP, depending on the presence of negative preconditions [22] and whether the actions in the domain and the given plan become delete-relaxed or just those in the domain [11]. The respective heuristics are the *Add Heuristic for POCL planning* [57], the *Relax Heuristic* [46], and a variant of the latter based on partial delete-relaxation, called *SampleFF* [11]. The second kind of heuristics is not just one single POCL heuristic, but a technique that allows us to *directly* use heuristics known from state-based planning in the POCL planning setting [10]. The technique encodes a plan into a classical (i.e., non-hierarchical) planning problem, where the POCL plan is encoded within the domain.

The idea of delete-relaxation has also been transferred to hierarchical planning. Here, the complexity of the plan existence problem is reduced from undecidable to NP or P, depending on various relaxations [2]. There is not yet an implementation of that idea, however. Instead, we developed the so-called *task decomposition graph* that is a relaxed representation of how the abstract tasks may be decomposed [24, 25]. That graph may both be used for pruning infeasible plans from the search space (i.e., plans that cannot be refined into a solution) [24] and for designing well-informed heuristics for hierarchical and hybrid planning [13, 25].

5.3 Plan Execution

In most real-world application domains, the effect of actions is not fully deterministic, though there is often an outcome that can be regarded as the intended or standard effect. Since *Companion*-Systems flexibly adapt to any changes in the user's situation and environment, they must be able to detect and deal with unforeseen effects. The sub system that monitors the environment and detects state changes that conflict with the current plan is called *execution monitor* and described in Sect. 5.3.1. When a state deviation is detected that may cause the current plan to fail, the *plan repair* component is started. The plan repair mechanism is introduced

[2]Delete-relaxation means ignoring negative literals in the effects and, optionally, in the preconditions of any action.

later on in Sect. 5.4. Solution plans are not totally ordered: they include only ordering constraints that are necessary to guarantee executability. Thus it is likely that there is more than one linearization of the solution. The *plan linearization* component is responsible for deciding which one is the most appropriate to be presented to a user. This functionality is described in Sect. 5.3.2. What it means to execute a single plan step, and how it may be done, is described in Sect. 5.3.3.

5.3.1 Monitoring

As stated above, the monitoring compares changes that have been detected in the environment with the intended effect of a started action. When differences are detected, it may not necessarily be a problem for the execution of the current plan, so the monitoring has to decide whether repair (see Sect. 5.4) is initiated or not. The decision may be based on the set of *active* causal links. A causal link is active if and only if its producer has been executed while the consumer has not. When there is an active link on a literal that has changed, repair is started (see Fig. 5.1).

Intuitively, this means that (a part of) the precondition of the consumer should have been fulfilled by the producer, but this has not been successful. Now there is no guarantee that the precondition of the consumer is fulfilled (i.e., a causal link that supports it) and plan execution may fail. There are special cases, however, where the currently executed sequence of actions is still executable although there is a

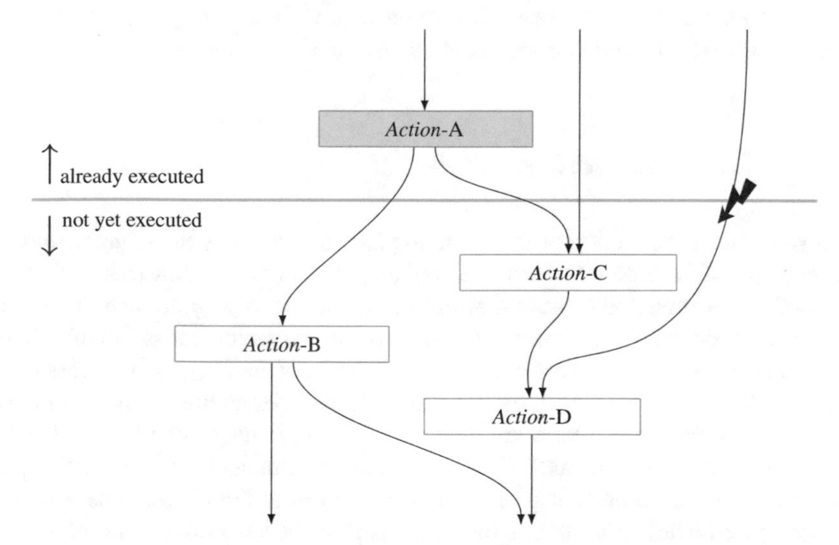

Fig. 5.1 The figure shows how an unexpected state deviation influences the execution of the remaining actions. The *horizontal line* indicates the execution horizon. An execution error flips the truth value of the literal protected by the right-most causal link. Because that causal link crosses the execution horizon, the causal link's consumer (Action-D) might not be executable anymore

causal link that is violated (a valid POCL plan could hence be found by simply choosing a new producer for the invalidated causal link within that plan). Although in that case the user could proceed executing that action sequence, plan repair must be initiated, since the respective causal link may be mandatory: if it has not been inserted by the planner as a means to ensure executability, but comes from the domain (specified within a plan referenced by a decomposition method) or from the initial plan, it may not be changed. Such links may be intended by the modeler to protect certain properties during execution (referred to as prevail conditions) and are thus not allowed to be removed.

So, whenever a condition of an active causal link is violated, the plan monitoring initiates plan repair. It creates an altered plan (if there is one) that is able to deal with unexpected changes and fulfills the constraints given in the model.

The approach given above is able to deal with unforeseen changes of the environment and minimizes the computational effort that is necessary. Plan repair is only started if active causal links are violated. However, there are situations where it would be beneficial to start the repair mechanism even in cases where no active causal links are violated and, hence, the plan is still executable. Consider, e.g., the case where the unforeseen change does not result in any violated causal link, but at the same time causes the original goal condition to become true. In that setting, the planning problem would be solved when no further actions are executed.[3] Without starting repair, the user would have had to continue executing the plan. Though this is a good reason to start plan repair, as often as there is enough time to wait for the new plan, there are also reasons to continue the execution of the original plan: if there is no notable problem with the plan currently executed, it might confuse the user when its execution is canceled to proceed with another plan. The question about when to repair could be answered by an empirical evaluation.

5.3.2 Plan Linearization

As given in the introduction of this section, plans generated by the planning system are only partially ordered. They include only the ordering constraints that are included in the model and those that have to be included to guarantee that a goal state is reached after execution. This makes the execution most flexible, since it commits only on necessary constraints. In many situations, it is necessary to choose a linearization of that partial order for plan execution. When plans are executed by a machine, like a smartphone or robot, it may not matter which of their linearizations is executed. However, whenever humans are involved in plan execution, the low commitment of the ordering given in the plan can be exploited to choose the linearization that is most suitable for the specific user in the current

[3]Assuming there are no not-yet executed actions that are inserted due to the underlying action hierarchy; cf. Solution Criterion 1a.

situation. Consider a user who has to achieve two tasks that are not related in any sense. This scenario is likely to result in a plan with two lines of action that are not interrelated. It is no problem to execute the first step of the first line, then the first step of the second line, and so on. However, it might be much more intuitive for the user to finish the first line before starting the second one (or vice versa). The overall process, committing on some ordering constraints during planning and determining the other ordering relations during post-processing, can be seen as a model that consists of two parts.

There are several objectives for the linearization that may be competing. This could be to consider the convenience of the user during execution, to optimize a metric that can be measured (e.g., execution time) or to imitate human behavior. Since *Companion*-Systems need to adapt to the specific user and his or her current situation, finding a user-friendly and maybe user- and situation-specific linearization is another point where adaptivity may come to light. As a starting point for situation- and user-specific strategies, we identified three domain-independent strategies to linearize plans [32].

All of them exploit knowledge that is included in the plan or the domain definition to linearize plans:

1. **Parameter Similarity.** In the home theater domain (see Chap. 24) it seems reasonable to complete all actions involving a specific device before starting on another. It is a feasible design decision of the modeler to pass on the devices as parameters to an action (though there are other ways to model the domain), as is the case for the example action in Eq. (5.1). A parameter-based strategy would exploit this: it orders plan steps in a way that maximizes successive actions that share constants in their parameter set [32, Section 4.1].
2. **Causal Link Structure.** The causal link structure of a plan represents which effect of a plan step fulfills a certain precondition of another. The planning procedure is problem-driven, i.e., there is no needless causal link in the plan. Therefore this is also a valuable source of linearization information, because the user may keep track of the causality behind steps that are executed. A strategy based on this structure orders the steps in a way that minimizes the distance between producer and consumer of a causal link. Besides the decisions of the domain modeler, this strategy also depends on the planning process [32, Section 4.2].
3. **Decomposition Structure.** Since the planning domain is commonly modeled by a human domain designer, it is reasonable to assume that tasks that are introduced by a single method are also semantically related. Generalizing this assumption, tasks that have a short distance in the tree of decompositions that spans from the initial task network to the actual plan steps are supposed to be semantically more closely related than tasks that have a long distance. This property can be used for plan linearization. In this form, it depends on both the domain and the planning process. When using the task decomposition graph instead, it only depends on domain properties [32, Section 4.3].

As given above, all strategies depend on the planning domain, the planning system, or both. Thus it is possible to model the same application domain in such a way that they work well or poorly. Consider, e.g., the strategy based on parameter similarity in a propositional domain—there is no information included that could be used for linearization.

The given strategies can be used to pick the next plan step from a set of possible next actions (those where all predecessors in the ordering relation have already been finished), i.e., for a local optimization. Another possibility is to optimize them globally over the linearization of the whole plan. They can also be used as a starting point for a domain-specific strategy.

5.3.3 Plan Step Execution

There are several possibilities of how to proceed when a single plan step has been selected for execution. In some cases, the action is just present due to technical reasons and nothing has to be done for its execution. Consider, e.g., the actions a_{init} and a_{goal}. Their purpose is to cause a certain change during the planning process and it is likely that they can be ignored by the execution system, although reaching action a_{goal} could trigger a notification that states the successful plan completion.

A second possibility is that actions control some part of the system. These are executed internally, but are not necessarily required to be communicated to the user. They may cause, for example, a light to be switched on/off, or a door to open/close, or a new entry to be added in a calendar.

Besides these possibilities, there are actions that have to be communicated to the user, as he or she is the one that carries them out or because the presentation itself is the desired purpose. Such actions can be easily communicated to the user by relying on additional system components taken from dialog management (Chap. 9) and user interaction (Chap. 10), as explained in Chap. 24. For that purpose, each action has an associated dialog model that specifies how it may be presented to a user [16, 47]. The dialog model may itself be structured in a hierarchical manner to enable the presentation of an action with a level of detail that is specific to the individual user. So, depending on the user's background knowledge, the action may be presented with more or less details [48]. The resulting information is sent to the fission component [34] that is responsible for selecting the adequate output modality (Chap. 10). For this, each action may have a standard text template associated with it, which can be used for visualization. Further, each constant used by an action parameter can be associated with respective graphics or videos. In Fig. 5.2 we see how the action given in Eq. (5.1) may be presented to a user.

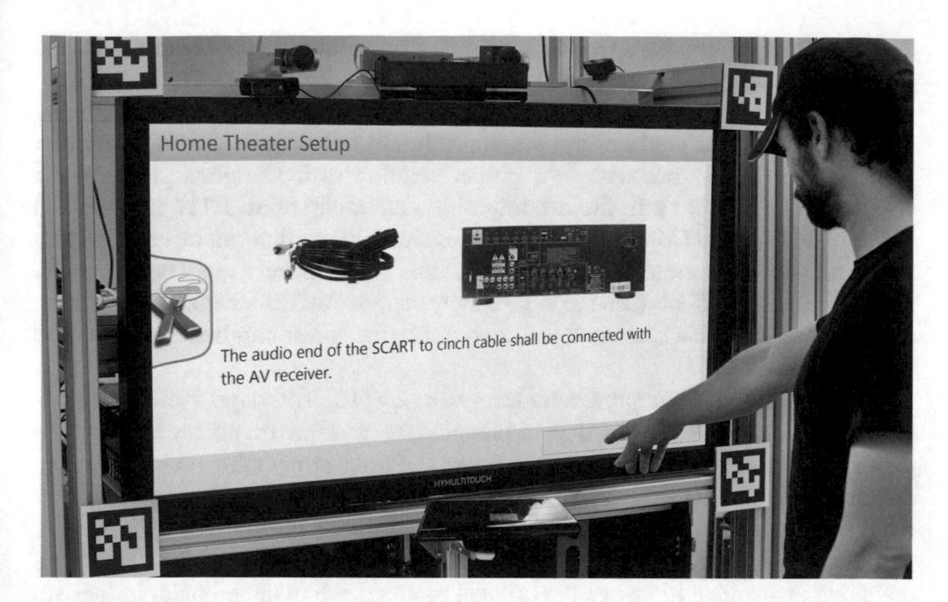

Fig. 5.2 Here, we see how a single planning action can be presented to a human user

5.4 Repairing Failed Plans

Companion-Systems have to adapt to changes in the current situation [21]. This is especially necessary when the execution of a plan fails. If the plan monitoring component (see Sect. 5.3.1) decides that—due to an execution failure—a new plan has to be found, there are two ways this can be done:

- **Re-planning.** The plan at hand is discarded and the planning process is done from scratch. The changed environment is used as initial state and a new plan is found that transfers it into a state that fulfills the goal criteria.
- **Plan Repair.** The original plan is re-used and adapted to the needs of the changed situation. Thereby the unexpected changes of the environment have to be considered and to be integrated into the new plan.

Both approaches have several advantages and disadvantages. *Re-planning* enables the use of sophisticated planning heuristics. For some cases in classical planning, Nebel and Koehler showed that plan repair might be computationally more expensive than planning from scratch [45]. The system could come up with a completely new solution that has nothing in common with the original one, albeit a minor change would have resulted in a valid solution. Presenting a very different solution to a human user might cause confusion and reduce the user's trust in the system.

When a plan is *repaired*, the new plan might be more similar to the original solution. However, this strategy might increase computational complexity [45], prevents the planning system from finding shorter/more cost-effective solutions; and an altered algorithm with effective heuristics that are able to deal with the altered planning problem has to be realized.

In HTN planning there is another aspect to consider: While in classical planning the already executed prefix of the original solution followed by a completely new plan that reaches a goal state is a proper solution to the original problem, the combination may not be in the decomposition hierarchy of an HTN problem and thus violate Solution Criterion 1. There are circumstances that can be encoded into the decomposition hierarchy of an HTN planning problem that cannot be assured by preconditions and effects (see the expressivity analysis by Höller et al. [31, 33]). So it has to be ensured that a repaired plan also fulfills the constraints that are introduced by the hierarchy.

We now introduce our approach for *re-planning* [12]. Although, from a theoretical point of view, it is classified as re-planning (because we do not try to repair the plan already found), it still combines aspects of both re-planning and plan repair. Aspects of repair are required to ensure that the plan prefix already executed is also part of any new solution that can cope with the execution error.

When the execution of a plan fails, a plan repair problem is created. Its domain definition is identical to that of the original problem, while the problem instance is adapted. It includes an additional set of obligations O, i.e., $\mathscr{I} = \langle C, s_{init}, P_{init}, O, g\rangle$. Obligations define which commitments that were made in the original plan have to be present in the new one. To ensure that they are fulfilled, we extend the solution criteria in such a way that all obligations need to be satisfied. There are obligations of the following kind:

- **Task Obligations.** These obligations ensure that a certain plan step (i.e., action) is present in the new solution. A task obligation is included in the problem for every step of the original plan that has already been executed. To overcome the unexpected environment change, a special task obligation is added to the repair problem. It makes sure that a new action is added that realizes the unforeseen changes of the environment. Therefore it has the detected change as its effect. This action is called *process* [17] and is introduced after the executed prefix of the original plan.
- **Ordering Obligations.** Ordering obligations define ordering constraints between the obligated task steps.

Obligations from the given classes are combined in a way ensuring that the executed prefix of the original plan is also a prefix of any new plan. The process is placed exactly behind this prefix to realize the detected change of the world. So the new plan can cope with the unforeseen changes of the environment.

The additional constraints (i.e., the obligations) require some small alterations of the planning procedure. Given an unsatisfied obligation, the algorithm needs to provide possible refinements: unsatisfied task obligations can be addressed via task insertion or decomposition and marking a task within a plan as one of those already executed. Ordering obligations are straightforward.

We have also developed a *repair* approach for hybrid planning [17, 20]. It starts with the original planning problem and the set of refinements applied to find the original solution. As is the case for our *re-planning* approach, the obligations are part of the planning problem as well to ensure that the execution error is reflected and the actions already executed are part of the repaired solution. In contrast to standard repair, the algorithm tries to re-apply all previously applied refinements. It only chooses different refinements where the particular choice leads to a part of the plan that cannot be executed anymore due to the execution failure.

5.5 Plan Explanation

Plans generated via automated planning are usually fairly complex and can contain a large number of plan steps and causal links between them. If the decisions of a *Companion*-System are based upon such plans, its user may not immediately understand the behavior of the system completely. In the worst case, he or she might even reject the system's suggestions outright and stop using it altogether. In general, unexpected or non-understandable behavior of a cognitive system may have a negative impact on trust in the human-computer relationship [42], which in turn is known to have adverse effects on the interaction with the user [49]. To avert this problem, a system should be able to explain its decisions and internal behavior [9, 40]. If a planner is the central cognitive component of the system, it has to be able to explain its decisions (i.e., the plan it has produced) to the user.

A first step towards user-friendly interaction and eliminating questions from the users even before they come up is an intelligent plan linearization component (see Sect. 5.3.2), which presents the whole plan in an easy to grasp step-by-step fashion. Obviously, this capability is not sufficient for complete transparency. Although the order in which actions are presented is chosen in such a way that it is intuitive for the user, he or she might still wonder about it or propose a rearrangement. The user might also be confused about the actual purpose of a presented action and ask why it is part of the solution in the first place. The hybrid plan explanation [53] technique is designed to convey such information to the user.

5.5.1 Generating Formal Plan Explanations

Usually, plan explanations are generated upon user request. Currently, the hybrid plan explanation technique supports two types of requests. The first inquires about the necessity of a plan step, i.e., "Why is action A in the plan?" or "Why should I do A?". The second requests information on an ordering of plan steps, i.e., "Why must action A be executed before B?" or "Why can't I do B after A?". In both cases the explanation is based upon a proof in an axiomatic system Σ, which encodes the plan, the way it was created, and general rules about how facts about the plan can be justified [53]. The request of the user is transformed into a fact F and an automated reasoner is applied to compute a proof for $\Sigma \vdash F$. This proof is regarded as the actual formal explanation of the fact the user has inquired about and is—subsequently—transformed into natural language by a dialogue management component and presented to the user [12, 53]. Obtaining such a proof in a *general* first-order axiomatic system is undecidable. In our case it is decidable, since all necessary axioms are Horn-formulas, i.e., disjunctions of literals with at most one being positive. This allows for the application of the well-known SLD-resolution [38] to find proofs.

We now describe which axioms are contained in Σ and how the inference, i.e., obtaining the formal proof, can be done. The plan itself is encoded in Σ by several axioms, using two ternary predicates cr and dr, which describe the causal and hierarchical relations, respectively. For every causal link $\langle ps, \varphi, ps' \rangle$ in the plan, the axiom $cr(ps, \varphi, ps')$ is added to Σ. As described in Sect. 5.2, the plan to be explained has been obtained by applying a sequence of modifications, i.e., by adding causal links, ordering constraints or tasks, or by decomposing abstract tasks. Each used method m is applied to decompose some abstract plan step ps'. It adds a set of new plan steps PS (and ordering constraints and causal links) to the plan. For every such plan step $ps \in PS$ the axiom $dr(ps, m, ps')$ is added to Σ.

5.5.1.1 Explaining the Necessity of Plan Steps

To answer the first kind of question, axioms proving the necessity of a plan step must be defined. That necessity is described using the unary predicate n. Note that by "necessity" we do not refer to an absolute or global necessity of a plan step. We do not answer the question whether the respective action has to be part of any solution (such actions are called *action landmarks* [51, 58]). Answering this question is in general as hard as planning itself. Instead, we explain the *purpose* of the action: we give a chain of arguments explaining for which purpose that action is used within the presented plan.

All plan steps of the initial plan P_{init} (which includes the action a_{goal} that encodes the goal condition) are necessary by definition, since P_{init} describes the problem itself. Thus, $n(ps)$ is included as an axiom for every plan step ps of P_{init}. If a plan step ps is contained in the plan in order to provide a causal link for another necessary plan step, ps is also regarded as necessary, as it establishes a precondition of a required action. A simple example application of this rule is the necessity of any action establishing one (or more) of the goal conditions. The information that a plan step establishes the precondition of another plan step is explicitly given in hybrid plans by causal links. Using the given encoding of causal links in Σ, we can formulate an axiom to infer necessity as follows:

$$\forall ps, \varphi, ps' : cr(ps, \varphi, ps') \land n(ps') \rightarrow n(ps) \tag{5.2}$$

A similar argument can be applied if a plan step ps has been obtained via decomposition. If a necessary abstract plan step ps' is decomposed into ps, then ps serves the purpose of refining ps'. Converted into an axiom this reads:

$$\forall ps, m, ps' : dr(ps, m, ps') \wedge n(ps') \rightarrow n(ps) \tag{5.3}$$

One can use both Axiom (5.2) and (5.3) to show the purpose of any plan step: it either is used to ensure the executability of another plan step (in this case, the first rule may be applied), or is part of the plan because of decomposition (then, the second rule applies). Any chain of arguments (i.e., rule applications) will subsequently root in a plan step of the initial plan, i.e., the number of proof steps is always finite.

So far, the explanations based on causal dependencies (cf. Axiom (5.2)) only rely on *primitive* plan steps. However, even these causality-based explanations could be improved when taking into account abstract tasks. E.g., the presence of the plan step *plugIn*(SCART-CABLE, AUDIO-PORT, AV-RECEIVER, AUDIO-PORT) should be explained as follows: it is necessary, as it is part of the abstract task *connect*(BLUERAY-PLAYER, AV-RECEIVER), which provides *signalAt*(AUDIO, AV-RECEIVER), which in turn is needed by the action *connect*(AV-RECEIVER, *TV*) to achieve the goal *signalAt*(AUDIO, *TV*). To obtain such explanations, *cr* predicates (i.e., causal links) involving abstract tasks must be inferred. Here, the idea is that if a plan step ps has an effect (or precondition) linked to some other plan step ps' that has been introduced into the plan by decomposing ps'', then ps'' is also linked to ps' as one of its primitive tasks generated the condition necessary for ps'. The axiomatic system Σ contains two further axioms inferring these *cr* relations. Figure 5.3 contains a visual representation of both axioms.

$$\forall ps, m, ps'', ps' : dr(ps, m, ps'') \wedge cr(ps', \varphi, ps) \rightarrow cr(ps', \varphi, ps'') \tag{5.4}$$

$$\forall ps, m, ps'', ps' : dr(ps, m, ps'') \wedge cr(ps, \varphi, ps') \rightarrow cr(ps'', \varphi, ps') \tag{5.5}$$

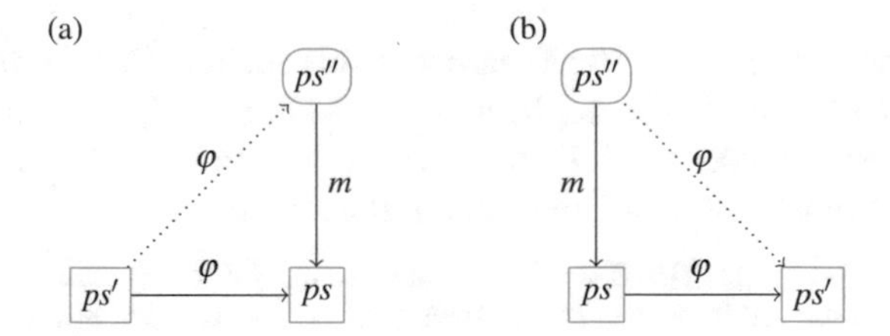

Fig. 5.3 The *rectangular boxes* depict primitive plan steps, the ones with *rounded corners* depict abstract plan steps. The *arrows* labeled with m indicate a performed decomposition using the method m. The *arrows* labeled with the literal φ indicate causal links, whereas the *dotted* ones are inferred by one of the Axioms (5.4) or (5.5). (**a**) Visualization of Axiom (5.4). (**b**) Visualization of Axiom (5.5)

5.5.1.2 Explaining the Order of Plan Steps

The second question a user might pose, i.e., why a plan step ps is arranged before some other plan step ps', has two possible answers. Either the order is contained in the plan presented to the user or it was chosen as part of the plan linearization process. In the latter case the system's answer could state that the order was chosen to obtain a plausible linearization and can be changed if the user so wishes. In the former case, again a proof for a fact is generated and conveyed to the user. Necessary order between plan steps is encoded using the binary relation $<$. If the user poses the said question, the fact $ps < ps'$ is to be proven in Σ and its proof constitutes the formal explanation for the order's necessity. A necessary order can be caused for several reasons, each of which is described by an axiom. For the sake of brevity, we will only provide an intuition on these axioms, while the interested reader is referred to the work of Seegebarth et al. [53] for further details.

Orderings can be contained in the plans referenced by decomposition methods; thus they are necessary if the respective abstract task is. Further, an ordering constraint may be added to a plan if a causal threat is to be dissolved. Here, the necessity is based on the threatening plan step of the threat (cf. Solution Criterion 2c). Order is also implicitly implied by every causal link in the plan, and its necessity is based on the necessity of the consuming plan step of the link.

5.5.2 *Verbalizing Plan Explanations*

After a formal plan explanation, expressed by a proof in first-order logic, is obtained, it has to be conveyed to the user in a suitable way. As a default approach, the explanation is transformed into text, which can be read to the user or displayed on a screen (see Fig. 5.2). To generate a natural language text, we use a pattern-based approach, an approach commonly used by automated theorem provers to present their proofs to humans [28, 30, 54]. Additionally, one could use techniques similar to those of the Interactive Derivation Viewer [56], which uses both verbal and visual explanations.

Consider the example mentioned earlier in this section. The formal explanation in this case consists of one application of Axiom (5.3), two applications of Axiom (5.2), one of Axiom (5.4), and two of Axiom (5.5). Resulting from this proof, the following natural language text is generated:

> Plug the audio end of the SCART-to-Cinch cable into the AV Receiver to connect the Blu-ray Player with the AV Receiver. This ensures that the AV Receiver has an audio signal, needed to connect the AV Receiver with the TV. This ensures that the TV has an audio signal, needed to achieve the goal.

We chose not to verbalize Axioms (5.4) and (5.5), as their application should be intuitively clear to the user. The remaining applications of Axioms (5.2) and (5.3) form a linear list. Each occurrence of Axiom (5.2) is translated into the text

"This ensures that $\langle\varphi\rangle$, needed to $\langle ps'\rangle$", where $\langle x\rangle$ denotes a domain-dependent verbalization of x. Likewise, each instance of Axiom (5.3) is translated into "Do this to $\langle ps'\rangle$". For the very first axiom in the explanation the beginnings of the sentences, "This" and "Do this", are replaced with the verbalization of the action to be explained.

5.6 Conclusion

Flexible system behavior is essential for realizing *Companion*-Systems [21]. We summarized how different system capabilities supporting this design goal can be implemented using the hybrid planning approach, starting with the generation process that might integrate the user and the execution and communication of generated solutions, as well as by discussing how to cope with unforeseen situations.

Though the current abilities of user-centered planning contribute valuable capabilities to the overall system, there are several promising lines of research for further improvements. Especially the problem of how plans are linearized [32] may offer further benefits for a convenient system. Another direction is a deeper explanation of system behavior [12, 53]. Here answers to questions like "Why can't I use this action/method?" or an explanation on why a problem at hand has no solution may help the user. The overall explanation quality might also be improved by further integrating ontology- as well as plan-based explanations [6]. Another important matter in real-world applications is the presentation of different alternatives to reach a goal [9].

Acknowledgements This work was done within the Transregional Collaborative Research Centre SFB/TRR 62 "*Companion*-Technology for Cognitive Technical Systems" funded by the German Research Foundation (DFG).

References

1. Alford, R., Shivashankar, V., Kuter, U., Nau, D.S.: HTN problem spaces: structure, algorithms, termination. In: Proceedings of the 5th Annual Symposium on Combinatorial Search (SoCS), pp. 2–9. AAAI Press, Palo Alto (2012)
2. Alford, R., Shivashankar, V., Kuter, U., Nau, D.: On the feasibility of planning graph style heuristics for HTN planning. In: Proceedings of the 24th International Conference on Automated Planning and Scheduling (ICAPS), pp. 2–10. AAAI Press, Palo Alto (2014)
3. Alford, R., Bercher, P., Aha, D.: Tight bounds for HTN planning. In: Proceedings of the 25th International Conference on Automated Planning and Scheduling (ICAPS), pp. 7–15. AAAI Press, Palo Alto (2015)
4. Alford, R., Bercher, P., Aha, D.: Tight bounds for HTN planning with task insertion. In: Proceedings of the 25th International Joint Conference on AI (IJCAI), pp. 1502–1508. AAAI Press, Palo Alto (2015)

5. Behnke, G., Höller, D., Biundo, S.: On the complexity of HTN plan verification and its implications for plan recognition. In: Proceedings of the 25th International Conference on Automated Planning and Scheduling (ICAPS), pp. 25–33. AAAI Press, Palo Alto (2015)
6. Behnke, G., Ponomaryov, D., Schiller, M., Bercher, P., Nothdurft, F., Glimm, B., Biundo, S.: Coherence across components in cognitive systems – one ontology to rule them all. In: Proceedings of the 25th International Joint Conference on AI (IJCAI), pp. 1442–1449. AAAI Press, Palo Alto (2015)
7. Behnke, G., Höller, D., Bercher, P., Biundo, S.: Change the plan - how hard can that be? In: Proceedings of the 26th International Conference on Automated Planning and Scheduling (ICAPS), pp. 38–46. AAAI Press, Palo Alto (2016)
8. Bercher, P., Biundo, S.: A heuristic for hybrid planning with preferences. In: Proceedings of the 25th International Florida AI Research Society Conference (FLAIRS), pp. 120–123. AAAI Press, Palo Alto (2012)
9. Bercher, P., Höller, D.: Interview with David E. Smith. Künstliche Intelligenz **30**, 101–105 (2016). doi:10.1007/s13218-015-0403-y
10. Bercher, P., Geier, T., Biundo, S.: Using state-based planning heuristics for partial-order causal-link planning. In: Advances in AI, Proceedings of the 36th German Conference on AI (KI), pp. 1–12. Springer, Berlin (2013)
11. Bercher, P., Geier, T., Richter, F., Biundo, S.: On delete relaxation in partial-order causal-link planning. In: Proceedings of the 25th International Conference on Tools with AI (ICTAI), pp. 674–681. IEEE Computer Society, New York (2013)
12. Bercher, P., Biundo, S., Geier, T., Hörnle, T., Nothdurft, F., Richter, F., Schattenberg, B.: Plan, repair, execute, explain - how planning helps to assemble your home theater. In: Proceedings of the 24th International Conference on Automated Planning and Scheduling (ICAPS), pp. 386–394. AAAI Press, Palo Alto (2014)
13. Bercher, P., Keen, S., Biundo, S.: Hybrid planning heuristics based on task decomposition graphs. In: Proceedings of the 7th Annual Symposium on Combinatorial Search (SoCS), pp. 35–43. AAAI Press, Palo Alto (2014)
14. Bercher, P., Richter, F., Hörnle, T., Geier, T., Höller, D., Behnke, G., Nothdurft, F., Honold, F., Minker, W., Weber, M., Biundo, S.: A planning-based assistance system for setting up a home theater. In: Proceedings of the 29th National Conference on Artificial Intelligence (AAAI), pp. 4264–4265. AAAI Press, Palo Alto (2015)
15. Bercher, P., Höller, D., Behnke, G., Biundo, S.: More than a name? On implications of preconditions and effects of compound HTN planning tasks. In: Proceedings of the 22nd European Conference on Artificial Intelligence (ECAI 2016), pp. 225–233. IOS Press (2016)
16. Bertrand, G., Nothdurft, F., Honold, F., Schüssel, F.: CALIGRAPHI-creation of adaptive dialogues using a graphical interface. In: 35th Annual Computer Software and Applications Conference (COMPSAC), pp. 393–400. IEEE, New York (2011)
17. Bidot, J., Schattenberg, B., Biundo, S.: Plan repair in hybrid planning. In: Advances in AI, Proceedings of the 31st German Conference on AI (KI), pp. 169–176. Springer, Berlin (2008)
18. Biundo, S., Schattenberg, B.: From abstract crisis to concrete relief (a preliminary report on combining state abstraction and HTN planning). In: Proceedings of the 6th European Conference on Planning (ECP), pp. 157–168. AAAI Press, Palo Alto (2001)
19. Biundo, S., Wendemuth, A.: *Companion*-technology for cognitive technical systems. Künstliche Intelligenz **30**, 71–75 (2016). doi:10.1007/s13218-015-0414-8
20. Biundo, S., Bercher, P., Geier, T., Müller, F., Schattenberg, B.: Advanced user assistance based on AI planning. Cogn. Syst. Res. **12**(3–4), 219–236 (2011); Special Issue on Complex Cognition
21. Biundo, S., Höller, D., Schattenberg, B., Bercher, P.: Companion-technology: an overview. Künstliche Intelligenz **30**, 11–20 (2016). doi:10.1007/s13218-015-0419-3
22. Bylander, T.: The computational complexity of propositional STRIPS planning. Artif. Intell. **94**(1–2), 165–204 (1994)
23. Edelkamp, S.: On the compilation of plan constraints and preferences. In: Proceedings of the 16th International Conference on Automated Planning and Scheduling (ICAPS), pp. 374–377. AAAI Press, Palo Alto (2006)

24. Elkawkagy, M., Schattenberg, B., Biundo, S.: Landmarks in hierarchical planning. In: Proceedings of the 20th European Conference on AI (ECAI), pp. 229–234. IOS Press, Amsterdam (2010)
25. Elkawkagy, M., Bercher, P., Schattenberg, B., Biundo, S.: Improving hierarchical planning performance by the use of landmarks. In: Proceedings of the 26th National Conference on Artificial Intelligence (AAAI), pp. 1763–1769. AAAI Press, Palo Alto (2012)
26. Erol, K., Hendler, J.A., Nau, D.S.: UMCP: a sound and complete procedure for hierarchical task-network planning. In: Proceedings of the 2nd International Conference on AI Planning Systems (AIPS), pp. 249–254. AAAI Press, Palo Alto (1994)
27. Erol, K., Hendler, J.A., Nau, D.S.: Complexity results for HTN planning. Ann. Math. Artif. Intell. **18**(1), 69–93 (1996)
28. Fiedler, A.: P.rex: an interactive proof explainer. In: Proceedings of the 1st International Joint Conference on Automated Reasoning (IJCAR), pp. 416–420. Springer, Berlin (2001)
29. Geier, T., Bercher, P.: On the decidability of HTN planning with task insertion. In: Proceedings of the 22nd International Joint Conference on Artificial Intelligence (IJCAI), pp. 1955–1961. AAAI Press, Palo Alto (2011)
30. Holland-Minkley, A.M., Barzilay, R., Constable, R.L.: Verbalization of high-level formal proofs. In: Proceedings of the 16th National Conference on AI and the 11th Innovative Applications of AI Conference (AAAI/IAAI), pp. 277–284. AAAI Press, Palo Alto (1999)
31. Höller, D., Behnke, G., Bercher, P., Biundo, S.: Language classification of hierarchical planning problems. In: Proceedings of the 21st European Conference on AI (ECAI), pp. 447–452. IOS Press, Amsterdam (2014)
32. Höller, D., Bercher, P., Richter, F., Schiller, M., Geier, T., Biundo, S.: Finding user-friendly linearizations of partially ordered plans. In: 28th PuK Workshop "Planen, Scheduling und Konfigurieren, Entwerfen" (PuK) (2014)
33. Höller, D., Behnke, G., Bercher, P., Biundo, S.: Assessing the expressivity of planning formalisms through the comparison to formal languages. In: Proceedings of the 26th International Conference on Automated Planning and Scheduling (ICAPS), pp. 158–165. AAAI Press, Palo Alto (2016)
34. Honold, F., Schüssel, F., Weber, M.: Adaptive probabilistic fission for multimodal systems. In: Proceedings of the 24th Australian Computer-Human Interaction Conference (OzCHI), pp. 222–231. ACM, New York (2012)
35. Honold, F., Bercher, P., Richter, F., Nothdurft, F., Geier, T., Barth, R., Hörnle, T., Schüssel, F., Reuter, S., Rau, M., Bertrand, G., Seegebarth, B., Kurzok, P., Schattenberg, B., Minker, W., Weber, M., Biundo, S.: Companion-technology: towards user- and situation-adaptive functionality of technical systems. In: International Conference on Intelligent Environments (IE), pp. 378–381. IEEE, New York (2014). http://companion.informatik.uni-ulm.de/ie2014/companion-system.mp4
36. Kambhampati, S., Mali, A., Srivastava, B.: Hybrid planning for partially hierarchical domains. In: Proceedings of the 15th National Conference on AI (AAAI), pp. 882–888. AAAI Press, Palo Alto (1998)
37. Keyder, E., Geffner, H.: Soft goals can be compiled away. J. Artif. Intell. Res. **36**, 547–556 (2009)
38. Kowalski, R.A.: Predicate logic as programming language. In: IFIP Congress, pp. 569–574 (1974)
39. Lin, N., Kuter, U., Sirin, E.: Web service composition with user preferences. In: Proceedings of the 5th European Semantic Web Conference (ESWC), pp. 629–643. Springer, Heidelberg (2008)
40. Lyons, J.B., Koltai, K.S., Ho, N.T., Johnson, W.B., Smith, D.E., Shively, R.J.: Engineering trust in complex automated systems. Ergon. Des. **24**(1), 13–17 (2016). https://doi.org/10.1177/1064804615611272
41. McAllester, D., Rosenblitt, D.: Systematic nonlinear planning. In: Proceedings of the 9th National Conference on AI (AAAI), pp. 634–639. AAAI Press, Palo Alto (1991)

42. Muir, B.M.: Trust in automation: part I. theoretical issues in the study of trust and human intervention in automated systems. Ergonomics **37**(11), 1905–1922 (1994)
43. Nau, D.S., Au, T.C., Ilghami, O., Kuter, U., Murdock, J.W., Wu, D., Yaman, F.: SHOP2: an HTN planning system. J. Artif. Intell. Res. **20**, 379–404 (2003)
44. Nau, D.S., Au, T.C., Ilghami, O., Kuter, U., Wu, D., Yaman, F., Muñoz-Avila, H., Murdock, J.W.: Applications of SHOP and SHOP2. IEEE Intell. Syst. **20**, 34–41 (2005)
45. Nebel, B., Koehler, J.: Plan reuse versus plan generation: a theoretical and empirical analysis. Artif. Intell. **76**(1-2), 427–454 (1995)
46. Nguyen, X., Kambhampati, S.: Reviving partial order planning. In: Proceedings of the 17th International Joint Conference on Artificial Intelligence (IJCAI), pp. 459–466. Morgan Kaufmann, San Francisco (2001)
47. Nothdurft, F., Bertrand, G., Heinroth, T., Minker, W.: GEEDI - guards for emotional and explanatory dialogues. In: 6th International Conference on Intelligent Environments (IE), pp. 90–95. IEEE, New York (2010)
48. Nothdurft, F., Honold, F., Zablotskaya, K., Diab, A., Minker, W.: Application of verbal intelligence in dialog systems for multimodal interaction. In: 10th International Conference on Intelligent Environments (IE), pp. 361–364. IEEE, New York (2014)
49. Parasuraman, R., Riley, V.: Humans and automation: use, misuse, disuse, abuse. Hum. Factors: J. Hum. Factors Ergon. Soc. **39**(2), 230–253 (1997)
50. Penberthy, J.S., Weld, D.S.: UCPOP: a sound, complete, partial order planner for ADL. In: Proceedings of the 3rd International Conference on Principles of Knowledge Representation and Reasoning (KR), pp. 103–114. Morgan Kaufmann, San Francisco (1992)
51. Porteous, J., Sebastia, L., Hoffmann, J.: On the extraction, ordering, and usage of landmarks in planning. In: Proceedings of the 6th European Conference on Planning (ECP), pp. 37–48. AAAI Press, Palo Alto (2001)
52. Russell, S., Norvig, P.: Artificial Intelligence – A Modern Approach, 1 edn. Prentice-Hall, Englewood Cliffs (1994)
53. Seegebarth, B., Müller, F., Schattenberg, B., Biundo, S.: Making hybrid plans more clear to human users – a formal approach for generating sound explanations. In: Proceedings of the 22nd International Conference on Automated Planning and Scheduling (ICAPS), pp. 225–233. AAAI Press, Palo Alto (2012)
54. Simons, M.: Proof presentation for Isabelle. In: Proceedings of the 10th International Conference on Theorem Proving in Higher Order Logics (TPHOLs), pp. 259–274. Springer, Berlin (1997)
55. Sohrabi, S., Baier, J.A., McIlraith, S.A.: HTN planning with preferences. In: Proceedings of the 21st International Joint Conference on AI (IJCAI), pp. 1790–1797. AAAI Press, Palo Alto (2009)
56. Trac, S., Puzis, Y., Sutcliffe, G.: An interactive derivation viewer. Electron. Notes Theor. Comput. Sci. **174**(2), 109–123 (2007)
57. Younes, H.L.S., Simmons, R.G.: VHPOP: versatile heuristic partial order planner. J. Artif. Intell. Res. **20**, 405–430 (2003)
58. Zhu, L., Givan, R.: Heuristic planning via roadmap deduction. In: IPC-4 Booklet, pp. 64–66 (2004)

Chapter 6
Addressing Uncertainty in Hierarchical User-Centered Planning

Felix Richter and Susanne Biundo

Abstract Companion-Systems need to reason about dynamic properties of their users, e.g., their emotional state, and the current state of the environment. The values of these properties are often not directly accessible; hence information on them must be pieced together from indirect, noisy or partial observations. To ensure probability-based treatment of partial observability on the planning level, planning problems can be modeled as Partially Observable Markov Decision Processes (POMDPs).

While POMDPs can model relevant planning problems, it is algorithmically difficult to solve them. A starting point for mitigating this is that many domains exhibit hierarchical structures where plans consist of a number of higher-level activities, each of which can be implemented in different ways that are known a priori. We show how to make use of such structures in POMDPs using the Partially Observable HTN (POHTN) planning approach by developing a Partially Observable HTN (POHTN) action hierarchy for an example domain derived from an existing deterministic demonstration domain.

We then apply Monte-Carlo Tree Search to POHTNs for generating plans and evaluate both the developed domain and the POHTN approach empirically.

6.1 Introduction

Companion-Systems offering decision making capabilities need to reason about dynamic properties of their environment. Most notably, this environment consists of the user of the system and its physical surroundings. Aspects of interest include the user's emotional state as well as attributes of relevant physical objects, both of which and may be difficult or costly to measure. It is, however, often possible to gain some knowledge about the aspects of interest by observing related, more easily accessible properties, even if these observations offer only noisy or partial information.

F. Richter (✉) • S. Biundo
Institute of Artificial Intelligence, Ulm University, Ulm, Germany
e-mail: felix.richter@alumni.uni-ulm.de; susanne.biundo@uni-ulm.de

S. Biundo, A. Wendemuth (eds.), *Companion Technology*, Cognitive Technologies,
DOI 10.1007/978-3-319-43665-4_6

A natural planning model that accounts for partial observability within a probability-based framework is given by Partially Observable Markov Decision Processes (POMDPs) [19]. A drawback of POMDPs, however, is that it is difficult to compute policies that prescribe good courses of action. This is especially true for large problems, which often arise when adequately modeling a task at hand requires factoring in many different user and world properties.

At the same time, many problems humans are confronted with typically consist of a number of higher-level activities. These activities can be hierarchically divided into smaller activities and eventually simple actions. Often, there is also a limited number of useful possibilities how a particular activity can be performed. This hierarchical structure opens opportunities for *Companion*-Systems to mitigate the computational burden of computing policies in partially observable environments.

For deterministic planning domains, Hierarchical Task Network (HTN) planning [5, 6] is a practical planning approach that allows modeling and exploiting such hierarchical structures. It allows for efficient plan generation and makes it easy to express expert knowledge about a planning domain. As such, it has been successfully applied to many real-world problems [14]. Typical such hierarchically structured real-world problems that can be tackled with HTN planning are, e.g., given by instances of the home theater setup domain developed for demonstration purposes in the context of intelligent assistance systems [1, 2, 9] and described in Chap. 24. The task in this domain is depicted in Fig. 6.1a: given an assortment of devices and cables, connect the devices such that in the end the user has a working home theater, as depicted in Fig. 6.1b.

A number of approaches exist that make use of hierarchical structure in planning domains that exhibit uncertainty of some kind, in either fully observable MDP or partially observable POMDP settings. Some approaches augment the original set of actions with macro-actions in the spirit of the options framework [20]: approaches in this category either have a one-to-one correspondence between macro-actions and their implementation [21] or try to generate restricted implementations at planning time [8]. Other hierarchical POMDP approaches such as PolCA+ [16] or MAXQ-

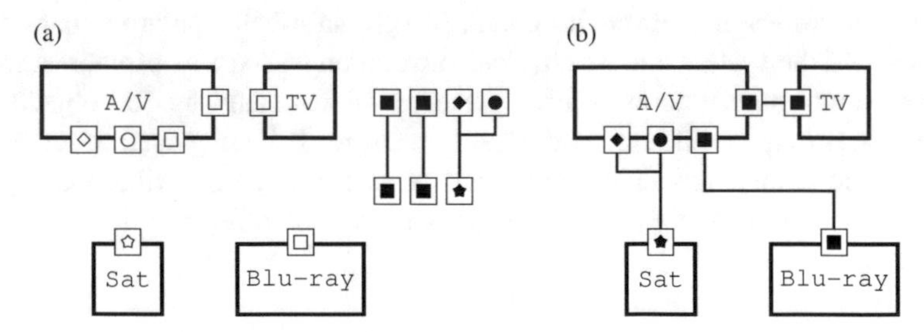

Fig. 6.1 Schematic representation of the home theater setup domain. The A/V receiver, TV, satellite receiver, and Blu-ray player each have various female ports, where *rectangle*, *star*, *diamond*, and *circle* denote HDMI, SCART, cinch video and cinch audio ports, respectively. Black ports on cables denote male ports. There are two HDMI cables and one SCART-to-cinch-AV cable. (**a**) The task: an assortment of unconnected devices and cables. (**b**) The goal: devices are properly connected

hierarchical policy iteration [7] define a fixed hierarchical decomposition of a task a priori, similar to the MAXQ decomposition [4] or HAM [15] in the fully observable MDP setting, and individually optimize sub-policies for each abstract action.

An alternative to the above approaches is to directly extend HTN planning to POMDPs, which results in the Partially Observable HTN (POHTN) approach that we describe in our earlier work [12, 13]. In this chapter, we develop a variant of the domain sketched in Fig. 6.1 that relaxes the full observability assumption and creates a suitable POHTN hierarchy. To demonstrate the effectiveness of the POHTN approach, we evaluate our approach empirically on several instances of the home theater setup domain.

The chapter is structured as follows. We first explain POMDP concepts and how the home theater domain is modeled using the Relational Dynamic Influence Diagram Language (RDDL) [17] in Sect. 6.2. Next, Sect. 6.3 reviews an existing popular non-hierarchical approach to POMDP planning, namely Monte-Carlo Tree Search (MCTS) on the basis of observable histories. Section 6.4 presents the POHTN approach and shows how a POHTN hierarchy can be constructed for the home theater setup domain. Section 6.4.4 describes the application of MCTS to POHTN planning. Our experiments in Sect. 6.5 compare MCTS-based POHTN planning and history-based MCTS planning and also show how modeling choices lead to domain variants with different computational difficulties and practical implications. Section 6.6 concludes with some final remarks.

6.2 The Home Theater Setup POMDP

A POMDP is an 8-tuple $(S, A, O, T, Z, R, b_0, H)$, where S, A, and O are finite sets of states, actions, and observations, respectively. The effects of executing actions on the system's environment are defined by the transition function T, in the sense that for a given state $s \in S$ and action $a \in A$, $T(s, a)$ defines a probability distribution over possible successor world states $s' \in S$. Similarly, the system's sensor model is determined by the observation function Z such that for a given action $a \in A$ and successor state $s' \in S$, $Z(a, s')$ defines a probability distribution over possible observations $o \in O$. The system alternatingly executes actions and receives observations as depicted in Fig. 6.2. Note that the successor world state s' is not visible to the system; it must infer information on the identity of s' from the observation o.

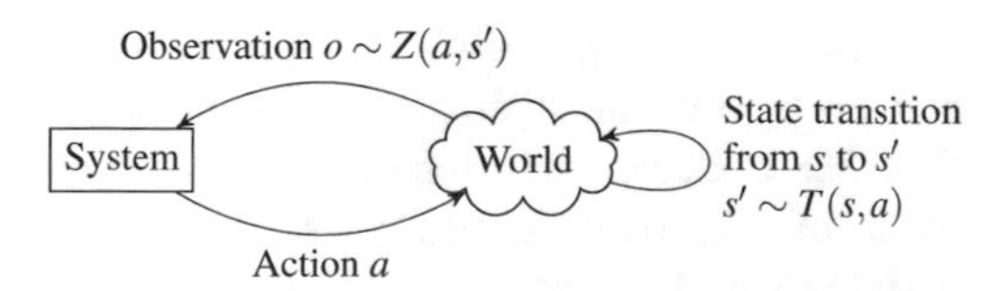

Fig. 6.2 The POMDP interaction cycle

What the system can see about its environment is the actions it executes and the observations it receives. After t steps, the system's entire knowledge about the evolution of its environment is thus the sequence $a^1o^1 \dots a^to^t$, called an *observable history*. Because of that, the system's policy can be represented as a function that maps observable histories to actions.

The system's goals are given in terms of a real-valued reward function $R(s, a, s')$ that determines how beneficial it is for the system when the result of executing $a \in A$ in $s \in S$ is $s' \in S$. The system's success is determined by the expected accumulated reward it is able to gather in a given number of time steps H, called the horizon, starting in a state that is sampled from the probability distribution b_0, the initial belief state. Formally, this can be captured as follows: for a given history $h = a^1o^1 \dots a^to^t$, let hao denote the history extended by a and o, i.e., $a^1o^1 \dots a^to^tao$, and let π denote the system's policy. Then the quality of π in a given state s is given by

$$V_\pi^h(s) = \begin{cases} \mathbb{E}_{s' \sim T(s,\pi(h))}[R(s, \pi(h), s')] & \text{if } h \text{ has } H \text{ steps} \\ \mathbb{E}_{s' \sim T(s,\pi(h))}[R(s, \pi(h), s') + \mathbb{E}_{o \sim Z(s',\pi(h))}[V_\pi^{h\pi(h)o}(s')]] & \text{else,} \end{cases} \tag{6.1}$$

where $\mathbb{E}_{x \sim X}[f(x)]$ denotes the expectation of $f(x)$ when x is distributed according to X. The quality of π in the initial belief state is $V_\pi^h(b_0) = \mathbb{E}_{s \sim b_0}[V_\pi^h(s)]$. The goal in POMDP planning is finding an optimal policy, i.e., a policy $\pi^* = \text{argmax}_\pi V_\pi^h(b_0)$.

In RDDL, POMDPs are defined using typed first-order logic. First, a set of types is defined that determines the relevant objects. For the home theater domain, these include types for devices, ports, and so on:

```
Device: object; // TV, Blu-ray player, etc.
SignalType: object; // audio, video, ...
// ultimate source of signal, e.g., Blu-ray player
SignalSource: object;
Port: object; // a port such as HDMI, cinch, ...
// numbers for counting the number of free ports
count: {@zero,@one,@two,@three};
// how tight a connection is
tightness: {@none,@loose,@tight};
```

For defining the set of states S, typed parametrized state fluents are used in RDDL, which are either predicates or functions in the first-order logic sense. The home theater domain uses four state fluents to capture relevant aspects of a given state. The first two, `freeFemalePorts(Device,Port)` and `freeMalePorts(Device,Port)`, determine the number of free female and male ports on a device, respectively, and can assume values from 0 to 3. Whether a device has received a signal of a certain source and type is kept track of via `hasSignal(Device,SignalType,SignalSource)`. E.g., `hasSignal(TV,audio,Sat)` means that the audio signal of the satellite receiver has reached the TV. The `connected(Device,Device,Port)` fluent models how tightly two devices are connected—not at all, loosely, or tightly. This represents a deviation from the original domain, where devices are connected tightly or not at all [1]. The difference between a tight and a loose connection is that while the cable is plugged in both cases, a loose connection does not transport a signal of any kind. This could be interpreted as a halfheartedly plugged-in cable, for example. We will later use this to add partial observability to the domain.

The set of states is then given by the set of possible interpretations of state fluents. E.g., suppose the available devices are a TV, a satellite receiver, and an HDMI cable. A state where both the satellite receiver and the TV have one free female HDMI port each, the HDMI cable has two free male HDMI ports, the satellite receiver creates an audio and a video signal, and nothing is connected yet is defined as follows:

```
freeFemalePorts(Sat,HDMI) = @one;
freeFemalePorts(TV,HDMI) = @one;

freeMalePorts(HDMI_Cable,HDMI) = @two;

hasSignal(Sat,audio,Sat);
hasSignal(Sat,video,Sat);
```

Actions and observations are defined in a similar manner, using specific action and observation fluents, respectively. The home theater setup domain features a `connect(Device,Device,Port)` action fluent for instructing the user to connect two devices via some port, a `tighten(Device,Device,Port)` action fluent for instructing the user to tighten loose connections, and a `checkSignals(Device)` action fluent for instructing the user to check the signals on a given device. The only observation fluent of the home theater setup domain is `hasSignalObs(Device,SignalType,SignalSource)`, which signifies whether a given device has a signal of a given type from a given source.

RDDL also has the possibility to define so-called intermediate fluents, whose values are calculated from the current state and which can be used to calculate the successor state. These non-observable fluents are useful when several successor state fluents depend on a single probabilistic outcome. The home theater setup domain uses this feature in `connectSucceeded(Device,Device,Port)` for determining whether a connection between two devices is tight or loose after they are connected. The last type of fluent, called non-fluents, is useful for static properties, such as whether a signal type can be transported through a certain kind of port. E.g., HDMI ports will transport both video and audio signals, but a cinch video port will only transport video.

State transitions are defined in terms of parametrized functions, one for each state fluent and intermediate fluent, called conditional probability functions. For each instantiation of a state fluent or intermediate fluent, they define a probability distribution over its value in the successor state. E.g., the value of `connectSucceeded(Device,Device,Port)` is defined as follows using RDDL syntax:

```
connectSucceeded(?d1,?d2,?p) =
 if (connect(?d1,?d2,?p) ^
  (freeFemalePorts(?d1,?p) ~= @zero ^
   freeMalePorts(?d2,?p) ~= @zero |
   freeMalePorts(?d1,?p) ~= @zero ^
   freeFemalePorts(?d2,?p) ~= @zero)) then
    Discrete(tightness,@none:0,@loose:0.2,@tight:0.8)
 else if (tighten(?d1,?d2,?p) ^
   connected(?d1,?d2,?p) == @loose) then
    KronDelta(@tight)
 else KronDelta(@none);
```

This means that when `connect` is executed and suitable ports are free on the devices that should be connected, the connect action will succeed. However, the connection will be loose with probability 0.2. This can be fixed by executing `tighten` to make sure the connection is tight.

The value of `connected(Device,Device,Port)` persists unless the connection is tighter than it was before. Similarly, the number of free ports on a device remains unchanged unless `connect` is executed. Signal availability is determined via `hasSignal(Device,SignalType,SignalSource)` by propagating signals over tight connections in every time step. Note that depending on the order in which connect actions are executed, it will take several time steps for a signal to propagate over a chain of devices. It would be preferable to have instantaneous signal propagation, but this would require computing the transitive closure over `hasSignal(Device,SignalType,SignalSource)`, which is not supported in RDDL.

Observation probabilities are defined analogously to state transition probabilities. The home theater domain has `hasSignalObs(Device,SignalType, SignalSource)` as its sole observation fluent. When `checkSignals (Device)` was executed on a device and the device can be checked for the type of signal in question (modeled using a non-fluent), it will reveal every signal available on the device to the system. The rationale behind this is that a TV can be checked for both video and audio signals by simply turning it on, as opposed to an HDMI cable. Checking for signals is the only way for the system to determine whether connections are tight.

The reward function in the model is constructed to fulfill several conditions:

1. The system's first priority should be to bring signals from source devices, such as a satellite receiver, to target devices, such as a TV. The target devices are identified using a non-fluent. In its simplest form, the reward for each time step is simply given by summing the number of signals that have already been brought to their target devices:

```
sum_{?d: Device, ?t: SignalType, ?s: SignalSource}
 DEVICE_NEEDS_SIGNAL(?d,?t,?s)*hasSignal'(?d,?t,?s)
```

A system trying to maximize its expected accumulated reward will therefore strive to bring all signals to their respective target devices in as few time steps as possible. It also urges the system to order connect actions such that cables are plugged in from source devices to target devices. This is more an artifact introduced by the signal propagation mechanism than intended. However, one could argue that this is in the interest of the user of the system, since the system's action recommendations can then be understood as "bringing" the signals to the target devices.

2. The system should avoid executing actions when their "preconditions" are not fulfilled (attempting connects when the required ports do not exist or are not free, attempting tightens when there is no connection at all, checking uncheckable devices). While these conditions are not observable in a strict sense, the system can in principle still derive whether the corresponding actions will succeed. E.g., the number of free ports evolves deterministically; therefore the system can always know whether a connect will fail to at least establish a loose connection. Similarly, whether a device can be checked for signals is known in advance, so checking uncheckable devices can be avoided.
3. The system should avoid executing unnecessary actions (tighten when a connection is already tight, checking for signals more than once on a given device). In contrast to the conditions described in (2), these conditions are harder to fulfill, because they depend on partially observable state properties: suppose a satellite receiver and a TV are connected via a cable and the system has determined that the signal does not reach the TV. This means that one or both of the connections is loose. In this case, the system cannot see which connection is loose, so it must risk tightening an already tight connection to make sure both connections are tight. Still, these conditions are useful since, e.g., tighten never needs to be executed more than once for a given connection.
4. Checking signals and then deciding whether tightening is necessary should be more attractive than simply tightening connections without looking. Some care needs to be taken considering the balance between the different rewards mentioned above. As argued, the system sometimes cannot avoid tightening an already tight connection. But doing so should still be unattractive; otherwise the system can simply ignore the possibility for checking signals and just tighten all connections directly after connecting.

We conclude the description of the home theater setup domain by noting that, in the initial state of every instance, all devices are unconnected and the only devices that have any signals are the signal source devices.

6.3 History-Based POMDP Planning

Next, we review a non-hierarchical POMDP planning approach, which will serve as baseline for our experiments in Sect. 6.5. Monte-Carlo Tree Search is a very popular approach to planning in uncertain environments [3], in particular UCT (Upper Confidence Bound applied to Trees [11]) and its variants, such as MaxUCT [10]. MCTS is a round-based anytime algorithm that incrementally constructs an explicit representation of the search space in memory. The tree contains alternating layers of decision nodes n_d and chance nodes n_c, where the root node is a decision node. The number of visits and estimated value of a node in the tree after k rounds of search are denoted by $C^k(n)$ and $V^k(n)$, respectively. Chance nodes also have estimates $R^k(n_c)$ for the immediate reward of executing their associated action. Applied to POMDPs, the search space is the space of observable histories [18] as depicted in Fig. 6.3.

Each round of search consists of two phases, tree traversal and backup. In the traversal phase, a generative model of the search space is used in conjunction with a tree traversal strategy to explore promising parts of the search space: starting from the root node, the tree is traversed by selecting actions and simulating their outcomes, until a terminal node is reached. In the case of search in the space of observable histories, the root node represents the empty history and terminal nodes are histories of length H. A typical tree traversal strategy is the UCT action selection formula [11], which, in a given decision node n_d, chooses the action (equivalently chance node n_c) that maximizes $V^k(n_c) + B\sqrt{(\log C^k(n_d))/C^k(n_c)}$, where B is a parameter that trades of between exploitative (favor high $V^k(n_c)$) and explorative (favor rarely visited n_c) behavior.

The backup phase updates the value estimates by incorporating the information gathered during the traversal, in reverse order of traversal. Chance node values are defined as

$$V^k(n_c) = R^k(n_c) + \frac{\sum_{n_d \in \mathrm{succ}^k(n)} C^k(n_d)V^k(n_d)}{C^k(n_c)}, \tag{6.2}$$

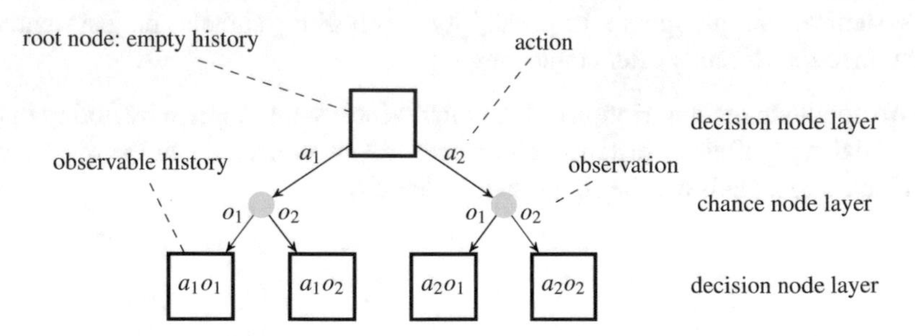

Fig. 6.3 Search tree for history-based POMDP planning. *Rectangular* nodes represent observable histories and are called decision nodes. *Grey round* nodes are called chance nodes

where the counters $C^k(n)$ are simply incremented in each visit, the immediate reward estimates $R^k(n_c)$ are averages over the experienced immediate rewards, and $\text{succ}^k(n)$ denotes the successor nodes of n in the search tree.

Several useful variants for backing up decision node values exist, most notably Monte-Carlo backup used in UCT [11], i.e.,

$$V^k(n_d) = \begin{cases} 0 & \text{if } n_d \text{ is a terminal node} \\ \frac{\sum_{n_c \in \text{succ}^k(n_d)} C^k(n_c) V^k(n_c)}{C^k(n_d)} & \text{else,} \end{cases} \tag{6.3}$$

and Max-Monte-Carlo backups used in MaxUCT[10]:

$$V^k(n_d) = \begin{cases} 0 & \text{if } n_d \text{ is a terminal node} \\ \max_{n_c \in \text{succ}^k(n_d)} V^k(n_c) & \text{else.} \end{cases} \tag{6.4}$$

Both have different strengths and weaknesses, the discussion of which is beyond the scope of this paper. We use a custom combination of both based on the weighted power mean which we call Soft Max-Monte-Carlo backups. For this, let $p = C^k(n_d)/\left|\text{succ}^k(n_d)\right|$ and define

$$V^k(n_d) = \begin{cases} 0 & \text{if } n_d \text{ is a terminal node} \\ \left(\frac{\sum_{n_c \in \text{succ}(n_d)} C^k(n_c) \left(V^k(n_c)\right)^p}{C^k(n_d)} \right)^{1/p} & \text{else.} \end{cases} \tag{6.5}$$

Until all actions have been tried once, this resembles Monte-Carlo backups when used in conjunction with UCT tree traversal, and converges to Max-Monte-Carlo backups as the number of samples grows.

For our experiments, we denote MCTS in the space of observable histories in conjunction with Soft Max-Monte-Carlo backups and UCT tree traversal by POUCT.

6.4 Partially Observable HTN Planning

Hierarchical Task Network planning, as originally proposed for deterministic planning domains [5], has been successfully applied to a range of real-world planning problems [14]. HTN planning domains are defined in terms of a hierarchy of actions: *abstract* actions are introduced to represent higher-level activities. Normal actions are then often called primitive to distinguish them from abstract actions. An HTN planning problem is given as a number of activities to be performed in a certain order. Since these activities are higher-level, they cannot be directly executed and act as placeholders. For each abstract action, several implementations, called methods, are given. They represent possible ways in which the activity can be performed in terms of a "sub-plan", in turn consisting of primitive or abstract actions. Plan generation in HTN planning means iteratively replacing the abstract actions of the initially specified abstract plan with suitable implementations until a solution plan is found that only contains primitive actions. Next, we review the definition of the POHTN approach given in our earlier work [12, 13].

6.4.1 HTN Planning

As a basis, we start by giving a formal description of a simple deterministic HTN planning framework based on the HTN planning formalization of Geier and Bercher [6]. The formalism only uses parameter-less fluents for representing states and actions to keep its description as simple as possible. However, the description can be easily generalized. Furthermore, we simplify the formalism by requiring that all task networks be totally ordered. We will therefore prefer to speak of action sequences instead of task networks, and denote the set of action sequences over action set X by TN_X. An HTN planning problem is a 6-tuple (L, C, A, M, c_I, s_I), where L is a finite set of state fluents. The primitive and abstract actions are given by the disjoint finite sets A and C, respectively. The available methods are given by $M \subseteq C \times \mathrm{TN}_{A \cup C}$, and $c_I \in C$ and $s_I \in 2^L$ denote the initial action and initial state, respectively. The dynamics of a primitive action a is defined in terms of precondition, add and delete lists, $(\mathrm{prec}(a), \mathrm{add}(a), \mathrm{del}(a)) \in 2^L \times 2^L \times 2^L$.

Let $\mathrm{ts}_1 = a_1, \ldots, a_{k-1}, c, a_{k+1}, \ldots, a_n$ be an action sequence, c an abstract action, and $m = (c, \mathrm{ts}_m)$ with $\mathrm{ts}_m = a_1^m, \ldots, a_l^m$ by a method for c. Applying m to ts_1 creates a new action sequence $\mathrm{ts}_2 = a_1, \ldots, a_{k-1}, a_1^m, \ldots, a_l^m, a_{k+1}, \ldots, a_n$ and is denoted $\mathrm{ts}_1 \rightarrow_m \mathrm{ts}_2$. When ts_2 can be created from ts_1 by applying an arbitrary number of methods from M, we write $\mathrm{ts}_1 \rightarrow_M^* \mathrm{ts}_2$.

An action sequence $a_1, \ldots, a_n$ is called executable in s if and only if every a_i is primitive and there exists a state sequence $s_0, \ldots, s_n$ such that $s_0 = s$, and $\mathrm{prec}(a_i) \subseteq s_{i-1}$ and $s_i = (s_{i-1} \setminus \mathrm{del}(a_i)) \cup \mathrm{add}(a_i)$ for all $1 \leq i \leq n$. Finally, an action sequence ts_S is a solution to an HTN problem if and only if it is executable in s_I and can be created from c_I, i.e., $c_I \rightarrow^* \mathrm{ts}_S$.

6.4.2 POHTN Planning

We will now describe the POHTN formalism in order to apply the HTN planning principles just described for POMDPs. We will again describe the POHTN formalism for parameter-less fluents only for brevity, but will give parametrized examples from the home theater domain.

First, we need an appropriate policy representation. We choose logical finite state controllers [12] for this purpose, since they can compactly represent POMDP policies and are a natural generalization of action sequences. A (logical) finite state

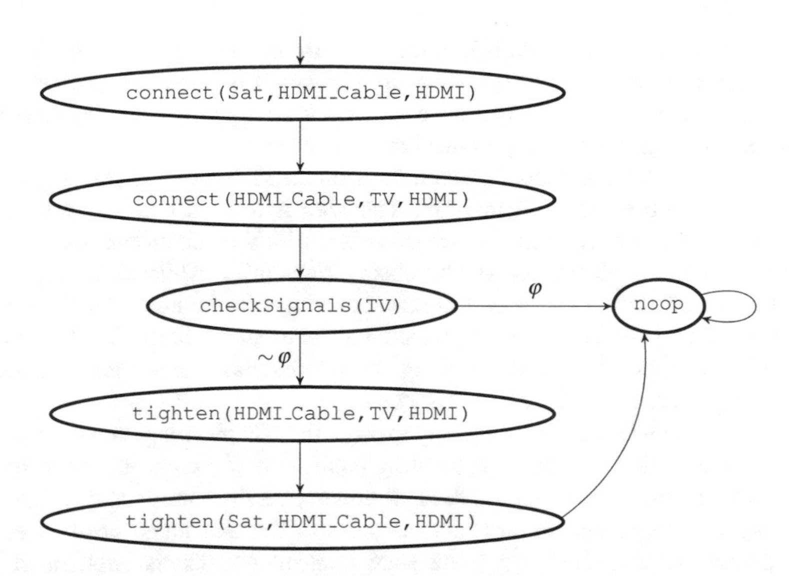

Fig. 6.4 A finite state controller for the home theater POMDP. The φ symbol is short for `exists_{?type,?source} hasSignalObs(TV,?type,?source)`. To reduce clutter, unlabeled edges denote `true` transition conditions; the transition condition between two unconnected nodes is `false`

controller fsc $= (N, \alpha, \delta, n_0)$ is a directed graph with node set N. Each node $n \in N$ is labeled with an action from action set X via $\alpha(n) \in X$. Edges are labeled with transition conditions $\delta(n, n')$, which are first-order formulas over a set of observation fluents Y. We call a controller well-defined if the outgoing transition conditions for every given node are mutually exclusive and exhaustive [13], i.e., if transitions define a distinct successor node for every possible observation. The controller is equipped with an initial node $n_0 \in N$. We denote the set of well-defined finite state controllers over X and Y by FSC(X, Y). Figure 6.4 shows a finite state controller for the home theater POMDP.

Executing a finite state controller in a given POMDP works by executing the action associated with the current node, starting with the initial node. It is then checked which transition condition φ is fulfilled by the received observation o, i.e., whether $o \models \varphi$, and the current node is updated to the target node of the transition whose condition is fulfilled. The process is repeated with the new current node until the horizon is reached. Given a history h, it is thus simple to determine the prescribed action of a controller by following observation edges.

Just as in HTN planning, we introduce a set of abstract actions C to complement the primitive actions A of the POMDP. For the home theater scenario, we will introduce the abstract action `connect_abstract(Device,Device)`.[1] The

[1]To be precise, this actually introduces n^2 parameter-less abstract actions when n is the number of devices.

intended meaning of the action is that it is an abstraction of bringing the signals of the first device to the second device via any number of intermediate devices, while also making sure that all intermediate connections are tight by checking signals on the target device and tightening connections if necessary.

Additionally, we also model observations on an abstract level and introduce a set of abstract observation fluents O^C. The idea is that, just as abstract actions are an abstraction of different courses of action with a common purpose, abstract observations are an abstraction of the observations made while executing such a course of action. The sole abstract observation fluent in the home theater domain is `loose_connection_found`, which is used to represent the fact that a loose connection was found somewhere along the intermediate connections created by `connect_abstract(Device,Device)`.

Methods in POHTN are, analogously to those in HTN planning, tuples consisting of an abstract action and an implementing controller. The controller representing the implementation part of the method is, however, a little more complicated due to the need to represent the abstract outcome of the associated abstract action. Such a method controller is a finite state controller which is augmented with a set of terminal nodes N_t, $N_t \cap N = \emptyset$. Each terminal node n_t is labeled with an interpretation $L(n_t)$ of the abstract observation fluents and can be the target of transitions. As an example, consider one of the implementations of `connect_abstract(Device,Device)` given in Fig. 6.5. We denote the set of method controllers over action set X, observation fluent set Y and abstract observation fluent set T as MFSC(X, Y, T).

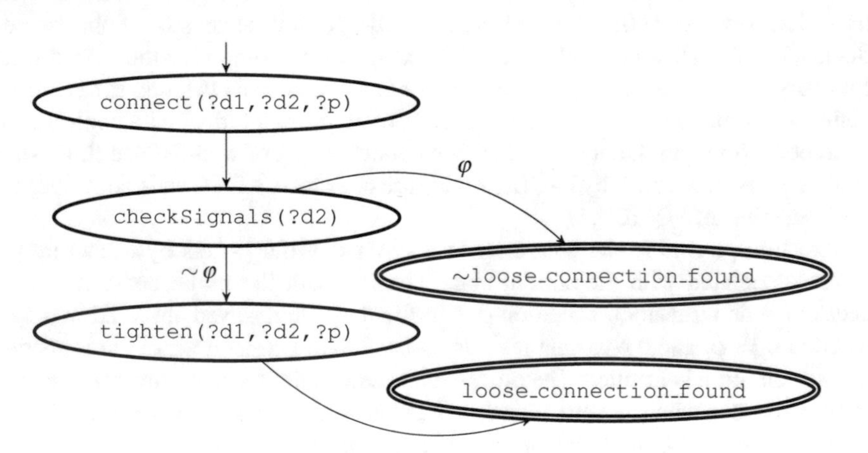

Fig. 6.5 The method controller of a method for `connect_abstract(?d1,?d2)`. The φ symbol is short for `exists_{?type,?source} hasSignalObs(?d2,?type,?source)`. The method implements `connect_abstract(?d1,?d2)` by connecting the devices directly, checking the connection between them, and creating an abstract observation with the result. To represent this method in our parameter-less framework, consider an instantiation of `connect_abstract(?d1,?d2)` with a fixed pair of values for `?d1` and `?d2`. This yields one method for each possible value for `?p`

With the above elements, we can syntactically define the POHTN planning problem as a tuple $(P, C, O^C, M, \mathrm{fsc}_I)$, where

- $P = (S, A, O, T, Z, R, b_0, H)$ is a POMDP,
- C is a finite set of abstract actions with $C \cap A = \emptyset$,
- O^C is a finite set of abstract observations with $O \cap O^C = \emptyset$,
- $M \subseteq C \times \mathrm{MFSC}(A \cup C, O \cup O^C, O^C)$ is the set of methods, and
- $\mathrm{fsc}_I \in FSC(A \cup C, O \cup O^C)$ is the partially abstract initial controller.

In spirit, method application in POHTN planning works very similarly to how it works in HTN planning, albeit a little more involved due to the multiple possible "results" of abstract actions. Let $\mathrm{pred}_{\mathrm{fsc}}(n) = \{n' \in N | \delta(n', n) \neq \texttt{false}\}$ be the set of predecessor nodes of a controller node n, i.e., the set of nodes from which n can be reached in one step. Let $\mathrm{fsc}^1 = (N^1, \alpha^1, \delta^1, n_0^1)$ be a partially abstract controller, $c \in C$ an abstract action, and $n_C^1 \in N^1$ with $\alpha^1(n_C^1) = c$ the node to decompose. Let further $m = (c, \mathrm{mfsc})$ be a decomposition method and let $\mathrm{fsc}^2 = (N^2, \alpha^2, \delta^2, n_0^2, N_t^2, L^2)$ be an isomorphic copy of mfsc for which $N^1 \cap (N^2 \cup N_t^2) = \emptyset$. Applying m to fsc^1 results in a new controller $\mathrm{fsc}^3 = (N^3, \alpha^3, \delta^3, n_0^3)$ which is defined as follows:

- The resulting node set $N^3 = (N^1 \cup N^2) \setminus \{n_C^1\}$ contains all nodes from N^1 and N^2 except the decomposed node n_C^1.
- Action labels are kept from the original controllers, i.e.,

$$\alpha^3(n) = \begin{cases} \alpha^1(n), & \text{if } n \in N^1 \\ \alpha^2(n), & \text{if } n \in N^2. \end{cases}$$

- Unless the initial node of the original controller was decomposed, the initial node remains unchanged:

$$n_0^3 = \begin{cases} n_0^1, & \text{if } n_0^1 \neq n_C^1 \\ n_0^2, & \text{if } n_0^1 = n_C^1. \end{cases}$$

- For the node transitions, inner transitions of fsc^1 and fsc^2 are kept. Transitions to the replaced node n_C^1 are converted to transitions to the initial node of the method controller n_0^2. Transitions to the terminal nodes of fsc^2 are redirected to the successor nodes of n_C^1 by checking whether the terminal node labels, i.e., interpretations of abstract observation fluents, fulfill the outgoing transition conditions of the decomposed node, which are formulas over abstract observation

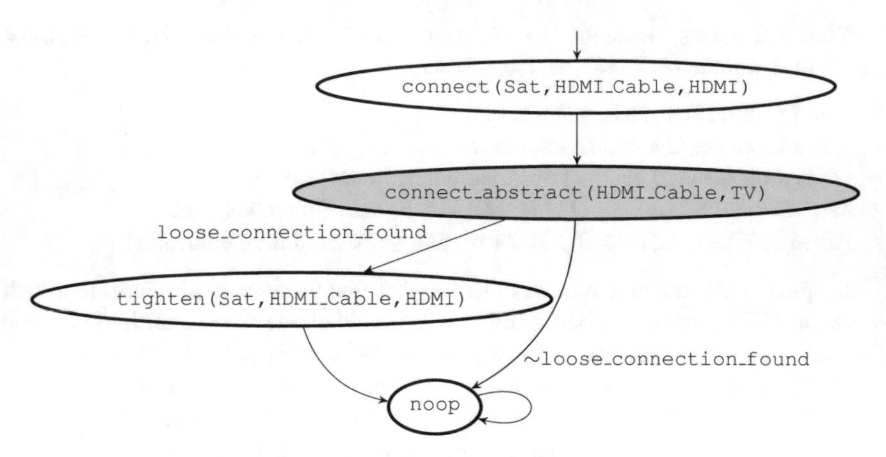

Fig. 6.6 A partially abstract policy. *Grey* nodes denote abstract actions

fluents:

$$\delta^3(n,n') = \begin{cases} \delta^1(n,n'), & n,n' \in N^1 \\ \delta^2(n,n'), & n,n' \in N^2, n' \neq n_0^2 \\ \delta^1(n,n_C^1), & n \in \text{pred}_{\text{fsc}^1}(n_C^1), n' = n_0^2 \\ \bigvee_{t \in N_t^2, L^2(t) \models \delta^1(n_C^1,n')} \delta^2(n,t), & n \in \text{pred}_{\text{fsc}^2}(t), n' \in N^1 \\ \bigvee_{t \in N_t^2, L^2(t) \models \delta^1(n_C^1,n_C^1)} \delta^2(n,t) \vee \delta^2(n,n'), & n \in \text{pred}_{\text{fsc}^2}(t), n' = n_0^2 \\ \texttt{false}, & \text{else} \end{cases}$$

It can be shown that if the transitions in fsc^1 and fsc^2 are well defined, then so are the transitions in fsc^3 [13].

As an example, consider the partially abstract controller in Fig. 6.6. Applying the method shown in Fig. 6.5 to the node labeled with `connect_abstract (HDMI_Cable,TV)` yields the primitive policy shown in Fig. 6.4.

Again, we write $\text{fsc}^1 \to_M^* \text{fsc}^2$ when fsc^2 can be created from fsc^1 by applying an arbitrary number of methods from M. Let $L(\text{fsc}_I, M) = \{\text{fsc}|\text{fsc}_I \to_M^* \text{fsc}, \text{fsc} \in \text{FSC}(A,O)\}$ be the set of primitive controllers that can be constructed from initial controller fsc_I by applying methods from M. The solution to a POHTN planning problem $\text{fsc}^* = \text{argmax}_{\text{fsc} \in L(\text{fsc}_I,M)} V_{\text{fsc}}^H(b_0)$ is then defined as the best primitive controller in $L(\text{fsc}_I, M)$.

For deterministic problems, the POHTN approach very closely resembles total-order HTN planning in the formalism described above. Intuitively, a total-order HTN problem can be converted into a POHTN problem by using a sufficiently large planning horizon and a reward function that is defined as -1 for violated preconditions and 0 else. If a policy with an accumulated reward of 0 exists, then it also represents an HTN solution. The task sequences in methods are converted to controllers where each element of a sequence is a node and there are `true` transition conditions between nodes.

6.4.3 A Hierarchy for Home Theater Setup

We are now ready to define a POHTN hierarchy for the home theater setup domain, which boils down to defining methods for `connect_abstract(?d1,?d2)`. The key assumption is that `connect_abstract(?d1,?d2)` is always used in such a way that `?d1` has a signal that needs to be transported to `?d2`, and that `?d2` is checkable for that signal. We also assume that `?d2` is checked for signals and that `loose_connection_found` is true whenever there is no signal at `?d2`.

Given that this is the case, we distinguish four cases. The simplest case is that the two devices can be connected directly, which leads to the method controller depicted in Fig. 6.5, i.e., the devices are connected, `?d2` is checked, and the connection is tightened if necessary.

When the devices cannot be connected directly, e.g., a `Sat` and `TV`, we can attempt connecting via a third device, e.g., an `HDMI_Cable`. This is realized by first executing `connect(?d1,?d3,?p)` for some port type `?p`, and afterwards using `connect_abstract(?d3,?d2)`. When the result of `connect_abstract(?d3,?d2)` is `loose_connection_found`, we execute `tighten(?d1,?d3,?p)` to make sure the connection between `?d1` and `?d3` is tight.

We also address the case that a connection between `?d1` and `?d2` was already established in the course of connecting two other devices. In this case, we only check `?d2` for signals and report the result in `loose_connection_found`.

Sometimes, it is necessary to create more than one connection between `?d1` and `?d2`, for example when the video and audio signals need to be transported via dedicated video and audio cables. Therefore, the last method contains `connect_abstract(?d1,?d2)` twice.

An initial controller for a given problem instance is always easily created: it simply consists of a sequence of `connect_abstract(?d1,?d2)` nodes for each pair of signal source device `?d1` and target device `?d2`.

6.4.4 Monte-Carlo Tree Search for POHTNs

MCTS can also be applied to POHTNs [13]. The search space in this case is $L(\text{fsc}_I, M)$, i.e., the set of controllers that can be generated by applying methods to the initial controller. The root node of the tree is consequently labeled with fsc_I, and the available "actions" are the methods applicable to a given controller, as illustrated in Fig. 6.7. Note that method application is deterministic; therefore each chance node only has a single successor. Also, it does not incur an immediate

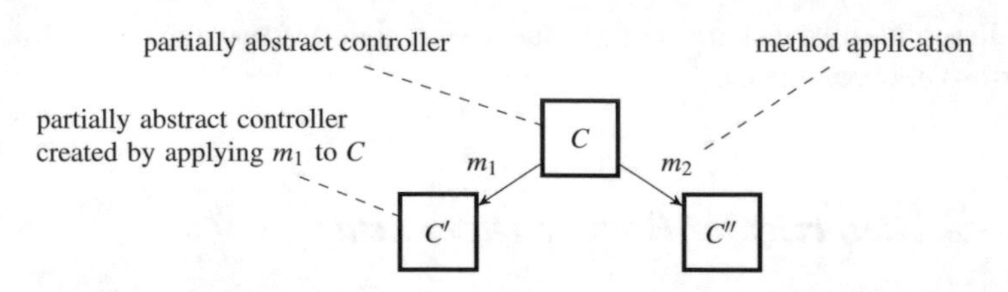

Fig. 6.7 A visualization of the MCTS search space when applied to POHTNs. The redundant chance node layer is omitted

reward. Terminal nodes correspond to primitive controllers, where both uncertainty and non-zero rewards occur: Once a primitive controller is reached, its execution is simulated once and the resulting accumulated reward is propagated up the tree.

If the hierarchy is chosen suitably, the number of decisions that need to be made during planning can be much smaller than in the case of MCTS in the space of histories. An important commonality of MCTS in the space of controllers and MCTS in the space of histories is that they make the same number of simulator calls in each round. This means that we can compare their performance on the basis of the number of iterations, i.e., their sample complexity.

6.5 Experiments

We conducted several kinds of experiments to (1) check whether our model satisfies our design goals stated at the end of Sect. 6.2, (2) compare the performance of the POUCT algorithm against MCTS applied to POHTNs planning (henceforth just POHTN), and (3) measure the influence of different modeling choices for the home theater domain.

6.5.1 *Quality of Hand-Crafted Policy*

In the first experiment, we checked whether our intuition of appropriate courses of action is in agreement with the results of automated planning methods. To this end, we considered a very simple instance of the domain, where there is only a satellite receiver with an HDMI port, a TV with an HDMI port, and an HDMI cable. The satellite receiver has a video and an audio signal, which the TV needs. The TV is also checkable for both signals. The planning horizon is set to 10. We expect that the policy shown in Fig. 6.4 is an optimal, or at least a near-optimal, policy for this instance. We therefore compare the expected accumulated reward of this policy with the result of running POUCT for a very long time. Figure 6.8 shows the result. It can be seen that POUCT eventually finds a policy of similar quality as the hand-crafted

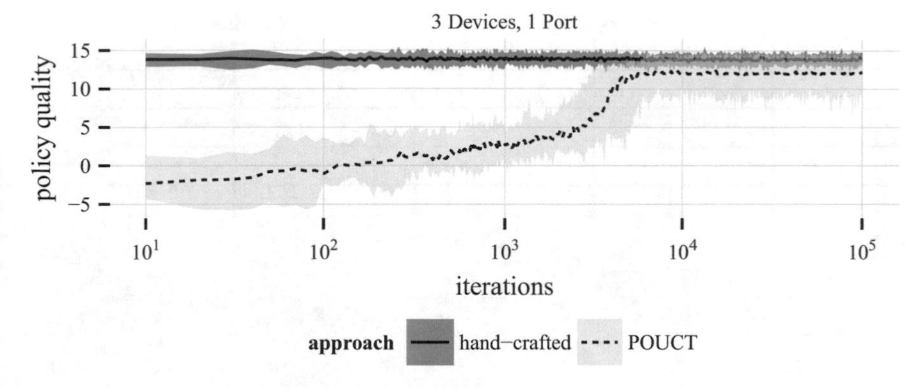

Fig. 6.8 A plot comparing the policy depicted in Fig. 6.4 and policies generated using POUCT on a small problem instance. It shows the median (curve) and the max./min. (ribbon) policy quality from 30 runs over the number of iterations. The value of a policy is estimated by simulating its execution 40 times and averaging over the accumulated rewards

policy in some runs. Also, there does not seem to be a policy of significantly higher accumulated reward than the hand-crafted policy. This indicates that our intuition of what the optimal policy should look like is correct.

6.5.2 POHTN vs. POUCT

For the second experiment, we created several instances of the home theater domain with an increasing number of devices and port types (i.e., difficulty), again with the horizon set to 10. We run both POUCT and POHTN on these instances. We expected the POHTN approach to scale better due to the smaller search space. Indeed, it can be seen in Fig. 6.9 that the POHTN approach outperforms POUCT on all instances and that the difference in performance grows for larger instances.

6.5.3 Alternative Reward Structure

In the third experiment, we consider a variant of the reward function: the system receives its reward for bringing signals to target devices only when *all* signals are at their respective target devices. This is closer to how the goal is defined in the original, deterministic domain.

It is not immediately apparent whether this improves or degrades planning performance. On the one hand, this makes it harder for the system to initially identify useful actions. On the other hand, it does not reward "dead-end" cable configurations that cannot be completed to solutions. We present a comparison of

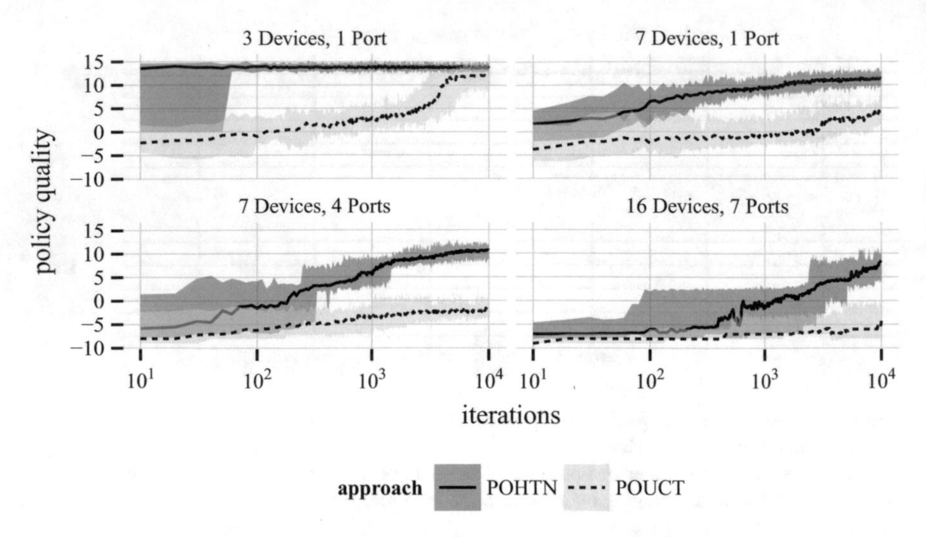

Fig. 6.9 A plot comparing POUCT and POHTN policy quality on different instances of the home theater setup domain. It shows the median (curve) and max./min. (ribbon) policy quality from 30 runs over the number of iterations

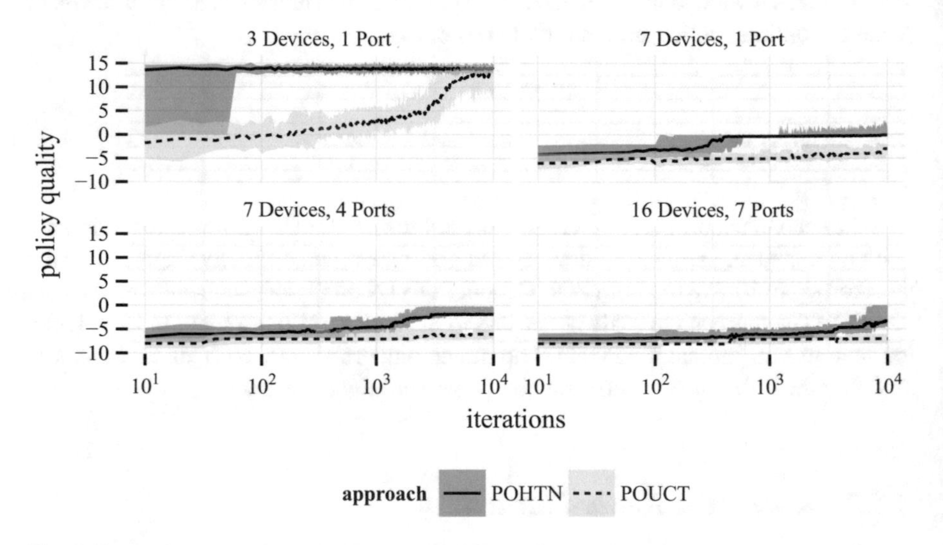

Fig. 6.10 A plot comparing POUCT and POHTN policy quality when the system receives a positive reward only when all devices are connected correctly. It shows the median (curve) and max./min. (ribbon) policy quality from 30 runs over the number of iterations

the two variants in Fig. 6.10 on the same instances as in Fig. 6.9. The policy qualities achieved by the planners are not directly comparable to the results in Fig. 6.9 due to the different reward structure. However, it can be seen that policy quality improves only very slowly over time for both approaches, except for very small instances. This indicates that rewarding the system for achieving intermediate goals is necessary for generating high-quality policies.

6.5.4 Eliminating Uncertainty

Lastly, we wanted to analyze the effect of eliminating uncertainty in the home theater setup task, i.e., when connecting devices always results in tight connections. The results are shown in Fig. 6.11. This plot shows that, at least for the smallest instance, the best hierarchical controller is not necessarily the best possible policy for the POMDP. In this case, we can even explain where the hierarchical policy is suboptimal: A controller generated with our hierarchy will always check whether the signals have reached the target devices. However, when connect always results in tight connections, this is unnecessary, and therefore punished. Apart from this, it can be seen from the collapsing ribbon that all POHTN runs on the first three instances eventually generated a policy of the same quality. While this does not guarantee that the best hierarchical controller has been found, it certainly is suggestive.

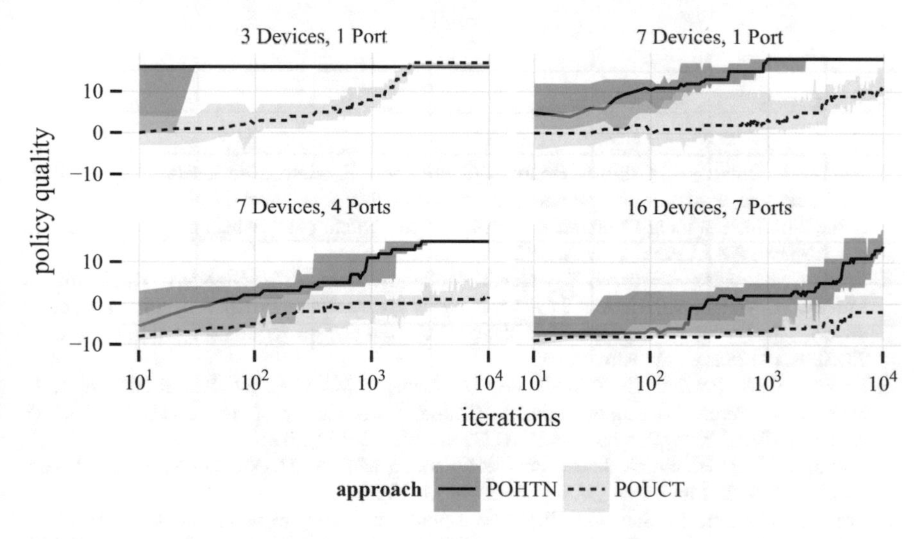

Fig. 6.11 A plot comparing POUCT and POHTN policy quality when connecting devices always results in tight connections. It shows the median (curve) and max./min. (ribbon) policy quality from 30 runs over the number of iterations

6.6 Conclusion

We described an approach for exploiting hierarchical structures in partially observable planning problems. To demonstrate its effectiveness, we developed a partially observable variant of the home theater setup domain and constructed a suitable action hierarchy. Our empirical evaluation pursued two goals. First, we tested our modeling decisions by examining the influence of various parameters on solution quality. Second, we compared the performance of the POHTN approach with that of non-hierarchical planning. The former experiments indicate that our model of the home theater setup domain is adequate. From the latter experiment, we can conclude that our approach is indeed able to exploit existing hierarchical structures in partially observable planning domains. Also, note that the same POHTN hierarchy is used in all experiments, regardless of the number devices in the problem instance, which means that the POHTN approach scales to POMDPs with large state, action, and observation spaces.

Acknowledgements This work was done within the Transregional Collaborative Research Centre SFB/TRR 62 "*Companion*-Technology for Cognitive Technical Systems" funded by the German Research Foundation (DFG).

References

1. Bercher, P., Biundo, S., Geier, T., Hoernle, T., Nothdurft, F., Richter, F., Schattenberg, B.: Plan, repair, execute, explain - how planning helps to assemble your home theater. In: Proceedings of the 24th International Conference on Automated Planning and Scheduling (ICAPS 2014), pp. 386–394. AAAI Press, Palo Alto (2014)
2. Bercher, P., Richter, F., Hörnle, T., Geier, T., Höller, D., Behnke, G., Nothdurft, F., Honold, F., Minker, W., Weber, M., Biundo, S.: A planning-based assistance system for setting up a home theater. In: Proceedings of the 29th National Conference on Artificial Intelligence (AAAI 2015). AAAI Press, Palo Alto (2015)
3. Browne, C.B., Powley, E., Whitehouse, D., Lucas, S.M., Cowling, P.I., Rohlfshagen, P., Tavener, S., Perez, D., Samothrakis, S., Colton, S.: A survey of monte carlo tree search methods. IEEE Trans. Comput. Intell. AI Games **4**(1), 1–43 (2012)
4. Dietterich, T.G.: Hierarchical reinforcement learning with the MAXQ value function decomposition. J. Artif. Intell. Res. (JAIR) **13**, 227–303 (2000)
5. Erol, K., Hendler, J., Nau, D.: UMCP: a sound and complete procedure for hierarchical task-network planning. In: Proceedings of the 2nd International Conference on Artificial Intelligence Planning Systems (AIPS 1994), pp. 249–254 (1994)
6. Geier, T., Bercher, P.: On the decidability of HTN planning with task insertion. In: Proceedings of the 22nd International Joint Conference on Artificial Intelligence (IJCAI 2011), pp. 1955–1961 (2011)
7. Hansen, E.A., Zhou, R.: Synthesis of hierarchical finite-state controllers for POMDPs. In: Proceedings of the Thirteenth International Conference on Automated Planning and Scheduling (ICAPS 2003), pp. 113–122 (2003)
8. He, R., Brunskill, E., Roy, N.: PUMA: planning under uncertainty with macro-actions. In: Proceedings of the Twenty-Fourth AAAI Conference on Artificial Intelligence, AAAI 2010 (2010)

9. Honold, F., Bercher, P., Richter, F., Nothdurft, F., Geier, T., Barth, R., Hörnle, T., Schüssel, F., Reuter, S., Rau, M., Bertrand, G., Seegebarth, B., Kurzok, P., Schattenberg, B., Minker, W., Weber, M., Biundo, S.: Companion-technology: towards user- and situation-adaptive functionality of technical systems. In: Proceedings of the 10th International Conference on Intelligent Environments (IE 2014), pp. 378–381. IEEE, New York (2014). doi:10.1109/IE.2014.60
10. Keller, T., Helmert, M.: Trial-based heuristic tree search for finite horizon MDPs. In: Proceedings of the 23rd International Conference on Automated Planning and Scheduling (ICAPS 2013), pp. 135–143. AAAI Press, Palo Alto (2013)
11. Kocsis, L., Szepesvári, C.: Bandit based monte-carlo planning. In: Proceedings of the 17th European Conference on Machine Learning (ECML 2006), pp. 282–293 (2006)
12. Müller, F., Biundo, S.: HTN-style planning in relational POMDPs using first-order FSCs. In: Bach, J., Edelkamp, S. (eds.) Proceedings of the 34th Annual German Conference on Artificial Intelligence (KI 2011), pp. 216–227. Springer, Berlin (2011)
13. Müller, F., Späth, C., Geier, T., Biundo, S.: Exploiting expert knowledge in factored POMDPs. In: Proceedings of the 20th European Conference on Artificial Intelligence (ECAI 2012), pp. 606–611. IOS Press, Amsterdam (2012)
14. Nau, D., Au, T.C., Ilghami, O., Kuter, U., Muñoz-Avila, H., Murdock, J.W., Wu, D., Yaman, F.: Applications of SHOP and SHOP2. IEEE Intell. Syst. **20**(2), 34–41 (2005)
15. Parr, R., Russell, S.J.: Reinforcement learning with hierarchies of machines. In: Advances in Neural Information Processing Systems (NIPS 1997), vol. 10, pp. 1043–1049. MIT Press, Cambridge (1997)
16. Pineau, J., Gordon, G., Thrun, S.: Policy-contingent abstraction for robust robot control. In: Proceedings of the 19th conference on Uncertainty in Artificial Intelligence (UAI 2003), pp. 477–484. Morgan Kaufmann, San Francisco (2003)
17. Sanner, S.: Relational dynamic influence diagram language (RDDL): language description (2010). http://users.cecs.anu.edu.au/~ssanner/IPPC_2011/RDDL.pdf
18. Silver, D., Veness, J.: Monte-carlo planning in large POMDPs. In: Lafferty, J., Williams, C., Shawe-Taylor, J., Zemel, R., Culotta, A. (eds.) Advances in Neural Information Processing Systems 23, pp. 2164–2172. Curran Associates, Red Hook (2010)
19. Sondik, E.: The optimal control of partially observable Markov decision processes. Ph.D. Thesis, Stanford University (1971)
20. Sutton, R.S., Precup, D., Singh, S.: Between MDPs and semi-MDPs: a framework for temporal abstraction in reinforcement learning. Artif. Intell. **112**(1), 181–211 (1999)
21. Theocharous, G., Kaelbling, L.P.: Approximate planning in POMDPs with macro-actions. In: Thrun, S., Saul, L., Schölkopf, B. (eds.) Advances in Neural Information Processing Systems (NIPS 2004), vol. 16, pp. 775–782. MIT Press, Cambridge (2004)

Chapter 7
To Plan for the User Is to Plan with the User: Integrating User Interaction into the Planning Process

Gregor Behnke, Florian Nielsen, Marvin Schiller, Denis Ponomaryov, Pascal Bercher, Birte Glimm, Wolfgang Minker, and Susanne Biundo

Abstract Settings where systems and users work together to solve problems collaboratively are among the most challenging applications of Companion-Technology. So far we have seen how planning technology can be exploited to realize Companion-Systems that adapt flexibly to changes in the user's situation and environment and provide detailed help for users to realize their goals. However, such systems lack the capability to generate their plans in cooperation with the user. In this chapter we go one step further and describe how to involve the user directly into the planning process. This enables users to integrate their wishes and preferences into plans and helps the system to produce individual plans, which in turn let the Companion-System gain acceptance and trust from the user.

Such a Companion-System must be able to manage diverse interactions with a human user. A so-called mixed-initiative planning system integrates several Companion-Technologies which are described in this chapter. For example, a—not yet final—plan, including its flaws and solutions, must be presented to the user to provide a basis for her or his decision. We describe how a dialog manager can be constructed such that it can handle all communication with a user. Naturally, the dialog manager and the planner must use coherent models. We show how an ontology can be exploited to achieve such models. Finally, we show how the causal information included in plans can be used to answer the questions a user might have about a plan.

G. Behnke (✉) • M. Schiller • P. Bercher • B. Glimm • S. Biundo
Institute of Artificial Intelligence, James-Franck-Ring, 89081 Ulm, Germany
e-mail: gregor.behnke@uni-ulm.de; marvin.schiller@uni-ulm.de; pascal.bercher@uni-ulm.de; birte.glimm@uni-ulm.de; susanne.biundo@uni-ulm.de

F. Nielsen • W. Minker
Institute of Communications Engineering, Albert Einstein-Allee 43, 89081 Ulm, Germany
e-mail: florian.nothdurft@alumni.uni-ulm.de; wolfgang.minker@uni-ulm.de

D. Ponomaryov
A.P. Ershov Institute of Informatics Systems, 6, Acad. Lavrentjev pr., Novosibirsk 630090, Russia
e-mail: ponom@iis.nsk.su

S. Biundo, A. Wendemuth (eds.), *Companion Technology*, Cognitive Technologies,
DOI 10.1007/978-3-319-43665-4_7

The given capabilities of a system to integrate user decisions and to explain its own decisions to the user in an appropriate way are essential for systems that interact with human users.

7.1 Introduction

Planning has proven to be a successful technology for problem solving in scenarios involving humans and technical systems [6, 7, 22, 26]. Usually, a planner generates a plan to solve a given problem, e.g., to set up a home theater system (see Chap. 24), and presents the generated plan to the user in a stepwise fashion, while providing additional advanced planning capabilities, like explanations or plan repair. In this process the user is only viewed as an operator who inputs an objective and subsequently executes the actions presented to him, while the planner is treated as a black-box system. This scheme is well suited if the task to be performed is combinatorially complex and a deeper understanding of the proposed solution's structure is not relevant as long as the goal is achieved.

However, if the problem at hand is of a more personal nature, e.g., creating a fitness-training plan, or the user has certain preferences and wishes about the plan to be executed, or the user has domain knowledge not easily encodable in terms of planning actions, a black-box approach is not adequate. The user may not accept the plan if it does not suit his individual needs or preferences. The same holds if the plan is associated with grave risks, e.g. in spaceflight [1] and military settings [25]. Here a human must be the final decider on what actions are actually executed. To circumvent these problems, the user has to be integrated into the planning process itself. Such planning systems are commonly called "mixed-initiative" as both the planner and the user propose courses of action and ask one another questions about the plan. As the result of their interplay the planner generates a final plan which solves the task and satisfies the user's wishes. We focus our discussion on the planning formalism *hybrid planning* [8], which is well suited for user-centered planning applications (see Chap. 5 and [4]). Most notably, it is similar to the way humans solve problems, i.e., in a top-down fashion [11].

Mixed-initiative planning (MIP) systems have already been studied by several researchers as they are often necessary to successfully deploy planning techniques to real-world problems. In the TRAINS/TRIPS project [13], a domain-specific planner for path finding and transportation tasks was extended with the capability to interact with a user. Similarly, MAPGEN [1, 5] was developed to support planning of operations of the Mars rovers Spirit and Opportunity while ensuring energy and safety constraints. Another approach, PASSAT [25], employs hierarchical planning. Here the user can "sketch" a plan, i.e., state some actions he wants to be part of the solution, and the system finds a plan containing these actions. Further, PASSAT guides the user through the search space by repeatedly asking how a current plan should be altered until an acceptable solution is found. In contrast, the techniques described in this chapter aim at integrating the user directly into the planning process.

In addition to a pure mixed-initiative planner a *Companion*-System [9] needs additional advanced capabilities to suitably interact with the user, including a dialog management system and explanation facilities. They need to be specifically tailored to the task of mixed-initiative planning, as they, e.g., must support dynamic changes in the current plan as well as changes in initiative from the system to the user. In the latter part of the chapter, we demonstrate how the planner and interaction components can be brought together with a mixed-initiative planner. An important part of such an integration is a shared model used by every component. We propose using an ontology to store this model and describe how a planning model can be suitably encoded and what additional benefits can be obtained.

In this chapter we study the design of such a mixed-initiative planning system. To begin with, we outline in Sect. 7.3 challenges a mixed-initiative planning system has to tackle in order to successfully cooperate with a user and consider whether the issue should be addressed by the planner or by a dialog manager. Next, we show how a *Companion*-System can be built atop a mixed-initiative planner in Sect. 7.4. Section 7.5 explains how the common model of all components of the *Companion*-System can be created and how relevant information can be accessed. More specifically, we describe how a planning domain can be encoded in the ontology and how parts of the domain, i.e., decomposition methods, can even be inferred automatically. In Sect. 7.6 we study how users react to strategies applied by the dialog manager to mediate between the planner and the user. Then we discuss how explanations for plans can be enhanced to make them more comprehensible as described in Sect. 7.7.

We will use a fitness training domain as a running example. It is the mixed-initiative planner's objective to develop an individualized training plan achieving some fitness objective, e.g., to have well-defined abdominal muscles. Exercises are grouped into workouts, which are to be performed on a single day and contribute to some fitness objective. We call a training a longer sequence of exercises that achieves a specific objective. A training is partitioned into several workouts, which are collections of exercises done on a single day of the training's schedule. The planner starts with a plan specifying a training task, representing the user's objective, and refines the plan repeatedly, first into suitable workout tasks and those in turn into concrete exercises. A similar application domain was considered by Pulido et al. [33], who described how a planner can be utilized to arrange physiotherapy exercises to rehabilitate people with upper limb injuries.

7.2 Preliminaries

The content of this chapter is based on the notions of hybrid planning and ontologies, which are both briefly introduced in this section. Chapter 5 explains hybrid planning in further detail and illustrates how it can be applied to user assistance in particular. In this chapter we show how the presented techniques can be complemented by integrating the user into the planning process itself.

Hybrid Planning Planning is an AI technique for solving complex combinatorial problems. The objective of the planner is to find a so-called plan, a (partially ordered) set of actions which, if executed in a given initial state, achieve some goal. States in planning are abstractions of the real world and are represented as sets of predicates. Actions are described in terms of their preconditions and effects, two formulae that must be true such that the action can be executed and that describe the change to the world state if the action is executed, respectively.

Hybrid planning is the fusion of two other planning approaches, namely Hierarchical Task Network (HTN [12]) and Partial Order Causal Link (POCL [24, 31]) planning. From the former it inherits the subdivision of actions into primitive and abstract ones. Primitive actions are assumed to be directly executable by an operator, e.g. a human user, while abstract actions represent more complex courses of action. The aim in HTN planning is, given an initial set of abstract actions, to refine them into a plan solely containing primitive actions. To do so, an HTN planning domain contains so-called *decomposition methods* mapping abstract actions to plans—not necessarily containing only primitive tasks—by which they may be replaced. We use the expression $\mathtt{A} \mapsto_{\prec} \mathtt{B}_1, \ldots, \mathtt{B}_n$ to denote such a method for the abstract task $\mathtt{A}$, decomposing it into a plan containing the subtasks $\mathtt{B}_1$ to $\mathtt{B}_n$ which is ordered w.r.t. the partial order $\prec$. If the partial order $\prec$ is omitted, we assume that no order is present, i.e., $\prec = \emptyset$. POCL planning introduces the notion of causal links to hybrid planning. A causal link describes the relation between two actions, i.e., that one is executed to achieve an effect needed by the other. Furthermore, in standard HTN planning, abstract actions have neither preconditions nor effects, whereas in hybrid planning they do. They enable causal reasoning about the plan at every level of abstraction and especially early during the planning process.

A planner for hybrid planning domains, PANDA, is presented in Chap. 5. Its purpose is to refine a given initial plan into a solution to the hybrid planning problem, i.e., to be an executable plan without abstract action that can be obtained via decomposition from the initial plan. It uses a heuristically guided plan-space search to find solutions. At each step during the search a so-called *flaw*, a property that keeps it from being a solution, is selected to be resolved. For instance, each abstract action in a plan constitutes a flaw, as well as preconditions of actions without supporting causal links. Thereafter, all possible modifications solving this flaw are applied to generate the plan's successors in the search space.

Ontologies Ontologies based on Description Logics (DLs) serve to model knowledge in an application domain, with a focus on the *concepts* of a given domain and the relations (*roles*) that hold between them. The formal representation of a knowledge base enables the use of reasoners to infer additional knowledge that is logically implied and provides a model-theoretic semantics for the contents of such a knowledge base. One application of ontologies is the Semantic Web, in whose context the web ontology language OWL was established as a W3C standard. In this chapter, we consider the fragment of OWL corresponding to the DL $\mathcal{ALC}$ [35]. Concepts are either primitive (represented by a set of concept names) or complex. Complex concept expressions are formed using the connectives $\sqcap$

(conjunction of concepts) and $\sqcup$ (disjunction), and quantifier-like $\exists/\forall$ constructs, which specify relationships between concepts with respect to a particular role. For instance, suppose that *includes* is a role name, and *HardExercise* is a concept name; then the expression $\exists$*includes.HardExercise* represents the concept of all things that include (at least) something that is a *HardExercise*. By contrast, the expression $\forall$*includes. HardExercise* represents the concept of all things that include nothing but *HardExercise*s. Two distinguished concepts, $\top$ and $\bot$, represent the universal concept (which encompasses every other concept) and the empty concept, respectively.

DLs are fragments of first-order logic and possess a model-theoretic semantics. Concepts are interpreted as subsets of a domain. A domain element that is in the interpretation of a concept is referred to as an instance of the concept. Roles are interpreted as binary relations in the domain. For example, to qualify as an instance of the concept $\exists$*includes.HardExercise*, a domain element needs to be related by the role *includes* to (at least) one instance of the concept *HardExercise*.

An *ontology* or *knowledge base* $\mathcal{O}$ specifies a finite set of axioms. Axioms of the form $C \sqsubseteq D$ are referred to as concept inclusion axioms (alternatively: subsumption axioms), and specify that every instance of the concept C is also an instance of the concept D. Equivalence axioms are of the form $C \equiv D$, and state that D subsumes C and vice versa. An interpretation that satisfies all axioms of a knowledge base $\mathcal{O}$ is called a *model* of $\mathcal{O}$. Using an ontology reasoner, a given knowledge base $\mathcal{O}$ can be queried about whether the subsumption relationship holds between two concepts C and D, namely whether the axiom $C \sqsubseteq D$ holds in any model of $\mathcal{O}$ (in which case the axiom is *entailed* by the knowledge base). Another type of query concerns whether a concept is *satisfiable* in $\mathcal{O}$, that is, whether a concept can have any instances (without leading to a contradiction).

The generation of natural-language output from ontologies (*ontology verbalization*) has traditionally focused on making the formalized content accessible to non-expert users (e.g. in the *NaturalOWL* system [2]). Recent work also aims at verbalizing ontology reasoning (in a stepwise manner), including [28] and [34].

7.3 Technology Concept and Design

In this section, we take a closer look at how automated planners and humans solve problems. We elucidate the major differences between them and describe how these can be handled either by altering the planner or by equipping a dialog manager with appropriate behavior.

User-Friendly Search Strategies As mentioned earlier, most automated planners employ efficient search strategies, like $A*$ or *greedy search*, guided by heuristics. These strategies visit search nodes, in our case yet unfinished plans, in an order determined by the heuristic starting with the most promising plan, e.g., the plan for which the heuristic estimate of the distance to a solution is minimal. Given a perfect heuristic, the planner would basically maintain always the same plan and alter it until a solution has been found. Since all heuristics computable in reasonable time are necessarily imperfect, the planner does not necessarily visit plans after each other which are neighbors in the search space, but may jump between completely separate parts of the search space. The considered plans may have nothing in common at all.

This alternating between plans is not in line with the human planning processes, as we tend to refine only one plan at a time. For example, in an experimental study on planning behavior, Byrne [11] found that subjects dealt with goals "one by one". Further research helped to put this finding in perspective, and postulates that the degree to which people adhere to a hierarchical top-down approach of sequential plan refinement depends on various factors: whether the problem domain can easily be recognized by people as hierarchically structured, whether human problem-solvers already dispose of expertise with hierarchical schemata to address problems in the domain, or—on the other hand—much how much they feel enticed to explore the domain by using bottom-up processes rather than a more straightforward top-down approach [17]. Thus, these empirical studies suggest that a structured planning process is a feature of skilled and goal-directed decision-making. In contrast, an automated planner's $A*$ strategies may seem erratic to the user. Such may result in the perception that user decisions exert only an arbitrary influence on the planning process and may promote a lack of subjective control and transparency. In order to prevent this perception—which would most probably lead to the user not using the system—the gap between the way human and automated planners solve a planning problem needs to be bridged in a mixed-initiative planning system. Instead of A* or greedy search, we propose using the search strategy *depth-first search* (DFS), which repeatedly refines a current plan until a solution has been found. If a plan is reached that cannot be further refined into a solution, e.g. a plan without a possible refinement, the decisions leading to this plan are reverted until another possible refinement is found. This continuous refinement of a single plan reflects the human process of problem solving much closer.

A drawback of DFS is that it is a blind search, i.e., it does not consider an estimate of how far a plan is away from being a solution when choosing a refinement. To remedy this problem, a mixed-initiative planner can consider a heuristic if the user is indifferent between options, or even weight the information of a heuristic against the decisions of the user. This scheme enables the user to inform the planner that it should use its best judgment to determine a solution for a subproblem, e.g. if it is combinatorially too complex to be solved by a human. DFS is also incomplete, meaning that it may not find a solution even if it exists, as it can "get stuck" in infinite parts of the search space not containing a solution.

Handling Unsolvable Plans Another difference is the way humans and planners deal with failed plans. During the search, the planner will explore plans that cannot possibly be refined into a solution anymore, either because the plan has a flaw without a possible modification to solve it or because a heuristic has determined this property. In this case, no successors are generated for the plan—it may still have resolvable flaws—and the search is continued. If the search procedure DFS is applied, this will lead to backtracking. As every practically usable computable heuristic is necessarily imperfect,[1] there are usually whole parts of the search space only containing unsolvable plans which are not recognized as such. If, either by chance or due to the user's decisions, the search enters such a region, a failed plan will eventually be obtained and the backtracking procedure will be started, leading to the complete exploration of this unsolvable part of the search space.

On the other hand, if humans recognize that a plan is unsolvable, they can most of the time determine a reason for failure and alter only relevant parts of the plan. Current planning systems often fail in determining which applied modification was the reason for the failure. Instead they use backtracking to resolve the problem. *Backtracking*, especially through a large search space, is very tedious and frustrating for a human user performing or witnessing these steps. A large search space requires extensive backtracking through unsolvable alternatives. This may result in the user interpreting the system's strategy as naive and impairs the trust in and perceived competence of the planner. Additionally, this strategy does not prevent the repetition of similar (unsuccessful) decisions, leading furthermore to frustration.

To remedy this problem, we can utilize the computational power of the planner. If options for a refinement are presented to the user he usually takes several seconds (if not longer) to decide on one of the options. During this time, the planner can start to explore the search spaces induced by the modifications presented to the user. The planner may determine, using a well-informed heuristic, that the search space induced by one of the options only leads to failed plans. We call such a modification a *dead-end*, and, if it occurs, the respective option can be removed from consideration and thus backtracking through this part of the search space can be averted. Here the important question is how this information should be communicated to the user, as simply disabling the respective option in the user interface without any apparent reason would be rather irritating and seems not to be appropriate. If the planner, on the other hand, has found a solution, the mixed-initiative planning system knows that backtracking is not necessary if the user agrees with the solution. We describe our approach in the next section and evaluate it in Sect. 7.6.

7.4 Integration of Planning and Dialog

The integration of automated planning and user-centered dialog begins with the statement of the user's goals. This first dialog between user and machine has the objective of defining the goals in a way understandable to the assisting automated planning system. This requires on the one hand a user-friendly and efficient task-selection dialog, and on the other hand the creation of a valid *planning problem*.

[1] Computing a perfect heuristic is as difficult as planning itself, e.g. in the case of HTN planning, undecidable [14].

Thus, the semantics of the dialog have to be coherent with the *planning domain*, resulting in a valid mapping between dialog result and *planning problem*.

Once the problem is passed on to the planner the interactive planning itself may start. Using the described DFS, the initial plan will be refined by selecting appropriate modifications for available flaws. In order to decide whether to integrate the user or not during this process, an elaborate decision model, integrating various information sources, is required. Relevant information sources are, for example, the *dialog history*, e.g., was the user's decision the same for all past similar episodes, the kind of *plan flaw*, e.g., is this flaw relevant for the user, the *user profile*, e.g., does the user have the competencies for this decision, and the current *situation*, e.g. is the current cognitive load of the user low enough for interaction. These sources illustrate that a decision model uses information from the *dialog management* and the *planner*, and is therefore located in a superordinate component.

In the case of user integration the information on the current *plan decision* has to be communicated to the user. This means that the *plan flaw* and the corresponding decision between the available *modifications* have to be represented in the dialog suitably. Hence, the corresponding plan information needs to be mapped to human-understandable dialog information. As this mapping potentially needs to exist for all plan information and for all dialog information, the requirement of coherent models between planner and dialog system becomes an existential factor for MIP systems. The thorough matching of both models would be an intricate and strenuous process, requiring constant maintenance, especially when a model needs to be updated. Thus, a more appropriate approach is the automatic generation of the respective models using one mutual model as source. This way, once the transformation functions work correctly, coherence is not an issue anymore, even when updating the domain. How these essential constituents of a conceptual MIP system architecture (depicted in Fig. 7.1) were implemented in our system is explained below.

The Decision Model This model is in charge of deciding when and how to involve the user in the planning process. It is composed of several subcomponents, acts as an interface to the planner and decides, upon planner requests, whether user involvement is useful, i.e., if this kind of flaw is understandable to a human user.

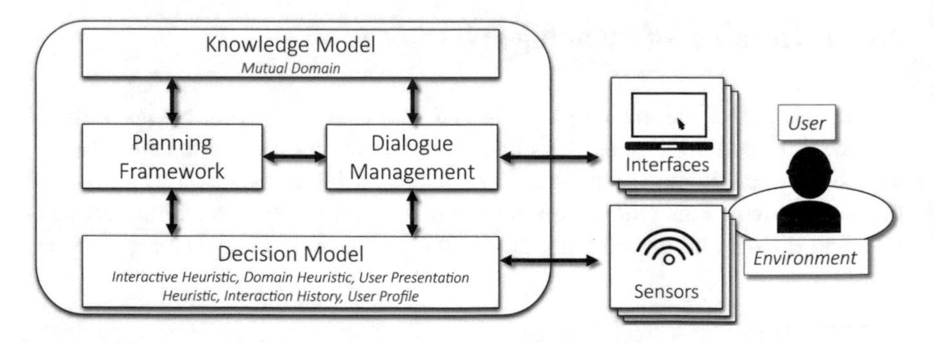

Fig. 7.1 Essential components of a mixed-initiative planning system integrating the user [30]

For this it also includes a list of essential domain decisions that are interesting and relevant for the user (e.g. for a training domain: day, workout, and exercises)—the rest is left for the fallback-heuristic and thus decided by the planner. If it is in favor of user involvement, the flaw and its corresponding modifications have to be passed on to the user. Then, the decision on the form of user integration is made. The dialog may provide the complete set of modifications, a pruned list, a sorted list, implicit confirmations, or explicit confirmations, for presentation or only to inform the user. This decision depends not only on the interaction history, but also on additional information (e.g. affective user states like overextension, interest, or engagement) stored in the user state.

The *Decision Model* also records the dialog and planning history. There are several reasons for that: The dialog history may enable a prediction of future user behavior (e.g. in selections), and additionally this knowledge is mandatory for *backtracking* processes, when the current plan does not lead to a solution. The history stores which decisions were made by the user. In the case of *backtracking* the decisions are undone step-by-step, with the goal of finding a solution by applying alternative modifications. Whenever a user-made decision is undone, the user is notified, because this system behavior would otherwise appear irritating.

Since *backtracking* as well as *dead-ends* are peculiar phenomena in an MIP system, the communication of these might have a critical influence on the user experience. Together with the *Dialog Management (DM)*, the *Decision Model* orchestrates the corresponding system behavior. The main difference between *backtracking* and *dead-ends* is the temporal ordering of the awareness of the unsolvable plan and made decision. For *backtracking* the awareness is achieved after the decision, and for *dead-ends* during the decision. As we assumed that *backtracking* will impair the user experience significantly, a parallel search for *dead-ends*, as described in Sect. 7.3, was implemented. The process itself is, of course, inherently different from *backtracking*, but may prevent it. Removing dead-ends from the search space when the relevant modification is not part of the current selection is a rather easy task. Otherwise, the current selection has to be modified to prevent the user from selecting a *dead-end*. However, removing it without any notification from the list seems like confusing behavior.

7.5 Coherent Models Across the System

The system described in the previous section relies on a shared vocabulary and a coherent description of the planning domain in both the dialog system (DS) and the planner. In this section we describe how this coherence can be achieved using an ontology as the central knowledge base component. To do so, relevant[2] parts of the planning domain are encoded in description logic. To obtain a unified view of

[2]Those which must be accessible by other systems.

the system's knowledge we also demonstrate how the remaining planning-specific information can be encoded in a way not interfering with the reasoning process, and thus how all information can be stored in the ontology. Further, we describe how planning and dialog domain can be automatically extracted from the ontology.

As an additional advantage, new decomposition methods for the planning domain can be inferred using ontology reasoning without the help of a human expert modeler, easing creating domains significantly. This is especially useful in our application scenario—individualized fitness training. Imagine a user found a new workout,[3] e.g. while browsing the Internet, and wishes to add it to his training plan.[4]

Using ontological reasoning, the system can infer which training objective the workout has and how it can be integrated into the existing planning domain, without additional input from the user. Furthermore, using plan and ontology explanations, the *Companion*-System can explain how and why it has integrated the new workout into the model in a certain way. If the workout does not comply with the user's objective, it could even explain why the user should not use the workout.

A few previous approaches have attempted to integrate ontological reasoning into planning. Most approaches targeting classical, i.e., non-hierarchical, planning attempt to either increase the performance of the planner or increase the expressivity of the planning formalism, e.g. by changing the notion of states. We refer to an article by Gil [15] for an extensive survey. Another approach by Sirin et al. [36, 37], called HTN-DL, uses ontology reasoning to solve web service composition problems encoded in an HTN. Their main objective is to determine which abstract tasks can be decomposed by a predefined plan, based on annotations to that plan. In that, they infer new decomposition methods, but only those for which the plan's task network was provided by the domain modeler. Their matching cannot take the actual content of the plan into account, but only an abstract description of the plan in terms of preconditions and effects. Furthermore, there is no guarantee on the relation of the plan's content and these descriptions. One could, e.g., use a legality criterion for decomposition methods [8] to determine whether the description is correct. Our approach, on the other hand, can infer completely new decomposition methods and is able to infer them based on the actual plan's steps to be contained in them.

7.5.1 Integrating Planning Knowledge into Ontologies

We start by describing how a planning domain can be encoded in an ontology. Tasks in the planning domain are represented as concepts in the ontology. For each planning task `T` there is a corresponding concept T in the ontology. Preconditions and effects of actions are encoded in the ontology using four distinct data properties:

[3] A partially ordered set of exercises.

[4] A system that supports this search-and-extraction, resulting in an extension to the ontology, has been developed but is not yet published.

needs for positive preconditions, *hindered-by* for negative preconditions, and *adds* and *deletes* for positive and negative preconditions, respectively. Here, only the predicates of preconditions and effects are contained in the ontology, while their parameters are omitted. Expressing them correctly, i.e. in a way amenable to logical reasoning, would require common references, e.g., for an action requiring $has(x)$ and resulting in $\neg has(x)$ where both instances of x must be equal. Unfortunately, description logics are not suited for such kinds of expressions, due to the tree model property [39]. One example of such an action is the `Day` action, describing the transition of a day to the next one, i.e. modeling time explicitly in the domain.

$$Day \equiv \exists deletes.\ \text{“}fatigue\text{”}$$

This concept describes an action without any preconditions, leading to a state where the predicate `fatigue` does not hold, after the action has been executed, for some parameters. In this case it models that a person is not fatigued after he has slept, i.e., a day has passed. Additionally our approach allows for domain-dependent extensions. They define that certain axioms $C \sqsubseteq E$, where C is a concept and E an expression, are interpreted as a set of preconditions and effects. In our application scenario, we used this mechanism to map declarative descriptions of exercises to actions. These descriptions are based on parts of the NCICB corpus [27], describing the musculoskeletal system of the human body. As an example, we provide the following definition of the exercise *BarbellHackSquat*, which describes that the action `BarbellHackSquat` has the precondition *warmedup(Hamstring)* and the effects *trained(Hamstring)*, *warmedup(Soleus)*, and *warmedup(GluteusMaximus)*.

$$\begin{aligned} BarbellHackSquat \sqsubseteq\ & \exists trains.Hamstring \sqcap \exists engages.Soleus \\ & \sqcap \exists engages.GluteusMaximus \end{aligned}$$

The most important part of a hierarchical planning domain are its decomposition methods. Each method $\mathtt{A} \mapsto_{\prec} \mathtt{B}_1, \ldots, \mathtt{B}_n$ describes that a certain abstract task `A` can be achieved by executing the tasks $\mathtt{B}_1, \ldots, \mathtt{B}_n$ under a restriction, $\prec$, on their order. To transform decompositions into ontological structures consistently, we state an intuition on the meaning of the concept-individual relation in the created ontology, by which our subsequent modeling decisions are guided. Individuals in the ontology should be interpreted as plans and a concept T, corresponding to some task `T`, as the set of all plans (i.e. individuals) that can be obtained by repeatedly decomposing `T`. A concept inclusion $B \sqsubseteq A$ thus states that every plan obtainable from `B` can also be obtained by decomposing `A`. To ease modeling, a special role—*includes*—is designated to describe that a plan contains some other plan. That is, if an individual a is an instance of a plan and $includes(a, b)$ holds, the plan described by the individual a also contains all actions of the plan b. First, so-called unit-methods $\mathtt{A} \mapsto \mathtt{B}$, which allow for replacing the task `A` with the task `B`, are translated into a simple concept inclusion, $B \sqsubseteq A$, for which the intuition clearly holds. Methods creating more than a single task must be translated into more elaborate constructs in the ontology.

Such a method, $\mathtt{A} \mapsto_{\prec} \mathtt{B}_1, \ldots, \mathtt{B}_n$, defines a set of tasks $\{\mathtt{B}_1, \ldots, \mathtt{B}_n\}$ which must be contained in the plan obtained from decomposing A, while also stating that they are sufficient. The relation "is contained in a plan" is expressed by the role *includes*. Following its definition, an expression $\exists includes.T$ describes the set of all plans containing the task T. Using this property, the decomposition method could be translated into the following axiom:

$$\bigsqcap_{i=1}^{n} \exists includes.B_i \sqsubseteq A$$

It is, however, not sufficient due to the open world assumption of description logics. The expression solely describes the required tasks, but not that only these tasks are contained in the plan. To express the latter, we use the syntactic *onlysome* quantifier, originally introduced for OWL's Manchester syntax [20].

Definition 1 Let r be a role and $C_1, \ldots, C_n$ concept expressions. Then we define the onlysome quantification of r over $C_1, \ldots, C_n$ by

$$\mathrm{O}r.[C_1, \ldots, C_n] := \sqcap_{i=1}^{n} \exists r.C_i \sqcap \forall r \left(\sqcup_{i=1}^{n} C_i\right).$$

As an example in the fitness domain, consider a method that specifies that a particular workout decomposes into the set of tasks represented by its exercises, for example `Workout1` $\mapsto$ `FrontSquat,BarbellDeadlift`. This can be specified in the ontology as $\mathrm{O}includes.[FrontSquat, BarbellDeadlift] \sqsubseteq Workout1$. In general, with this definition we can describe a decomposition method $\mathtt{A} \mapsto_{\prec} \mathtt{B}_1, \ldots, \mathtt{B}_n$ with the following axiom, fulfilling the stated requirement.

$$\mathrm{O}includes.[B_1, \ldots, B_n] \sqsubseteq A$$

The intuition on the interpretation of concepts and individuals implies a criterion for when two concepts A and B, described by such axioms, should subsume each other. That is, A should subsume B based on the axioms in the ontology if and only if every plan described by B is also a plan for A. We have stated this criterion previously and proven that it holds with only minor restrictions to the ontology for expressions defined in terms of *onlysome* restrictions [3]. So far, we have not mentioned possible ordering constraints imposed on the tasks in a method. Representing them s.t. that they can be accessed by a DL reasoner poses a problem similar to that of representing variables, as a single task may be referred to several times in describing a partial order. To circumvent this problem, ordering is encoded only syntactically, i.e., in a way ignored by any reasoner while it still can be retrieved from the ontology by analyzing its axioms. If a plan contains some task A before some task B, then any occurrence of B in an onlysome restriction is substituted with $B \sqcup (\bot \sqcap \exists after.A)$.

7.5.2 Generating New Decomposition Methods Using DL Reasoning

Having integrated the planning domain into an ontology, we can utilize DL reasoning to infer new decomposition methods. Based on the interpretation of concept subsumption, each inferred subsumption $E \sqsubseteq A$ can be interpreted as the fact that every plan obtainable from the expression E can also be obtained from the abstract task A. If it is possible to associate E with a distinct plan P, we can add a method `A` $\mapsto$ P to the planning model. The task of ontology classification is to find the subsumption hierarchy of the given ontology, i.e., all subsumptions between named concepts occurring in the ontology. They have the form $B \sqsubseteq A$ and can be easily transformed into unit-methods `A` $\mapsto$ `B`. Taking *Workout1* as an example, if it can be inferred that *Workout1* $\sqsubseteq$ *StrengthTraining* (*Workout1* is a strength training), a method `StrengthTraining` $\mapsto$ `Workout1` is created. A more challenging task is to find more complex decomposition methods, described by concept inclusions between named concepts (i.e. the task to be decomposed) and expressions describing plans. For example, is the combination of *Workout1* with another workout classified as a *StrengthTraining*?

Since generating all possible subsumptions between concepts and arbitrary expressions is impossible in practice, only a certain set of candidate expressions should be considered. The easiest way to do so is to add new named concepts $C \equiv E_C$ for every candidate expression E_C to the ontology. This enables a uniform scheme to generate new decomposition methods. First, a set of candidate expressions is generated and added as named concepts to the ontology. Second, ontology classification is used to determine all subsumptions between named concepts in the ontology. Third, these subsumptions are translated into decomposition methods. This is done for subsumptions that connect two named concepts from the original ontology, and subsumptions between a named concept and the concepts in a newly generated candidate expression.

Behnke et al. [3] described which expressions should be considered as potential candidate concepts. Most notably, they argued that a syntactic combination of concepts defined by onlysome expressions should be defined. Our fitness training domain initially contains 310 tasks and only a few methods, while the corresponding ontology contains 1230 concepts (of which 613 are imported from the NCICB corpus) and 2903 axioms (of which 664 are from NCICB). Using DL reasoning—provided by the OWL reasoner FaCT++ [38]—206 new decomposition methods are created. On an up-to-date laptop computer (Intel Core i5-4300U) it takes 3.6 s to compute the whole extended planning domain.

7.5.3 Dialog Domain

In order to integrate the user into the planning process and to communicate the generated solution, a dialog management component is needed to control the flow and the structure of the interaction. In order to communicate a solution, all planned tasks have to be represented in the dialog domain, while integrating the user requires the ongoing presentation of planning decisions. This includes, most notably, the choice of a decomposition method if an abstract task is to be refined. The use of shared knowledge considerably facilitates coherency of the interaction. Although the planning knowledge stored in the ontology alone is not sufficient for the generation of the dialog domain, it contributes to its structure and enables an unisono view on the domain, eliminating inconsistency and translation problems.

The integrated planning knowledge, used to infer new decompositions for existing planning domains, can be used to create a basic dialog structure as well. As in Sect. 7.5.1, a dialog $\mathtt{A}$ can be decomposed into a sequence of subdialogs containing the dialogs $\mathtt{B}_1, \ldots, \mathtt{B}_n$ by an axiom O *includes* $[B_1, \ldots, B_n] \sqsubseteq A$. For example, in our application scenario a strength training can be conducted using a set of workouts $\mathtt{A}_1, \ldots, \mathtt{A}_m$, each of which consists of a set of exercises $\mathtt{B}_1, \ldots, \mathtt{B}_n$. This way a dialog hierarchy can be created, using the topmost elements as entry points for the dialog between user and machine. Nevertheless, this results only in a *valid* dialog structure, but not in a *most suitable* one for the individual user. For this, concepts of the ontology can be excluded from the domain generation or conjugated to other elements in an XML configuration file. This way elements can be hidden or rearranged for the user. The dialogs are also relevant during the MIP process. When selecting between several *Plan Modifications*, these have to be translated to a format understandable by the user. Hence, in addition to the knowledge used to generate plan steps, resources are required for communicating these steps to the user. Therefore, texts, pictures, or videos are needed, which can be easily referenced from an ontology. Using this information, dialogs suitable for a well-understandable human-computer interaction can be created and presented to the user.

One key aspect of state-of-the-art DS is the ability to individualize the ongoing dialog according to the user's needs, requirements, preferences, or history of interaction. Coupling the generation of the dialog domain to the ontology enables us to accomplish these requirements using ontological reasoning and explanation in various ways as follows: Dialogs can be pruned using ontological reasoning according to the user's needs (e.g. "show only exercises which do not require gym access") and requirements (e.g. "show only beginner exercises") or adapted to the user's dialog history (e.g. "preselect exercises which were used the last time") and preferences (e.g. "present only exercises with dumbbells").

7.6 Explanations for Plans and Planning Behavior

Integrating proactive as well as requested explanations into the interaction is an important part of imparting used domain knowledge and clarifying system behavior. Using a coherent knowledge source to create dialog and planning domains enables us to use predefined declarative explanations [29] together with the dynamically generated plan explanations described in Chap. 5 and explanations for ontological inferences, without dealing with inconsistency issues. This way *Plan Steps* (e.g. exercises) can be explained in detail, dependencies between plan steps can be explained to exemplify the necessity of tasks (i.e. plan explanation), and ontology explanations can justify inferences from which the planning model and the dialog domain were generated, all of which increase the user's perceived system transparency. In the scope of mixed-initiative planning, events like *backtracking* may confuse the user. Especially in this context we deem the integration of explanations a very valuable system capability. In order to investigate the effects of the use of explanations in MIP we designed an experiment comparing different kinds of explanations. As context a typical *backtracking* situation was chosen, initiated by external information.

7.6.1 Methodology

Participants were presented a scenario where they were tasked to create individual strength training workouts. They were guided through the process by the system, which provided a selection of exercises for training each specific muscle group necessary for the workout. Figure 7.2 shows an exemplary course of interaction for the introductory round. Here, the user had to plan a *Full Body Workout* to train the listed body parts. For each body part a dialog was presented, providing the user with a selection of exercises to train the specific body part. For example, when training the *legs* the user could choose from exercises such as the *barbell squat* or the *dumbbell squat* (see Fig. 7.3).

The experimental design compared two conditions, distinguished by the use of explanations. During the session, *backtracking* was initiated due to an artificially induced event. In the first condition (backtracking with notification, BT-N), a context-independent high-level explanation was provided:

> The system has detected that the previously presented options do not lead to a solution. Therefore, you have to decide again.

In the second condition (backtracking with explanation, BT-E), a more concrete description of the reason for backtracking (the external event) was provided:

> The system has detected that the gym is closed today due to a severe water damage. Therefore, you have to decide again and select exercises suitable for training at home.

Participants were assigned randomly to conditions. Due to incomplete data some had to be removed, resulting in 43 participants (25 to BT-N, 18 to BT-E). Measures were obtained after the interaction for the following variables:

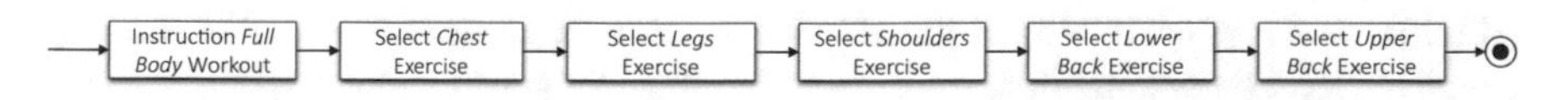

Fig. 7.2 Sequence of (sub-)dialogs for planning the introductory workout

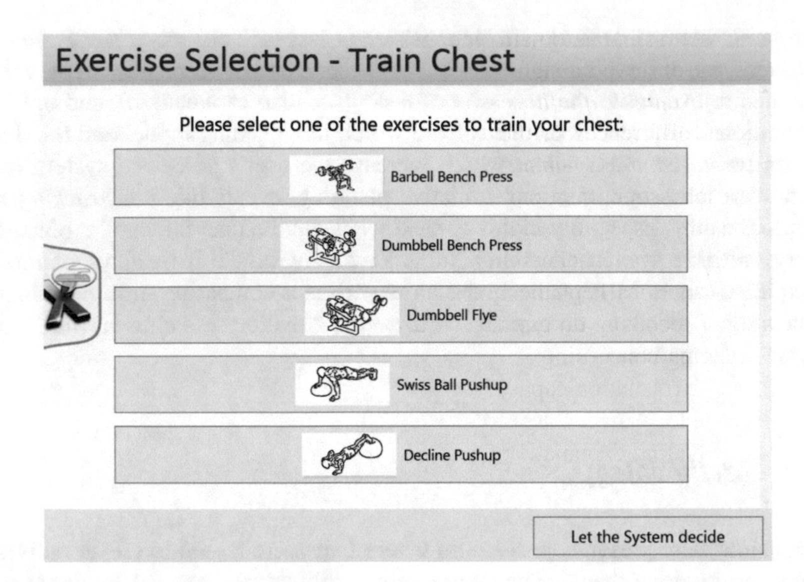

Fig. 7.3 A screenshot of a typical selection dialog (we employ a system developed by Honold et al. [18]). The user is prompted to select an exercise to train the legs or to "let the system decide", in which case the exercise is selected at random

Human-Computer Trust (HCT) describes the trust relationship between human and computer and was assessed using the questionnaire by Madsen and Gregor [23], measuring five dimensions (*Perceived Understandability, Perceived Reliability, Perceived Technical Competence, Personal Attachment, Faith*).

AttrakDiff assesses the perceived pragmatic quality, the hedonic qualities of stimulation and identity, and the general attractiveness of dialog systems (or software in general). We used the questionnaire developed by Hassenzahl et al. [16].

Cognitive Load is assessed using an experimental questionnaire developed by Pichler et al. [32] which measures all three types of cognitive load (intrinsic, extraneous and germane cognitive load) separately, along with the overall experienced cognitive load, fun and difficulty of the tasks.

7.6.2 *Results*

A pairwise t-test on the uniformly distributed data revealed significantly higher scores for the HCT item's *perceived reliability* ($t(3.0) = 57, p = 0.004$), *perceived understandability* ($t(3.99) = 57, p = 0.000$) and *perceived technical competence* ($t(2.06) = 41, p = 0.045$) in the BT-E condition as compared to BT-N. For the cognitive load questionnaires, we only found that the germane load was higher for the BT-E condition, but only to a marginally significant degree ($t(1.99) = 41, p = 0.053$). Note that germane load is positive load—it occurs in the processes inherent in the construction and automation of schemas (e.g. building mental models). In the AttrakDiff we observed a significantly higher score for BT-E in the dimension of experienced *pragmatic qualities* ($t(2.37) = 41, p = 0.022$), which can be attributed to significant differences in the subscales *unpredictable–predictable, confusing–clearly structured, unruly–manageable* and *unpleasant–pleasant*.

These results strengthen our hypothesis that providing explanations of system behavior, in this case of backtracking, does indeed help to perceive the system as more reliable, more understandable, and more technically competent. It seems that the explanations kept the user motivated, compared to the more frustrating experience of receiving no explanation for the impairing system behavior for BT-N. The findings concerning AttrakDiff provide evidence for the conjecture that systems with explanation capabilities seem to be perceived as not so complicated, more predictable, manageable, more clearly structured and in general more pleasant. However, providing explanations of internal system processes, which increase the transparency of the system, requires corresponding reasoning capabilities. This includes the ability to explain causal dependencies between tasks and their related hierarchical structure (i.e. decomposition methods), using *plan explanations* (cf. Chap. 5) and extensions thereof discussed in the following.

7.7 Extending Plan Explanations with Ontology Explanations

Plan explanations focus on elaborating on the structure of a generated plan, namely the relationships between preconditions and effects and how primitive tasks are obtained by decomposing abstract tasks. However, as described in Sect. 7.5.1, task decompositions are inferred from domain knowledge in the ontology, which lends itself to more detailed explanations. For example, rather than presenting a task decomposition only in the form of a statement such as "Task A was necessary, since it must be executed to achieve B", the assumptions and the reasoning behind this decomposition can be used to further elucidate *why* such a task A serves to achieve B. One advantage of using a description logics formalism for this knowledge is the opportunity to make use of *ontology verbalization* techniques that have been devised to generate texts and explanations from ontologies that are understandable to lay people. A considerable part of related work in this field has so far concentrated on how selected facts from an ontology can be presented to users in a fluent manner, and how well they are understood, for example [2, 21]. Approaches to the generation of explanations for reasoning steps have been developed [10, 28, 34]. The work presented here is in line with the principles common to these three approaches. Whereas in the following we illustrate our approach using the explanation of a task decomposition as an example, also other logical relationships between facts modeled in the ontology can be explained with the help of the presented techniques.

The presented mechanism has been implemented as a prototype—developing the prototype into a mature system remains a topic for future work.

Explanations are generated for facts that can be inferred from the axioms in the ontology. For a short example, consider a workout that includes two tasks, front squat and barbell deadlift (each to be performed for a medium number of repetitions), called *Workout1*. Further assume that with the background knowledge of the ontology, *Workout1* is classified as a strength training (i.e. $Workout1 \sqsubseteq StrengthTraining$ holds), which—as discussed in Sect. 7.5.1—introduces a decomposition, `StrengthTraining` $\mapsto$ `Workout1`. The user may now ask for a justification for the decomposition, which is provided based on the axioms and the logical relationships that served to classify *Workout1* as a strength training. These explanations are generated in a stepwise fashion from a formal proof, such that the individual inference steps (in particular, including intermediate facts) are translated to simple natural language texts. In our running example, the resulting explanation is the following:

> A strength training is defined as something that has strength as an intended health outcome. Something that includes an isotonic exercise and a low or medium number of repetitions has strength as an intended health outcome, therefore being a strength training. Workout1 is defined as something that strictly includes a medium number of repetitions of front squat and a medium number of repetitions of barbell deadlift. In particular Workout1 includes a medium number of repetitions of front squat. Given that front squat is an isotonic exercise, a medium number of repetitions of front squat is an isotonic exercise. Thus, we have established that Workout1 includes an isotonic exercise and a low or medium number of repetitions. Given that something that includes an isotonic exercise and a low or medium number of repetitions is a strength training, Workout1 is a strength training.

In the following, we discuss the two submechanisms involved in generating such an explanation—reasoning and explanation generation—in more detail.

Reasoning The first step consists of identifying those axioms that are logically sufficient to show (and thus explain) the fact in question. This task is known as *axiom pinpointing*, for which an efficient approach has been developed by Horridge [19]. The set of necessary axioms (which need not be unique) is called a *justification* for the inferred fact. We simply use Horridge's mechanism as a preprocessing step to obtain a set of relevant axioms to infer the fact in question, and then use inference rules implemented in the prototype to build a stepwise (also called consequence-based) proof. As of current, the prototype uses inference rules from various sources (e.g. [28]) and has not been optimized for efficiency (working on the justifications instead of the full ontology makes using such a simple mechanism feasible).

Explanation Generation The generated proofs have a tree structure, which is used to structure the argument. Each inference rule provides a template, according to which textual output is produced. As a general rule, first the derivation of the premises of an inference rule needs to be explained before the conclusion may be presented (this corresponds to a post-order traversal, with the conclusion at the root of the tree). Inference rules with more than one premise specify the order in which the premises are to be discussed, which is determined by the logical form of the premises. For example, for the following inference rule, the derivation for the left premise is presented before the derivation of the right premise, though logically the order of the premises is irrelevant for the validity of the proof.

$$\frac{\begin{array}{c}\vdots \\ X \sqsubseteq \exists r.B\end{array} \qquad \begin{array}{c}\vdots \\ B \sqsubseteq C\end{array}}{X \sqsubseteq \exists r.C} \; R_{\exists}^{+}$$

Lexicalisation of facts is done with the help of the ontology. For each concept name or role name (or other named elements in the ontology) the value of the *label* attribute is used as a lexical entry. Thus, the domain modeler is required to specify adequate names for the elements of the domain when modeling them in the ontology, for example, "isotonic exercise" for the concept name *IsotonicExercise*. A special case is represented by concepts that can be used as attributes in complex concept expressions. For instance, consider *MediumNumberOfRepetitions* (things that are repeated a moderate number of times) and *FrontSquat*, which can be combined to *MediumNumberOfRepetitions*⊓*FrontSquat*. Since *MediumNumberOfRepetitions* can be used on its own, the label specifies a name that can stand on its own, such as "medium number of repetitions". However, the concept needs to be combined adequately when in combination, for which a second type of label is used, in this case "medium number of repetitions of", such that the combination reads as "medium number of repetitions of front squat". In formulae, concept names are generally represented in indeterminate form, e.g., "a medium number of repetitions of front squat". Connectives in formulae are translated as $\sqsubseteq$: "is"; $\sqcap$: "and" (unless the concepts can be combined as described above); $\sqcup$: "or"; $\exists$: "something that", etc. This way, texts are generated similar to those studied in [21] and [28]. While this template-based approach is in general similar to previous related work (e.g. [28]), some mechanisms and templates are specific to our approach. For example, our modeling relies on the *onlysome* macro, which represents a rather lengthy formula when expanded. When generating output, class expressions that correspond to the form of *onlysome* are identified and treated using a succinct text template to avoid being repetitive; in the example above, this is done for the third sentence, in which the onlysome statement O*includes*.[*MediumNumberOfRepetitions* ⊓ *FrontSquat*, *MediumNumberOfRepetitions* ⊓ *BarbellDeadlift*] is verbalized as: "something that strictly includes a medium number of repetitions of front squat and a medium number of repetitions of barbell deadlift". Without this macro, the verbalization of the logically equivalent statement would state both the existence of and the restriction to these two exercises separately, and can thus be considered repetitive.

7.8 Conclusion

This chapter discussed a nexus of considerations and techniques that can form the basis of a planning system in which the user actively takes part in the planning process. For this purpose, we put priority on technologies that serve to make the planning process *comprehensible* and that emphasize its structure, by identifying adequate search strategies and techniques for handling flaws and dead-ends. Furthermore, we addressed the requirements of such a system to provide a coherent view on its domain knowledge. To this end, we developed an integration of the planning domain with an ontology to provide a central knowledge component for such an interactive system, such that planning is suitably linked with reasoning and dialog management. A further aspect considered important for the empowerment of the user concerns the provision of explanations. In addition to enabling the user to effectively participate in the problem-solving process, explanations increase the perceived reliability, understandability and competence of such a system, as was shown in the presented experiment. We discussed how different kinds of explanations (in particular, of reasoning steps) help to realize a dedicated user-oriented approach to planning, and outlined the scope for individualizing the planning process and the system's communication to address users' preferences.

Acknowledgements This work was done within the Transregional Collaborative Research Centre SFB/TRR 62 "*Companion*-Technology for Cognitive Technical Systems" funded by the German Research Foundation (DFG).

References

1. Ai-Chang, M., Bresina, J., Charest, L., Chase, A., Hsu, J.J., Jonsson, A., Kanefsky, B., Morris, P., Rajan, K., Yglesias, J., Chafin, B., Dias, W., Maldague, P.: MAPGEN: mixed-initiative planning and scheduling for the Mars exploration rover mission. IEEE Intell. Syst. **19**(1), 8–12 (2004)
2. Androutsopoulos, I., Lampouras, G., Galanis, D.: Generating natural language descriptions from OWL ontologies: the NaturalOWL system. J. Artif. Intell. Res. **48**, 671–715 (2013)
3. Behnke, G., Ponomaryov, D., Schiller, M., Bercher, P., Nothdurft, F., Glimm, B., Biundo, S.: Coherence across components in cognitive systems – one ontology to rule them all. In: Proceedings of the 24th International Joint Conference on Artificial Intelligence (IJCAI), pp. 1442–1449. AAAI Press, Palo Alto, CA (2015)
4. Behnke, G., Höller, D., Bercher, P., Biundo, S.: Change the plan – how hard can that be? In: Proceedings of the 26th International Conference on Automated Planning and Scheduling (ICAPS). AAAI Press, Palo Alto, CA (2016)
5. Bercher, P., Höller, D.: Interview with David E. Smith. Künstl. Intell. (2016). doi:10.1007/s13218-015-0403-y. Special Issue on Companion Technologies
6. Bercher, P., Biundo, S., Geier, T., Hoernle, T., Nothdurft, F., Richter, F., Schattenberg, B.: Plan, repair, execute, explain - how planning helps to assemble your home theater. In: Proceedings of the 24th International Conference on Automated Planning and Scheduling (ICAPS), pp. 386–394. AAAI Press, Palo Alto, CA (2014)
7. Bercher, P., Richter, F., Hörnle, T., Geier, T., Höller, D., Behnke, G., Nothdurft, F., Honold, F., Minker, W., Weber, M., Biundo, S.: A planning-based assistance system for setting up a home theater. In: Proceedings of the 29th National Conference on Artificial Intelligence (AAAI). AAAI Press, Palo Alto, CA (2015)
8. Biundo, S., Schattenberg, B.: From abstract crisis to concrete relief (a preliminary report on combining state abstraction and HTN planning). In: Proceedings of the 6th European Conference on Planning (ECP), pp. 157–168. AAAI Press, Palo Alto, CA (2001)

9. Biundo, S., Höller, D., Schattenberg, B., Bercher, P.: Companion-technology: an overview. Künstl. Intell. (2016). doi:10.1007/s13218-015-0419-3. Special Issue on Companion Technologies
10. Borgida, A., Franconi, E., Horrocks, I.: Explaining ALC subsumption. In: Proceedings of the 14th European Conference on Artificial Intelligence (ECAI), pp. 209–213. IOS Press, Palo Alto, CA (2000)
11. Byrne, R.: Planning meals: problem solving on a real data-base. Cognition **5**, 287–332 (1977)
12. Erol, K., Hendler, J.A., Nau, D.S.: UMCP: a sound and complete procedure for hierarchical task-network planning. In: Proceedings of the 2nd International Conference on Artificial Intelligence Planning Systems (AIPS), pp. 249–254. AAAI Press, Palo Alto, CA (1994)
13. Ferguson, G., Allen, J.F.: TRIPS: an integrated intelligent problem-solving assistant. In: Proceedings of the 15h National Conference on Artificial Intelligence (AAAI), pp. 567–572. AAAI Press, Palo Alto, CA (1998)
14. Geier, T., Bercher, P.: On the decidability of HTN planning with task insertion. In: Proceedings of the 22nd International Joint Conference on Artificial Intelligence (IJCAI), pp. 1955–1961. AAAI Press, Palo Alto, CA (2011)
15. Gil, Y.: Description logics and planning. AI Mag. **26**(2), 73–84 (2005)
16. Hassenzahl, M., Burmester, M., Koller, F.: AttrakDiff: Ein Fragebogen zur Messung wahrgenommener hedonischer und pragmatischer Qualität. In: Mensch & Computer 2003: Interaktion in Bewegung, pp. 187–196. Teubner, Wiesbaden (2003)
17. Hayes-Roth, B., Hayes-Roth, F.: A cognitive model of planning. Cogn. Sci. **3**, 275–310 (1979)
18. Honold, F., Schüssel, F., Weber, M.: Adaptive probabilistic fission for multimodal systems. In: Proceedings of the 24th Australian Computer-Human Interaction Conference (OzCHI), pp. 222–231. ACM, New York (2012)
19. Horridge, M.: Justification based explanations in ontologies. Ph.D. thesis, University of Manchester, Manchester (2011)
20. Horridge, M., Drummond, N., Goodwin, J., Rector, A., Stevens, R., Wang, H.H.: The Manchester OWL syntax. In: Proceedings of the OWLED'06 Workshop on OWL: Experiences and Directions, vol. 216 (2006). CEUR Workshop Proceedings
21. Kuhn, T.: The understandability of OWL statements in controlled English. Semantic Web **4**(1), 101–115 (2013)
22. Lin, N., Kuter, U., Sirin, E.: Web service composition with user preferences. In: The Semantic Web: Research and Applications. Lecture Notes in Computer Science, vol. 5021, pp. 629–643. Springer, Berlin (2008)
23. Madsen, M., Gregor, S.: Measuring human-computer trust. In: Proceedings of the 11th Australasian Conference on Information Systems (ACIS), pp. 6–8 (2000)
24. McAllester, D., Rosenblitt, D.: Systematic nonlinear planning. In: Proceedings of the 9th National Conference on Artificial Intelligence (AAAI), pp. 634–639. AAAI Press, Palo Alto, CA (1991)
25. Myers, K.L., Jarvis, P., Tyson, M., Wolverton, M.: A mixed-initiative framework for robust plan sketching. In: 13th International Conference on Automated Planning and Scheduling (ICAPS), pp. 256–266. AAAI Press, Palo Alto, CA (2003)
26. Nau, D.S., Au, T.C., Ilghami, O., Kuter, U., Muñoz-Avila, H., Murdock, J.W., Wu, D., Yaman, F.: Applications of SHOP and SHOP2. IEEE Intell. Syst. **20**(2), 34–41 (2005)
27. NCICB (NCI Center for Bioinformatics): http://ncicb.nci.nih.gov/xml/owl/EVS/ (2015). Thesaurus.owl. Accessed 9 Feb 2015
28. Nguyen, T.A.T., Power, R., Piwek, P., Williams, S.: Predicting the understandability of OWL inferences. In: The Semantic Web: Semantics and Big Data. Lecture Notes in Computer Science, vol. 7882, pp. 109–123. Springer, Berlin (2013)
29. Nothdurft, F., Richter, F., Minker, W.: Probabilistic human-computer trust handling. In: Proceedings of the Annual Meeting of the Special Interest Group on Discourse and Dialogue, pp. 51–59. ACL, Menlo Park, CA (2014). http://www.aclweb.org/anthology/W14-4307

30. Nothdurft, F., Behnke, G., Bercher, P., Biundo, S., Minker, W.: The interplay of user-centered dialog systems and AI planning. In: Proceedings of the 16th Annual Meeting of the Special Interest Group on Discourse and Dialogue (SIGDIAL), pp. 344–353. ACL, Menlo Park, CA (2015)
31. Penberthy, J.S., Weld, D.S.: UCPOP: a sound, complete, partial order planner for ADL. In: Proceedings of the 3rd International Conference on Principles of Knowledge Representation and Reasoning (KR), pp. 103–114. Morgan Kaufmann, Los Altos, CA (1992)
32. Pichler, M., Seufert, T.: Two strategies to measure cognitive load. In: EARLI Conference 2011. Education for a Global Networked Society, pp. 928–929. European Association for Research on Learning and Instruction, Leuven (2011)
33. Pulido, J.C., González, J.C., González-Ferrer, A., García, J., Fernández, F., Bandera, A., Bustos, P., Suárez, C.: Goal-directed generation of exercise sets for upper-limb rehabilitation. In: Proceedings of the 5th Workshop on Knowledge Engineering for Planning and Scheduling (KEPS), pp. 38–45 (2014)
34. Schiller, M., Glimm, B.: Towards explicative inference for OWL. In: Proceedings of the 26th International Description Logic Workshop, vol. 1014, pp. 930–941. CEUR (2013)
35. Schmidt-Schauß, M., Smolka, G.: Attributive concept descriptions with complements. Appl. Artif. Intell. **48**, 1–26 (1991)
36. Sirin, E.: Combining description logic reasoning with AI planning for composition of web services. Ph.D. thesis, University of Maryland at College Park (2006)
37. Sirin, E., Parsia, B., Wu, D., Hendler, J., Nau, D.: HTN planning for web service composition using SHOP2. Web Semant. **1**(4), 377–396 (2004)
38. Tsarkov, D., Horrocks, I.: FaCT++ description logic reasoner: system description. In: Proceedings of the 3rd International Joint Conference on Automated Reasoning (IJCAR), pp. 292–297. Springer, Berlin (2006)
39. Vardi, M.Y.: Why is modal logic so robustly decidable? Descriptive Complex. Finite Models **31**, 149–184 (1997)

Chapter 8
Neurobiological Fundamentals of Strategy Change: A Core Competence of *Companion*-Systems

Andreas L. Schulz, Marie L. Woldeit, and Frank W. Ohl

Abstract *Companion*-Systems interact with users via flexible, goal-directed dialogs. During dialogs both, user and Companion-System, can identify and communicate their goals iteratively. In that sense, they can be conceptualized as communication partners, equipped with a processing scheme producing actions as outputs in consequence of (1) inputs from the other communication partner and (2) internally represented goals. A quite general core competence of communication partners is the capability for strategy change, defined as the modification of action planning under the boundary condition of maintaining a constant goal. Interestingly, the biological fundamentals for this capability are largely unknown. Here we describe a research program that employs an animal model for strategy change to (1) investigate its underlying neuronal mechanisms and (2) describe these mechanisms in an algorithmic syntax, suitable for implementation in technical Companion-Systems. It is crucial for this research program that investigated scenarios be sufficiently complex to contain all relevant aspects of strategy change, but at the same time simple enough to allow for a detailed neurophysiological analysis only obtainable in animal models. To this end, two forms of strategy change are considered in detail: Strategy change caused by modified feature selection, and strategy change caused by modified action assignment.

A.L. Schulz (✉) • M.L. Woldeit
Leibniz Institute for Neurobiology, Brenneckestraße 6, 39118 Magdeburg, Germany
e-mail: andreas.schulz@lin-magdeburg.de; marie.woldeit@lin-magdeburg.de

F.W. Ohl
Leibniz Institute for Neurobiology, Brenneckestraße 6, 39118 Magdeburg, Germany

Otto-von-Guericke University, 39118 Magdeburg, Germany

Center for Behavioral Brain Sciences (CBBS), 39118 Magdeburg, Germany
e-mail: frank.ohl@lin-magdeburg.de

S. Biundo, A. Wendemuth (eds.), *Companion Technology*, Cognitive Technologies,
DOI 10.1007/978-3-319-43665-4_8

8.1 Introduction

The exchange between a user and a *Companion*-System should ideally take the form of a dialog. Dialogs are characterized by the fact that the output of one dialog partner becomes the input of the other dialog partner. As the dialog unfolds, a cyclic exchange of information occurs between partners, ideally towards reaching a common goal. In a very simplified scheme, dialog partners can be considered symmetric. In order to produce the outputs on the basis of the represented goal and current input information, the newly incoming information has to be evaluated. This evaluation is a central part for establishing flexibility of dialogs (see Fig. 8.1a).

Successful dialogs require a certain amount of cognitive flexibility in both partners. Hereby the interaction of several cognitive processes is required, e.g. attention shift, conflict monitoring and perception [32].

A specific form of cognitive flexibility is the capability for strategy change. This term refers to adaptation of action or action planning in response to changing conditions while the overarching individual goal remains constant. Dialog partners need to realize that actions taken so far or actions planned for the immediate future have a low likelihood of reaching an individual or shared goal of the dialog partners. Even in the simplified scheme sketched above, changed action or action planning can manifest itself in many different forms, but becomes particularly well observable if it is manifested in altered analysis of the input ("set shifting") or the altered production of the output ("task switching"). The former case occurs, for example, when a dialog partner realizes that previously attended aspects of the sensory input have become useless for attaining the existing goal and other sensory features have to be attended to. This could happen while the actual actions

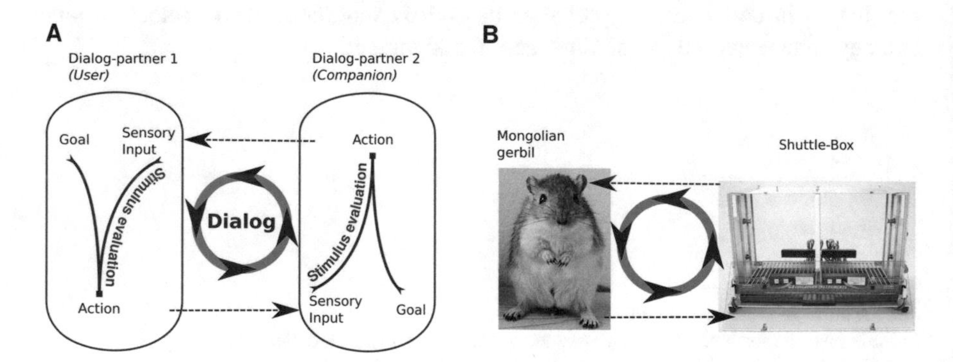

Fig. 8.1 **(a)** Stimulus evaluation and strategy change within the dialog model. Both dialog partners, the user and the *Companion*-System, evaluate incoming sensory information and choose actions to achieve an individual goal. Feature selection and stimulus-action contingencies are central to determining the subject's strategy within this process. Actions of one partner are fed back to the other partner and become evaluated; therefore this schema can be understood as a cyclic exchange. **(b)** In the laboratory the animal subject acts within a shuttle-box as one dialog partner, while the computational control of the experiment represents the second partner

themselves remain unchanged, but are now recruited for different reasons. The latter case occurs when the collected sensory inputs now trigger different behavioral outputs.

Technical systems can be designed with an a priori world model and built-in strategies, with the ability to switch between existing strategies according to an inbuilt algorithm with defined criteria for switching. Any designed world model, however, is limited in real-life scenarios because once the built-in strategies fail, the *Companion*-System should be able to continue the dialog. This requirement implies that a *Companion*-System is able to learn strategies, recognize unknown situations and apply already learned strategies, but also learn new strategies without overwriting the old ones. Furthermore, once the *Companion*-System has stored a small number of different strategies, it should be able to switch between them swiftly.

Humans and higher vertebrates have the capability to adapt their behavior to environmental and context-induced demands through cognitive flexibility. In human subjects disruptions or lack of cognitive flexibility represent a common feature of neuropsychiatric disorders, such as schizophrenia or obsessive-compulsive disorder (OCD). In the adaptation process behavioral flexibility always has to be contrasted with stability of already incorporated rules and conditions from the past [32]. That implies that existing strategies are not overwritten but updated and refined for the larger context, possibly serving as a foundation to create new strategies. To guarantee a most convenient and efficient user-*Companion* dialog/interaction, it is desirable that a *Companion*-System be capable of developing such flexibility.

Neuroscience can add valuable information about: (a) the mechanistic view of how biological systems enforce cognitive flexibility at a neuronal network level, and (b) the holistic view of when to switch to a new strategy or when to stay with the existing rules by assessing behavioral flexibility in defined learning-paradigms. Results from both neuroscientific areas of research can serve as a basis to create principles, parameters to algorithms for such biologically inspired artificial intelligence. In this chapter we will:

- give an overview of how technical learning principles and learning in biological systems have converged through the concept of reinforcement learning
- sketch briefly the neuronal substrates found to underlie the described processes
- flesh out why learning through negative reinforcement is crucial for a behaviorally and cognitively flexible *Companion*-System
- provide physiological data from our own animal model to illustrate functional interactions between brain areas involved in the read-out of goal-determined behavior
- present the reader with the concept of serial reversal learning, a task used experimentally to assess cognitive flexibility
- provide our own findings on serial reversal learning with our animal model
- give an overview of the already known neuronal substrates of serial reversal learning.

8.2 Reinforcement Learning: A Technical and Biological Learning Model

Reinforcement learning (RL) is a widely established conceptional approach in the field of both artificial intelligence and bioscience to describe learning by trial and error. An overview of the technical aspects of reinforcement learning is given in [64].

RL-methods have their roots in the field of psychology of animal learning behavior:

> The history of reinforcement learning has two main threads, both long and rich, that were pursued independently before intertwining in modern reinforcement learning. One thread concerns learning by trial and error and started in the psychology of animal learning. This thread runs through some of the earliest work in artificial intelligence and led to the revival of reinforcement learning in the early 1980s. The other thread concerns the problem of optimal control and its solution using value functions and dynamic programming. For the most part, this thread did not involve learning. All threads came together in the late 1980s to produce the modern field of reinforcement learning. [64]

RL ideally results in the optimal mapping of environmental situations to an agent's behavior (called actions) in order to maximize a numerical reward signal. The learner is not told which actions to take, in contrast to many forms of machine learning, but instead must discover which actions yield the largest reward by trial-and-error. The overarching goal for the agent is to choose such actions which maximize (or minimize) the overall reinforcements by minimizing the prediction error. The prediction error is the difference between obtained and predicted reward. Several approaches for solving this problem have been suggested. In the field of animal learning, however, the suitability of the so called Temporal-Difference (TD) learning algorithm has been widely applied, particularly the actor-critic variant. The mathematical description is given in [64].

Advantageous to TD learning is the fact that the agent is not required to maintain a model of the environment (e.g. a representation of different states and transitions between them) but only needs a prediction of a potential reinforcement. This model class allows for two distinctive submodules: one for evaluating the prediction of the reward and the other for action selection.

Despite reinforcement learning's roots in animal behavior, it was not until the 1990s that this concept attracted the attention of neuroscientists again [27, 61]. Dopaminergic neurons in midbrain structures (*substantia nigra pars compacta, ventral tegmental area, dorsolateral substantia nigra*) of animals produce action potentials (short electrical pulses) in response to rewarding conditions. Interestingly, the frequency of these action potentials appears to directly correlate with the magnitude of prediction errors (Fig. 8.2).

In the striatum, a large subcortical structure, neurons encode cue-induced reward predictions. This cue might for example correspond to environmental events signaling reward. Furthermore, neurons in the midbrain, striatum and nucleus accumbens show phasic responses, an increased production of action potentials,

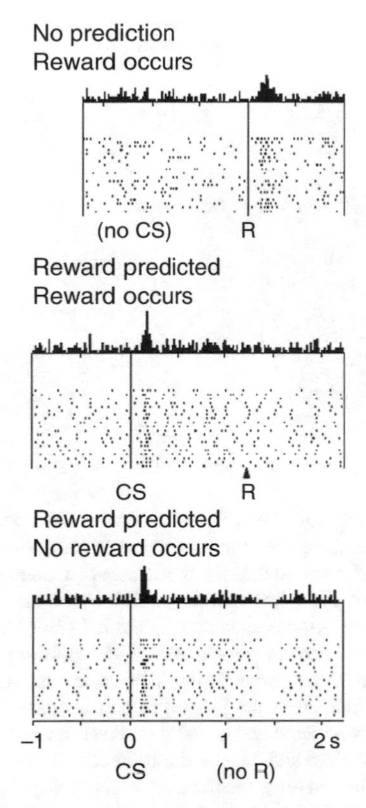

Fig. 8.2 Reward prediction error responses of single dopamine transmitting neurons. Changes in these neurons' output code for an error in the prediction of appetitive events. (*Top*) Before learning, a drop of appetitive fruit juice is given in the absence of prediction; hence a positive error in the prediction of reward results. The dopaminergic neuron is activated by this unpredicted occurrence of juice. (*Middle*) After learning, the conditioned stimulus predicts reward, and the reward occurs according to the prediction; hence no error in the prediction of reward is calculated. The dopamine transmitting neuron is activated by the reward-predicting stimulus but fails to be activated by the predicted reward (*right side*). (*Bottom*) After learning, the conditioned stimulus predicts a reward, but due to a mistake of the monkey is not given to the subject. The activity of the dopaminergic neuron is depressed at the exact time when the reward would have occurred. The depression occurs more than 1 s after the conditioned stimulus without any intervening stimuli, revealing an internal representation of the time of the predicted reward. Neuronal activity is aligned to the electronic pulse that drives the solenoid valve delivering the reward liquid (*top*) or the onset of the conditioned visual stimulus (*middle and bottom*). Each panel shows the peri-event time histogram and raster of impulses from the same neuron. Horizontal distances of dots correspond to real-time intervals. Each *line of dots* represents one trial. The original sequence of trials is plotted *from top to bottom*. CS: conditioned reward-predicting stimulus; R: primary reward (from [61])

at the onset of primary rewards such as food, juice or water (Fig. 8.2). With these three elements (prediction of reward, reward magnitude and prediction error) the main elements for an RL-circuit are available in the brain. Based on the experimental work of Schultz and colleagues, as well as others [48, 59, 60, 65],

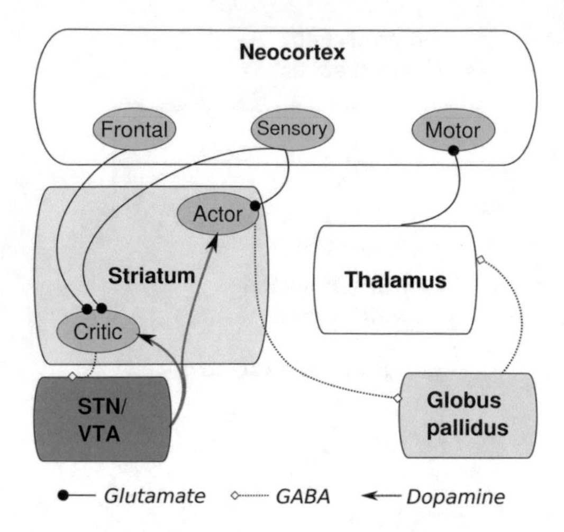

Fig. 8.3 Critic-Actor Model of the Basal Ganglia. The sensory areas in the neocortex receive sensory and contextual information from the environment. Neurons in these sensory areas project to the striatum, consisting of critic and actor modules via glutamatergic (excitatory) synapses. The activation of neurons in the actor module inhibits neurons in the *globus pallidus* (GP) via GABAergic synapses. The suppression of activation in the GP leads to a disinhibition of thalamic neurons which project to the motor cortex. A specific population activity results in a certain motor response or action respectively. The critic module receives the same sensory information as the actor module from the neocortex and encodes the reward prediction related to the sensory cues. With the predicted and actually received reward, the dopaminergic neurons in the midbrain (SNc/VTA) calculate and encode the prediction error. Their firing rate is supposed to be proportional to the prediction error (see Fig. 8.2). These dopaminergic neurons innervate the entire striatum (actor and critic modules). The released dopamine acts modulatory at the cortico-striatal glutamatergic synapses of critic and actor modules and alters the synaptic strength of these synapses. The modified synaptic weights result in a different mapping from the sensory cues coded in the sensory cortical areas to the actor and critic modules in the striatum

the actor-critic implementation of TD learning was mapped to the basal ganglia and dopaminergic midbrain neurons, already known for their role in learning and goal-directed behavior. Several models of neuronal realizations of this mapping have been deployed to date [19, 20, 26, 27, 44, 61], but are under continuous revision.

The main aspects of the actor-critic model are depicted in Fig. 8.3. The sensory areas of the neocortex receive sensory and contextual information from the environment. This information is transmitted to neurons in the striatum, the main input hub of the basal ganglia, via glutamatergic synapses. In this model the striatum incorporates a critic and an actor module. The actor module selects an appropriate action. This action selection command is then transmitted to *globus pallidus* (GP), which contains projections to thalamic motor areas that directly connect to the motor cortex. Here, motor commands are effected for the chosen action. The critic module receives the same sensory information as the actor module from the neocortex and encodes the reward prediction related to the sensory cues. With the predicted and actually received rewards, the dopaminergic neurons in the midbrain (SNc/VTA) calculate and encode the prediction error. Their firing rate is proportional to the prediction error (see Fig. 8.2). These dopaminergic neurons project to the entire striatum (actor and critic modules). Note that dopaminergic neurons project to other brain structures as well, e.g. the neocortex. The released dopamine acts modulatory at the cortico-striatal glutamatergic synapses of the critic and actor modules. As a result, the synaptic strength of these synapses is altered. The modified synaptic weights result in a different map from the sensory cues coded in the sensory cortical areas to the actor and critic modules in the striatum.

The actual physiological mechanisms are still under investigation. One hypothesis states that the critic module is located in the ventral part of the striatum whereas the actor module is located in dorso-medial and dorso-lateral divisions of the striatum (Fig. 8.3; [26, 27]).

Temporal difference learning (RL), without an additional a priori defined knowledge base (world models) or additional elements (like task-specific expert networks with only one network active for every different task), is not suitable to explain cognitive flexibility. With constant parameters a trained RL agent will adapt to a changing, non-stationary environment with the same dynamic as during initial learning, and shows phenomena similar to interference or task switch cost.

8.3 Learning with Negative Reinforcement

Previous and current research related to RL concepts have preferentially investigated the effects of appetitive rewards and positive reinforcement. However, in real-world scenarios a *Companion*-System may not only learn from positive reinforcement, meaning supporting feedback for a chosen action, but also from negative reinforcement. This is a key point concerning basal competences of a *Companion*-System. With negative feedback the *Companion*-System has to learn which actions are not advantageous from a set of different actions to resolve which actions minimize the negative feedback. In our laboratory we investigate the RL-model in experiments with aversive reinforcement in a two-way active avoidance (2-WAA) conditioning task using Mongolian gerbils (*Meriones unguiculatus*) as experimental animals (Fig. 8.1b).

The 2-WAA procedure is performed in a shuttle-box (Fig. 8.4). It consists of two compartments separated by a small hurdle. Mongolian gerbil subjects are free to cross this hurdle and change compartments. Electrical foot-shocks can be delivered through a metal grid floor, serving as negative reinforcement. A trial in a 2-WAA task typically begins with the onset of a stimulus (e.g. a tone or a light). This stimulus is termed conditioned stimulus (CS). The animal subject now has the opportunity to cross to the contralateral side of the shuttle-box within a certain time window (in the range of 6 s). After the time has elapsed, the foot-shock is turned on. The subject consequently shuttles to the other compartment in order to escape the shock, also called unconditioned stimulus (US). This leads to a simultaneous termination of shock and tone. If the animal shuttles within the

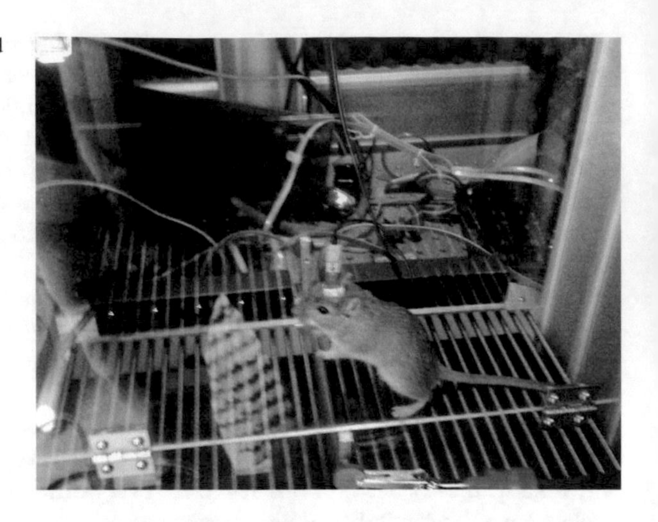

Fig. 8.4 Shuttle-box as used in the laboratory. Two compartments are separated by a small hurdle. Foot shocks can be delivered through the metal grid-floor. The setup is enclosed in an acoustically and electrically shielded chamber

specified time-window, the Go response is scored as a *hit*, otherwise as a *miss*. After an inter-trial interval, usually 15–30 s, the next trial commences. Typically, the animal subjects learn that the sensory stimulus (CS) predicts the foot-shock (US) after a few trials. During CS presentation, they now start orienting towards the hurdle and quickly cross into the other compartment at US onset. The subjects are now escaping both shock and tone. Further, during the training, subjects begin shuttling into the contralateral compartment before the shock has been turned on. This shuttling still terminates the tone and the animals now have successfully avoided the shock. Trials are started independently of the animal's position within the shuttle-box: it can shuttle from the left compartment to the right or vice versa. The shuttling behavior is necessary to avoid the shock, not a specific location (egocentric perspective).

According to Mowrer's two-factor theory of avoidance learning [46], the termination of the tone (or another sensory stimulus), which has been closely associated with the foot-shock, results in release of fear, which in turn acts as an internal positive reinforcement signal during successful avoidance trials. This proceeding is in contrast to the immediate reinforcement signal of appetitive rewards like water or food.

The paradigm can be expanded to a discrimination paradigm by introducing a second stimulus (CS–). One kind of stimulus (now called CS+) still requires the animal to cross within a specified time window. During CS- trials the animals have to stay in the ipsilateral compartment for a certain amount of time (e.g. at least 10 s). If they cross the hurdle they will receive a foot-shock on the contralateral side (negatively reinforced). Shuttling during a CS- presentation is called a *false alarm* and residing ipsilaterally a correct rejection.

In our experiments we used linearly frequency modulated tones (FM; e.g. 2–4 kHz vs. 4–2 kHz) as CS+ or CS- respectively. The tones have a duration of 200 ms. The entire CS+ period of 6 s consists therefore of up to 12 such tones (12 times 0.2 s

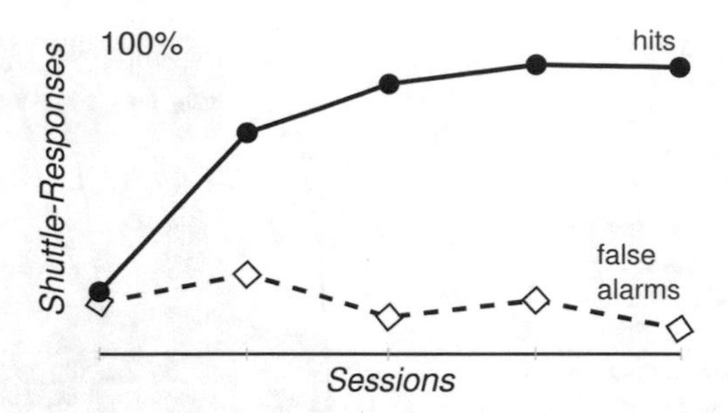

Fig. 8.5 Typical learning curve of discrimination learning. The *straight line* indicates the percentage of shuttling behavior as response to the Go tone per session. This is regarded as correct response and dubbed *hit*. The *dashed line* shows the percentage of shuttling per session as response to the NoGo tone. This is incorrect behavior and called a *false alarm*. The higher the number of *hits* and the lower the number of *false alarms*, the better the discrimination performance of the animal

tone duration +0.2 s pause between tones). A CS- period of 10 s consists of 20 tones. Figure 8.5 shows a typical individual learning curve. The animal learns to shuttle during the CS+ tones before shock onset and to stay in the compartment for at least 10 s during the CS- period.

As we have described earlier, all neural actor-critic models assume functional connections between cortical sensory areas, coding the environment, and the striatal structures containing the actor and critic modules. Therefore, actor-critic models base the process of learning on the modification of these functional connections (or mappings from sensory areas and the striatum). Studies about anatomical connections found connections from cortical sensory areas in different parts of the basal ganglia [6, 42, 50, 62, 67, 71]. Interestingly, there is not much experimental data available investigating the learning-related modifications of functional strength of these connections (see [71]).

Therefore, we employed our behavioral task to measure the modulation of functional strength of the mapping from the auditory cortex as sensory cortical structure to the ventral striatum (actor and critic modules) during learning.

To this end gerbils were implanted with multi-site electrodes in the auditory cortex and the ventral striatum (Fig. 8.6, [70]). A small plug was fixed on the skull of the animal and allowed tethering of a cable to a recording system. Due to this preparation we are capable of recording neuronal signals while the animals perform a Go/NoGo behavioral task in the shuttle-box (e.g. [49]).

The coherence of the local field potentials of both brain structures can be used as a measure of their functional coupling. The coherency C_{ij} (Eq. (8.1)) between two time series x_i and x_j is their complex-valued normalized cross-spectrum S_{ij} (Eq. (8.2)).

$$C_{ij} = \frac{S_{ij}(f)}{(S_{ii}(f)(S_{jj}(f))^{1/2}} \tag{8.1}$$

$$S_{ij} = \langle x_i(f) x_j(f)^* \rangle \tag{8.2}$$

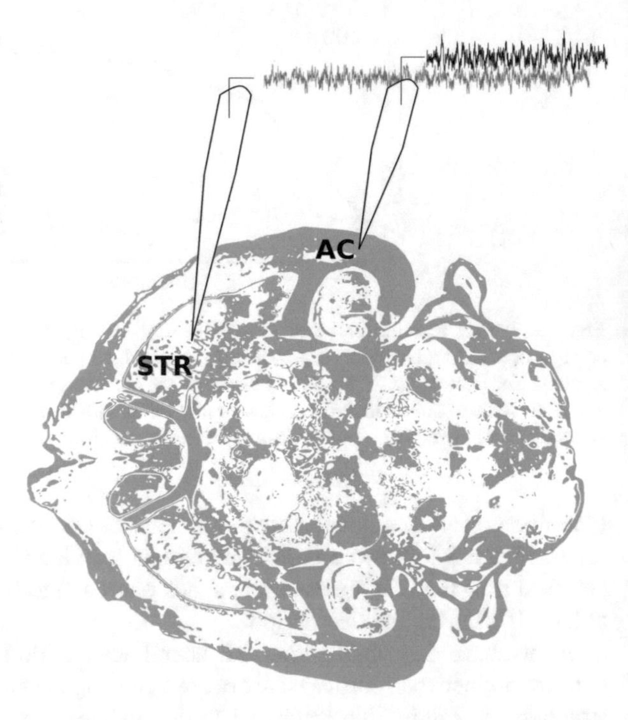

Fig. 8.6 Scheme of positions of recording electrodes. Electrodes were chronically implanted in the auditory cortex (AC) and the ventral striatum (STR). Note that the AC is at the surface of the brain, whereas the STR is a deep brain structure beneath the neocortex. Measured was the electrical local field potential in both brain structures

where * means complex conjugation and denotes the expectation value. S_{ii} and S_{jj} represent the auto spectra of the respective time-series x_i and x_j.

A commonly studied measure is the coherence which is defined as the absolute value (here referred to as mCoh) of the coherency (see [47]).

$$mCoh = |C_{ij}| \tag{8.3}$$

$$iCoh = imag(C_{ij}) \tag{8.4}$$

where $imag(C_{ij})$ represents the imaginary part of the complex coherency C_{ij}.

The coherency values were calculated for the onsets of the first six tones of a trial. For the comparison between naïve and trained animals, the coherency values during the first six CS- and CS+ presentations in every trial were averaged over the entire session. This resulted in one mCoh value for every session per animal, separated for the CS+ and the CS- conditions (Fig. 8.7a).

All animals showed significant improvement of discrimination performance over the course of training. In order to increase statistical power, coherence values from sessions 1 and 2 were pooled into "early sessions" and coherence values from sessions 4 and 5 were pooled into "late sessions". The mCoh values during CS+ presentations increased significantly over training sessions. The mCoh during CS- presentations, however, showed no significant changes (Fig. 8.7b). Altogether, these results indicated a specific increase of functional cortico-striatal coupling in frequency bands up to 12 Hz during CS+ presentations accompanied by increased discrimination performance during learning of the Go/NoGo-task (Fig. 8.7c). The

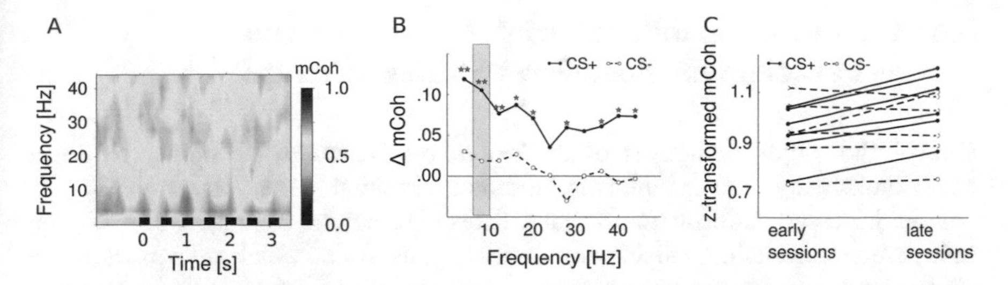

Fig. 8.7 Coherence during training. (**a**) *Gray scale* coded stimulus evoked coherence between AC and ventral striatum during presentation of tones. The time axis shows a sequence of six tones, duration 200 ms, indicated by *small black bars*. The coherence was averaged over 30 of these sequences. Each sequence starts at t = 0. The strength of the coherence was modulated with the repetition rate of the tones. Coherence values at the onset of each tone are dubbed stimulus evoked coherence (**b**) Coherence values of the 8 Hz frequency band for six investigated animals. CS+ evoked coherence increased from early to late session but this was not so for CS-. (**c**) Differences between late and early session for all coherence frequency bands for CS+ (*solid*) and CS- (*dashed*). $^{**}p < 0.01$, $^{*}p < 0.05$ (paired t-test)

increased functional coupling cannot be merely explained by the motor response (shuttling behavior) *hit rate* [62].

These findings are in support of the theory that learning differential behavioral responses to conditioned stimuli requires the formation of neuronal populations that are then differentially recruited during the task accordingly [1, 20, 60].

One possible mechanism of population-specific modulations can be achieved by pairing Hebb's rule with the action of a neuromodulatory substance, e.g. dopamine. The CS+ first becomes associated with the foot shock and elicits fear in the subject. Later, during learning avoidance, responses are shown and become associated with the termination of the CS+. The resulting subsequent relief of fear produces an internal reinforcement that is associated with dopaminergic signaling during successful avoidance [15, 28–31]. If at a cortico-striatal synapse the presynaptic neuron (in the cortex) and the postsynaptic neuron (in the striatum) were active contingently (Hebb's rule) and dopamine was released in the vicinity of this synapse by dopaminergic midbrain neurons after reward or relief from fear, this synapse will be modified [59, 60, 64]. As a result a different cortico-striatal coherence can be measured. In the case of the CS- the pre- and postsynaptic neuron could have been active as with well. However, due to the lack of dopaminergic neuromodulatory action, because there was no relief from fear, the coherence remains constant. Note that while monosynaptic cortico-striatal coupling represents a parsimonious rationale behind the reported changes in coherence between auditory cortex and ventral striatum, indirect polysynaptic connections or a common neuronal input could account for our findings, as well. Candidates for this common source are thalamic nuclei and midbrain structures [24, 43, 63].

Taken together, results from neuroscience support the principle of Reinforcement Learning, particularly the Temporal Difference Learning, with plastic functional connections from neocortex, coding the environments, and the striatum representing actor and critic modules, at least for initial learning problems.

8.4 Cognitive Flexibility: Serial Reversal Learning as Paradigm for Behavioral Strategy Change

One of the major challenges of a *Companion*-System is the ability to adapt behavioral strategies to different circumstances. Preferably the *Companion*-System has all necessary information to take decisions, but in the real world not all information is available, reliable or relevant, e.g. decisions which are correct in one context or to one user might be incorrect in another context or to a different user.

Reinforcement learning can only sub-optimally solve this problem. Specifically, temporal difference learning, without an additional a priori defined knowledge base (world model) or additional elements (like task-specific expert networks with only one network active for every different task), are not suitable to explain the found cognitive flexibility in biological systems. With constant parameters a trained RL agent will adapt to a changing, non-stationary environment with the same dynamics as during initial learning. Moreover, such conflicting situations would require a complete re-learning. This re-learning would take longer (with identical parameter sets) than the initial learning phase due to interference phenomena from the previously learned policies and strategies ("switch cost").

In biological systems such interferences can come into existence from learned non-reward if in the new contingencies the subject has to respond to cues previously without consequences. Cues that lose their previous meaning in the new contingencies, on the other hand, can lead to high perseverance and ultimately to longer learning durations in the new contingencies. There is a thin line between usefulness of cognitive flexibility and habitual responding, the latter allowing the system to efficiently and rapidly respond to environmental challenges and freeing cognitive resources at the same time.

Therefore, another mechanism would be advantageous that enables a technical *Companion*-System to hold multiple sets of policies or strategies and to switch between these. Ideally, the *Companion*-System's performance increases after a few negative feedbacks, immediately after a reversal; hence an additional mechanism would enact and facilitate the switch between the two different strategies (or RL sets/frames).

An experimental method to enforce the formation of these different strategies is serial reversal learning. Within such tasks test subjects learn two state-action pairs, and once learned, the mapping of these pairs is reversed. After a re-learning period, which usually takes longer than the initial learning, the contingencies are swapped back again to the original mapping. These switches are repeated several times.

Many studies suggest that at least all vertebrates show improved performances over successive reversals. However, there are differences in the performance between species and the properties of the learning task. Bitterman [3] classified different species according to their ability to solve visual and spatial serial reversal tasks. Turtles, for instance, showed improvement on spatial but not on visual problems, whereas pigeons, rats and monkeys were able to show improved performance on spatial and visual tasks [2, 4, 9, 10, 12, 16, 21–23, 34, 36, 38–40, 45, 52, 53, 55].

The improved performance after multiple reversals is expressed by faster learning dynamics after contingency reversal. As already pointed out above, biologically inspired RL-models are not able to capture this nature of improved performance. Various animal studies show that subjects are principally able to improve their performance after multiple successive reversals with increased learning dynamics. Under certain circumstances some species are even able to perform a single error reversal suggesting switching between strategies without any new low-level learning. A *Companion*-System should feature the exact same behavior of switching strategies without new low-level learning.

To assess cognitive flexibility during serial contingency reversals in terms of strategy development in our animal model of 2-WAA in the shuttle-box and to analyze different reinforcement schedules, we have expanded the discrimination task described under Sect. 8.3. For this purpose we have designed two experimental animal groups, based on the CS-assigned reinforcements. In the first group (FA-FS group) the animals had to stay in the ipsilateral compartment for at least 10 s to avoid a foot-shock on the contralateral side of the shuttle-box during CS- trials (negatively reinforced *false alarms*). In the other group (FA-no FS group) animals were free to shuttle or stay during CS- trials ("neutral" CS-). Reinforcement during CS+ trials was equal in both groups.

In either paradigm, shuttling during a CS- presentation is called a *false alarm* and residing ipsilaterally a *correct rejection*. After the subjects had acquired the task, the stimulus-action associations were be reversed and the former CS+ now became the CS- and vice versa. The serial reversal task consisted of multiple such switching contingencies following each other (Figs. 8.5 and 8.8).

As in the discrimination task, we have trained the animals using frequency-modulated tones with falling and rising modulation directions as conditioned stimuli. Animals were trained in daily sessions with 50% CS+ and 50% CS- trials in pseudo-randomized order. In both groups multiple contingency reversals were possible, although the dropout quote of animals failing a reversal criterion (three sequential sessions with a d' > 1, d' from signal detection theory, data not shown) increased with each reversal phase (approximately 30% in each phase). This was due to different overall strategy development that typically took place within each experimental group (Fig. 8.8).

We observed that subjects from the FA-FS group, when trained with multiple contingency reversals, mostly relapsed from avoidance to escape behavior, thereby increasing the number of *misses* and *false alarms*, but also the number of foot-shocks they received (Fig. 8.8, top right). For this group this appeared to be the preferential strategy to cope with multiple stimulus-action association switches.

In the FA-no FS group animals typically ceased discriminating between the two CS classes with their behavioral responses. They resumed crossing the hurdle during each trial independent of CS+ or CS- presentation. This led to an overall decrease in the number of shocks these animals received (Fig. 8.8, bottom right). In both groups serial auditory discrimination learning led to a strategy that bypassed frequent re-learning. Surprisingly, both groups displayed an overall strategy development that addressed the serial reversal training in its entirety, considering more switches to

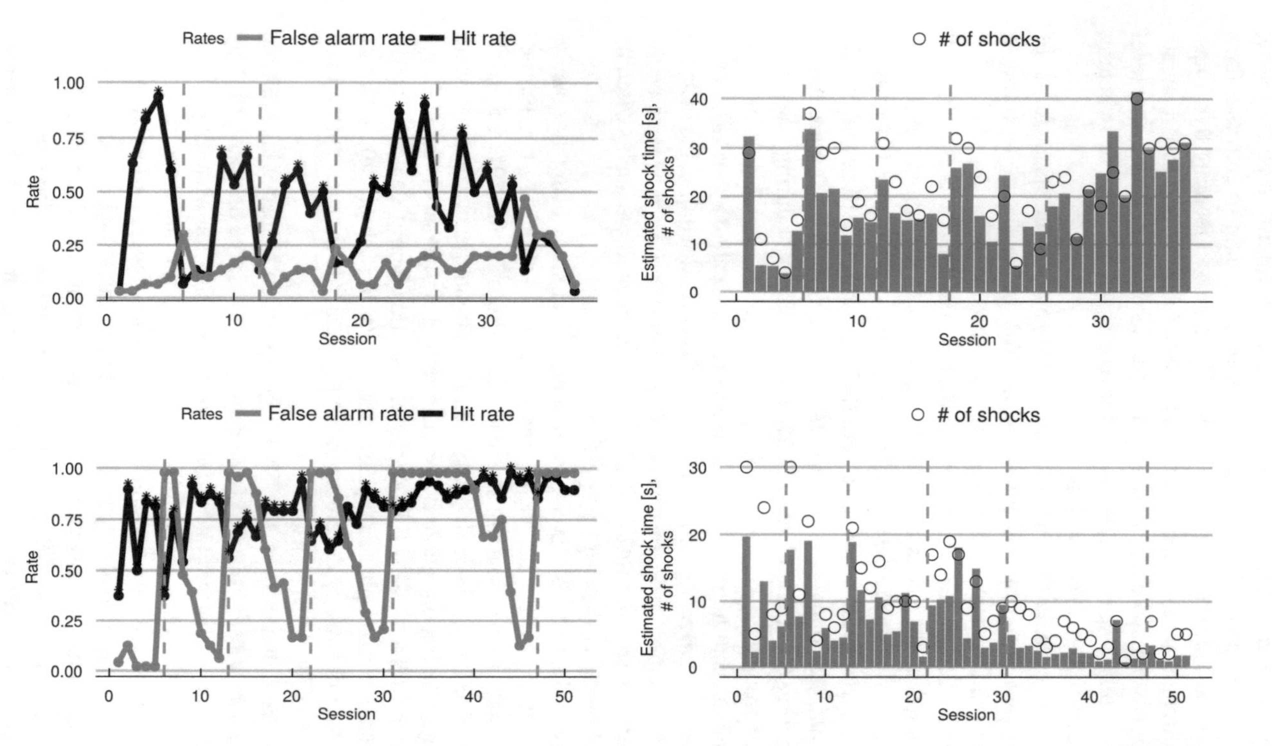

Fig. 8.8 Exemplary individual learning curves of animals trained in serial reversals of auditory discrimination learning in the shuttle-box. The subject trained with a foot-shock on *false alarm* responses (symmetric reinforcement schedule, *upper row*) fell back to an escape strategy after multiple contingency reversals. These animals generally showed an overall increase in the number of foot-shocks they received during each reversal phase. The exemplary animal trained in an asymmetric reinforcement schedule did not receive foot-shocks on *false alarms* (*lower row*). This subject evolved the overall strategy to cross the shuttle-box hurdle in every trial indifferent to CS type. It can be seen that this led to an overall decrease in the number of shocks this subject received during every reversal phase

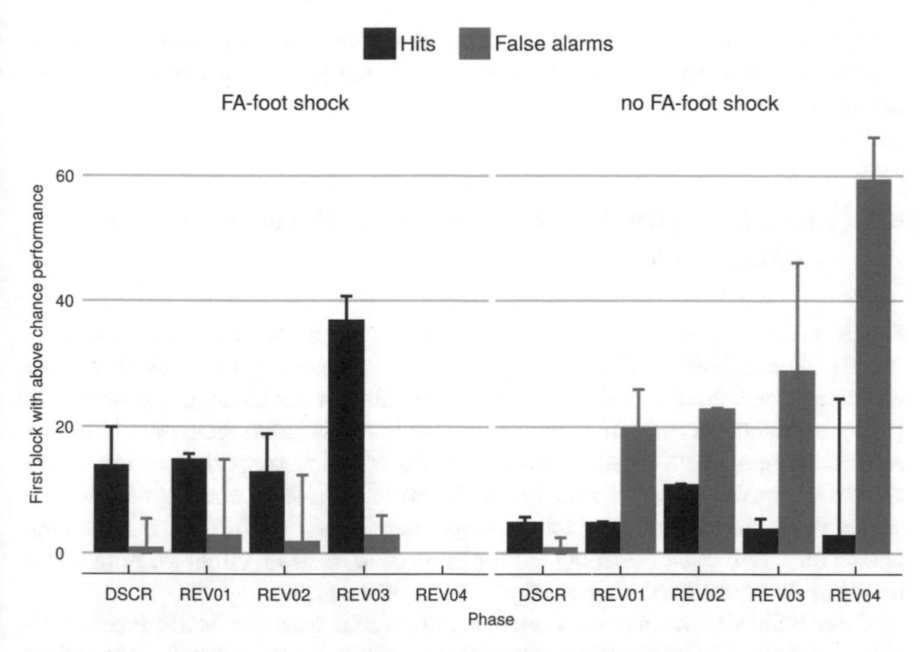

Fig. 8.9 First training blocks with above chance Go and NoGo performance. Animals trained with a *false alarm* foot shock (*left*, n = 27) quickly showed above chance performance in NoGo trials (decreased FA) but compared to animals trained without FA-foot shock (*right*, n = 8) took longer to perform *hits*. These animals subsequently needed more blocks to decrease FA behavior in successive contingency reversals

come. This superordinate strategy switches from a solution that best serves the contingencies in the present reversal phase to a policy that optimizes the serial aspect of the training.

Comparing animals that have been trained with a *false alarm* foot shock regimen to a group that has been trained in serial 2-WAA reversal task with a neutral CS- yielded also prominent differences in terms of increased learning dynamics. Assessing block wise performance (one block = 24 trials) yielded tendencies for increased go behavior on Go conditions in every sequential reversal phase in the FA-no FS group (Fig. 8.9, right panel), with simultaneous increases of the block numbers to show above chance crossing responses in NoGo conditions. In the FA-foot shock group (Fig. 8.9, left) subjects decreased crossings in NoGo conditions constantly fast while displaying the tendency to increase the number of blocks needed to perform *hits* above chance.

Inferring from the above results, while we could see faster learning dynamics especially in the FA-no FS group, switching contingencies was superseded by the implementation of the respective superordinate strategy in both groups. Interestingly, we observed a large inter-individual variance for the training lengths to which animal subjects started using a superordinate strategy.

To identify such thresholds will be a future challenge when designing *Companion*-Systems. But the neurobiological results also call for some expert modules that flexibly switch policies on different hierarchical context-levels.

8.5 Neurobiological Substrates of Serial Reversal Learning (Cognitive Flexibility)

Several studies agreed on the role of the orbito-frontal cortex (OFC) for reversal learning (for review, see [25, 33, 68]). The OFC is, however, not crucial for initial discrimination learning. Lesions of the OFC impair reversal learning of positive and negative outcome contingencies, while lesions in the striatum seem only to impair reversal learning with negative reinforcement, which is supported by measuring electrical signals of striatal neurons (reviewed in [35, 41]). A study with human subjects suffering from frontal lobe damages confirmed that the OFC is in fact not crucial for initial choice learning but for its reversal. Instead, initial learning seems to rely critically on the dorsomedial prefrontal cortex [17].

Other brain structures are also involved in reversal learning. Many experiments showed that lesions or electrical stimulation disrupting normal functionality of the dorso-medial striatum do not impair initial discrimination learning but its reversal [54]. Reversal learning is impaired by the dopaminergic-receptor D2 agonist in the medial striatum, which consequently reduces inputs from the PFC to the medial striatum; the same is true for serotonergic depletion in the orbitofrontal cortex (OFC). Furthermore, the ability for reversal learning appears to correlate with availability of D2 receptors in the caudate nucleus and the putamen, whereas it has no influence on initial discrimination learning or retention of discrimination performance (for a review see, [51]).

The basolateral amygdala (BLA) is involved in discrimination learning with negative outcomes. Based on lesion and electrophysiological studies, there are good indications that the BLA plays a role in specific types of reversal learning, particularly if different outcomes (reward and punishment) are reversed [69].

In serial reversal discrimination tasks the same brain structures seem to be involved as in reversal learning. A number of studies have identified the medial prefrontal cortex (mPFC), including prelimbic and infralimbic cortices and the orbitofrontal cortex (OFC), but also parts of the striatum and other brain structures to be involved in the improved performance dynamics of serial reversal learning [5, 7, 8, 11, 13, 14, 18, 37, 57, 58, 66, 72]. In one experiment Rodgers and DeWeese [56] trained rats to discriminate between different sounds according to pitch or location (left vs. right). Rats were able to switch to the appropriate selection rule in multiple reversals just after a few error trials. The authors showed that not only neurons in the mPFC encode this selection rule, but also neurons in the sensory cortical areas like the auditory cortex. Furthermore, electrical disruption of the mPFC impaired performance significantly.

Referring to the neurobiologically mapped actor-critic model, a simple discrimination learning can be explained with synaptic changes in sensory cortico-striatal loops. A first reversal might trigger another change of the synaptic strength within this loop. This might explain the slower dynamics of re-learning. However, the improved performance dynamics in serial reversals contradict this rewiring model and call for another neural mechanism. The cited studies suggest that mainly the mPFC and OFC, but also other brain areas such as the anterior cingulate cortex, the striatum and sensory cortical areas, take part in serial reversal learning. Similarly to reinforcement learning experiments, most of the serial reversal learning studies were all positively reinforced, utilizing water or food rewards. As mentioned above, within the *Companion*-user interaction, negative feedback from the user appears to be critical for the system to switch strategies. Such a negative feedback could be the user's indication that the dialog went in a wrong direction, for instance.

However, current knowledge about reversal learning is mainly based on lesion studies and pharmacological manipulation of neurotransmitters and to a lesser extent on electrophysiological and fMRI measurements. Detailed functions of the underlying circuitry for reversal learning remain still poorly understood.

8.6 Conclusion and Outlook

In this chapter we exploited the advantage of animal research to investigate mechanisms enabling cognitive systems to flexibly adapt their behavioral strategies to changing environments. Reinforcement learning models alone offer a poor solution to this problem, because they have to relearn the changed contingencies, even if it they had learned the rules before. An optimal solution can be seen in a system which consists of a low-level RL module and another module which can acquire different strategies. By top-down control of the RL modules, this system would be able to switch flexibly between these strategies, needing only a few unsuccessful outcomes.

With the experimental method of serial reversal learning, the ability of animals to develop strategic changes can be investigated. The results of these experiments depend on the setup of the task and especially on the reinforcement schedule. In some experiments with positively reinforced behavior (reward) well-trained animals are able to adapt their behavior after a few non-rewarded trials. In the context of the *Companion*-System, as the system may select its actions by learning which actions are not preferable, we were particularly interested in strategy change capabilities of animals learning with a punishment schedule (or avoidance of such) instead of with a reward schedule. We used two different reinforcement schedules. In one condition a previously correct, not punished action was negatively reinforced after a reversal (*false alarm shock*). In the other condition the previously correct action had no negative consequences after reversal if chosen. In both conditions all animals developed a strategy which is not the optimal solution as described above, however, which cannot be explained by a strict RL model. In the first case, after each reversal the animals experience strong interference effects by the negative reinforcement of the previously successful action. Consequently, after several reversals they did not take any actions to avoid the negative reinforcement, but awaited the shock due to the CS and escaped at its onset immediately. This strategy is a suboptimal solution (the optimal solution would be to receive no shock, or only very few at the start of each reversal switch), by minimizing the amount of shock received. This strategy can be applied independently of the (reversal task, switch) contingencies, however.

In the reinforcement condition without *false alarm* shock the animals developed an almost optimal strategy. During the first reversal switches, animals demonstrated a strategy change as desired for a *Companion*-System. With each reversal switch fewer numbers of trials were necessary to avoid the shocks, and this cannot be explained by RL. In consequence, after several reversals only very few trials would be required to change the strategy. However, eventually the animals developed another almost optimal solution to avoid the shock. They shuttled to the other compartment regardless of the presented stimulus (CS+ or CS-). This represents a valid strategy in terms of number of received shocks. The reason it can be regarded as suboptimal is the fact that the animals made an increased physical effort to shuttle, even if it would not be required to avoid the shock.

A conclusion of these experiments is that the behavior of animals in a serial reversal task cannot be explained by RL alone and additional mechanisms of strategy building and changes are involved, even if strategy changes lead only to suboptimal solutions. The challenge in the future is to develop biological-inspired theoretical concepts and identify neural brain mechanisms of these strategy changes.

A *Companion*-System, which is supposed to be able to adapt to changing environments and dialog courses and has not an a priori "hard-wired" program, might be based on these future theoretical concepts, consisting of RL modules and additional modules for different strategies, derived from animal behavioral and neural mechanisms. Furthermore, animal behavior illustrates that it is not always the optimal solution that needs to be achieved. Certain contexts will favor a suboptimal solution, as well.

Acknowledgements This work was done within the Transregional Collaborative Research Centre SFB/TRR 62 "*Companion*-Technology for Cognitive Technical Systems" funded by the German Research Foundation (DFG).

References

1. Amemori, K.I., Gibb, L.G., Graybiel, A.M.: Shifting responsibly: the importance of striatal modularity to reinforcement learning in uncertain environments. Front. Hum. Neurosci. **5**, 47 (2011). doi:10.3389/fnhum.2011.00047. http://dx.doi.org/10.3389/fnhum.2011.00047
2. Bathellier, B., Tee, S.P., Hrovat, C., Rumpel, S.: A multiplicative reinforcement learning model capturing learning dynamics and interindividual variability in mice. Proc. Natl. Acad. Sci. USA **110**(49), 19950–19955 (2013). doi:10.1073/pnas.1312125110. http://dx.doi.org/10.1073/pnas.1312125110

3. Bitterman, M.E.: The comparative analysis of learning. Science **188**(4189), 699–709 (1975). doi:10.1126/science.188.4189.699. http://dx.doi.org/10.1126/science.188.4189.699
4. Bond, A.B., Kamil, A.C., Balda, R.P.: Serial reversal learning and the evolution of behavioral flexibility in three species of North American corvids (Gymnorhinus cyanocephalus, Nucifraga columbiana, Aphelocoma californica). J. Comp. Psychol. **121**(4), 372–379 (2007). doi:10.1037/0735-7036.121.4.372. http://dx.doi.org/10.1037/0735-7036.121.4.372
5. Boulougouris, V., Dalley, J.W., Robbins, T.W.: Effects of orbitofrontal, infralimbic and prelimbic cortical lesions on serial spatial reversal learning in the rat. Behav. Brain. Res. **179**(2), 219–228 (2007). doi:10.1016/j.bbr.2007.02.005. http://dx.doi.org/10.1016/j.bbr.2007.02.005
6. Budinger, E., Laszcz, A., Lison, H., Scheich, H., Ohl, F.W.: Non-sensory cortical and subcortical connections of the primary auditory cortex in mongolian gerbils: bottom-up and top-down processing of neuronal information via field ai. Brain Res. **1220**, 2–32 (2008). doi:10.1016/j.brainres.2007.07.084. http://dx.doi.org/10.1016/j.brainres.2007.07.084
7. Bussey, T.J., Muir, J.L., Everitt, B.J., Robbins, T.W.: Triple dissociation of anterior cingulate, posterior cingulate, and medial frontal cortices on visual discrimination tasks using a touchscreen testing procedure for the rat. Behav. Neurosci. **111**(5), 920–936 (1997)
8. Castañé, A., Theobald, D.E.H., Robbins, T.W.: Selective lesions of the dorsomedial striatum impair serial spatial reversal learning in rats. Behav. Brain. Res. **210**(1), 74–83 (2010). doi:10.1016/j.bbr.2010.02.017. http://dx.doi.org/10.1016/j.bbr.2010.02.017
9. Clayton, K.N.: The relative effects of forced reward and forced nonreward during widely spaced successive discrimination reversal. J. Comput. Physiol. Psychol. **55**, 992–997 (1962)
10. Dabrowska, J.: Multiple reversal learning in frontal rats. Acta Biol. Exp. (Warsz) **24**, 99–102 (1964)
11. Deco, G., Rolls, E.T.: Synaptic and spiking dynamics underlying reward reversal in the orbitofrontal cortex. Cereb. Cortex **15**(1), 15–30 (2005). doi:10.1093/cercor/bhh103. http://dx.doi.org/10.1093/cercor/bhh103
12. Dias, R., Robbins, T.W., Roberts, A.C.: Primate analogue of the Wisconsin card sorting test: effects of excitotoxic lesions of the prefrontal cortex in the marmoset. Behav. Neurosci. **110**(5), 872–886 (1996)
13. Dias, R., Robbins, T.W., Roberts, A.C.: Dissociable forms of inhibitory control within prefrontal cortex with an analog of the Wisconsin card sort test: restriction to novel situations and independence from "on-line" processing. J. Neurosci. **17**(23), 9285–9297 (1997)
14. Divac, I.: Frontal lobe system and spatial reversal in the rat. Neuropsychologia **9**(2), 175–183 (1971)
15. Dombrowski, P.A., Maia, T.V., Boschen, S.L., Bortolanza, M., Wendler, E., Schwarting, R.K.W., Brandão, M.L., Winn, P., Blaha, C.D., Cunha, C.D.: Evidence that conditioned avoidance responses are reinforced by positive prediction errors signaled by tonic striatal dopamine. Behav. Brain Res. **241**, 112–119 (2013). doi:10.1016/j.bbr.2012.06.031. http://dx.doi.org/10.1016/j.bbr.2012.06.031
16. Feldman, J.: Successive discrimination reversal performance as a function of level of drive and incentive. Psychon. Sci. **13**(5), 265–266 (1968). doi:10.3758/BF03342516. http://dx.doi.org/10.3758/BF03342516
17. Fellows, L.K.: Orbitofrontal contributions to value-based decision making: evidence from humans with frontal lobe damage. Ann. N. Y. Acad. Sci. **1239**, 51–58 (2011). doi:10.1111/j.1749-6632.2011.06229.x. http://dx.doi.org/10.1111/j.1749-6632.2011.06229.x
18. Ferry, A.T., Lu, X.C., Price, J.L.: Effects of excitotoxic lesions in the ventral striatopallidal–thalamocortical pathway on odor reversal learning: inability to extinguish an incorrect response. Exp. Brain Res. **131**(3), 320–335 (2000)
19. Frank, M.J.: Dynamic dopamine modulation in the basal ganglia: a neurocomputational account of cognitive deficits in medicated and nonmedicated parkinsonism. J. Cogn. Neurosci. **17**(1), 51–72 (2005). doi:10.1162/0898929052880093. http://dx.doi.org/10.1162/0898929052880093

20. Frank, M.J., Seeberger, L.C., O'Reilly, R.C.: By carrot or by stick: cognitive reinforcement learning in parkinsonism. Science **306**(5703), 1940–1943 (2004). doi:10.1126/science.1102941. http://dx.doi.org/10.1126/science.1102941
21. Garner, H., Wessinger, W., McMillan, D.: Effect of multiple discrimination reversals on acquisition of a drug discrimination task in rats. Behav. Pharmacol. **7**(2), 200–204 (1996)
22. Gossette, R.L., Hood, P.: Successive discrimination reversal measures as a function of variation of motivational and incentive levels. Percept. Mot. Skills **26**(1), 47–52 (1968). doi:10.2466/pms.1968.26.1.47. http://dx.doi.org/10.2466/pms.1968.26.1.47
23. Gossette, R.L., Inman, N.: Comparison of spatial successive discrimination reversal performances of two groups of new world monkeys. Percept. Mot. Skills **23**(1), 169–170 (1966). doi:10.2466/pms.1966.23.1.169. http://dx.doi.org/10.2466/pms.1966.23.1.169
24. Haber, S.N., Calzavara, R.: The cortico-basal ganglia integrative network: the role of the thalamus. Brain Res. Bull. **78**(2-3), 69–74 (2009). doi:10.1016/j.brainresbull.2008.09.013. http://dx.doi.org/10.1016/j.brainresbull.2008.09.013
25. Hamilton, D.A., Brigman, J.L.: Behavioral flexibility in rats and mice: contributions of distinct frontocortical regions. Genes Brain Behav. **14**(1), 4–21 (2015). doi:10.1111/gbb.12191. http://dx.doi.org/10.1111/gbb.12191
26. Houk, J.C.: Agents of the mind. Biol. Cybern. **92**(6), 427–437 (2005). doi:10.1007/s00422-005-0569-8. http://dx.doi.org/10.1007/s00422-005-0569-8
27. Houk, J.C., Wise, S.P.: Distributed modular architectures linking basal ganglia, cerebellum, and cerebral cortex: their role in planning and controlling action. Cereb. Cortex **5**(2), 95–110 (1995)
28. Ilango, A., Wetzel, W., Scheich, H., Ohl, F.W.: The combination of appetitive and aversive reinforcers and the nature of their interaction during auditory learning. Neuroscience **166**(3), 752–762 (2010). doi:10.1016/j.neuroscience.2010.01.010. http://dx.doi.org/10.1016/j.neuroscience.2010.01.010
29. Ilango, A., Shumake, J., Wetzel, W., Scheich, H., Ohl, F.W.: Effects of ventral tegmental area stimulation on the acquisition and long-term retention of active avoidance learning. Behav. Brain Res. **225**(2), 515–521 (2011). doi:10.1016/j.bbr.2011.08.014. http://dx.doi.org/10.1016/j.bbr.2011.08.014
30. Ilango, A., Shumake, J., Wetzel, W., Scheich, H., Ohl, F.W.: The role of dopamine in the context of aversive stimuli with particular reference to acoustically signaled avoidance learning. Front. Neurosci. **6**, 132 (2012)
31. Ilango, A., Shumake, J., Wetzel, W., Ohl, F.W.: Contribution of emotional and motivational neurocircuitry to cue-signaled active avoidance learning. Front. Behav. Neurosci. **8**, 372 (2014). doi:10.3389/fnbeh.2014.00372. http://dx.doi.org/10.3389/fnbeh.2014.00372
32. Ionescu, T.: Exploring the nature of cognitive flexibility. New Ideas Psychol. **30**(2), 190–200 (2012). doi:10.1016/j.newideapsych.2011.11.001. http://dx.doi.org/10.1016/j.newideapsych.2011.11.001
33. Jonker, F.A., Jonker, C., Scheltens, P., Scherder, E.J.A.: The role of the orbitofrontal cortex in cognition and behavior. Rev. Neurosci. **26**(1), 1–11 (2015). doi:10.1515/revneuro-2014-0043. http://dx.doi.org/10.1515/revneuro-2014-0043
34. Kangas, B.D., Bergman, J.: Repeated acquisition and discrimination reversal in the squirrel monkey (Saimiri sciureus). Anim. Cogn. **17**(2), 221–228 (2014). doi:10.1007/s10071-013-0654-7. http://dx.doi.org/10.1007/s10071-013-0654-7
35. Kehagia, A.A., Murray, G.K., Robbins, T.W.: Learning and cognitive flexibility: frontostriatal function and monoaminergic modulation. Curr. Opin. Neurobiol. **20**(2), 199–204 (2010). doi:10.1016/j.conb.2010.01.007. http://dx.doi.org/10.1016/j.conb.2010.01.007
36. Kulig, B.M., Calhoun, W.H.: Enhancement of successive discrimination reversal learning by methamphetamine. Psychopharmacologia **27**(3), 233–240 (1972)
37. Li, L., Shao, J.: Restricted lesions to ventral prefrontal subareas block reversal learning but not visual discrimination learning in rats. Physiol. Behav. **65**(2), 371–379 (1998)
38. Mackintosh, N., Cauty, A.: Spatial reversal learning in rats, pigeons, and goldfish. Psychon. Sci. **22**, 281–282 (1971)

39. Mackintosh, N.J., McGonigle, B., Holgate, V., Vanderver, V.: Factors underlying improvement in serial reversal learning. Can. J. Psychol. **22**(2), 85–95 (1968)
40. McAlonan, K., Brown, V.J.: Orbital prefrontal cortex mediates reversal learning and not attentional set shifting in the rat. Behav. Brain Res. **146**(1-2), 97–103 (2003)
41. McDannald, M.A., Jones, J.L., Takahashi, Y.K., Schoenbaum, G.: Learning theory: a driving force in understanding orbitofrontal function. Neurobiol. Learn. Mem. **108**, 22–27 (2014). doi:10.1016/j.nlm.2013.06.003. http://dx.doi.org/10.1016/j.nlm.2013.06.003
42. McGeorge, A.J., Faull, R.L.: The organization of the projection from the cerebral cortex to the striatum in the rat. Neuroscience **29**(3), 503–37 (1989). http://www.ncbi.nlm.nih.gov/pubmed/2472578
43. McHaffie, J.G., Stanford, T.R., Stein, B.E., Coizet, V., Redgrave, P.: Subcortical loops through the basal ganglia. Trends Neurosci. **28**(8), 401–407 (2005). doi:10.1016/j.tins.2005.06.006. http://dx.doi.org/10.1016/j.tins.2005.06.006
44. Montague, P.R., Dayan, P., Sejnowski, T.J.: A framework for mesencephalic dopamine systems based on predictive Hebbian learning. J. Neurosci. **16**(5), 1936–1947 (1996)
45. Mota, T., Giurfa, M.: Multiple reversal olfactory learning in honeybees. Front. Behav. Neurosci. **4** (2010). doi:10.3389/fnbeh.2010.00048. http://dx.doi.org/10.3389/fnbeh.2010.00048
46. Mowrer, O.H.: Two-factor learning theory reconsidered, with special reference to secondary reinforcement and the concept of habit. Psychol. Rev. **63**(2), 114–128 (1956)
47. Nolte, G., Bai, O., Wheaton, L., Mari, Z., Vorbach, S., Hallett, M.: Identifying true brain interaction from eeg data using the imaginary part of coherency. Clin. Neurophysiol. **115**(10), 2292–2307 (2004). doi:10.1016/j.clinph.2004.04.029. http://dx.doi.org/10.1016/j.clinph.2004.04.029
48. O'Doherty, J., Dayan, P., Schultz, J., Deichmann, R., Friston, K., Dolan, R.J.: Dissociable roles of ventral and dorsal striatum in instrumental conditioning. Science **304**(5669), 452–454 (2004). doi:10.1126/science.1094285. http://dx.doi.org/10.1126/science.1094285
49. Ohl, F.W., Scheich, H., Freeman, W.J.: Change in pattern of ongoing cortical activity with auditory category learning. Nature **412**(6848), 733–736 (2001). doi:10.1038/35089076. http://dx.doi.org/10.1038/35089076
50. Pennartz, C.M.A., Berke, J.D., Graybiel, A.M., Ito, R., Lansink, C.S., van der Meer, M., Redish, A.D., Smith, K.S., Voorn, P.: Corticostriatal interactions during learning, memory processing, and decision making. J. Neurosci. **29**(41), 12831–12838 (2009). doi:10.1523/JNEUROSCI.3177-09.2009. http://dx.doi.org/10.1523/JNEUROSCI.3177-09.2009
51. Piray, P.: The role of dorsal striatal d2-like receptors in reversal learning: a reinforcement learning viewpoint. J. Neurosci. **31**(40), 14049–14050 (2011). doi:10.1523/JNEUROSCI.3008-11.2011. http://dx.doi.org/10.1523/JNEUROSCI.3008-11.2011
52. Pubols, B. Jr.: Successive discrimination reversal learning in the white rat: a comparison of two procedures. J. Comput. Physiol. Psychol. **50**(3), 319–322 (1957)
53. Pubols, B.H.: Serial reversal learning as a function of the number of trials per reversal. J. Comput. Physiol. Psychol. **55**, 66–68 (1962)
54. Ragozzino, M.E.: Acetylcholine actions in the dorsomedial striatum support the flexible shifting of response patterns. Neurobiol. Learn. Mem. **80**(3), 257–267 (2003)
55. Remijnse, P.L., Nielen, M.M.A., Uylings, H.B.M., Veltman, D.J.: Neural correlates of a reversal learning task with an affectively neutral baseline: an event-related fMRI study. Neuroimage **26**(2), 609–618 (2005). doi:10.1016/j.neuroimage.2005.02.009. http://dx.doi.org/10.1016/j.neuroimage.2005.02.009
56. Rodgers, C.C., DeWeese, M.R.: Neural correlates of task switching in prefrontal cortex and primary auditory cortex in a novel stimulus selection task for rodents. Neuron **82**(5), 1157–1170 (2014). doi:10.1016/j.neuron.2014.04.031. http://dx.doi.org/10.1016/j.neuron.2014.04.031
57. Schoenbaum, G., Nugent, S.L., Saddoris, M.P., Setlow, B.: Orbitofrontal lesions in rats impair reversal but not acquisition of go, no-go odor discriminations. Neuroreport **13**(6), 885–890 (2002)

58. Schoenbaum, G., Setlow, B., Nugent, S.L., Saddoris, M.P., Gallagher, M.: Lesions of orbitofrontal cortex and basolateral amygdala complex disrupt acquisition of odor-guided discriminations and reversals. Learn. Mem. **10**(2), 129–140 (2003). doi:10.1101/lm.55203. http://dx.doi.org/10.1101/lm.55203
59. Schultz, W.: The reward signal of midbrain dopamine neurons. News Physiol. Sci. **14**, 249–255 (1999)
60. Schultz, W.: Reward signaling by dopamine neurons. Neuroscientist **7**(4), 293–302 (2001)
61. Schultz, W., Dayan, P., Montague, P.R.: A neural substrate of prediction and reward. Science **275**(5306), 1593–1599 (1997)
62. Schulz, A.L., Woldeit, M.L., Gonçalves, A.I., Saldeitis, K., Ohl, F.W.: Selective increase of auditory cortico-striatal coherence during auditory-cued go/nogo discrimination learning. Front. Behav. Neurosci. **9**(368) (2016). doi:10.3389/fnbeh.2015.00368
63. Smith, Y., Surmeier, D.J., Redgrave, P., Kimura, M.: Thalamic contributions to basal ganglia-related behavioral switching and reinforcement. J. Neurosci. **31**(45), 16102–16106 (2011). doi:10.1523/JNEUROSCI.4634-11.2011. http://dx.doi.org/10.1523/JNEUROSCI.4634-11.2011
64. Sutton, R.S., Barto, A.G.: Reinforcement Learning: An Introduction. MIT, Cambridge, MA (1998)
65. Tremblay, L., Hollerman, J.R., Schultz, W.: Modifications of reward expectation-related neuronal activity during learning in primate striatum. J. Neurophysiol. **80**(2), 964–977 (1998)
66. von der Gablentz, J., Tempelmann, C., Münte, T.F., Heldmann, M.: Performance monitoring and behavioral adaptation during task switching: an fMRI study. Neuroscience **285**, 227–235 (2015). doi:10.1016/j.neuroscience.2014.11.024. http://dx.doi.org/10.1016/j.neuroscience.2014.11.024
67. Voorn, P., Vanderschuren, L.J.M.J., Groenewegen, H.J., Robbins, T.W., Pennartz, C.M.a.: Putting a spin on the dorsal-ventral divide of the striatum. Trends Neurosci. **27**(8), 468–74 (2004). doi:10.1016/j.tins.2004.06.006. http://www.ncbi.nlm.nih.gov/pubmed/15271494
68. Walton, M.E., Behrens, T.E.J., Noonan, M.P., Rushworth, M.F.S.: Giving credit where credit is due: orbitofrontal cortex and valuation in an uncertain world. Ann. N. Y. Acad. Sci. **1239**, 14–24 (2011). doi:10.1111/j.1749-6632.2011.06257.x. http://dx.doi.org/10.1111/j.1749-6632.2011.06257.x
69. Wassum, K.M., Izquierdo, A.: The basolateral amygdala in reward learning and addiction. Neurosci. Biobehav. Rev. **57**, 271–283 (2015). doi:10.1016/j.neubiorev.2015.08.017. http://dx.doi.org/10.1016/j.neubiorev.2015.08.017
70. Woldeit, M.L., Schulz, A.L., Ohl, F.W.: Phase de-synchronization effects auditory gating in the ventral striatum but not auditory cortex. Neuroscience **216**, 70–81 (2012). doi:10.1016/j.neuroscience.2012.04.058. http://dx.doi.org/10.1016/j.neuroscience.2012.04.058
71. Xiong, Q., Znamenskiy, P., Zador, A.M.: Selective corticostriatal plasticity during acquisition of an auditory discrimination task. Nature (2015). doi:10.1038/nature14225. http://dx.doi.org/10.1038/nature14225
72. Xue, G., Xue, F., Droutman, V., Lu, Z.L., Bechara, A., Read, S.: Common neural mechanisms underlying reversal learning by reward and punishment. PLoS One **8**(12), e82169 (2013). doi:10.1371/journal.pone.0082169. http://dx.doi.org/10.1371/journal.pone.0082169

Chapter 9
Assistive and Adaptive Dialog Management

Florian Nielsen and Wolfgang Minker

Abstract One of the most important challenges in the field of human-computer interaction is maintaining and enhancing the willingness of the user to interact with the technical system. This willingness to cooperate provides a solid basis which is required for a collaborative human-computer dialog. For the dialog management this means that a *Companion*-System adapts the course and content of human-computer dialogs to the user and assists during the interaction through individualized help and explanation. In this chapter we elucidate our dialog management approach, which provides user- and situation-adaptive dialogs, and our explanation management approach, which enables the system to provide assistance and clarification for the user during run-time.

9.1 Introduction

The usual task of the dialog management (DM) is to control the structure, content, and flow of the dialog between human and computer. It communicates with the application and gathers from as well as provides to the user all necessary information, which is needed to accomplish a specific task in cooperation with a technical system. As an individual cognitive technical system should be able to adapt to a user's capabilities, preferences, and current goals and take into account the situation and emotional state, the dialog management is one of the most influential components to foster the realization of these system properties. These features and properties qualify a system as a so-called *Companion*-System (cf. Chap. 1). *Companion*-Systems are by definition "continually available, co-operative, reliable and trustworthy assistants which adapt to a user's capabilities, preferences, requirements, and current needs" [23].

From this *Companion*-System definition the authors derived those user characteristics (see Fig. 9.1) which are most important for the use in a dialog management component, fostering the realization of the *Companion*-properties. Here, *emotional state* is exchanged with *affective state*, because state-of-the-art research focuses on

F. Nielsen (✉) • W. Minker
Institute of Communications Engineering, Ulm University, Ulm, Germany
e-mail: florian.nothdurft@alumni.uni-ulm.de; wolfgang.minker@uni-ulm.de

S. Biundo, A. Wendemuth (eds.), *Companion Technology*, Cognitive Technologies,
DOI 10.1007/978-3-319-43665-4_9

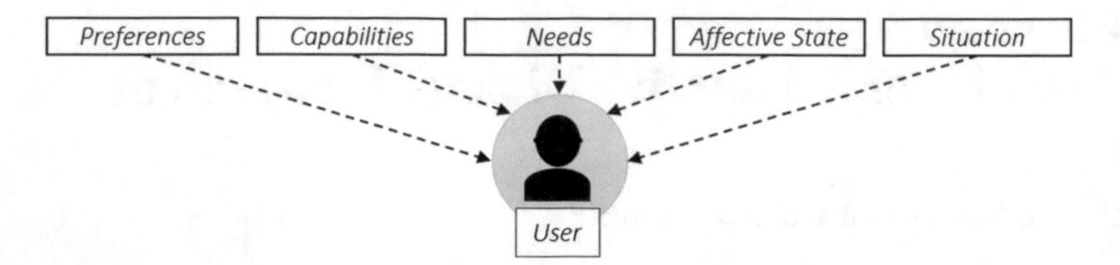

Fig. 9.1 The main user characteristics: A user's *capabilities* represent, for example, the user's knowledge or handicaps. *User needs* are derived from the user's current goals while the *situation* describes the user's present environment. *Preferences* are user made or learned adaptation criteria, and the *affective state* represents emotions as well as dispositions towards the technical system

recognizing and reacting not only to a user's primary emotions, but also to a user's disposition towards a technical system. Those dispositions are secondary emotions and can describe a user's stance towards a system better, by using terms like confusion, interest or frustration, rather than primary emotions like anger, sadness or joy [19]. Systems adapting their functionalities to users' general characteristics (e.g. capabilities or preferences), users' affective states or the present situation may help to lead to a cooperative and effective dialog between man and machine. Adapting the human-computer interaction (HCI) to the individual user does yield possible advantages. For example, the DM may adapt the flow of a dialog between human and machine to the user (e.g. to the user's knowledge) and by that help to prevent overextension or mental overload. However, individual systems that act upon implicit user information and adapt their system behavior accordingly may confuse or disturb the user's perception of the system. Especially proactive behavior (e.g. reacting autonomously to the user's affective state) might not be understandable to the user, or in other words incongruent to the user's mental model of the system. One of the main aspects in *Companion*-Systems is their ability to evolve the capabilities of solving problems collaboratively with the user. However, solving complex problems together with a *Companion*-System is only feasible if the human user interprets and understands the machine correctly.

These situations are critical and may have a negative impact on the human-computer trust (HCT) relationship [12]. The main problem is that if the user does not trust the system and its actions, advice or instructions, the way of interaction may change up to complete abortion of future interaction [18]. Glass et al. [6] observed that the capabilities to provide explanations in adaptive agents can reduce most of the trust concerns identified by the user. In addition, several other studies showed that explanations are a way to prevent the decrease of trust (e.g. [3, 4]). Therefore, a *Companion*-System should also be able to integrate explanations about the system's proactive behavior instantaneously in the dialog. This means that, depending on the current user model and situation, the *Companion*-System should be able to induce the appropriate kind of explanation to increase the chances of a successful and trustworthy HCI. Therefore, in the following we will elucidate how the DM may adapt the flow and structure of the HCI using explanations to individual user characteristics, but also why and how a dialog system may cope with problems resulting from misunderstood system behavior.

The chapter is structured as follows. In Sect. 9.2 we provide introductory information on dialog management approaches. Section 9.3 describes the general concept and design ideas, including requirements for the user's knowledge and mental model. In Sect. 9.4 the implementation of the adaptive and assistive capabilities of the DM in a rule-based and probabilistic fashion is described, followed by the conclusion in Sect. 9.5.

9.2 Background

In the past decades several approaches to the DM task have been developed that can be classified in four basic categories [9]. First of all, basic *finite-state-machine* approaches (e.g. [25]), where a set of states is defined with a set of moves for each state which transitions to a new state in the automaton. Second, *frame-based* approaches (e.g. [8]), where the DM is monitoring the current so-called frame, which is specified by a set of needed information (slots), the context for the utterance interpretation and the context for the dialog progress. This is more advanced, since it allows for mixed-initiative interaction and allows multiple paths to acquire the information. Thirdly, stochastic-based approaches (e.g. [24]), which apply reinforcement learning techniques to the DM by determining the best policy or choice of actions from all available actions a system can take in a dialog, which will optimize the system's performance as measured by a utility function, such as the user's evaluation of the system [11]. The last main group of approaches is *agent-based approaches* (e.g. [14]) where the dialog is controlled by several intelligent agents capable of reasoning about themselves (e.g. in the BDI (beliefs, desires, intentions) approach [20]) using Artificial Intelligence techniques. The agent-based approach subsumes plan-based dialog models, where preconditions, actions and effects are used to control the ongoing dialog. Here, the structure of the dialog (i.e. the flow) is determined at run-time by the different agents using the current world state and the goals left to achieve. Thus, the agent-based approach is the closest to the one used here.

9.3 Concept and Design

For an individual *Companion*-System, the DM integrates into a complex architecture and splits its responsibilities with the *planning framework* (PF). Due to the fact that DM and PF are closely linked together, a dialog management approach related to the planning approach used in the PF was chosen. This means that, as already described in Chap. 7, the course of plan steps representing the solution for the given planning problem is generated and executed by the planning-framework (i.e. the *plan generation* and *plan execution* components). In case that the proposed plan steps require user interaction, plan steps are passed from the plan execution to the DM. The main purpose of this component is to decompose and refine the plan step (if necessary) into a user-adaptive dialog. If the dialog for the passed-on plan step requires adaptation to the individual user, the structure and the content of the dialog can be adapted. Therefore, our combination of DM and PF is closely related to the split into task level and dialog level used in agent-based approaches.

During this work two different approaches were used. First, a rule-based approaches using preconditions, thresholds and predefined rules to adapt the structure and flow of the dialog. Second, a hybrid approach using rule-based approaches for the task-oriented part of the interaction, and a probabilistic approach for dealing with adaptivity to user and situation characteristics based on uncertain or noisy information sources (e.g. user knowledge, user understanding or affective user states) (cf. Sect. 9.4.3). The framework for the integration of user- and situation-adaptive dialog management capabilities was a cognitive knowledge-based technical system, which extends the architectures of classic unimodal (e.g. spoken dialog systems) or multimodal dialog systems. This system [7] can be called a prototypical *Companion*-System [2]. To be able to provide *Companion*-functionalities in a completely individualized way, several interacting components (see Fig. 9.2) are necessary. The main differences with a classic architecture are the underlying knowledge base, and the cognitive capabilities by the use of a sensor system, and that the dialog management and the planning framework share the tasks of planning, controlling and structuring the HCI.

9.3.1 Required User Model

As already mentioned, the following adaptation criteria are needed for the integration of the desired explanation capabilities and stored in the user model: the general user knowledge, the fine-grained user knowledge and the user's mental model.

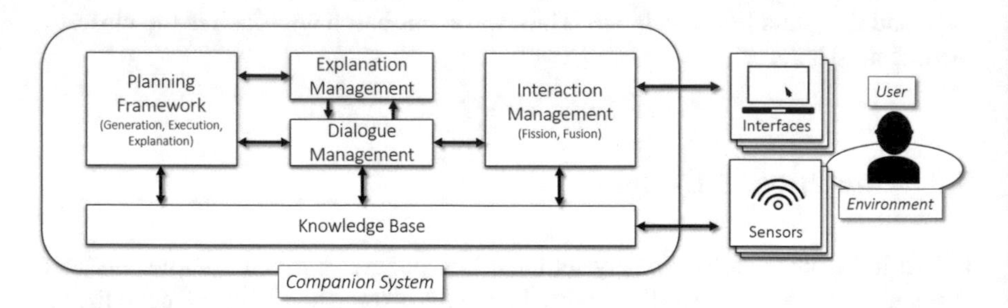

Fig. 9.2 This figure shows the prototypical *Companion*-System. This system serves as framework for the modules developed or extended: *explanation* and *dialog management*. Embedded between the *planning framework*, *interaction management*, and the underlying *knowledge base*, they control the user-adaptive dialog between human and machine

9.3.1.1 General and Fine-Grained User Knowledge

One of the first design decisions was to model user knowledge in a fine-grained way, using the same level of detail as used in the domain model of planning and dialog. For example, in the example domain used here, the task represented as relation *connect(TV, Receiver, HDMI)* models the knowledge of the action *connect* as well as the concepts *TV*, *Receiver*, and *HDMI*. Additionally, we draw a distinction between declarative knowledge and procedural knowledge. The former can be used to describe the being of things (i.e. appearance, purpose). Possessing declarative knowledge about something does not necessarily mean begin able to use this knowledge for a task or action. In comparison to that, procedural knowledge can be applied to a task. Procedural knowledge provides the knowledge on how to execute a task or on how to solve a problem (Fig. 9.3).

Those knowledge constituents are modelled as probability distributions over a five-step knowledge scale ranging from *novice* to *expert* for every constituent (see Fig. 9.7). This means that knowledge is modelled in small pieces instead of in only one general level. This helps the model to allow for a more realistic and exact individualization and adaptation to the user. Though we concentrate on a fine-grained knowledge representation, we also build a mean overall knowledge value for coarse adaptations of the dialog flow (cf. Fig. 9.4). This means that a mean of all domain knowledge constituents is calculated, representing the overall knowledge of the user. Hence, the categorization into groups or levels of expertise is based not only on a general assumption, but also on a more thoroughly combination of all domain elements. The development of the user's knowledge is based on observations made during the interaction and on past interaction episodes (cf. Fig. 9.5). Therefore, the knowledge levels are system-made assumptions about the user, requiring a probabilistic representation. Occurring events during the HCI influence the probability distribution of relevant contained knowledge objects.

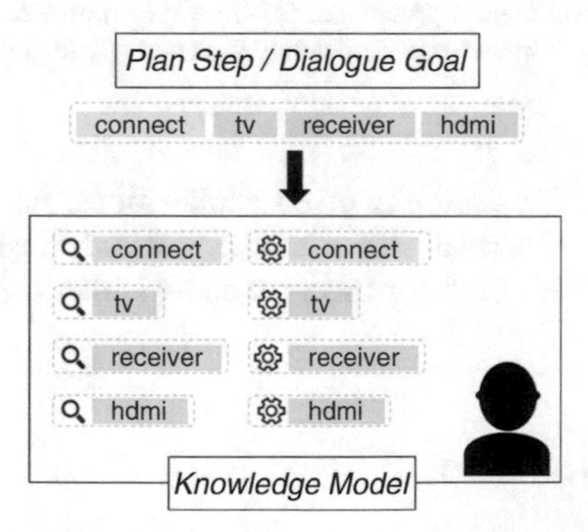

Fig. 9.3 The relation coming from either the dialog goal or the plan step directly can be analyzed separately by the explanation management, because they are stored in a fine-grained way in the *User's Knowledge Model*. Here, *the magnifying glass* represents declarative knowledge elements and the cogwheel procedural knowledge elements

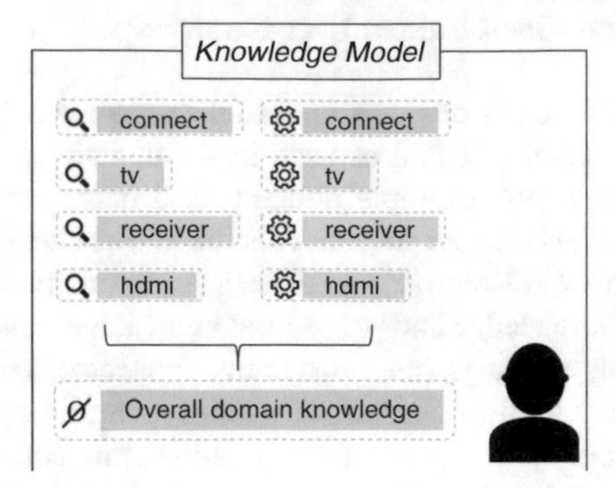

Fig. 9.4 The user's overall knowledge of a domain is used for the coarse adaptation of the dialog flow. It consists of a combination of all domain elements and therefore represents the mean domain knowledge

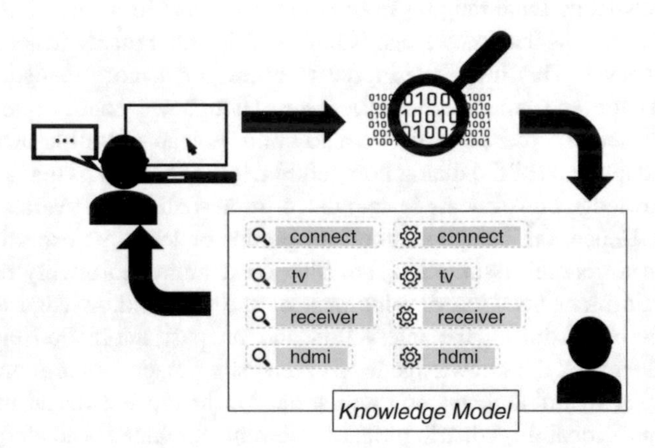

Fig. 9.5 The fine-grained user's knowledge evolves over time. This means that in a first step user interactions or other observations are processed. This may then influence the knowledge model and its constituents. The changed knowledge model may subsequently lead to adaptation and influence the HCI

These events may be, for example, given explanations, failed actions or, plainly elapsed time. Contained means, for example, not only tasks the user executed, but also entities that were used for this task and whether the task completion was successful.

9.3.1.2 User's Mental Model

Our main interest in mental models is the consequence that they are built by the user and only imitate the perceived system behavior. This means that the perceived mental model may not correspond to the real-world model and may therefore be incorrect. This process of adapting existent mental models or "fine-tuning" them is very important. Whenever some feedback is observed which is incongruent to the present mental model, the model has to be adapted. Additionally, these situations are those in which the user does not understand the system behavior, because of incongruent models, and the risk increases that the user's perception of the system's trust components (e.g. understandability, reliability) might be impaired. To foster effective mental model correction it showed that transparency explanations worked best [17]. However, as mental models cannot be directly observed we are left to estimate whether incongruence in mental and actual system models is present. However, is we look for situations in which not only the mental model incorrect, but also the receptor is not sure whether the mental model applies.

Therefore, due to the lack of direct observability, the mental model is hard to build. However, what can be used, in order to estimate a mental model, is affect recognition combined with context information like the interaction history and current state of the dialog. For this we are keeping track of the HCI in a so-called dialog history. This history records the decisions and actions of the user and the technical system. The history is used to note when, for example, something was explained to the user, when the user executed a task or requested help from the system. Here, information of specific system actions or decisions that have been performed in the past is stored in the history as well. For example, the estimated mental model includes whether explanations were given for system decisions or tasks and whether the transparency on decisions was upheld by providing information on the reasoning process that led to the decision.

9.3.2 *Consequences for the Design*

While the description of the underlying architecture already gave some hints on where the main adaptation criteria, derived from the design requirement, are applied, this section will elucidate the theoretical intent behind these design decisions. Where, why and how those criteria (*general user knowledge*, *fine-grained user knowledge*, and *user's mental model*) may be used in the present system will be described in the following.

9.3.2.1 General User Knowledge

This model represents the user's general knowledge about a domain or sub-domain. Therefore, it is an arithmetic mean of user's knowledge about a specific topic, for example, the user's knowledge about cooking, technical systems or cars. Analogous to human-human interaction (HHI), a person's knowledge is based on the subjective opinions and objective criteria of another party. Humans tend to assign competencies to other persons, like "he is a good cook", "she knows how to repair cars" or "he knows everything about computers". However, we do not only assign abstract competencies like that, but also very specific competencies, like “she makes a great lasagna”, “he knows everything about Linux”, or “she can change the tires of her VW Golf”. Now, if a human seeks help or clarification, the most important classification criteria are the abstract ones. However, if the human already possesses a certain amount of knowledge in that domain, the more specific competencies tend to become more important. This process of selecting the most competent help for the user’s current situation gets more and more complex the more specific his or her problem and existing knowledge in the domain is. Transferred to technical systems, this means that a user’s general knowledge is suitable for a coarse adaptation of the dialog flow and structuring of HCI. However, if more specific competencies in a domain are required, a more fine-grained estimation of the user’s knowledge is needed for an effective and sound HCI.

Therefore, in our present architecture, the dialog management and its *guard-based* dialog model are suitable for adapting to a user’s general knowledge. The dialog model is able to adapt to *guards*, which control the flow of the dialog. The drawback of increasing complexity for higher numbers of *guards* is not relevant if the user’s general knowledge is regarded as one of only a small number of sub-concepts.

9.3.2.2 Fine-Grained User Knowledge

As already explained in the former paragraph, a detailed estimation of the user’s knowledge is needed to assist the user in a more specific and more effective manner. Since the *guard-based* dialog model is not suitable for a fine-grained adaptation to the user’s knowledge model, an automatic solution not based on predefined dialog paths seems preferable. Hence, an automatic, proactive augmentation of the already roughly adapted dialog has to be done. Analogously to HHI, where, for example, “chef apprentices” get, on top of their already adapted instructions, more detailed information based on their specific competencies (e.g. “cut the chicken filet in this direction”), in HCI this has to be done as well. While observing the user’s behavior and interaction history, one has to decide whether additional clarification or instruction is needed in order to achieve the desired goal. In the present architecture the explanation management has the purpose of dealing with these kinds of adaptations at run-time. It observes the ongoing dialog and decides based upon the history of interaction and the upcoming dialogs whether additional explanations are needed to match the user’s fine-grained knowledge model. Therefore, dialogs have to be modelled in a way that they can be assessed in terms of required knowledge. For example, the task “prepare the béchamel sauce” requires knowledge about béchamel sauce in general, as well as knowledge about the ingredients and of course about the procedure for preparing it. In order to enable the system to assess the requirements, the knowledge constituents have to be modelled for the dialogs.

9.3.2.3 User's Mental Model

The user's mental model cannot be observed directly. As explained before, only symptoms rather than the disease combined with present context information can be used to estimate whether the user's mental model and the actual system behavior do align. Hence, affective states that indicate, for example, *frustration* or *confusion* have to be used in combination with information on the dialog history and the surroundings to estimate mental model incongruences. However, the recognition of affective user states is error-prone and burdened with uncertainty, especially considering non-acted data. Thus, to treat the uncertainty-based affective states in a proper way, a probabilistic model is needed.

The main idea of the probabilistic model is the integration of proper uncertainty treatment for information based on uncertain observations. Apart from factors like uncertain environments and decision-making, incorporating mental model adaptivity directly into the *guard-based* DM is not good idea. Modelling the incongruence of mental models as a guard variable would be possible, but predefining the adaptations is not.

In previous research we showed that especially transparency explanations, which aim for explaining the system processes, lead to a higher perception of understandability of the system's behavior [15]. However, this goal of explanation cannot be predefined for every situation by the designer. That's why the adaptivity to a user's mental model has to be done during run-time, and for our systems this means in the *explanation management*. In Sect. 9.4.3 we explain how the estimated mental model can trigger explanations. Though we can decide on whcther explanations are needed, the explanation itself has to be generated by the modules responsible for the decision-making in terms of the dialog and plan, which will be elucidated in the next section.

9.4 Implementation

As already mentioned, different approaches and facets were developed, which will be explained in the following. In Sect. 9.4.1 the dialog flow adaptation of a task-oriented dialog is exemplarily explained on the user's knowledge level. Section 9.4.2 describes how a fine-grained user knowledge model can be used to provide user-adaptive assistivity by augmenting a task-oriented dialog in a rule-based way without interfering in its original dialog approach. How a probabilistic approach can be used and why such an approach is necessary to integrate explanation capabilities in a realistic and effective fashion is then explained in Sect. 9.4.3.

9.4.1 Rule-Based Adaptivity

As described in Sect. 9.3, the generated plan of the planning framework serves as skeleton of the user-adaptive dialog. The provided plan steps are decomposed one by one into a hierarchical dialog structure which consists of so-called *dialog goals* [16]. Each dialog goal represents a single interaction step between human and system (e.g. one screen on a device filled with specific information). The term dialog goal arises from the fact that every step in the interaction pursues a goal. This goal is, in this case, to achieve one or several of the desired plan step effects. Therefore, the term dialog goal is to be distinguished from the term goal used in the PF components. This means that a plan step may be decomposed into several dialog goals and that for every desired plan step effect a set of similar dialog goals may exist. These similar dialog goals usually have so-called *guards* which formulate conditions that need to be fulfilled in order for the dialog goal to be entered at run-time. These guards may, for example, take into account general user knowledge levels and therefore help to adapt the dialog to user characteristics stored in the knowledge base as marginal.

Goals can be arranged in a vertical structure and also in a horizontal structure. In a vertical structure each goal may yield several sub-goals. In a horizontal structure each goal may have a fixed successive goal that is next in the dialog. This implies that the dialog may be roughly structured like a finite state machine, but there is enough room left to dynamically arrange the sub-goals to satisfy the user's needs. Such an arrangement is handled by the way the guards for the goals are defined. Guards are preconditions protecting the related dialog goal of inappropriate execution, leading most likely to dialog failure (e.g. due to inadequate user knowledge). Those roughly made dialog structure adaptations are later augmented by assistive behavior manifested by additional explanations during run-time, adaptive to the user's fine-grained knowledge, which will be explained in Sect. 9.4.2.

During the interaction, the dialog management traverses through its dialog structure (see Fig. 9.6) to select the path of dialog goals most suitable for the

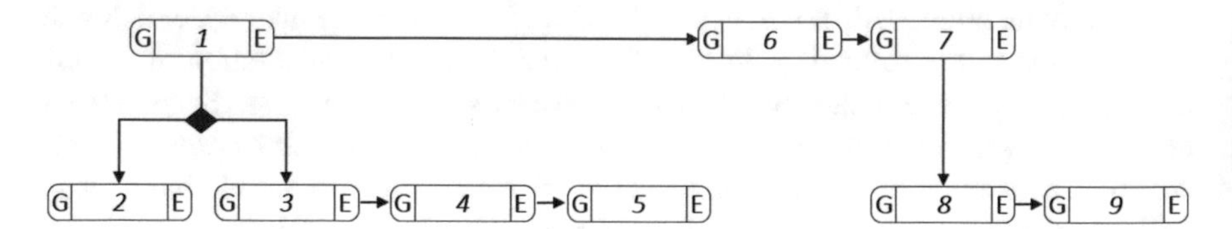

Fig. 9.6 *Dialog Goals* may have sub-goals (cf. 2 and 3 for goal 1) or link to the next goal in the sequence (cf. 6 to 7). More abstract goals (e.g. 1 and 7) are not directly executable, but have to be refined. The dialog content and flow may change according to predefined variables used as *guards*, depicted as *G*. For example, if one *guard* of goal number 2 requires expert knowledge, the resulting sequence of dialog steps, in case of fulfilment, would include a simplified instruction (i.e. 2,6,8,9). Contrary to that, a novice user would receive a more detailed as well as longer sequence of dialog steps with additional extent of assistance (i.e. 3,4,5,6,8,9). Dialog Goals have effects as well, describing the effects of execution (depicted as *E*)

current user. The selection of the next dialog goal is therefore made in a user-adaptive manner and leads to an individual dialog appropriate for the current user. In order to conduct the selection of the next appropriate dialog goal, a constraint solving algorithm is used. Constraint programming has proved to be especially useful in problems on finite domains where many conditions limit the possible variable value configurations [5]. It is a technique to find solutions to problems by backtracking and efficient reasoning. As a dialog in our case is limited to a certain number of possibilities how the system can traverse through the dialog structure, it is reasonable to use a finite domain for the variables that constrain the execution of the dialog goals. These conditions make the dialog model suited for applying a finite domain constraint-solving mechanism.

The decision about which dialog goals can be executed is based on the values that variables specified in the guard conditions are allowed to have. Applied to constraint programming, we consider the conditions in the guards as constraints, and, based on the current values of the variables, the constraint solver tells which guard conditions can be fulfilled. Based on this variable configuration we can select a number of dialog goals which can be executed at a certain time in the dialog. The described adaptation of the dialog flow is only suitable for roughly customizing the dialog using the mean user knowledge due to complexity issues. Using the fine-grained knowledge model would lead rapidly to overly complex dialog structures. This means that knowledge is here regarded as one mean value for the complete domain or coarse subdomains. For example, in the sample domain of connecting a new home theater system, the user's mean technical knowledge is considered the main adaptation variable. The dialog content and flow will change if two different users interact with the system. For example, if the first user is an *expert* (i.e. has a mean technical knowledge ≥ 4.5), the sequence of dialog steps should include simplified instructions, sufficient for an expert. Contrary to that, a novice (i.e. mean technical knowledge <1.5) should receive a more detailed as well as a longer sequence of dialog steps with additional extent of assistance. The user variable *mean technical knowledge* is designed to be the arithmetic mean of the combination of the user's default knowledge model and the evolution of its fine-grained constituents during past episodes of HCI (see Fig. 9.5).

The mean technical knowledge is used as a *guard* to formulate conditions necessary for the execution of the dialog goal. Due to the rule-based DM approach using links to sub-goals or to proceeding goals, combined with the requirement to define those links by hand, the overhead vastly explodes in the case of a large number of guard variables. Therefore, the *user knowledge adaptation* is only used for few selected variables, which influence the dialog flow to a great extent.

9.4.2 Rule-Based Assistivity

Adaptation of structure and flow of the dialog to levels of user characteristics (e.g. mean technical knowledge) yields only coarse individualization results. Due to the effort necessary to design a specific dialog course or flow for every level of user characteristic and the resulting complexity, it is uneconomical to use the former approach for more fine-grained adaptation and individualization. Therefore, the course of interaction steps is adapted and extended for a more individualized dialog to the individual user's knowledge during run-time.

The assistive part of the DM, the *dialog augmentation*, deals with the automatic extension of the ongoing dialog with additional dialog steps, helping the user to accomplish tasks ahead. Contrary to the rule-based adaptation, which chose the most appropriate already predefined dialog path for the user, this approach integrates independent dialog steps into the running dialog. This means the designer does not have to cope with the knowledge-based fine-grained individualization, but the explanation management makes sure that the upcoming dialog steps are conform to the user model.

As already explained in Sect. 9.3, the course of steps the user has to fulfil in order to solve the task he wants to accomplish is first planned by the planning framework and later decomposed further by the DM into so-called dialog steps. Each of these steps is represented as a relation with the name of the task and its appendant concepts as its arguments. As our main goal is to prevent task failure we have to ensure that the upcoming or current tasks and appendant concepts do not exceed the user's knowledge. Therefore, prior to sending the dialog steps for presentation to the *multimodal fission* (see Chap. 10), they are first sent to the explanation management. Here, the content of the predefined dialog is analyzed and compared to the user's knowledge model, stored in the knowledge base as marginal. If the user's knowledge is probably not sufficient, the course of the dialog is updated by including additional dialog steps to better fit the user's knowledge model. Those additional dialog steps are meant to explain missing knowledge to the user.

As previously reported, in our knowledge model we distinguish between declarative knowledge and procedural knowledge. For example, in the domain used in the demonstrator, the task represented as relation *connect(TV, Receiver, HDMI)* models the knowledge of the action *connect* as well as the concepts *TV*, *Receiver* and *HDMI*. This means that compared to other systems the user's knowledge is modeled in small pieces instead of in only one general level. Of course, the resulting model is very complex, but also more realistic and conforming more to the idea of human knowledge modelling.

The knowledge constituents are modelled as probability distributions over a five-step knowledge scale ranging from *novice* to *expert* (see Fig. 9.7). As the

```
<proceduralGoalKnowledge goalName="connect" goalID="3.2">
  <knowledgeValue value="novice" probability="0.5"/>
  <knowledgeValue value="advanced novice" probability="0.3"/>
  <knowledgeValue value="intermediate" probability="0.1"/>
  <knowledgeValue value="advanced intermediate" probability="0.1"/>
  <knowledgeValue value="expert" probability="0"/>
</proceduralGoalKnowledge>
```

```
<declarativeConceptKnowledge conceptName="hdmi">
  <knowledgeValue value="novice" probability="0.5"/>
  <knowledgeValue value="advanced novice" probability="0.3"/>
  <knowledgeValue value="intermediate" probability="0.1"/>
  <knowledgeValue value="advanced intermediate" probability="0.1"/>
  <knowledgeValue value="expert" probability="0"/>
</declarativeConceptKnowledge>
```

Fig. 9.7 In this excerpt of the user's knowledge model the procedural knowledge of the action *connect* and the declarative knowledge of the concept *HDMI* as well as their respective probability distributions over the knowledge levels are listed

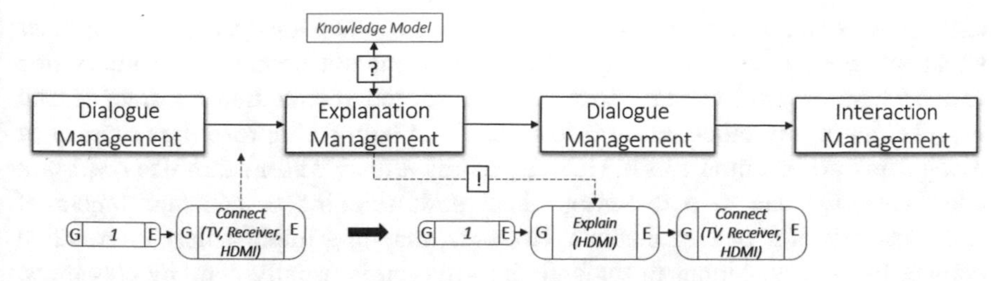

Fig. 9.8 Here we can see that the original predefined dialog on the *left* is augmented during runtime by an additional explanation dialog. The current dialog *1* is completed and before execution the next dialog is analysed by the explanation management. By checking the knowledge model it concludes that the user's knowledge about HDMI is not sufficient and decides to augment the dialog with an additional explanation dialog

user's knowledge is based on observations made during the interaction and on past interaction episodes and the user's knowledge is based on system-made assumptions about the user, a probabilistic representation is required. During the interaction we check if the user's knowledge for the current dialog step constituents is most probably high enough. The concepts and the action are analyzed by the explanation management, and if needed the dialog flow is augmented with additional dialog steps (see Fig. 9.8). If the user's knowledge is most probably too low, the explanation manager generates an additional explanation dialog, which tries to impart the missing knowledge for the user to execute the dialog step successfully.

The explanation management then selects which type of explanation is the appropriate one for the current lack of knowledge. As the dialog step consists of several parts, so does the explanation. Each task and concept are analyzed to generate a summarized explanation. The explanation management sends the content of the explanation to the knowledge base to be stored in the *information model*. This explanation may consist of pictures, text or text meant to be spoken. Afterwards, a dialog step is generated, which references the content of the explanation stored in the information model and sent to the DM to be included in the course of interaction. The DM will then proceed to send the additional dialog to the multimodal fission component, which then decides by modality arbitration on which device and how the content should be presented.

As mentioned earlier, the explanation management not only verifies the dialog step's contained knowledge, but also coordinates and processes explicitly stated explanation requests from the user and generates, if necessary, additional dialog steps to be included in the ongoing dialog.

One of the main drawbacks of the rule-based approach is the improper treatment of uncertainty. Though we are using probability distributions to model the user's knowledge levels from novice to expert, the processing of events does not meet all wished for requirements for uncertainty treatment. The development of the probability distribution is a rather imprecise rule-based approach, only incorporating explicit human-computer interaction. However, implicit information coming from the user (e.g. affective states) is an important indicator for the development of user knowledge (i.e. did the user understand or not). Obviously, recognizing those situations cannot be done solely by using information coming from interaction and its history. Multimodal input such as speech recognition accuracy, facial expressions or any other sensor information can help to improve the accuracy of recognizing critical moments in HCI. Especially affective user states like *confusion* and *frustration* can help to reason about understanding or accomplishment of tasks, instructions, or explanations. However, mapping implicit user information coming from sensor input to semantic information is usually done by classifiers, and those classifiers convey a certain amount of probabilistic inaccuracy, which has to be handled. Therefore, a decision model has to be able to handle probabilistic information in a suitable manner. How this kind of implicit user information is integrated to foster consistent and proper uncertainty treatment will be explained in the next section.

9.4.3 Probabilistic Assistivity

The former sections concentrated on the adaptation of the dialog to high-level user variables such as the user's mean knowledge or verbal intelligence, and on the augmentation of the task-oriented dialog with conceptual or procedural explanation dialogs. However, one of the most important factors in HCI is the user's affective state or disposition towards the system. This means that situations in which the user does not understand the system's actions or instructions are very critical for a trustworthy and sound HCI. Incomprehensible, not understandable system behavior may lead to the loss of trust and in the worst case to the abortion of interaction.

As elucidated in [17], the different goals of explanation are suitable for different situations in HCI. Therefore, we need to estimate the user's state and reason about the interaction state in order to decide upon the most appropriate and most effective system reaction strategy for the current situation and user. For this, not only explicit, but also implicit interaction information has to be used. Apart from explicit, data (e.g. touch, speech, or click), especially implicit data (e.g. affective state, location, or user profile) can help in estimating the user's state, for example the user's mental model, in a better way.

The main idea of the *probabilistic assistivity* is the integration of proper uncertainty treatment for information based on uncertain observations. However, since decision-making under uncertainty requires complex and elaborate models, and tends to get unsolvable fast, only the augmentation process is controlled by a probabilistic model. For the problem representation of when and how to react, a so-called partially observable Markov decision process (POMDP) was chosen and formalized in the Relational Dynamic Influence Diagram Language (RDDL) [21]. RDDL is a uniform language which allows an efficient description of POMDPs by representing its constituents (actions, observations, belief state) with variables. On the one hand we are using a classic dialog approach as described in Sect. 9.4.1 for the task-oriented part of the dialog. On the other hand a planner [13] is integrated

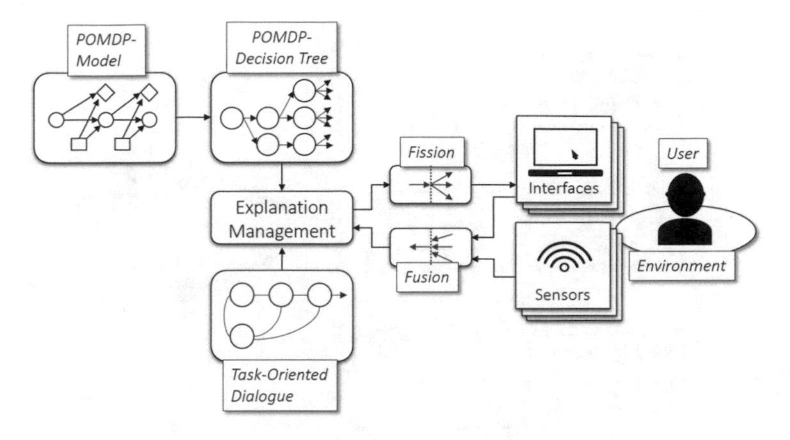

Fig. 9.9 The architecture consists of two dialog models, a fission and fusion engine, sensors as well as the multimodal interface representation to interact with the user. The dialog models can be separated in a task-oriented dialog model and into a POMDP-based decision tree for explanation augmentation. This decision tree is generated from a POMDP-model by a planner

to generate from a POMDP a decision tree. This POMDP is used only for the augmentation of the task-oriented part of the dialog (Fig. 9.9).

9.4.3.1 Dialog Augmentation Process

The task-oriented dialog is modelled in a classic dialog approach. Each dialog action has several interaction possibilities, each leading to another specified dialog action. Each of those dialog actions is represented as a POMDP action a as part of C (*communicative function*(c)). As already mentioned, only the communicative function is modelled to reduce the complexity in the POMDP.

The HCI is started using the classic dialog approach and uses the POMDP to check whether the user's trust or components of the user's trust are endangered. At run-time the next action in the task-oriented dialog is compared to the one determined by the POMDP (see Fig. 9.10). This means that if the next action in the task-oriented dialog is not the same as the one planned by the POMDP, the dialog flow is interrupted, and the ongoing dialog is augmented by the proposed action. For example, if the user is currently presented a communicative function of type *inform* and the decision tree recommends providing a transparency explanation because the understanding and reliability are probably false, the originally next step in the dialog is postponed and first the explanation is presented. The other way around, if the next action in the task-oriented dialog is subsumed by the one scheduled by the POMDP, the system does not need to intervene. For example, if the next dialog step is to instruct the user about how to connect *amplifier* and *receiver* and the POMDP would recommend an action of type communicative function *instruct*, no dialog augmentation is needed.

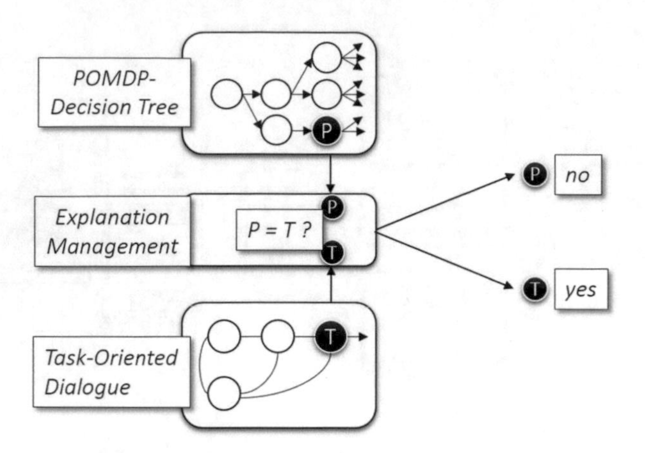

Fig. 9.10 This figure shows the comparison of task-oriented dialog to the POMDP-generated Decision Tree. If the next action T in the task-oriented dialog does not correspond to the one endorsed by the POMDP Decision Tree (P), the dialog will be augmented by the POMDP action. However, if they align, no intervention is necessary, and the planned action T is executed

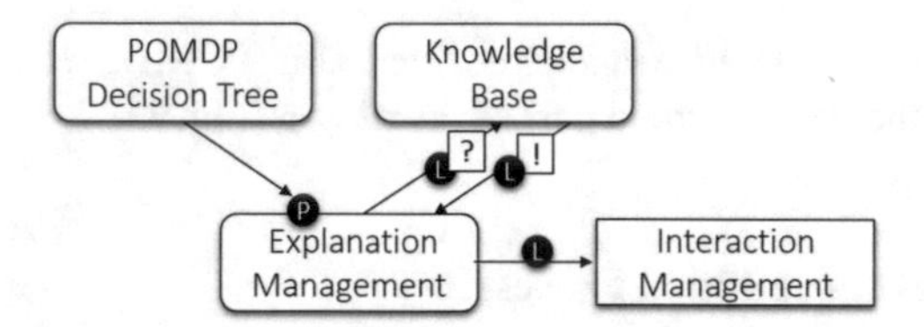

Fig. 9.11 The proposed decision P of the POMDP model is in this case a *learning* explanation request. Hence, the explanation management looks up the matching learning explanation (the *question mark L*) for the current topic in the knowledge base and returns the explanation content. This is then converted to a dialog step (the *exclamation mark L*), which is then passed to the interaction management for presentation

While the selection of the explanation is one thing, the integration of the real explanation is another. The POMDP only tells us whether and what kind of explanation should be integrated.

9.4.3.2 Explanation Selection

Basically there are four goals of explanation we are interested in covering. The first two are imparting declarative and procedural knowledge by using *conceptual* and *learning* explanations respectively. These typical explanations realised in contemporary systems by tutorials (learning) or plain help texts (conceptual) are usually explanations defined by experts. When the POMDP recommends integrating one of these explanations, the content of the explanations is gathered from the database, converted to a dialog step and integrated into the dialog (see Fig. 9.11). In our system these explanations are also stored in a predefined way. There are information fragments available, which can be used for explanations, for all contained concepts of the used planning and dialog domains. These information fragments can then be combined to explanation dialog steps, which can then be integrated in the ongoing dialog flow. For the actions themselves, information fragments exist as well, representing learning or parts of learning explanations. These fragments are requested from the explanation management and subsequently processed into dialog steps and passed to the interaction management.

However, incomprehensible system behavior cannot be explained using these goals of explanation. Incomprehensible situations are prone to influencing the relationship between human and technical system negatively [12]. Especially if the user does not trust the system and its actions, advice or instructions anymore, the way of interaction may change up to complete abortion of future interaction [18]. Therefore, not understandable system behavior is critical for a sound and trustworthy HCI. However, the risk of negative consequences can be reduced by providing specific types of explanations [6, 10]. The most effective explanation goals to handle incomprehensible system behavior are *transparency* and *justification* explanations [15, 17]. Hence, these are the other two goals of explanations we want to handle.

Justifications are the most obvious goal an explanation can pursue in HHI discussions. The main idea of this goal is to provide support for and increase confidence in given advice or actions. For example, "you have to eat an apple a day, because that keeps the doctor away" would be some sort of justification. The goal of transparency is to increase the user's understanding of how the interlocutor's behavior (e.g. providing advice or instructions) was attained, in terms of the interlocutor's inner processes; for example, "I just heard that your car's exhaust is rattling, which is a bad sign, therefore you have to bring it to the car repair shop". If a technical system can provide these kinds of explanations, it can help the user to change his perception of the system as a black-box to a system the user can comprehend. Thereby, the user can build a correct mental model of the system and its underlying reasoning processes. In general, one can remark that justifications tend to be abstract explanations, which do not necessarily directly relate to system behavior or processes. Thus, justifications can be predefined, at least to a certain degree, by experts. If the user requests *why* a dialog step *ds* (e.g. "connect(TV, Receiver, HDMI)") has to be done, an explanation prepared beforehand (e.g. "You have to connect TV and receiver with an HDMI cable to transmit audio and video signals") can be presented to the user. Alternatively, one could provide an explanation based on the hierarchy in the domain, such as "this has to be done to connect your home theater", a combination of both, or even something not related to the structure or dialog concepts, such as "to be able to watch movies". The integration of predefined justification explanations is the same as depicted in Fig. 9.11 for *learning* and *conceptual* explanations. A drawback of using predefined explanations is of course the vast amount of work necessary to cover all task available in the domain. Additionally, the content of justifications is limited to information known beforehand to the expert, and especially in incomprehensible situations transparency explanations might be the better option. Transparency explanations are, though, always dependent on the inner processes of a system and a distinct transformation of those, for the user, to a comprehensible form. Hence, in contrast to justifications, transparency explanations have to be generated dynamically during run-time.

While previously conducted experiments (see [15]) showed that providing transparency explanations is the best way to deal with incomprehensible situations in HCI, this is also the most complicated way of explanation. Here, this means that the decision-making processes performed in the planning framework to generate a plan, in turn leading to the coarse structure of the dialog, might have to be explained to the user. Fortunately, the PF does include a *plan explanation* module [22], which focuses on so-called *Why* explanations, which in this case correspond to some form of transparency explanations. Their explanations describe, for example, why a certain plan step p is part of the plan P or the ordering of two plan steps in P. This module generates a logically sound explanation, which is guaranteed by using a technique based on formal proofs. Specifically, they use the task in question p_e, the plan P, its decomposition structure, to generate a set Σ of first-order axioms, which in the end enables a logical interference. Basically, the plan explanation consists of an explanation on used causal links and their matching preconditions, as well as the used decomposition methods. However, the generated formal explanation still has to be transformed by the explanation management, in order to be understandable for human users. Hence, it is translated by the explanation manager using template-based natural language generation into human-readable text. This new generated content is then transmitted to the information model in the knowledge base. The output the explanation manager generates is in either way a new dialog step, which references content from the information model and is sent to the DM to be included in the ongoing interaction.

9.4.4 Intertwining with the Architecture

For information presentation the dialog goal is passed on to the IM (cf. Fig. 9.9) and by that transferred to an XML-based format called *dialog output*. Hereby, the dialog content is composed of multiple information objects referencing so-called *information IDs* in the information model. Each information ID can consist of different types (e.g. text, audio, and pictures) which are selected and combined at run-time by a fission sub-component to compose the user interface in a user- and situation-adaptive way (see Chap. 10 for more details).

After the user interaction, the DM receives the interaction results from the multimodal fusion. The results are then analyzed and if the results are related to the desired plan step effects implemented by the dialog step, these effects are transmitted to the knowledge base as observations. Additionally, the plan execution component is notified that the current plan step has been processed by the dialog management.

9.5 Conclusion

In this chapter we presented different approaches for a user-adaptive and assistive dialog management approach. Although some attempts were made to make the rule-based dialog model more flexible, for example by using so-called dialog widgets, which implement reusable parts of a dialog (e.g. confirmations) [1], the main parts of the rule-based dialog are still rigidly predefined finite-state automatons. Additionally, due to the knowledge base storing information a major part of which originates from sources prone to uncertainty, another approach using probabilistic methods was developed. This general probabilistic DM approach contributes to a more coherent and flexible *Companion*-System.

The modelling of the user's knowledge is made in a realistic fashion, considering the uncertainty of one's knowledge distribution, though the process of updating the knowledge over time is currently not. While presenting an explanation to the user does increase the chances of understanding, it does not guarantee it. Therefore, the update process of one's knowledge values should integrate uncertainty as well, e.g. using information indicating understanding (e.g. user affective states like engagement, interest, and disposition). Nevertheless, we presented an approach which may integrate such information in the future and by doing that facilitate a more individual, co-operative and reliable system that is a trustworthy and understandable partner for the user.

Acknowledgements This work was done within the Transregional Collaborative Research Centre SFB/TRR 62 "*Companion*-Technology for Cognitive Technical Systems" funded by the German Research Foundation (DFG).

References

1. Bertrand, G., Nothdurft, F., Minker, W.: "What do you want to do next?" Providing the user with more freedom in adaptive spoken dialogue systems. In: 2012 8th International Conference on Intelligent Environments (IE), pp. 290–296 (2012)
2. Biundo, S., Wendemuth, A.: *Companion*-technology for cognitive technical systems. Künstl. Intell. **30**(1), 71–75 (2016). Special Issue on Companion Technologies
3. Cheverst, K., Byun, H.E., Fitton, D., Sas, C., Kray, C., Villar, N.: Exploring issues of user model transparency and proactive behaviour in an office environment control system. User Model. User Adap. Inter. **15**, 235–273 (2005)
4. Dzindolet, M.: The role of trust in automation reliance. Int. J. Hum. Comput. Stud. **58**(6), 697–718 (2003)
5. Fernandez, A.J., Hortala-Gonzalez, T., Saenz-Perez, F., Del Vado-Virseda, R.: Constraint functional logic programming over finite domains. Theory Pract. Log. Program. **7**(5), 537–582 (2007). doi:10.1017/S1471068406002924. http://dx.doi.org/10.1017/S1471068406002924
6. Glass, A., McGuinness, D.L., Wolverton, M.: Toward establishing trust in adaptive agents. In: IUI '08: Proceedings of the 13th International Conference on Intelligent User Interfaces, pp. 227–236. ACM, New York (2008)

7. Honold, F., Bercher, P., Richter, F., Nothdurft, F., Geier, T., Barth, R., Hoernle, T., Schüssel, F., Reuter, S., Rau, M., Bertrand, G., Seegebarth, B., Kurzok, P., Schattenberg, B., Minker, W., Weber, M., Biundo, S.: Companion-technology: towards user- and situation-adaptive functionality of technical systems. In: 10th International Conference on Intelligent Environments (IE 2014), pp. 378–381. IEEE, New York (2014). doi:10.1109/ie.2014.60
8. Larsson, S., Traum, D.R.: Information state and dialogue management in the trindi dialogue move engine toolkit. Nat. Lang. Eng. **6**(3&4), 323–340 (2000)
9. Lee, C.J., Jung, S.K., Kim, K.D., Lee, D.H., Lee, G.G.B.: Recent approaches to dialog management for spoken dialog systems. J. Comput. Sci. Eng. **4**(1), 1–22 (2010)
10. Lim, B.Y., Dey, A.K., Avrahami, D.: Why and why not explanations improve the intelligibility of context-aware intelligent systems. In: Proceedings of the SIGCHI Conference on Human Factors in Computing Systems, CHI '09, pp. 2119–2128. ACM, New York (2009)
11. McTear, M.F.: Spoken dialogue technology: Enabling the conversational user interface. ACM Comput. Surv. **34**(1), 90–169 (2002). doi:10.1145/505282.505285. http://doi.acm.org/10.1145/505282.505285
12. Muir, B.M.: Trust in automation: Part I. Theoretical issues in the study of trust and human intervention in automated systems. In: Ergonomics, pp. 1905–1922. Taylor & Francis, London (1992)
13. Müller, F., Späth, C., Geier, T., Biundo, S.: Exploiting expert knowledge in factored POMDPs. In: Proceedings of the 20th European Conference on Artificial Intelligence (ECAI 2012), pp. 606–611 (2012)
14. Nguyen, A., Wobcke, W.: An agent-based approach to dialogue management in personal assistants. In: Proceedings of the 10th international conference on Intelligent user interfaces, pp. 137–144. ACM, New York (2005)
15. Nothdurft, F., Minker, W.: Justification and transparency explanations in dialogue systems to maintain human-computer trust. In: Proceedings of the 4th International Workshop On Spoken Dialogue Systems (IWSDS). Springer, Berlin (2014)
16. Nothdurft, F., Bertrand, G., Heinroth, T., Minker, W.: GEEDI - guards for emotional and explanatory dialogues. In: 6th International Conference on Intelligent Environments (IE'10), pp. 90–95 (2010)
17. Nothdurft, F., Richter, F., Minker, W.: Probabilistic human-computer trust handling. In: Proceedings of the 15th Annual Meeting of the Special Interest Group on Discourse and Dialogue (SIGDIAL), pp. 51–59. Association for Computational Linguistics, Philadelphia, PA (2014). http://www.aclweb.org/anthology/W14-4307
18. Parasuraman, R., Riley, V.: Humans and automation: use, misuse, disuse, abuse. Hum. Factors J. Hum. Factors Ergon. Soc. **39**(2), 230–253 (1997)
19. Picard, R.W., Picard, R.: Affective Computing, vol. 252. MIT, Cambridge (1997)
20. Rao, A.S., Georgeff, M.P.: BDI agents: from theory to practice. In: Proceedings of the First International Conference on Multi-Agent Systems, ICMAS-95, pp. 312–319 (1995)
21. Sanner, S.: Relational dynamic influence diagram language (RDDL): language description (2010). http://users.cecs.anu.edu.au/ ssanner/IPPC2011/RDDL.pdf
22. Seegebarth, B., Müller, F., Schattenberg, B., Biundo, S.: Making hybrid plans more clear to human users – a formal approach for generating sound explanations. In: Proceedings of the 22nd International Conference on Automated Planning and Scheduling (ICAPS 2012), pp. 225–233 (2012)
23. Wendemuth, A., Biundo, S.: A companion technology for cognitive technical systems. In: Esposito, A., Vinciarelli, A., Hoffman, R., Müller, V.C. (eds.) Proceedings of the EUCogII-SSPNET-COST2102 International Conference (2011). Lecture Notes in Computer Science. Proceedings on Cognitive Behavioural Systems, Dresden (2012)
24. Williams, J.D., Young, S.: Partially observable Markov decision processes for spoken dialog systems. Comput. Speech Lang. **21**(2), 393–422 (2007)
25. Zeigler, B., Bazor, B.: Dialog design for a speech-interactive automation system. In: Second IEEE Workshop on Interactive Voice Technology for Telecommunications Applications, 1994, pp. 113–116. IEEE, New York (1994)

Chapter 10
Management of Multimodal User Interaction in *Companion*-Systems

Felix Schüssel, Frank Honold, Nikola Bubalo, Michael Weber, and Anke Huckauf

Abstract While interacting, human beings continuously adapt their way of communication to their surroundings and their communication partner. Although present context-aware ubiquitous systems gather a lot of information to maximize their functionality, they predominantly offer rather static ways to communicate. In order to fulfill the user's communication needs and demands, ubiquitous sensors' varied information could be used to dynamically adapt the user interface. Considering such an adaptive user interface management as a major and relevant component for a *Companion*-Technology, we also have to cope with emotional and dispositional user input as a source of implicit user requests and demands. In this chapter we demonstrate how multimodal fusion based on evidential reasoning and probabilistic fission with adaptive reasoning can act together to form a highly adaptive and model-driven interactive system component for multimodal interaction. The presented interaction management (IM) can handle uncertain or ambiguous data throughout the complete interaction cycle with a user. In addition, we present the IM's architecture and its model-driven concept. Finally, we discuss its role within the framework of the other constituents of a *Companion*-Technology.

10.1 Introduction

A *Companion*-System appears to the user as different device types and different input and output modalities. Device types could be classical computers, small mobile devices like PDAs or smartphones, larger wall-mounted displays equipped with touch input technology, or even physical objects that are connected to their digital representations. Therefore, interaction modalities vary from mouse and

F. Schüssel (✉) • F. Honold • M. Weber
Institute of Media Informatics, Ulm University, 89081 Ulm, Germany
e-mail: felix.schuessel@uni-ulm.de; frank.honold@uni-ulm.de; michael.weber@uni-ulm.de

N. Bubalo • A. Huckauf
General Psychology, Ulm University, 89081 Ulm, Germany
e-mail: nikola.bubalo@uni-ulm.de; anke.huckauf@uni-ulm.de

S. Biundo, A. Wendemuth (eds.), *Companion Technology*, Cognitive Technologies,
DOI 10.1007/978-3-319-43665-4_10

keyboard input, touchscreen gestures and speech interaction to physical interactions with objects detected by sensors. The combination of these modalities leads to multimodal interaction with multiple devices.

Users will use *Companion*-Systems predominantly in a task-oriented manner. For the task fulfillment, it is thereby not important which devices and interaction modalities are available; this will be determined by the situation and the current state of the user and the surroundings. Therefore, there can be no fixed a priori user interface (UI) between a *Companion*-System and a user. What is needed is an abstract model that allows the different components within a *Companion*-System to express their requirements and delegate their concretion and realization to an interaction management (IM).

In the following sections we introduce a possible architecture for *Companion*-Systems and give an insight into the interaction management by explaining the two main components, namely multimodal fission and fusion. The last part of this chapter focuses on the interplay of fission and fusion.

10.2 Basic System Architecture

The architectural view of a *Companion*-System depicted in Fig. 10.1 naturally follows Don Norman's *human action cycle* [35] of execution and evaluation.[1] Inspired by Norman's work, our system contains components that can plan a sequence of actions in order to achieve a goal (task planning), execute the single actions (tasks via dialog steps) towards the user (upper outgoing process flow), and evaluate the user's reaction (lower incoming process flow).

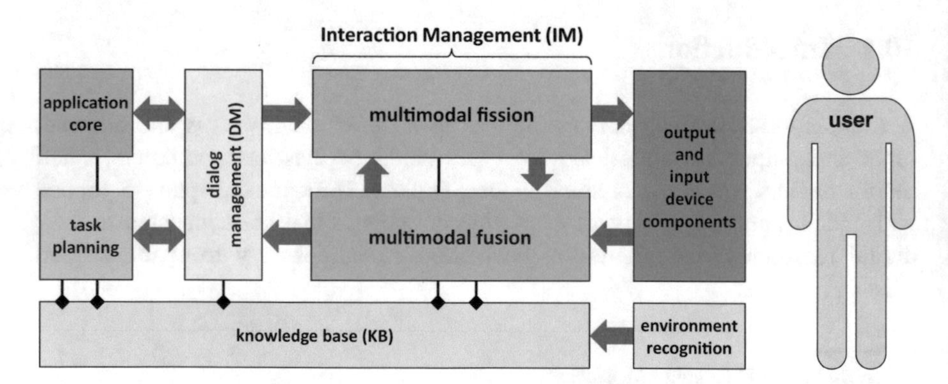

Fig. 10.1 The basic components of a *Companion*-System have a cyclic structure. The IM consists of multimodal fission and fusion components [22]

[1]Simplified action cycle process based on [35]: *form a goal* → *plan an action* → *execute an action* → *perceive feedback* → *evaluate the new state of the world* → ....

The main functional areas of the IM, namely fission and fusion, on the one hand permanently exchange information with the dialog management (DM), which serves as interface to the application core and the task planning. On the other hand, they realize and evaluate the adaptive communication and interaction with the user through the currently available output and input device components. Adaptions of the UI to the user, the surroundings, as well as available devices are influenced by information stored in the knowledge base (KB).

10.3 The Interaction Management

The following sections name the hows, whys and wherefores of the chosen approaches to adaptive output generation and semantic combination of user inputs. The last part describes the interplay of both components, fission and fusion.

10.3.1 Modality Arbitration in Multimodal Fission

Adaptive multimodal interaction is linked to the goal of providing the user with an adequate user interface (UI) with respect to the user's current context of use (CoU). The interaction management's fission component (see Fig. 10.1) has to decide which uni- or multimodal output configuration has to be realized as optimal UI. The fission's main input is an XML-based description of a dialog step, which is provided form the dialog manager as a single part of a more complex dialog structure [3].

This section focuses on the processes of modality arbitration and model-driven UI generation. The first part in this section gives an overview of the state of the art in modality arbitration and references related work. The second part considers the modeling of the CoU and describes how knowledge can be organized to support the fission's probabilistic reasoning. Next, the third part gives an introduction to the concept of model-driven UI evolution, as we use such an approach to realize an adaptive interaction management. The fourth part completes this section with a description of the fission's reasoning process for modality arbitration.

10.3.1.1 Related Work

As stated in [9], systems that combine different output modalities like text and speech have evolved since the early 1990s. The allocation of output modalities of these early multimodal systems was hard-coded rather than based on intelligent algorithms. To summarize the findings from [17] and [42], a fission component should give answers to the following four WWHT questions. First, *What* is the information to present? Second, *Which* modalities should be used to present this information? Third, *How* to present the information using these modalities? And then, as the forth question, *How* to handle the evolution of the resulting presentation?

An important survey on multimodal interfaces, principles, models, and frameworks is provided in [15]. Beyond that, the authors mention the idea of machine learning approaches for multimodal interaction. Their given example focuses on machine learning in multimodal fusion on the feature level. Another interesting approach is presented in [19]. The authors present a multi-agent system, where past interactions are taken into account to reason about the new output. They recommend a machine learning approach for case-based reasoning.

To reason about the best UI configuration in a certain CoU is a challenging task. Some approaches provide meta UIs where the user can specify a particular UI configuration, e.g. via an additional touch device as presented in [41]. Their system is able to respect the user's demands and can distribute the UI with the use of the referenced device components.

Based on our investigations, rule-based approaches can be seen as established practice. Recent work goes together with model-driven UI generation to realize adaptive UIs. Additional user input can be used to support the system's decision process. The awareness of uncertain knowledge is practically non-existent. In our current approach for modality arbitration [21], we use real-world data, which can even be afflicted with uncertainty. The presented approach is able to perform a continuous model-driven UI adaptation that respects the ongoing changes in the CoU.

10.3.1.2 Adaptivity Based on Changes in the Context of Use

In our work we refer to context as given by the definition in [11]. The authors define context as "...any information that can be used to characterize the situation of an entity. An entity is a person, place, or object that is considered relevant to the interaction between a user and an application, including the user and applications themselves."

According to [4], a *Companion*-System shall be adaptive and always available. That implies that the interaction management (IM) has to ensure an ongoing adaptation of the UI with respect to each sensed change in the CoU (see p. 192). The IM gains its knowledge from the knowledge base (see Fig. 10.1). Here we focus on six separate models, which describe knowledge about the user (1), the environment (2), the known device components (3) with their capabilities, the current dialog step (4), information (5) that can be referenced from within a dialog step, as well as different distance relations (6) that describe user-to-device and device-to-device distances [21, 22].

Since we see *Companion*-Systems as distributed systems, we decided to use a message-oriented middleware to connect the different components from our architecture. Information exchange is realized via XML-messages that can be sent and received via different topics from and to diverse components [24]. Our Schema-based definition does not only allow us to describe simple key-value pairs. It also allows us to map multiple values with an additional probability distribution to the keys [21]. This enables the KB (see Chap. 2) to express and communicate even uncertain observations from the CoU.

Fig. 10.2 Emulation by tangible interaction (*left*) and later real situation with various sensors (*right*). Moving the toy figures on the table results in almost equal signals as the person tracking via laser range finders as described in [18]

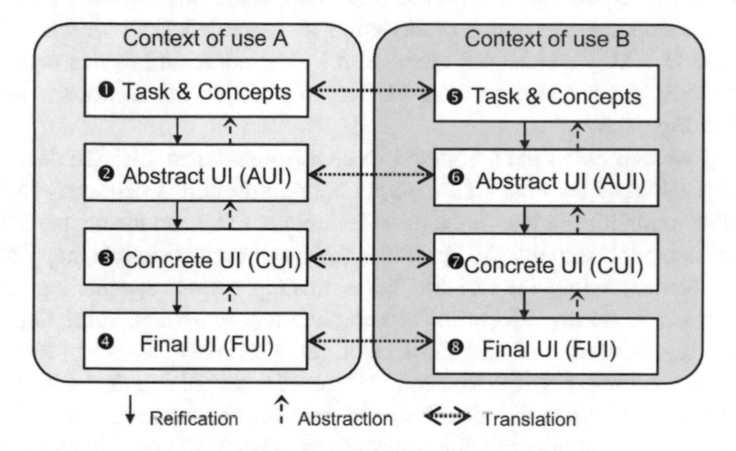

Fig. 10.3 The CAMELEON reference framework's process of UI refinement (taken from [47]). Starting from a simple task description, with the use of reification transformations, finally a Final UI (FUI) can be rendered on specific device components

The applied middleware approach comes with another advantage. It supports an easy exchange and scalability of divers device components and sensors. In [23] for example, we show how we are able to emulate the sensory system for diverse test settings by manipulating user and device models at runtime using tangible interaction (see Fig. 10.2). In that way, the fission's reasoner can be tested without the need of any further sensory system.

10.3.1.3 Evolution of the User Interface

The self-imposed requirements of adaptivity and availability for a *Companion*-System can be met best with a flexible model-driven approach for UI-Generation as characterized by the CAMELEON reference framework (CRF) (see [7] and Fig. 10.3). The CRF's approach comes with the idea of step-by-step refinement of an output description, whereas the last step of the process provides the final user interface (FUI). The refinement process itself can be influenced by external decisions. In that way it is possible to derive various FUIs with even different layouts (e.g. browser web-interface vs. front end of a smartphone app vs. augmented head-up display (see Chap. 23)) from only one identical task description.

The CRF can easily be mapped to the presented architecture of a *Companion*-System (see Fig. 10.1). In our current approach an AI planner decides about the sequence of tasks (see Chap. 5). The dialog management (see Chap. 9) decides about the abstract and modality-independent sequence of dialog steps as abstract user interface (AUI). The interaction management's fission component uses the dialog output as a description of an AUI as input. Based on that, it reasons about the most adequate output configuration as described in the next section. The fission's output represents the description of the output as concrete UI (CUI). It is a mark-up description in XML, which gets interpreted by the addressed device components. This results in the rendering of the FUI using multiple modalities for input and output (cf. Fig. 10.8).

In [21] we addressed the mapping problem from AUI to CUI. To decide about the configuration of the later CUI's output configuration it is necessary to identify all possible modalities (a.k.a. 'interaction technique' [34]). On the on hand, the used information model from the CoU provides the system with possible mappings from abstract information (e.g. the variable 'hdmi' that represents a specific type of cable) to all its possible concrete forms of representation (e.g. textual, aural, depicted, or otherwise representative of an HDMI cable (cf. Fig. 10.8) (see also [36]). On the other hand, the system is able to identify those of the available device components which are able to realize the identified and possible concrete ways of information representation. In combination, this allows us to identify all possible modalities and their multimodal combination. The next section describes the reasoning process that identifies the most adequate CUI as output.

10.3.1.4 Reasoning for Modality Arbitration

As mentioned before, the fission's reasoning process has to decide about the optimal output configuration in a given CoU, as depicted in Fig. 10.4. The fission's main goal is to solve the mapping problem from the modality-independent AUI, as provided by the dialog manager, to the modality-specific description of the CUI. In that way, the abstract output directly answers the basic *What*-question from the related work section (p. 189). The following reasoning process depends on the available knowledge about the CoU (see p. 190) and provides answers to the *Which*- and *How*-questions.

As a first step the fission module analyzes the received dialog output in order to cluster similar information items. Items for similar use (e.g. when reasoning about the rendering of a cable selection prompt) and with equal concrete ways of representation (e.g. selection items for 'cinch', 'scart' and 'hdmi') can be clustered for later case-based reasoning (CBR). The evaluation process for modality arbitration would treat all of them in the same way. So their mapping problem has to be solved only once and can be applied in terms of CBR for the equal others.

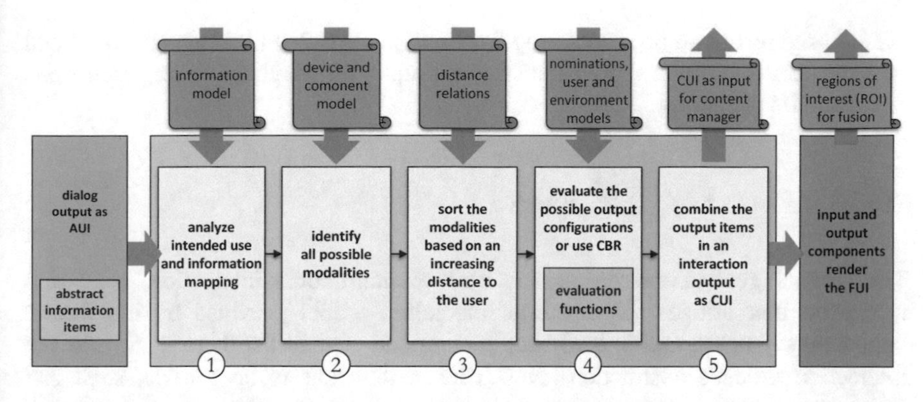

Fig. 10.4 The main steps within our fission process. The dialog output is a description of the abstract user interface (AUI). The possible output configurations for each information item are evaluated with the use of context knowledge and a set of evaluation functions. The set built by the highest rated output configurations (one from each information item) forms the interaction output as CUI

The second step inspects the available device models and identifies all possible modalities for each information item. The so identified list of possible modalities is sorted in the third step, based on their increasing distance from the user. This is of special importance when the number of possible multimodal combinations increases so much that not all of these output configurations can be evaluated. In such a case heuristical methods can be applied to define a best-so-far solution based on the distance criterion.

Modality arbitration takes place in the fourth step. A set of reward and punishment functions is used as a rule set to rate each possible output configuration. A function's reward is biased based on the probability of its activating knowledge items. Each function's meaning stems from common design principles or is influenced by study results, e.g. from [43].

This approach of probabilistic fission is described in detail in [21]. Besides knowledge from the CoU, user-given nominations in terms of explicitly stated desires or dislikes can also activate specific functions, which are especially provided for this purpose. New solutions to the mapping problem are kept in memory to be used for similar problems in the future. That way, known solutions from the CBR-base can be applied for similar items.

In the fifth and last step, the highest rated output configurations are assembled to form the modality-specific output description. This so-called interaction output is passed as CUI towards the involved device components for rendering. In addition to that, the CUI is also passed to the content manager in order to configure the fusion module for possible interaction inputs. If the FUI contains any visual elements (cf. Fig. 10.8), their rendering components send information about these regions of interest to the fusion. This will allow the fusion's preprocessing components to map pointing gestures to referenced objects on the screen.

To answer the *Then*-question (cf. related work on p. 189), the fission continuously evaluates its output according to possibly occurring changes in the CoU. These event-based re-fission processes loop from sub-process 2–5 (see Fig. 10.4) and end with the occurrence of a new abstract dialog output, for which the entire process can start from its beginning.

10.3.2 Fusion of User Inputs

The general goal of the fusion, in terms of multimodal interaction, is to find a method that utilizes the concrete interaction model provided by the fission component, gathers multimodal user inputs, and transforms them back into the interaction model's abstraction level. This method has to be generic, so it can incorporate the various sensor components of a *Companion*-System; and it has to deal with uncertainty to offer flexible as well as robust input options for the user.

Viewing the input processing of a *Companion*-System as a whole, it appears as a multi-level or rather hybrid fusion process (cf. [2]), where data-level, feature-level and decision-level fusions are performed at different stages (Fig. 10.5). In the domain of human-computer interaction (HCI), usually some kind of decision or hybrid level fusion is applied. The following gives a brief overview of current approaches and states their capabilities and limitations, before the actual input fusion process is described in greater detail. Parts of the following sections are based on [44].

10.3.2.1 Related Work

Within the decision-/hybrid-level fusion of HCI, the applied strategies can be further categorized into procedural, frame-based, unification-based, and statistical/hybrid ones [29, 30].

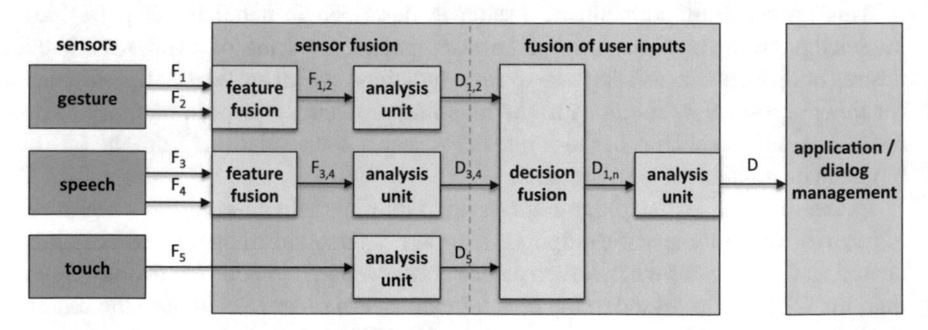

Fig. 10.5 The input processing of a *Companion*-System as a hybrid fusion process with different fusion levels (*from left to right*). Preprocessed sensor data is forwarded to the processing as features (F_n). These can be combined by feature fusion units (to $F_{n,m}$), before analysis units make decisions on these ($D_{n,m}$). The fusion of multimodal user inputs consists of a final decision fusion and analysis unit that forwards a conclusive decision (D) to the application/dialog management components

The procedural strategy is one of the oldest found in the literature. It uses state-based algorithms, finite state machines, petri nets and alike to perform an algorithmic management of input events to be combined. Examples can be found from the late 1980s to the present, as in [1, 10, 26, 32, 33], to name but a few.

Strategies are categorized as frame-based when tabular data structures and attribute value pairs (referred to as frames) are filled successively until a complete structure is obtained. Instances of this category are given in [14, 27, 28, 31, 34].

Unification-based approaches use more complex data structures (also called typed feature structures), upon which a unification operator or some form of production rules is applied. Among other well-known examples are the XTRA system [48], Quickset [8], and PATE [38].

The fourth category, statistical/hybrid strategies, is comprised of approaches combining unification-based methods with statistical processing and machine learning techniques. While machine learning is most prominent in feature-level fusion, it seems to make its way into decision-level fusion of multimodal interaction as well [15, 16]. Statistical processing can be found in [12, 39].

The most obvious drawback of procedural approaches is their increasing complexity when the multimodal state space grows large. Additionally, it remains unclear how they harmonize with dedicated dialog models of more sophisticated systems. Known frame-based and unification-based strategies seem to focus on combining two modalities at any one time (according to [30]), whereas multimodal input fusion within *Companion*-Systems must have the theoretical ability to combine an arbitrary number of inputs. Statistical/hybrid strategies often come with the burden of mandatory training before reasonable results are to be expected. As sensory inputs, which tend to be ambiguous in nature, are of utmost importance for *Companion*-Systems, another aspect of input fusion must be taken into account: its ability to deal with uncertainty. This aspect is neglected by most of the known approaches as they either expect clear-cut decisions from sensors or simply make decisions according to specific rules. Approaches that have a notion of uncertainty, mostly from the statistical/hybrid strategy, rely on simple n-best lists to make a final decision and lack a formally well-defined mechanism. This is especially true when it comes to support reinforcement and disambiguation of inputs, aside from semantic combination.

In the following, we describe our own approach for multimodal input fusion for *Companion*-Systems. Regarding the four strategies explained above, it represents a form of unification-based method enhanced by evidential reasoning and is roughly based on [40]. It worked within the related domain of human robot interaction.

10.3.2.2 Evidential Reasoning Applied to the Fusion of User Inputs

The *transferable belief model* (TBM) and its basic concept, Dempster-Shafer's theory of evidential reasoning, are generalizations of probability theory and quantify beliefs of sources in events. The TBM was proposed by Philippe Smets in the early 1990s [45] and differs from the classic probability in that a belief does not state the actual probability that an event happened, but only the confidence of the sensor about the event. The biggest advantage over classical probability is its ability to explicitly represent uncertainty in the form of a disjunction like *"event A or event B happened with a belief of m"*, without the necessity to assign probabilities to individual events. A complete overview and the basic concepts of TBM are given in [46]. The theory can be used to combine evidences from different sensors to make decisions on individual events described in the Frame Of Discernment Ω (FOD). Using their rule of combination, reinforcement, disambiguation, and detection of conflicting evidences are possible. What is missing is the ability to form actually fused extended concepts. For example, if a user wants to select an object o using a verbal deictic reference like "this one" and a pointing gesture, this would lead to beliefs about A = 'select' and B = 'object o'. The fusion system should be able to produce a belief about the combined concept AB = 'select object o'. However, the traditional rule of combination results in belief values for the individual concepts A and B, but not taken together. As described in [40], the TBM theory can be modified to allow a real fusion of multimodal inputs.

Key to the modification of TBM is the introduction of tuples as a representation of a combined concept [40]. In the above example, instead of having two propositions a ('select') and b ('object o'), the tuple (a, b) is used to denote the combined concept. While $m(\{a, b\})$ is the basic belief that the event was either a or b, $m(\{(a, b)\})$, or for short $m(\{ab\})$, is the belief in the combined event ab. Mathematically speaking, tuples are the result of a set product $\{a\} \times \{b\} = \{(a, b)\}$. Using this Cartesian product, TBM's original rule of combination can be altered and written as:

$$m(C) = \sum_{C:C=(A\times B)\cap\Omega} m_1(A) \cdot m_2(B) \tag{10.1}$$

where $(A\times B)\cap\Omega$ allows only those combinations out of the Cartesian product $A\times B$ that are defined in the FOD. The amount of conflict within evidences is defined as:

$$m(\emptyset) = \sum_{(A\times B)\cap\Omega=\emptyset} m_1(A) \cdot m_2(B) \tag{10.2}$$

One benefit of evidential reasoning is its easy scalability, as both equations can easily be extended to an arbitrary number of sources of evidence [44].

Using the strict mathematical definition of a Cartesian product, this new combination rule would only produce beliefs over combined concepts (tuples) and not elementary propositions anymore. To avoid this, the authors of [40] introduce a neutral evidence, '*', that must be part of every sensor's input. So the belief distribution of a sensor must in any case at least contain a belief for '*', meaning that nothing has been detected. Additionally, the combination of evidences obeys the following rules:

- Combination of an evidence with the neutral evidence '*' results in that evidence itself. So $\{a\} \times \{*\}$ results in $\{a\}$.

- The combination of an evidence with itself results in that evidence itself. So $\{a\} \times \{a\}$ results in $\{a\}$.
- The order of evidences in a combined concept is not stressed. That is, $\{ab\}$ is the same as $\{ba\}$.

While beliefs are useful to combine different sources of evidence as described above, they are not directly suited to derive decisions. A decision is better justified by probabilities for each elementary proposition out of the FOD. For this reason the so-called *pignistic transformation* (from Latin 'pignus': a bet) is applied. It transforms a belief function into a probability function over Ω, denoted by *BetP* and given by:

$$BetP(\omega) = \sum_{A:\omega\in A\subseteq\Omega} \frac{m(A)}{|A|} \qquad \forall\omega \in \Omega \tag{10.3}$$

where $|A|$ denotes the number of elementary propositions in set A. Simply stated, the pignistic transformation distributes the belief mass of a set A on all its single elements.

10.3.2.3 The Process of Input Fusion

When deploying the fusion technique described above in an actual system in order to provide the dialog management (and the application itself) with inputs, a complete processing chain is needed. The relevant components of this process and their connections are illustrated in Fig. 10.6. The overall processing is twofold. First, an abstract interaction model is sent to the fusion system that is used to dynamically reconfigure the fusion process for the current possible interactions. Second, there

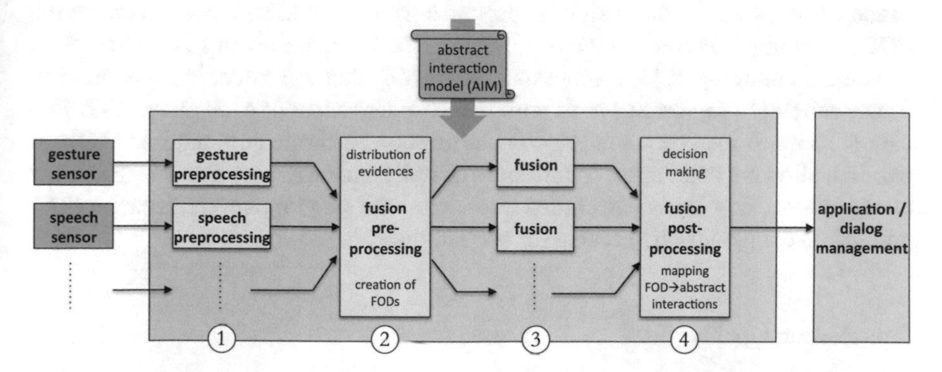

Fig. 10.6 The main components of the user input fusion. Configured by an abstract information model, multiple parallel fusions based on evidential reasoning are performed. A final decision is made on the fused results and interaction events are passed over to the application/dialog management components

is the input fusion pipeline (① to ④) that receives recognition events from the connected input sensors and passes them through the actual fusion process. Finally, fused input events are sent to the application/dialog management when indicated by the fusion result. The following paragraphs describe the two parts of processing, namely the configuration via an abstract interaction model and the input fusion pipeline, in detail.

Abstract Interaction Model

In order to realize a multimodal input fusion for a specific task, a description of all possible interactions that are allowed in the current state is needed. The description itself can be independent of the modalities used for input, as only the semantics of interactions and their mapping to abstract interactions is the principal task of the fusion system (cf. Sect. 10.3.2). Instead of using a proprietary language like SMUIML [13], finite state machines [10], or complex rules [20, 34, 38], we specify the fusion of input events as undirected graph. The nodes of the graph represent all possible events (the FOD of evidential reasoning), whereas the edges denote the semantic fusion of multiple events. Using GraphML as standardized XML format [6] with integrated XSL transformations, a maximum of flexibility and expressive power is achieved. The so-called abstract interaction model (AIM) is not relying on any specific model of inputs or modalities, as we are convinced that no predefined model of whatever nature can cover all possible interactions. Details on where such an AIM comes from within a *Companion*-System are given in Sect. 10.3.3.

Whenever the fusion system receives an AIM, it dynamically reconfigures to account for the change in possible interactions. For each sensor that is connected to the system, there exists a dedicated preprocessing component. The AIM is used to restrict the sensor messages reported to the inner fusion system to those relevant in the current state. In the fusion preprocessing part, the AIM is used to create the FOD, or multiple FODs to be more precise. The rationale for this is that an AIM can contain multiple disjoint subgraphs, something that was not considered in [40]. Those subgraphs represent events completely independent of each other. So fusion cannot be performed on a single FOD but must be performed on multiple FODs in parallel, allowing concurrent occurrences of independent events. For this reason, in our approach, the AIM is partitioned into its disjoint subgraphs and a separate FOD and fusion component are created for each of them.

Input Fusion Pipeline

At this point the system is configured for the current input possibilities and is ready to receive and fuse inputs. The following elucidates the different steps of the pipeline as indicated by ① to ④ in Fig. 10.6.

Currently we use an off-the-shelf speech recognition component, and gesture and location detection sensors from consortium partners. All sensors communicate with the system via XML messages. These messages usually contain recognition results together with beliefs or confidence scores. The different sensor preprocessing components ①, which have already been configured for the current AIM, create events over all possible interactions and assign beliefs. So they are 'translating' modality-specific recognitions into elementary events of the FODs. This can happen on an application-specific basis. E.g. the speech preprocessing could create an undetermined selection event when the user says "this one", or input parameters could be mapped to attributes of events.

Whenever a sensor preprocessing signals a new sensor event, the fusion preprocessing ② redistributes the incoming belief distributions among the currently available fusion components. That is, whenever a signaled belief of a sensor matches an element of a fusion's FOD, the belief is added to the input of that fusion. In order to preserve valid belief distributions that satisfy $\sum_{A:A\subseteq\Omega} m(A) = 1$ (basic beliefs must sum up to 1), every dropped-out belief mass gets assigned to the neutral belief, '*'. This can happen when a sensor reports an event that is currently not possible as stated by the AIM.

Using the constructed belief distributions over their FODs, the fusion components ③ perform the combination of evidences using Eqs. (10.1) and (10.2) to compute combined beliefs and the overall conflict. After that, the pignistic transformation *BetP* from Eq. (10.3) is applied and the resulting list of probabilities is handed over to the fusion postprocessing (cf. Fig. 10.6 ④). Here, the event with the highest probability is selected to be sent back to the application/dialog management. If the winning event is a combined event, as stated in the AIM, and therefore has an XSL transformation defined, this transformation gets applied to all constituent events to produce the final semantic result. There are two cases where no event is sent to the application/dialog management: either the winning event is incomplete (i.e. it is defined only as part of a combined event), or the neutral evidence is the one with the highest probability. Conflicting evidences can be communicated to the user, to make him become aware of the system's inability to decide on a particular action, as described in the following section.

When the user performs inputs in a sequential manner (using *integration patterns* [37]) or a sensor's processing takes some while, it is quite rarely (if ever) the case that different sensors raise recognition events simultaneously. Since the evidential reasoning is ad hoc, there is a need to extend the temporal scope of such events. Thus our approach uses *temporal fading*, similar to the way of using activation values as presented in [38]. After new evidences are assigned their initial belief values, they continuously get decreased over time, while at the same time the belief in the neutral evidence '*' is increased. Once all the beliefs of a sensor output have reached 0 (and $m(*) = 1$), the output is finally removed from the fusion system. In other words, the fusion system has a form of memory that fades over time.

10.3.3 Interplay of Fission and Fusion

Now that both parts of the IM have been described separately and mostly independently from one another, the question remains about how they interact in a complete *Companion*-System as depicted in Fig. 10.1. Reviewing the literature in recent years, there have been multiple approaches considering both fission and fusion [5, 12, 15, 37, 49]. While they make use of more or less sophisticated ways of fission and fusion, their interplay is mostly application-specific and hard-coded. A fully model-driven approach with a direct interplay of both components, as presented here, offers additional and application-independent possibilities. The following sections are based on the work in [25], first presenting a more detailed view on the architectural aspects of the interplay between the two components and the other parts of a *Companion*-System's architecture. Then, the introduced linking points are described separately with their respective advantages.

10.3.3.1 The Architectural View

The architectural view on the interplay of fission and fusion, where possible linking points are emphasized, is depicted in Fig. 10.7.

The dialog management (DM) initiates the HCI loop with a dialog output, D_{out}. It contains a modality-independent description of what to present to the user. Based on this, the interaction management's fission component infers a modality-specific user interface description as described in Sect. 10.3.1. The resulting interaction

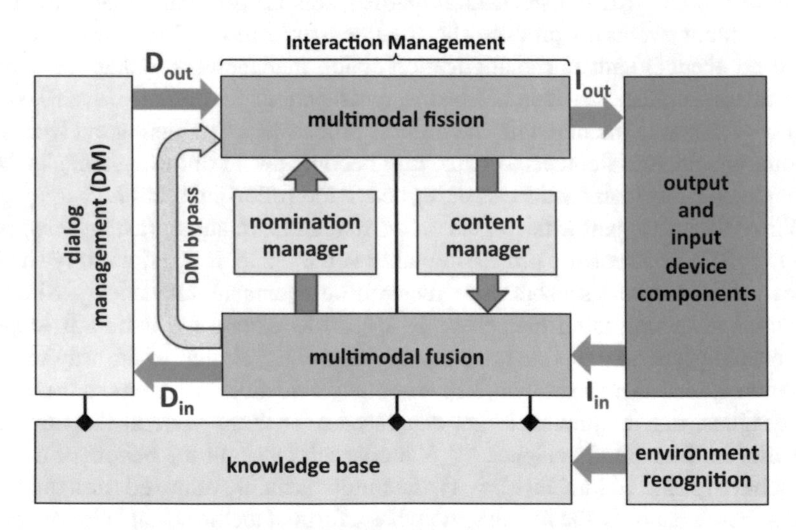

Fig. 10.7 Architectural detail of the interplay of fission and fusion components emphasizing collaborative aspects (nomination manager, content manager, and the dialog management bypass)

output I_{out} is then passed to the available output device components that render the user interface, including auditive information, e.g. using speakers for text-to-speech synthesis (TTS). The system is now ready to receive any kind of user input via the available input device components. Once they recognize a user input, they interpret it up to a certain decision level and provide the fusion component with their modality-specific interaction input I_{in}. After combining these inputs as described in Sect. 10.3.2, the fusion component passes the semantic of the most probable event (if there is any) back to the dialog management as dialog input D_{in}. This marks the end of the current HCI cycle and the system can continue with the next dialog output.

In addition, any context information change that is sensed by the environment recognition component(s) and maintained in the knowledge base can influence the UI generation of the fission component, e.g. user dispositions or location changes.

10.3.3.2 Content Manager: Tailoring the Fusion to An Adaptive UI

Fission and fusion both work on different models and abstraction levels that best fit their respective purposes. Once the fission component has decided on how to present an interface to the user, the fusion component needs to be informed on the resulting interaction possibilities. Based on that, the fusion is able to decide if different user inputs are ambiguous, are conflicting, or reinforce each other as described in Sect. 10.3.2.

Therefore the fusion component needs to be provided with an AIM that states all actions in the domain at hand the user could possibly trigger via the available input device components. In addition, the AIM must contain all domain-specific knowledge on how different inputs should be semantically combined. Within our approach, the content manager (cf. Fig. 10.7) is responsible for providing this kind of information. After the fission has reasoned about the concrete output configuration, the content manager inspects the resulting description of the concrete UI and identifies all objects that can be part of a user interaction. Using this information, the content manager creates a specific AIM that tailors the fusion for the current input possibilities.

The main benefit of using a dedicated component like the content manager to create the AIM is based on the fact that it allows the input fusion to be domain-independent and reusable in completely different applications. Domain-specific knowledge concerning, for example, what kind of inputs exist and how these have to be combined is completely stated in the AIM. This allows for a flexible and dynamic system output, while preserving a robust input recognition tailored to the current output configuration. Using a dedicated component for this task facilitates domain independence, helps to separate responsibilities, provides clear interfaces, and facilitates debugging.

10.3.3.3 Nomination Manager: User Demands for Information Presentation

We introduce the nomination manager (NM) (depicted in Fig. 10.7) to tailor the interaction's information flow towards the user. We assume that respecting a user's explicit demands for any modality can increase the credibility and reliability of an intelligent system. It is the fission component which is able to respect those user-initiated UI demands at runtime. But it is the fusion component which is able to identify those configuration nominations on a semantic level.

As an example, a user might utter the wish: "Show me more information about the HDMI cable." The fusion component analyzes its input and is able to recognize the distinct nomination for the visual channel. Based on the different confidence values, the fusion decides that the current input represents a request concerning HDMI. Accordingly, the fusion informs the NM with a new nomination containing only one identified desire. The probability, which is linked with the desire for the visual channel, is set to 1.0. The fusion does not have any further indications for other desired output channels concerning the requested HDMI explanation. The NM is able to aggregate different nominations for any referenced dialog output or information item via specific dialog or information IDs [25].

Right after sending the nomination message, the fusion passes the modality-independent HDMI explanation request to the dialog manager. In turn, the dialog manager responds with a suitable dialog output containing the HDMI explanation. The fission inspects the output and identifies the dialog ID. While reasoning about the dialog's modality-specific representation, the fission consults the nomination manager concerning nominations for the actual dialog output. The fission's reasoning algorithm is able to respect any stored nomination with a certain dialog or information ID that occurs within the processed dialog output description.

In that way the nomination manger supports the fission with additional knowledge for its reasoning process. Beyond the given example of enforcing a specific output channel, the nomination manager is able to handle nominations with a reference to certain desired device components or even expressing a certain dislike.

10.3.3.4 Resolution of Ambiguities Using a Bypass

In contrast to conventional GUI input technologies like mouse and keyboard, emerging technologies used for multimodal interaction, such as speech or gesture recognition, often provide inputs affected with uncertainty. And it is the interaction management that must deal with this new kind of ambivalent data, be it from sensors reporting false or unclear interpretations or users providing ambiguous or even conflicting inputs. Accidentally or on purpose, it can easily happen that input fusion cannot clearly decide on the user's intentions. In such cases, an intelligent system should deal with such a situation by providing helpful feedback to the user, and offer him the possibility to resolve existing ambiguities.

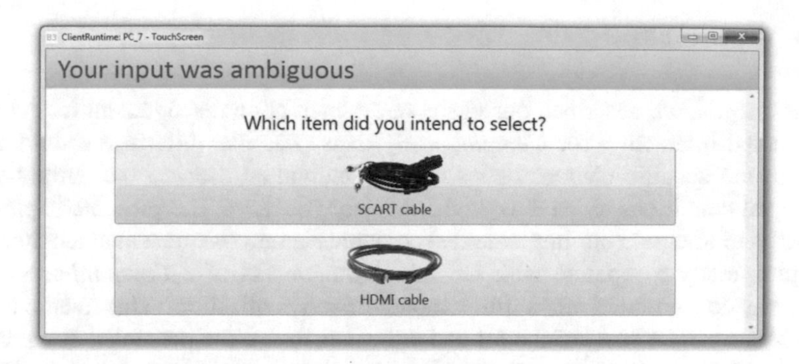

Fig. 10.8 Ad hoc conflict resolution by the interaction management. The inferred UI after detecting conflicting inputs (here: 'scart' vs. 'hdmi'), as displayed on a touch screen. In addition, the user is able to use speech input

If the input fusion component identifies conflicting inputs, it is not able to derive a definite user input. In order to deal with such a situation, we propose a direct cooperation of fusion and fission to resolve the ambivalent input. This allows bypassing the dialog management, resulting in a shortened HCI loop (cf. Fig. 10.7). As all necessary information is already present within the fusion module, a generic dialog output can automatically be constructed and forwarded to the fission module to be rendered, resulting in an output like the one shown in Fig. 10.8. This relieves the dialog management from the necessity to explicitly model such additional dialogs, which do not contribute to the overall dialog flow.

At the time when the ambiguity is resolved by the user, the fusion detects a valid input and sends a dialog input to the dialog management. In turn the normal dialog sequence is continued. This bypass approach addresses interaction-related conflicts within the interaction management according to the principle: solve the problems where they occur. This generic approach does not depend on the dialog management's ability to resolve conflicts. It turns out that this bypass concept works very reliably and is user-friendly. Furthermore, separating the dialog management from the used input techniques dismantles the obligation to explicitly model these additional dialogs that do not contribute to the overall dialog flow. Currently, the generated dialogs are quite simple and just present a list of possible alternatives the user is supposed to select from. Additional information might be useful that reveals the reasons for the system's current indecisiveness. This might also lead to a system's increasing credibility and perceived trustworthiness. Further investigation is needed to check if this is appropriate for all kinds of misunderstandings or if an additional component (like the nomination manager) that exclusively handles such queries could be advantageous.

10.4 Summary

In this chapter, we described our vision of an interaction management for adaptive multimodal interaction for *Companion*-Systems. The user interface should adapt itself to the specific characteristics of each individual user, to the current environmental conditions as well as to the available devices. The presented approach allows us to integrate our implemented components into the presented architectural structure, ready to operate with the other components of a *Companion*-System. Here, we concentrated on a fully model-driven realization. This methodology correlates with the state of the art and allows us to use the presented components in different scenarios and use-cases. The interaction management's context-adaptive components for fusion and fission can be updated at runtime and support an ongoing adaptation to the context of use. With the use of the applied middleware concept, it doesn't matter whether such context data is provided by external modules from other research groups or from our own context data emulators. Since real-world sensory data may be affected with uncertainty, both fusion and fission can deal with uncertain inputs.

Furthermore, we worked on a possible interplay of fusion and fission. The literature treats these components as rather isolated. However, we identified and presented three possible connecting factors where both components can benefit from each other. The presented bypass to the dialog manager can be used to clarify minor ambiguities within the interaction management. In addition, we presented a content manager that provides UI-specific configuration input for the fusion component. This facilitates domain independence, helps to separate responsibilities, provides clear interfaces, and simplifies debugging. We also introduced the nomination manager. This component allows us to tailor the interaction information flow towards the user by respecting his explicit demands for possible modalities. In that way, user-made nominations support the fission with additional knowledge for its reasoning process.

Acknowledgements This work was done within the Transregional Collaborative Research Centre SFB/TRR 62 "*Companion*-Technology for Cognitive Technical Systems" funded by the German Research Foundation (DFG).

References

1. Ameri Ekhtiarabadi, A., Akan, B., Çürüklu, B., Asplund, L.: A general framework for incremental processing of multimodal inputs. In: Proceedings of the 13th International Conference on Multimodal Interface, ICMI '11, pp. 225–228. ACM, New York (2011). doi:10.1145/2070481.2070521
2. Atrey, P., Hossain, M.A., El Saddik, A., Kankanhalli, M.: Multimodal fusion for multimedia analysis: a survey. Multimedia Systems **16**, 345–379 (2010)

3. Bertrand, G., Nothdurft, F., Honold, F., Schüssel, F.: CALIGRAPHI - Creation of Adaptive Dialogues Using a Graphical Interface. In: 2011 IEEE 35th Annual Computer Software and Applications Conference (COMPSAC), pp. 393–400. IEEE, New York (2011). doi:10.1109/COMPSAC.2011.58
4. Biundo, S., Wendemuth, A.: Companion-technology for cognitive technical systems. Künstl. Intell. (2016). doi:10.1007/s13218-015-0414-8
5. Blumendorf, M., Roscher, D., Albayrak, S.: Dynamic user interface distribution for flexible multimodal interaction. In: ICMI and the Workshop on Machine Learning for Multimodal Interaction. ACM, New York (2010). doi:10.1145/1891903.1891930
6. Brandes, U., Eiglsperger, M., Herman, I., Himsolt, M., Marshall, M.: GraphML progress report: structural layer proposal. In: Mutzel, P., Jünger, M., Leipert, S. (eds.) Graph Drawing. Lecture Notes in Computer Science, vol. 2265, pp. 501–512. Springer, Berlin, Heidelberg (2002). doi:10.1007/3-540-45848-4_59
7. Calvary, G., Coutaz, J., Thevenin, D., Bouillon, L., Florins, M., Limbourg, Q., Souchon, N., Vanderdonckt, J., Marucci, L., Paternò, F., Santoro, C.: The Cameleon reference framework. Technical Report 1.1. CAMELEON Reference Framework Working Group (2002)
8. Cohen, P.R., Johnston, M., McGee, D., Oviatt, S., Pittman, J., Smith, I., Chen, L., Clow, J.: QuickSet: multimodal interaction for distributed applications. In: Proceedings of the Fifth ACM International Conference on Multimedia, MULTIMEDIA '97, pp. 31–40. ACM, New York (1997). doi:10.1145/266180.266328
9. Costa, D., Duarte, C.: Adapting multimodal fission to user's abilities. In: Proceedings of the 6th International Conference on Universal Access in Human-Computer Interaction: Design for All and EInclusion - Volume Part I, UAHCI'11, pp. 347–356. Springer, Berlin (2011)
10. Cutugno, F., Leano, V.A., Rinaldi, R., Mignini, G.: Multimodal framework for mobile interaction. In: Proceedings of the International Working Conference on Advanced Visual Interfaces, AVI '12, pp. 197–203. ACM, New York (2012). doi:10.1145/2254556.2254592
11. Dey, A.K., Abowd, G.D.: Towards a better understanding of context and context-awareness. In: HUC '99: Proceedings of the 1st International Symposium on Handheld and Ubiquitous Computing, pp. 304–307. Springer, Berlin (1999)
12. Duarte, C., Carriço, L.: A conceptual framework for developing adaptive multimodal applications. In: IUI '06: Proceedings of the 11th International Conference on Intelligent User Interfaces, pp. 132–139. ACM, New York (2006). doi:10.1145/1111449.1111481
13. Dumas, B., Lalanne, D., Ingold, R.: Prototyping multimodal interfaces with the SMUIML modeling language. In: CHI 2008 Workshop on User Interface Description Languages for Next Generation User Interfaces, CHI 2008, Frienze, pp. 63–66 (2008)
14. Dumas, B., Lalanne, D., Guinard, D., Koenig, R., Ingold, R.: Strengths and weaknesses of software architectures for the rapid creation of tangible and multimodal interfaces. In: TEI '08: Proceedings of the 2nd International Conference on Tangible and Embedded Interaction, pp. 47–54. ACM, New York (2008). doi:10.1145/1347390.1347403
15. Dumas, B., Lalanne, D., Oviatt, S.: Multimodal interfaces: a survey of principles, models and frameworks. In: Lalanne, D., Kohlas, J. (eds.) Human Machine Interaction – Research Results of the MMI Program. Lecture Notes in Computer Science, vol. 5440/2009, chap. 1, pp. 3–26. Springer, Berlin, Heidelberg (2009). doi:10.1007/978-3-642-00437-7_1
16. Dumas, B., Signer, B., Lalanne, D.: Fusion in multimodal interactive systems: an HMM-based algorithm for user-induced adaptation. In: Proceedings of the 4th ACM SIGCHI Symposium on Engineering Interactive Computing Systems, EICS '12, pp. 15–24. ACM, New York (2012). doi:10.1145/2305484.2305490
17. Foster, M.E.: State of the art review: multimodal fission. Public Deliverable 6.1, University of Edinburgh (2002). COMIC Project
18. Geier, T., Reuter, S., Dietmayer, K., Biundo, S.: Goal-based person tracking using a first-order probabilistic model. In: Proceedings of the Ninth UAI Bayesian Modeling Applications Workshop (UAI-AW 2012) (2012)

19. Hina, M.D., Tadj, C., Ramdane-Cherif, A., Levy, N.: A Multi-Agent based Multimodal System Adaptive to the User's Interaction Context. Multi-Agent Systems – Modeling, Interactions, Simulations and Case Studies, chap. 2, pp. 29–56. InTech (2011)
20. Holzapfel, H., Nickel, K., Stiefelhagen, R.: Implementation and evaluation of a constraint-based multimodal fusion system for speech and 3D pointing gestures. In: Proceedings of the 6th International Conference on Multimodal Interfaces, ICMI '04, pp. 175–182. ACM, New York (2004). doi:10.1145/1027933.1027964
21. Honold, F., Schüssel, F., Weber, M.: Adaptive probabilistic fission for multimodal systems. In: Proceedings of the 24th Australian Computer-Human Interaction Conference, OzCHI '12, pp. 222–231. ACM, New York (2012). doi:10.1145/2414536.2414575
22. Honold, F., Schüssel, F., Weber, M., Nothdurft, F., Bertrand, G., Minker, W.: Context models for adaptive dialogs and multimodal interaction. In: 2013 9th International Conference on Intelligent Environments (IE), pp. 57–64. IEEE, New York (2013). doi:10.1109/IE.2013.54
23. Honold, F., Schüssel, F., Munding, M., Weber, M.: Tangible context modelling for rapid adaptive system testing. In: 2013 9th International Conference on Intelligent Environments (IE), pp. 278–281. IEEE, Athens (2013). doi:10.1109/IE.2013.9
24. Honold, F., Bercher, P., Richter, F., Nothdurft, F., Geier, T., Barth, R., Hörnle, T., Schüssel, F., Reuter, S., Rau, M., Bertrand, G., Seegebarth, B., Kurzok, P., Schattenberg, B., Minker, W., Weber, M., Biundo, S.: Companion-technology: towards user- and situation-adaptive functionality of technical systems. In: 2014 10th International Conference on Intelligent Environments (IE), pp. 378–381. IEEE, New York (2014). doi:10.1109/IE.2014.60
25. Honold, F., Schüssel, F., Weber, M.: The automated interplay of multimodal fission and fusion in adaptive HCI. In: IE'14: Proceedings of the 10th International Conference on Intelligent Environments, pp. 170–177. IEEE, Shanghai (2014). doi:10.1109/IE.2014.32
26. Johnston, M., Bangalore, S.: Finite-state multimodal integration and understanding. Nat. Lang. Eng. **11**, 159–187 (2005). doi:10.1017/S1351324904003572
27. Koons, D.B., Sparrell, C.J., Thorisson, K.R.: Integrating simultaneous input from speech, gaze, and hand gestures. In: Maybury, M.T. (ed.) Intelligent Multimedia Interfaces, chap. 11, pp. 257–276. American Association for AI, Menlo Park, CA (1993)
28. Krahnstoever, N., Kettebekov, S., Yeasin, M., Sharma, R.: A real-time framework for natural multimodal interaction with large screen displays. In: Proceedings of the 4th IEEE International Conference on Multimodal Interfaces, ICMI '02, pp. 349–354. IEEE Computer Society, Washington, DC (2002). doi:10.1109/ICMI.2002.1167020
29. Lalanne, D., Nigay, L., Palanque, P., Robinson, P., Vanderdonckt, J., Ladry, J.F.: Fusion engines for multimodal input: a survey. In: Proceedings of the 2009 International Conference on Multimodal Interfaces, ICMI-MLMI '09, pp. 153–160. ACM, New York (2009). doi:10.1145/1647314.1647343
30. LaViola, J.J. Jr., Buchanan, S., Pittman, C.: Multimodal Input for Perceptual User Interfaces, chap. 9, pp. 285–312. Wiley, New York (2014). doi:10.1002/9781118706237.ch9
31. Mansoux, B., Nigay, L., Troccaz, J.: Output multimodal interaction: the case of augmented surgery. In: Bryan-Kinns, N., Blanford, A., Curzon, P., Nigay, L. (eds.) People and Computers XX – Engage. BCS Conference Series, vol. 5, pp. 177–192. Springer/ACM, London/New York (2006). doi:10.1007/978-1-84628-664-3_14
32. Martin, J.C.: Tycoon: theoretical framework and software tools for multimodal interfaces. In: Lee, J. (ed.) Intelligence and Multimodality in Multimedia Interfaces. AAAI Press, Palo Alto, CA (1998)
33. Neal, J.G., Thielman, C.Y., Dobes, Z., Haller, S.M., Shapiro, S.C.: Natural language with integrated deictic and graphic gestures. In: Proceedings of the Workshop on Speech and Natural Language, HLT '89, pp. 410–423. Association for Computational Linguistics, Stroudsburg, PA (1989). doi:10.3115/1075434.1075499
34. Nigay, L., Coutaz, J.: A generic platform for addressing the multimodal challenge. In: CHI '95: Proceedings of the SIGCHI Conference on Human Factors in Computing Systems, pp. 98–105. ACM, New York (1995). doi:10.1145/223904.223917
35. Norman, D.A.: The Design of Everyday Things. iBasic Books, New York (2002)

36. Nothdurft, F., Honold, F., Zablotskaya, K., Diab, A., Minker, W.: Application of verbal intelligence in dialog systems for multimodal interaction. In: 2014 10th International Conference on Intelligent Environments (IE), pp. 361–364. IEEE, Shanghai (2014). doi:10.1109/IE.2014.59. (Short paper)
37. Oviatt, S.: Multimodal Interfaces, 2nd edn., chap. 21, pp. 413–432. CRC, Boca Raton (2007). doi:10.1201/9781410615862.ch21
38. Pfleger, N.: Context based multimodal fusion. In: ICMI '04: Proceedings of the 6th International Conference on Multimodal Interfaces, pp. 265–272. ACM, New York (2004). doi:10.1145/1027933.1027977
39. Portillo, P.M., García, G.P., Carredano, G.A.: Multimodal fusion: a new hybrid strategy for dialogue systems. In: Proceedings of the 8th International Conference on Multimodal Interfaces, ICMI '06, pp. 357–363. ACM, New York (2006). doi:10.1145/1180995.1181061
40. Reddy, B.S., Basir, O.A.: Concept-based evidential reasoning for multimodal fusion in human-computer interaction. Appl. Soft Comput. **10**(2), 567–577 (2010). doi:10.1016/j.asoc.2009.08.026
41. Roscher, D., Blumendorf, M., Albayrak, S.: A meta user interface to control multimodal interaction in smart environments. In: Proceedings of the 14th International Conference on Intelligent User Interfaces, IUI '09, pp. 481–482. ACM, New York (2009). doi:10.1145/1502650.1502725
42. Rousseau, C., Bellik, Y., Vernier, F., Bazalgette, D.: A framework for the intelligent multimodal presentation of information. Signal Process. **86**(12), 3696–3713 (2006). doi:10.1016/j.sigpro.2006.02.041
43. Schüssel, F., Honold, F., Weber, M.: Influencing factors on multimodal interaction during selection tasks. J. Multimodal User Interfaces **7**(4), 299–310 (2013). doi:10.1007/s12193-012-0117-5
44. Schüssel, F., Honold, F., Weber, M.: Using the transferable belief model for multimodal input fusion in companion systems. In: Schwenker, F., Scherer, S., Morency, L.P. (eds.) Multimodal Pattern Recognition of Social Signals in Human-Computer-Interaction. Lecture Notes in Computer Science, vol. 7742, pp. 100–115. Springer, Berlin, Heidelberg (2013). doi:10.1007/978-3-642-37081-6_12
45. Smets, P.: The combination of evidence in the transferable belief model. IEEE Trans. Pattern Anal. Mach. Intell. **12**(5), 447–458 (1990). doi:10.1109/34.55104
46. Smets, P.: Data fusion in the transferable belief model. In: Proceedings of the Third International Conference on Information Fusion. FUSION 2000, vol. 1, pp. PS21–PS33. IEEE, New York (2000). doi:10.1109/IFIC.2000.862713
47. Vanderdonckt, J., Limbourg, Q., Michotte, B., Bouillon, L., Trevisan, D., Florins, M.: USIXML: a user interface description language for specifying multimodal user interfaces. In: Proceedings of W3C Workshop on Multimodal Interaction WMI'2004, pp. 1–7 (2004)
48. Wahlster, W.: User and discourse models for multimodal communication. In: Sullivan, J.W., Tyler, S.W. (eds.) Intelligent User Interfaces, pp. 45–67. ACM, New York (1991). doi:10.1145/107215.128691
49. Wahlster, W. (ed.): SmartKom: Foundations of Multimodal Dialogue Systems. Springer, Berlin (2006). doi:10.1007/3-540-36678-4

Chapter 11
Interaction with Adaptive and Ubiquitous User Interfaces

Jan Gugenheimer, Christian Winkler, Dennis Wolf, and Enrico Rukzio

Abstract Current user interfaces such as public displays, smartphones and tablets strive to provide a constant flow of information. Although they all can be regarded as a first step towards Mark Weiser's vision of ubiquitous computing they are still not able to fully achieve the ubiquity and omnipresence Weiser envisioned. In order to achieve this goal these devices must be able to blend in with their environment and be constantly available. Since this scenario is technically challenging, researchers simulated this behavior by using projector-camera systems. This technology opens the possibility of investigating the interaction of users with always available and adaptive information interfaces. These are both important properties of a Companion-technology. Such a Companion system will be able to provide users with information how, where and when they are desired. In this chapter we describe in detail the design and development of three projector-camera systems(UbiBeam, SpiderLight and SmarTVision). Based on insights from prior user studies, we implemented these systems as a mobile, nomadic and home deployed projector-camera system which can transform every plain surface into an interactive user interface. Finally we discuss the future possibilities for Companion-systems in combination with a projector-camera system to enable fully adaptive and ubiquitous user interface.

11.1 Introduction to Ubiquitous User Interfaces

Traditionally, user interfaces are part of a physical device such as a laptop, a tablet or a smartphone. To be able to interact with such user interfaces fluidly throughout the day, users have to actually carry those devices with them. In [19], Mark Weiser describes his vision on technology which will blend into the user's environment and offer omnipresent interfaces. Current systems are not yet able to offer these characteristics Mark Weiser envisioned. Researchers started to use projection to simulate these types of interfaces.

J. Gugenheimer (✉) • C. Winkler • D. Wolf • E. Rukzio
Institute of Media Informatics, Ulm University, James-Frank-Ring, 89081 Ulm, Germany
e-mail: jan.gugenheimer@uni-ulm.de; christian.winkler@uni-ulm.de; dennis.wolf@uni-ulm.de; enrico.rukzio@uni-ulm.de

S. Biundo, A. Wendemuth (eds.), *Companion Technology*, Cognitive Technologies,
DOI 10.1007/978-3-319-43665-4_11

As already introduced earlier in Chap. 1, a *Companion*-System complies with several abilities such as individuality, adaptability, flexibility, cooperativeness and trustworthiness. This chapter focuses particularly on two abilities of a *Companion*-System [1]: *availability* and *adaptability*. Both these characteristics are investigated using Projector-Camera Systems.

One essential part of availability is the capability to access large information displays at any given time at any given location. The basic concept of an office environment offering these capabilities was introduced by Raskar et al. [16] using Projector-Camera Systems. Depth cameras were used to enable interaction with the projected interfaces. The cameras are adjusted in the same direction as the projector, thus allowing us to sense interactions such as touch on top of the projected image. Touch interaction was implemented using either infrared-based tracking [11, 22], color-based tracking [14] or marker-less tracking [8, 18, 20]. This basic concept was furthermore enhanced by attaching motors to the Projector-Camera Systems, allowing us to reposition the interactive projection almost everywhere inside the room [15, 21]. Raskar et al. [16] furthermore leveraged the tracking capabilities to adapt the image of the projection on to the projection surface, allowing us to project onto non-planar surfaces. Nowadays, basic Projector-Camera Systems can be built solely using consumer available products [7]. The necessary software to calibrate and implement the interaction on the projection can be developed by using toolkits such as WorldKit [24] or UbiDisplays [7]. Such toolkits offer quick and easy calibration and installation of projectors and depth cameras resulting in a touch sensitive projection interface created using solely consumer products.

Despite this progress in Projector-Camera Systems, researchers mainly focused on technical improvement and big laboratory setups resulting in little knowledge about the use of Projector-Camera Systems inside a real life environment. However, home deployment and real life usage open new questions about the design, interaction and use-cases of Projector-Camera Systems. Furthermore, there is currently a lack of small, portable and easily deployable Projector-Camera Systems which can be used for an in-situ study. In this chapter, we present in-situ user studies exploring the design-space of Projector-Camera Systems. Based on this study we are going to present three prototypes (*UbiBeam*, *smarTVision* and *SpiderLight*) which each focus on one of the use-cases/interaction concepts derived from the study results.

11.2 In-Situ User Study Using Projector-Camera Systems

To the best of our knowledge, no exploratory in-situ study was conducted focusing on the use and interaction with Projector-Camera Systems in a home environment. Huber et al. [10] did a qualitative user study by interviewing several HCI (Human-Computer Interaction) researchers on interaction techniques of pico projectors. The interviews however took place in a public environment and were focused solely on the interaction with small projectors. Hardy [6] deployed a Projector-Camera System at his working desk and used it for over 1 year. He reported valuable experiences and insights in the long term use of a Projector-Camera System inside an office environment. To investigate the use of home deployed Projector-Camera Systems, we conducted an in-situ user study using a mockup prototype in the home of 22 participants. The goal was to gain a deeper understanding of how the participants would use and interact with a Projector-Camera System in their own homes.

11.2.1 Method

To collect data, we conducted semi-structured interviews in 22 households (10 female, 12 male) and participants being between 22 and 58 years of age (M=29). We decided to interview participants in their homes since they were aware of the whole arrangement of the rooms and could therefore provide detailed insights into categories such as placement. Furthermore this helped to create a familiar environment for the participant which led to a pleasant atmosphere. This also allowed us to cover a variety of different use cases and rooms such as: the living room, bedroom, bathroom, working room, kitchen and corridor. The study was conducted using a mockup prototype consisting of an APITEK Pocket Cinema V60 projector inside of a card box mounted on a Joby Gorillapod. The cardboard box provided illustrations of non-functional input and output possibilities such as a touchpad, several buttons, a display and a depth camera. This low-fidelity mockup helped the participants to imagine how a future Projector-Camera System could look and what capabilities it could have.

The interviews were conducted in three parts. First, participants were briefed on the concept of ubiquitous computing/ubiquitous interfaces and introduced to Projector-Camera Systems. The second part was a semi-structured interview on the use and capabilities of Projector-Camera Systems. In the third part, participants had to go through each room in which they stated they wanted to use a Projector-Camera System and create and explain potential set-ups (Fig. 11.1). This resulted in participants actually challenging their own creations and led to a fruitful discussion with the interviewer.

The data gathered was analyzed using a grounded theory approach [17]. Two authors independently coded the data using open, axial and selective coding. The research questions for this exploratory study were: “How would people use a small and easy deployable projector-camera system in their daily lives? When and how would they interact with such a device, and how would they integrate it into their home?”

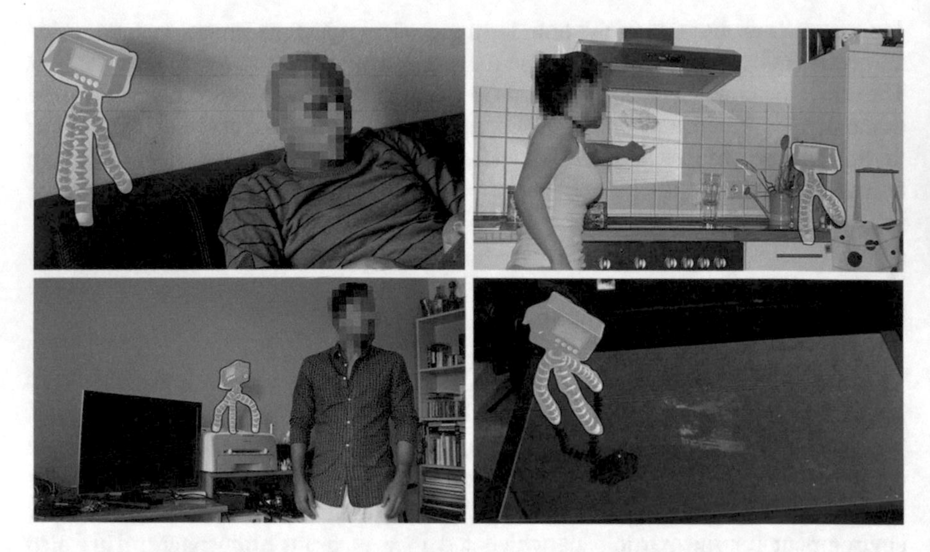

Fig. 11.1 Users building and explaining their setups (mockup highlighted for better viewability)

11.2.2 Results and Findings

We discovered four main categories [3] the participants were focusing on when they handled Projector-Camera Systems in their home environment:

- **Projector-Camera System placement**: *Where was the Projector-Camera System mounted inside the room?*
- **Projection surface**: *What projection surfaces did the participant choose?*
- **Interaction modalities**: *What modalities were mentioned for the input and why?*
- **Projected Content/Use Cases**: *What content did the participant want to project for each specific room?*

11.2.2.1 Content and Use Cases

Specific use cases were highly dependent on which room the participants were referring to. Nevertheless, two higher concepts derived from the set-ups the participants created: *information widgets* and *entertainment widgets*. The focus of *information widgets* was mainly to aggregate data. The majority of the use cases focused around an aid in finishing a certain task characteristic to the room. *Entertainment widgets* were mostly created in the living room, bedroom and bathroom. The focus of these was to enhance the free time spent in one of these rooms and make stays more enjoyable.

11.2.2.2 Placement of the Projector-Camera System

The placement was also classified into two higher concepts: the placement of the device *in reach* and *out of reach*. During the study, participants placed the Projector-Camera System within their reach and at waist height in the bedroom, bathroom and in the kitchen. The reasoning behind it was they could effortlessly remove it and carry it to a different room. In the living room, working room and corridor participants could imagine a permanent mounting and therefore placed the Projector-Camera System *out of reach*. The placement was done in a way so that the device could project on most surfaces and was "not in the way" (P19).

11.2.2.3 Orientation and Type of Surface

Even though it was explained to participants that projection onto non-planar surfaces is possible (due to certain distortion techniques), they always preferred flat and planar surfaces. Only one participant wanted to project onto a couch. The classification made for the projection surfaces was *horizontal* (e.g. table) or *vertical* (e.g. wall) orientation. Both types were used equally often throughout all setups inside the kitchen, bedroom, working room and living room. Only in the corridor and bathroom did the majority create vertical surfaces due to the lack of large horizontal spaces.

11.2.2.4 Interaction Modalities

In terms of modalities all participants focused mostly on speech recognition, touch, and a remote control. Techniques such as gesture interaction, shadow interaction or laser pointers were mentioned occasionally but were highly dependent on a very specific use case. The main influence on the preferred modality was the room and the primary task in there. *Out of reach* placements were mainly controlled using a remote control and *in reach* using touch interaction. One participant explained his choices are mostly driven by convenience: "You see, I am lazy and I don't want to leave my bed to interact with something" (P22).

11.3 The *UbiBeam* System

We designed *UbiBeam* based on the insights from the in-site study [5]. The focus was to create a small and portable Projector-Camera System which can be deployment in the majority of the rooms. In terms of a *Companion*-System, *UbiBeam* should offer *availability* in terms of everywhere available user interfaces and *adaptability* in the form of adapting the location of the interface and the interaction modality, depending on the use case. The system consists of several components such as a projector, a depth camera and two servomotors to be able to transform every ordinary surface into a touch-sensitive information display. In the future such a device could have different form factors such as a light bulb [12] or a simple small box [13] which can be placed inside the user's environment. The design of these devices will therefore focus on aspects such as deployment and portability and not solely on interaction. *UbiBeam* was a first step towards a home deployed Projector-Camera System which can work as a research platform to gather more insights on everyday usage of Projector-Camera Systems.

11.3.1 Implementation

The goal was to create a platform which can be easily rebuilt. The proposed architecture describes a compact and steerable stand-alone Projector-Camera System.

11.3.1.1 Hardware Architecture

We decided to use the ORDROID-XU as the processing unit for *UbiBeam* (Fig. 11.2) which offers a powerful eight-core system basis chip (SBC). As a depth camera *UbiBeam* uses the Carmine 1.08 of PrimeSence. Its advantages over similar Time-of-Flight cameras are its higher resolution and its good support by the OpenNI framework. The projector is the ultra-compact LED projector ML550 by OPTOMA (a 550 lumen DLP projector combined with an LED light source). The projection distance is between 0.55 and 3.23 m. Pan and tilt is enabled using two HS-785HB servo motors by HiTEC (torque of 132 N cm). The auto focus is realized similar to [23] by attaching a SPMSH2040L linear servo to the focusing unit of the projector

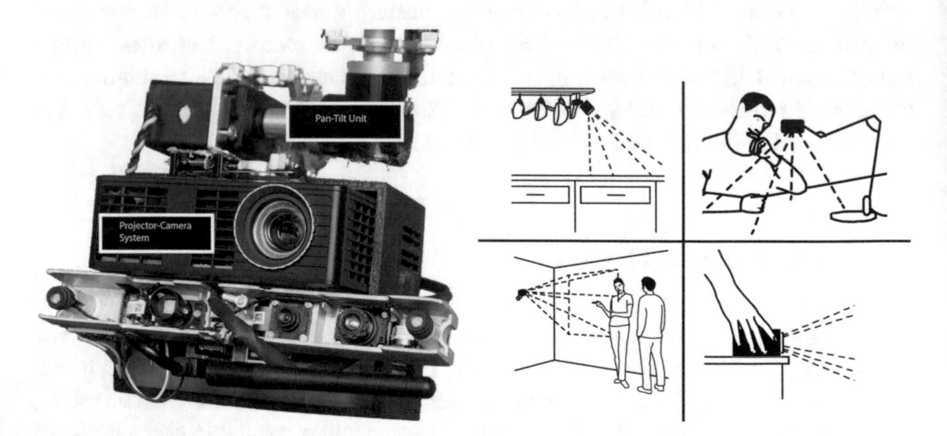

Fig. 11.2 The UbiBeam system in combination with the envisioned use cases for a home deployable Projector-Camera System

and refocusing based on the depth information. The control of the actuators is done by an Arduino Pro Mini. All the hardware components can be bought and assembled for less than 1000 USD.

11.3.1.2 Software Implementation

Since the goal was to create a stand-alone Projector-Camera System we tried to use lightweight and resource saving software. As an operating system we decided to use Ubuntu 12.04. The depth and RGB images were read and processed using OpenNI and OpenCV. UI widgets were implemented in QT, a library for UI development using C++ and QML. This allowed us to use an easy markup language (QML) to allow developers to create their own widgets.

UbiBeam was designed with the concept of an easy deployable system. Therefore, after the deployment at one particular location, the system automatically calibrates itself and enables touch interaction on the projection. The user can then create simple widgets using touch (e.g. calendar, clock, image frame) over the whole projection space. The orientation of the projection can either be controlled using the smart phone as a remote or dynamically by certain widgets (*adaptability*). After moving the device to a new space the auto focus and touch detection recalibrates automatically and creates a new interaction space.

11.3.2 Evaluation

To validate the quality of the proposed UbiBeam a technical evaluation was conducted. In particular, the precision and speed of the pan-tilt unit were examined as well as the touch accuracy.

11.3.2.1 Pan-Tilt Unit Performance

The task of the pan-tilt unit is to move the UbiBeam fast and accurately to a desired location. The two properties accuracy and pace were assessed in a laboratory study.

Alignment Accuracy The accuracy approaching a previously stored position was determined by placing the UbiBeam at a distance of 1 m to a wall. The projector was displaying a red cross to indicate the centre of the projection. Then the pan-tilt unit was commanded to approach the stored position from eight defined starting points. The position where the red cross came to a standstill was marked on the wall. Starting points were up, up-right, right, right-down, down, down-left, left, and left-up. Where up and down indicates a vertical shift by 45° from the stored position. Accordingly left and right indicates a horizontal shift by 90°. The measured distances in horizontal and vertical direction between the marked and

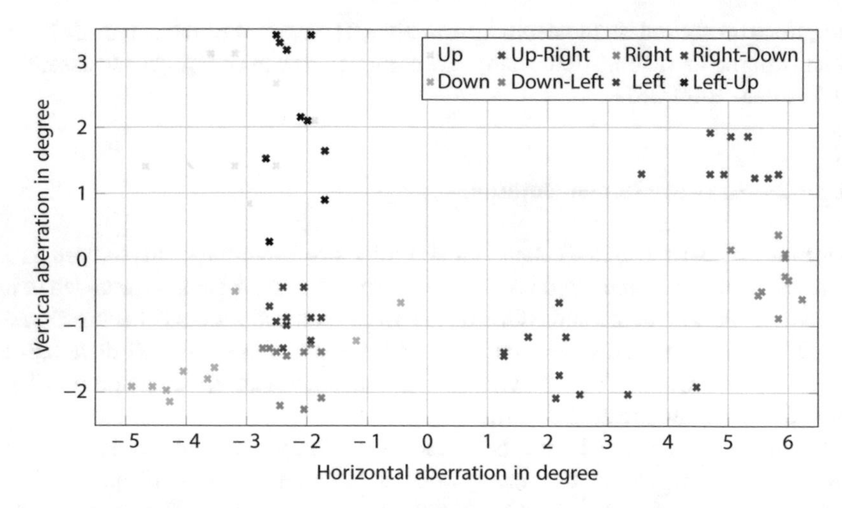

Fig. 11.3 Results of the positioning task for the pan-tilt motors

stored position lead to an angle of aberration. The stored position was approached ten times from each starting point. Thus 80 data points were obtained. A plot of the data is shown in Fig. 11.3. The average horizontal misalignment is 3.29°. For vertical alignment, the average error is 1.48°. Hence, the misalignment in an arbitrary direction is 3.74°. This accords to a shift of less than 10 cm if the surface is 150 cm away from the projector. A likely reason for the smaller misalignment in the vertical direction is caused by an accelerometer additionally used to control the servo for horizontal alignment. Since for horizontal alignment, no secondary sensor is used, the alignment is not as good. Overall the alignment is good enough to re-project a widget at almost the same location in the physical world, but is not sufficient enough to augment small tangible objects, for example a light switch. A more accurate alignment could be achieved by more powerful servos with a high-fidelity potentiometer.

Alignment Speed The pace of the pan-tilt unit was evaluated in a separate study. Therefore, the time needed for 164° horizontal pan and a 110° tilt was measured. Each movement was repeated ten times from both directions. Since panning and tilting is performed simultaneously, no combinations of tilt and pan were executed. On average the pan-tilt unit needed 3.5 s for the horizontal pan task. For the tilt task, the unit needed 4.8 s. A reason for the slower tilt movement could be the higher force needed for tilting compared to the rotation force. Overall the Projector-Camera System can reach every position in less than 6 s (worst-case: move 135° vertically). This seems to be a sufficient amount of time. Of course, there are faster servos available, but higher acceleration forces could damage the printed case holding the Projector-Camera System.

11.3.2.2 Touch Performance

Touch performance was evaluated in a similar laboratory study. The system was mounted over a desk at a distance of 75 cm. It was tilted down 70° from horizontal, pointing at the desk illuminating an interaction space of 40 cm×30 cm. The set-up is shown in Fig. 11.4. Four red crosses surrounded by a white circle posed as a target. They were distributed on three different surfaces. Two targets at the desk, one at the cardboard box on the left side and one on a ramp composed of a red notebook. In all cases, the diameter of the red cross was 18 mm.

During the study participants had the task to touch the targets as accurately as possible. Participants were instructed to take as much time as needed. Overall, 40 targets were presented in a counterbalanced order, one at a time. A detected touch was indicated by a green border. After touching the target, it disappeared and a new target appeared at one of the three other positions. Time as well as touch position in the projector and world coordinate system were recorded. From that data, the error in mm in the world coordinate system can be derived. Ten participants (all right handed) between 24 and 27 years took part in this study. Hence, 400 touch events were monitored. On average participants needed around 2 min to touch all 40 targets. In less than 1% the touch was not detected on the first approach. This was counted manually. The targets are labeled as follows: cardboard box (T1), ramp (T2), left desk (T3) and right desk (T4).

The mean touch error, variance and standard deviation for the different targets is specified in Table 11.1. Each target had a mean error of less than 20 mm. This requires large buttons for pleasant interaction. However, the small standard deviation for all targets indicates that the offset could be fixed by shifting the input by a few pixels. However, more studies must be conducted to verify this assumption.

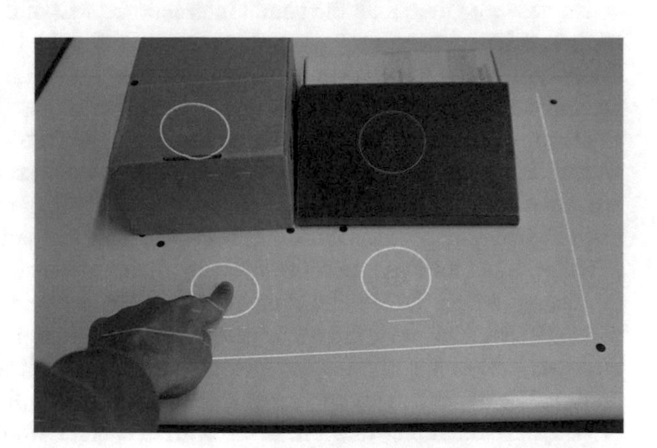

Fig. 11.4 Evaluation setup for the touch interaction

Table 11.1 Statistical data for the touch accuracy in mm

Target	T1	T2	T3	T4
Mean	14.11	19.32	16.58	17.82
Variance	8.48	12.50	8.50	7.57
SD	2.91	3.54	2.92	2.75

11.3.3 Discussion

The technical evaluation of *UbiBeam* has shown that the current setup is fast and accurate enough to support the use cases mentioned by participants in the earlier described user studies. Especially since users exclusively wanted to project on planar surfaces instead of augmenting specific items of their household, the current accuracy of below 2 cm seems sufficient in this regard. Further on, our evaluation of the touch accuracy showed that touches are robustly (99%) recognized and with a deviation below 2 cm. While the latter would clearly be too much for touch recognition on handheld devices one has to consider that the projected widgets of *UbiBeam* are much larger, typically having at least a size of 30×30 cm when projecting from only one meter away. Touch-Guidelines for smartphones typically agree on a minimal 1 cm bounding box required for touchable targets. Considering the at least four times larger displays generated by *UbiBeam*, a 2 cm deviation seems acceptable, although this accuracy should be further improved in the future.

11.4 The *SpiderLight* System

The focus of *SpiderLight* was to explore a body-worn Projector-Camera System and thereby investigate the interaction of a *Companion*-System which is always at hand (*availability*) and generate short cuts to context relevant information (*adaptability*). By observing smartphone users, we see that oftentimes getting hold of the device consumes more time than the actual interaction. Most of the time, the phone is used for micro-interactions such as looking up the time, the bus schedule, or to control a service like the flashlight or the music player [2]. With the recent emergence of wearable devices, such as smartwatches, users can access these kinds of information at all times without having to reach into their pockets. However, most of these wearable devices are merely equipped with a small screen so that only a little amount of content can be displayed and the user's finger is occluding most of the display during interaction (fat-finger problem). At the same time, the development of pico-projectors is progressing, allowing them to be incorporated in mobile phones (Samsung Beam), Tablets (Lenovo Yoga Pro 2) and even wearable devices such as a watch (Ritot). This way, the user overcomes the limitation of a small screen as pico-projectors allow the creation of comparably large displays from very small form factors. The larger display further enables sharing the displayed content with a group of people. Combining this projector with a camera would allow for interactions using the shadow of the fingers (Fig. 11.5). This would lead to having a large information display always available at the push of a finger.

The purpose of the *SpiderLight* is to facilitate micro-interactions that are too short to warrant getting hold of and possibly unlocking smartphone. Consequently, the *SpiderLight* is not meant to replace the user's smartphone. Instead, we understand

Fig. 11.5 The interaction space of the *SpiderLight*, which delivers quick access to context-aware information using a wrist-worn projector

SpiderLight as an accessory to the user's mobile phone that has more limited in and output capabilities in favor of a much shorter access time.

11.4.1 Implementation

The implemented system must be able to sense finger movements in line of sight of the projection, sense inertial movements, and project preferably with a wide angle to not excessively constrain the minimum distance between projecting hand and palm or wall. In addition, these components were supposed to be part of a single standalone system, with processing power and power supply on-board. The easiest hardware decision was for the projector to be a Microvision SHOWWX+ HDMI, as it was the smallest laser-beam-steering projector available on the market, providing the widest projection angle, too. The decision for a laser projector seemed inevitably to support quickly changing the projection surface and the projection distance, which would require constant adjustment of the focus using a DLP-based solution and even then could not provide the dynamic focus range required to project on the uneven human palm.

For the central processing unit we considered different commercially available systemboards like Raspberry Pi, Beaglebone, or Cupieboard and small smartphones that provide video output. However, they all seemed too bulky by themselves, considering that projector, camera, battery, and potentially additional sensors would all add to the overall size of the system. Our decision thus fell on an Android TV stick that would provide the same functionality at a much smaller size. In particular, we chose a system based on the Rockchip GT-S21D, that in addition to HDMI out and USB host, as all TV sticks offer, also provides a camera that is originally meant to be used with teleconferencing. Finding suitable cameras of the desired size that work well together with Android is often a very difficult challenge and by choosing a system that already integrated the camera we achieved the smallest possible footprint of the camera. However, the decision also implied two consequences: We decided against a depth camera, which at the time of engineering was not available at the required size and with the required support for mobile platforms like Android. Furthermore, the default placement of the camera required adding a surface mirror to the system to make the camera point in the direction of the projector. As the stick did not provide inertial sensors and inertial sensors of mobile platforms often are not very accurate we added the X-IMU to the overall setup that would allow us to accurately measure the device's orientation and translation for pre-warping the projected image against distortion and recognizing rotational device gestures. Finally, a battery supporting two USB ports with at least 1 A current output on each port was integrated to power the projector and the TV stick, which in turn powers the X-IMU (Fig. 11.6).

The SpiderLight system runs on Android with its UIs created in Java and rendered through OpenGLES. The computer vision and sensor fusion algorithms are written in C++ and integrated using JNI and Android's NDK interface. Apart from the decisions that were already taken regarding the interaction metaphors, we finally had to decide which type of menu interaction we wanted to support. Since more users of a pre-study preferred the approach using finger shadows for menu selection, we used the top menu that was designed with finger shadows in mind and supports absolute pointing (Fig. 11.7a). Conversely, for scroll selection, we selected rotational device gestures that were answered the most in the pre-study. For item selection, again finger shadow selection is used, whereby the first of four top segments returns to the menu selection and the other 2–3 menu items provide selection commands (Fig. 11.7b).

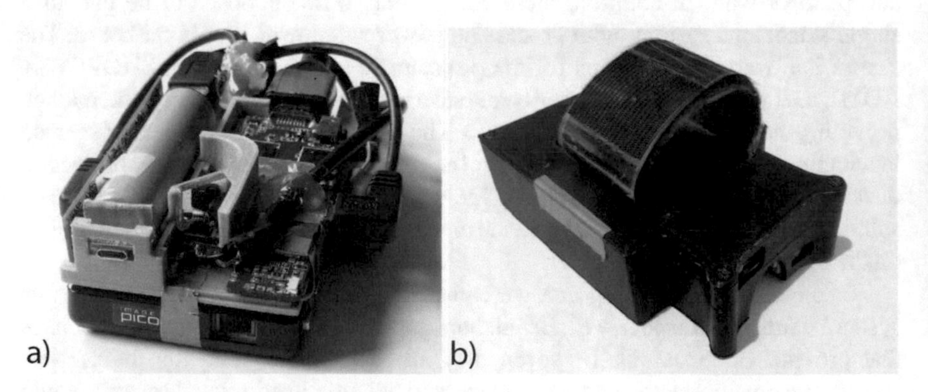

Fig. 11.6 The closure of *SpiderLight* (**b**) and the interior design (**a**) showing the projector at the bottom, the Android TV stick with the camera mirror on the right, and the battery on the left side. Not visible is the X-IMU which sits behind the projector on the lower side

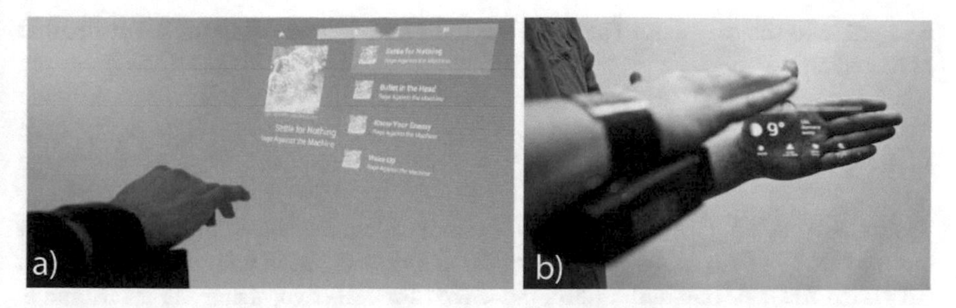

Fig. 11.7 Apart from the always available palm (**b**), any nearby surface (**a**) can be used for better clarity and single-hand interaction. Finger shadows facilitate button selections

11.4.2 Evaluation

To evaluate the performance and usability of *SpiderLight* we conducted a user study using the actual prototype.

11.4.2.1 Method

We recruited 12 participants (6 female) who were all right handed (since the prototype was optimized for the right hand) with an average age of 26 (range: 21–30). Except for two participants all had at least 2 years experience in using a smartphone. The goal of the study was to compare *SpiderLight* with a current smartphone in terms of access time, and usability in three applications that depict typical daily activities. Furthermore, we wanted to collect first impressions of participants using *SpiderLight*. The first task was to look up either the current weather or what time a certain bus is going to the train station. The second task was to scan an AR code and gather certain information (e.g. nutrition facts). The third task was to select a certain song in a music player. Each task was executed twice with a slight modification but stayed the same in term of complexity (e.g. only the piece of information to look up changed).

The study started with the participants being introduced to *SpiderLight*. Afterwards, they had time to practice and explore the system until they felt comfortable. Participants were encouraged to think aloud and give immediate feedback, which was written down. Participants were instructed to stand in front of a white wall and project onto it but without extending the arm to avoid exhaustion. After the introduction participants were using the smartphone and *SpiderLight* to finish the three tasks (tasks and systems were both counterbalanced). Every task started with taking the phone out of the pocket and unlocking it, respectively enabling the projection of the *SpiderLight* system. Once all tasks were finished, the users were asked to complete several questionnaires about their experiences using *SpiderLight*. During the study an error using the music application resulted in participants not being able to select a song. Therefore the third task was not used for the evaluation of the results.

11.4.2.2 Results

Task Completion Time On average it took participants 12.47 s (sd=3.7) for task one and 19.94 s (sd=9.72) for task two, using *SpiderLight*, in comparison to 12.00 s (sd=2.46) for task one and 14.80 s (sd=3.25) for task two, using the smartphone. We assume that the surprisingly high task completion time for *SpiderLight* resulted from the misdetections of input. Despite our efforts, the implementation of *SpiderLight* had sometimes problems in detecting a finger correctly. Therefore some participants resulted in taking longer using *SpiderLight* due to misdetection of input (which was manually recorded during the study). Nevertheless, looking at participants using *SpiderLight* without misdetection of input, the times show that most were able to finish the tasks with times below each smartphone time. We therefore argue that with a better implementation, *SpiderLight* would perform faster compared to smartphones.

Qualitative Feedback In the questionnaires about the usage of *SpiderLight*, participants reported that rotation interaction was simpler to conduct, less physically demanding and had a higher accuracy compared to finger input. This could partly be influenced by the misdetection of fingers, but also from the fact that using the shadow of a finger to interact with a device was more novel and challenging to participants compared to rolling their arm. In a last question participants were asked in what scenarios they would prefer to use *SpiderLight* instead of a smartphone. *SpiderLight* was highly preferred for sharing content and using the camera to scan AR codes, whereby it was less preferred to control the media player. We can explain this through the interaction concept of *SpiderLight*, which is designed for small interactions and quick lookups. Controlling a media player however is a task which can be considered longer and requires several selections such as browsing for a song.

11.4.3 Discussion

The unique advantage of projectors being able to create large displays from very small device form factors makes these devices very suitable for future wearable technology and for supporting micro-interactions. With *SpiderLight* we presented an approach to user interfaces for micro-interactions with wrist-worn projectors. We created a prototype that afforded most of the requirements in a standalone device, addressing several hardware and

(continued)

software challenges. Compared to other "smart devices" *SpiderLight* inhibits distinct advantages: it provides a much larger display than smart watches and can easily be shared in contrast to the display of smart glasses. Although our final evaluation could not entirely prove the superiority of *SpiderLight* over smartphone usage, we have to take the familiarity of users with smartphones and the described tracking issues we faced in the evaluation into account.

11.5 The *smarTVision* System

To evaluate the interaction with a Projector-Camera System in a stationary (home deployed) scenario, we decided to create a prototype for the use case of watching television. This television use case was often mentioned during the in-situ study (Sect. 11.2) and creates new challenges in terms of *availability* and *adaptability*. The interaction should now be possible using a remote control (*out of reach*) and touch (*in reach*). Initially, we analyze the current television setups in users' homes. The traditional setup of one television as the center of the living room is still widespread. However, a current trend shows that users tend to use second screens such as smartphones additionally to the content displayed on the main TV screen. Yet, the current setup does not allow for sharing additional content with others without interrupting the current content. With *smarTVision* [4], we present a concept which allows us to place any number of additionally projected screens inside the living room. We explored the space of input and output options and implemented several applications to investigate different interaction concepts.

The basic concepts allows the user to create several projected interfaces on the floor, the wall and the ceiling (Fig. 11.8b). Each location can be suited for a different use case (e.g. scoring information to a basketball game at the ceiling) and can either be controlled by the user or by the system (*adaptability*). The interaction with *smarTVision* is done either using the smartphone application (e.g. share player information to a basketball game on the floor) or via touch (e.g. scroll through different basketball players at the table). These should explore the two categories of *in reach* and *out of reach* projected user interfaces.

11.5.1 Implementation

To study the *smarTVision* concept we designed and implemented a prototype system. The hardware was attached to a stage lighting rig mounted on two tripods (Fig. 11.8a). The rig itself was positioned above a touch and a couch table. Three BenQ W1080ST projectors were mounted on the rig to be able to project onto the space from the couch up to the wall. A fourth projector placed below

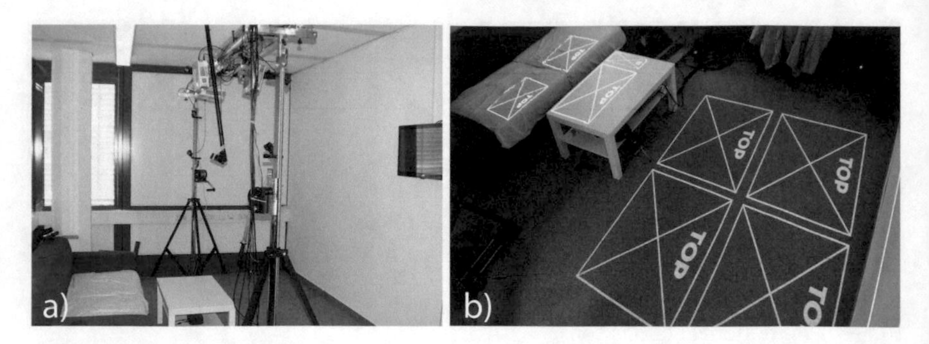

Fig. 11.8 The prototype hardware setup: a traverse mounted on two tripods spans across the room, holding a depth camera and projectors (**a**). The projected display space of the prototype allows us to create several surfaces (**b**)

the couch table created the projection onto the ceiling. This setup allowed for rendering any visual content on this continuous display space. The interaction was implemented using a Microsoft Kinect attached above the table and the couch. Using the UbiDisplays toolkit of Hardy et al. [7] allowed us to create a touch sensitive projection on the couch and on the couch table. In addition to the touch interaction, a LeapMotion placed on the couch allowed for controlling *out of reach* projections using gestural interaction (e.g. swipe through content). To manage complex applications *smarTVision* used a central Node.js server for the coordination between the internal application logic.

To illustrate and research the benefits of *smarTVision* we implemented several demo applications. In this section we will focus on four of these applications, namely *second screen manager*, *sharing mobile phone content*, *sports play application* and *quiz application*.

Second Screen Manager The second screen manager allows the users to extend and augment the traditional setup by placing additional content in the projection space. Initially, a subset of available television channels is presented to the user on the couch table. By selecting one channel via touch input, the user can assign the position of this channel to any new location. In addition to different camera perspectives, the user can also place related content such as social media feeds. The second screen manager provides a straightforward interface for placing and managing second screens inside the projection space.

Sharing Mobile Phone Content As already mentioned by participants of the in-situ study in the first section, interaction with Projector-Camera Systems should not only create new modalities but also blend with currently used technologies such as smartphones. Therefore, we implemented the functionality to share content such as images, videos and URLs from a personal device (e.g. smartphone, tablet, laptop). The *smarTVision* mobile application allows the user to connect to the backend and share his content on any surface inside the projection space. The interaction with the content is then controlled using the personal device. This reflects the feedback of users on interaction with *out of reach* interfaces using remote controls.

Sports Play Application To evaluate the concept of *adaptability*, we implemented a content specific application which supports watching a basketball game broadcast. The main screen will show the main camera of the game, whereby the system blends in additional information such as player highlights, scores and detailed statistics. The user is still able to control the content using touch interaction on the couch table. The adaption is currently only based on the action inside the broadcast and not based on the user's emotional state. However, this could easily be added when the user's emotional state is sensed in a good manner.

Quiz Application To explore multi-user interaction we additionally implemented an application which allows the user to play along with a quiz show broadcast (Fig. 11.9). Users are provided with a projected second screen next to them on the couch. Using touch, they can select their answer to the question currently discussed in the quiz show. Corresponding to the revealing of the correct answer a user is either illuminated in green (correct) or red (wrong). This application highlights the concept of *availability* of a *Companion*-System. The system is able to project a user interface next to and even on to the user to enable input in a comfortable position.

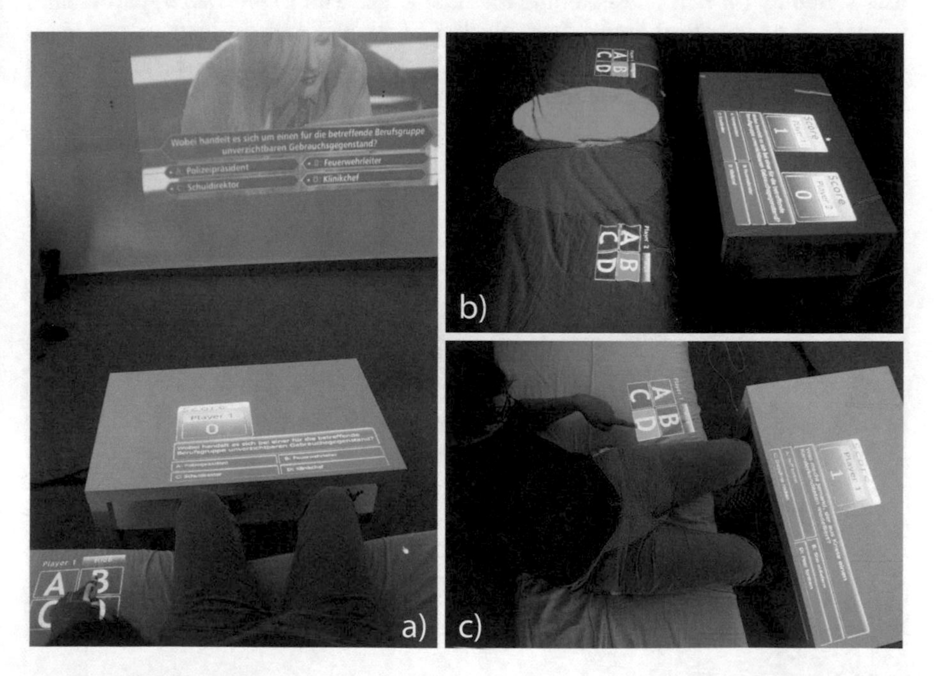

Fig. 11.9 The quiz application. Answer options can be selected via a small interface next to the user (**a**). Depending on the selection, the user is illuminated in a *red* (wrong) or *green* (correct) light (**b,c**)

11.5.2 Evaluation

We gathered qualitative feedback of users regarding the interaction and concepts of *smarTVision*. Therefore, we recruited 12 participants (7 male, 5 female). Participants were always seated in the same spot on the couch and first introduced to the general interaction concept of *smarTVision* and projected user interfaces in general. After the initial introduction, users were given specific training tasks to get familiar with the interaction of *smarTVision* and to explore the applications. After the practice participants had to fill out two questionnaires, one on specific questions regarding subjective feedback and one AttrakDiff questionnaire [9]. The study was video recorded and interactions and reactions were analyzed based on the video recordings. All participants rated *smarTVision* mildly positive in regards to the interface clarity and overview. Participants also agreed on the "good overview over the distributed second screens", showing that the interaction space itself (couch, floor, wall and ceiling) were chosen appropriately for the scenario of watching television. Regarding the readability of content, participants rated *smarTVision* more heterogeneous. Reading text on the wall and ceiling was considered a rather strenuous task. This should be considered for designing home deployed Projector-Camera Systems, so that user interfaces with a higher text density should be presented in the user's vicinity (*in reach*). Regarding the interaction with *smarTVision*, participants mentioned positively the effortless placing of second screen displays and were satisfied with their created results. The AttrakDiff questionnaire resulted in the prototype being a "rather desired" product.

11.5.3 Discussion

The majority of participants rated *smarTVision* as straightforward and easy to use. We focused on the effortless interaction so the system can blend into the user's environment and support him when necessary. This is particularly important if such a device will be deployed inside the homes of participants, so the frequency of use does not decrease over time. These design decisions were further confirmed with the positive result of the AttrakDiff questionnaire. Participants also praised the benefit of *smarTVision*, in being able to work solely without a physical remote control. This emphasizes the degree of *availability* a Projector-Camera System can offer and also the level of *adaptability*, since in certain use-cases (e.g. sharing pictures) participants preferred using a personal device such as a smartphone to interact with the interfaces.

(continued)

In this section we presented *smarTVision*, a continuous projected display space system that enables users to create any number of second screens and place them in their environment. We presented several applications which were implemented to utilize this novel interaction space. Finally, we showed the results of a preliminary user study collecting a first impression of users interacting with a Projector-Camera System combined with a television scenario.

11.6 Conclusion for *Companion*-Systems

In this chapter we presented first an in-situ user study on home deployable Projector-Camera Systems and explored the requirements such a system needs to fulfill to be accepted and used inside a user's home. Based on these insights we presented three implemented prototypes *UbiBeam*, *SpiderLight* and *smarTVision*. Each individual prototype focused on a certain interaction space and explored the particular scenario in the context of *availability* and *adaptability* of a *Companion*-System. *UbiBeam* showed how a small and portable Projector-Camera System must be implemented to conduct user studies at the participant's home. In the future we are planning to deploy the *UbiBeam* system for a longer period of time inside participants' homes and collect data on the frequency of use and on the type of use. The *SpiderLight* system explored how a highly available *Companion*-System can look and how the interaction with a portable Projector-Camera System must be designed to meet user requirements. Finally, with *smarTVision* we explored the interaction of a fixed Projector-Camera System inside a user's living room. The initial focus of the work was on building the prototype and collecting first user impressions. In the future we will focus on conducting a bigger user study and exploring the level and type of *adaptability* such a system can offer to the user. Currently all the prototypes use a simple form of *adaptability*, based on certain events. In collaboration with researchers in the fields of *adaptive planning and decision-making* and *knowledge modeling* a more sophisticated level of *adaptability* can be created. Furthermore, integrating knowledge from projects in the field of "Situation and Emotion" would result in being able to adapt the prototype not only on states of the system but also on the emotional state of the user.

Acknowledgements This work was done within the Transregional Collaborative Research Centre SFB/TRR 62 "*Companion*-Technology for Cognitive Technical Systems" funded by the German Research Foundation (DFG).

References

1. Biundo, S., Wendemuth, A.: *Companion*-technology for cognitive technical systems. Künstl. Intell. **30**(1), 71–75 (2016). Special issue on companion technologies
2. Ferreira, D., Goncalves, J., Kostakos, V., Barkhuus, L., Dey, A.K.: Contextual experience sampling of mobile application micro-usage. In: Proceedings of the 16th International Conference on Human-Computer Interaction with Mobile Devices and Services, MobileHCI '14, pp. 91–100. ACM, New York (2014). doi:10.1145/2628363.2628367. http://doi.acm.org/10.1145/2628363.2628367
3. Gugenheimer, J., Knierim, P., Seifert, J., Rukzio, E.: Ubibeam: an interactive projector-camera system for domestic deployment. In: Proceedings of the Ninth ACM International Conference on Interactive Tabletops and Surfaces, ITS '14, pp. 305–310. ACM, New York (2014). doi:10.1145/2669485.2669537. http://doi.acm.org/10.1145/2669485.2669537
4. Gugenheimer, J., Honold, F., Wolf, D., Schüssel, F., Seifert, J., Weber, M., Rukzio, E.: How companion-technology can enhance a multi-screen television experience: a test bed for adaptive multimodal interaction in domestic environments. KI-Künstl. Intell. **30**, 1–8 (2015)
5. Gugenheimer, J., Knierim, P., Winkler, C., Seifert, J., Rukzio, E.: Ubibeam: exploring the interaction space for home deployed projector-camera systems. In: Human-Computer Interaction–INTERACT 2015, pp. 350–366. Springer, Berlin (2015)
6. Hardy, J.: Reflections: a year spent with an interactive desk. Interactions **19**(6), 56–61 (2012). doi:10.1145/2377783.2377795. http://doi.acm.org/10.1145/2377783.2377795
7. Hardy, J., Alexander, J.: Toolkit support for interactive projected displays. In: Proceedings of MUM 2012, pp. 42:1–42:10. ACM, New York (2012). doi:10.1145/2406367.2406419. http://doi.acm.org/10.1145/2406367.2406419
8. Harrison, C., Benko, H., Wilson, A.D.: Omnitouch: wearable multitouch interaction everywhere. In: Proceedings of UIST 2011, pp. 441–450. ACM, New York (2011). doi:10.1145/2047196.2047255. http://doi.acm.org/10.1145/2047196.2047255
9. Hassenzahl, M.: The thing and I: understanding the relationship between user and product. In: Funology, pp. 31–42. Springer, Berlin (2005)
10. Huber, J., Steimle, J., Liao, C., Liu, Q., Mühlhäuser, M.: Lightbeam: interacting with augmented real-world objects in pico projections. In: Proceedings of MUM 2012, pp. 16:1–16:10. ACM, New York (2012). doi:10.1145/2406367.2406388. http://doi.acm.org/10.1145/2406367.2406388
11. Karitsuka, T., Sato, K.: A wearable mixed reality with an on-board projector. In: Proceedings of the 2nd IEEE/ACM International Symposium on Mixed and Augmented Reality, ISMAR '03, pp. 321. IEEE Computer Society, Washington, DC (2003). http://dl.acm.org/citation.cfm?id=946248.946820
12. Linder, N., Maes, P.: Luminar: portable robotic augmented reality interface design and prototype. In: Adjunct Proceedings UIST 2010, pp. 395–396. ACM, New York (2010). doi:10.1145/1866218.1866237. http://doi.acm.org/10.1145/1866218.1866237
13. Lumo interactive projector. http://www.lumoplay.com/ Accessed 18 April 2015
14. Mistry, P., Maes, P.: Sixthsense: a wearable gestural interface. In: ACM SIGGRAPH ASIA 2009 Sketches, SIGGRAPH ASIA '09, pp. 11:1–11:1. ACM, New York (2009). doi:10.1145/1667146.1667160. http://doi.acm.org/10.1145/1667146.1667160
15. Pinhanez, C.S.: The everywhere displays projector: a device to create ubiquitous graphical interfaces. In: Proceedings of UbiComp 2001, pp. 315–331. Springer, London (2001). http://dl.acm.org/citation.cfm?id=647987.741324
16. Raskar, R., van Baar, J., Beardsley, P., Willwacher, T., Rao, S., Forlines, C.: iLamps: geometrically aware and self-configuring projectors. In: ACM SIGGRAPH 2003 Papers, SIGGRAPH '03, pp. 809–818. ACM, New York (2003). doi:10.1145/1201775.882349. http://doi.acm.org/10.1145/1201775.882349
17. Strauss, A.L., Corbin, J.M., et al.: Basics of Qualitative Research, vol. 15. Sage, Newbury Park, CA (1990)

18. Tamaki, E., Miyaki, T., Rekimoto, J.: Brainy hand: an ear-worn hand gesture interaction device. In: CHI '09 Extended Abstracts on Human Factors in Computing Systems, CHI EA '09, pp. 4255–4260. ACM, New York (2009). doi:10.1145/1520340.1520649. http://doi.acm.org/10.1145/1520340.1520649
19. Weiser, M.: The computer for the 21st century. SIGMOBILE Mob. Comput. Commun. Rev. **3**(3), 3–11 (1999). doi:10.1145/329124.329126. http://doi.acm.org/10.1145/329124.329126
20. Wilson, A.D.: Using a depth camera as a touch sensor. In: Proceedings of ITS 2010, ITS '10, pp. 69–72. ACM, New York (2010). doi:10.1145/1936652.1936665. http://doi.acm.org/10.1145/1936652.1936665
21. Wilson, A., Benko, H., Izadi, S., Hilliges, O.: Steerable augmented reality with the Beamatron. In: Proceedings of UIST 2012, pp. 413–422. ACM, New York (2012). doi:10.1145/2380116.2380169. http://doi.acm.org/10.1145/2380116.2380169
22. Winkler, C., Reinartz, C., Nowacka, D., Rukzio, E.: Interactive phone call: synchronous remote collaboration and projected interactive surfaces. In: Proceedings of the ACM International Conference on Interactive Tabletops and Surfaces, ITS '11, pp. 61–70. ACM, New York (2011). doi:10.1145/2076354.2076367. http://doi.acm.org/10.1145/2076354.2076367
23. Winkler, C., Seifert, J., Dobbelstein, D., Rukzio, E.: Pervasive information through constant personal projection: the ambient mobile pervasive display (AMP-D). In: Proceedings of CHI 2014, pp. 4117–4126. ACM, New York (2014). doi:10.1145/2556288.2557365. http://doi.acm.org/10.1145/2556288.2557365
24. Xiao, R., Harrison, C., Hudson, S.E.: Worldkit: rapid and easy creation of ad-hoc interactive applications on everyday surfaces. In: Proceedings of CHI 2013, pp. 879–888. ACM, New York (2013). doi:10.1145/2470654.2466113. http://doi.acm.org/10.1145/2470654.2466113

Chapter 12
Interaction History in Adaptive Multimodal Interaction

Nikola Bubalo, Felix Schüssel, Frank Honold, Michael Weber, and Anke Huckauf

Abstract Modern *Companion*-Technologies provide multimodal and adaptive interaction possibilities. However, it is still unclear which user characteristics should be used in which manner to optimally support the interaction. An important aspect is that users themselves learn and adapt their behavior and preferences based on their own experiences. In other words, certain characteristics of user behavior are slowly but continuously changed and updated by the users themselves over multiple encounters with the *Companion*-Technology. Thus, a biological adaptive multimodal system observes and interacts with an electronic one, and vice versa. Consequently, such a user-centered interaction history is essential and should be integrated in the prediction of user behavior. Doing so enables the *Companion* to achieve more robust predictions of user behavior, which in turn leads to better fusion decisions and more efficient customization of the UI. We present the development of an experimental paradigm based on visual search tasks. The setup allows the induction of various user experiences as well as the testing of their effects on user behavior and preferences during multimodal interaction.

12.1 *Companion*-Systems

An integral part of the research and development of an intelligent adaptive *Companion*-System is the testing and validation of the current models and algorithms which represent and predict user behavior. To this end, an experimental setup needs to be developed, which enables such verification as well as the identification and quantification of factors relevant to human computer interaction (HCI). Given that a *Companion*-System is aimed to be a generally applicable module which improves

N. Bubalo • A. Huckauf
General Psychology, University Ulm, Albert-Einstein-Allee 47, 89081 Ulm, Germany
e-mail: nikola.bubalo@uni-ulm.de; anke.huckauf@uni-ulm.de

F. Schüssel (✉) • F. Honold • M. Weber
Institute for Media Informatics, University Ulm, James-Franck-Ring, 89081 Ulm, Germany
e-mail: felix.schuessel@uni-ulm.de; frank.honold@uni-ulm.de; michael.weber@uni-ulm.de

S. Biundo, A. Wendemuth (eds.), *Companion Technology*, Cognitive Technologies,
DOI 10.1007/978-3-319-43665-4_12

the interaction processes between the user and the system, the experimental setup needs to be an abstract representation of the majority of contemporary user interfaces and interaction patterns.

Key components for such an experimental setup are the task to be completed by the user, the multimodal user interface and interaction possibilities, the monitoring and evaluation of relevant factors of the interaction, as well as the possibility to manipulate these factors in order to create specific interaction situations.

12.1.1 The Task

The adaptive multi-modal user interfaces (MMUIs) used in corresponding research studies are, in their design, focusing on browsing through and sorting databases or websites. Examples are archives of corporate meeting data [1], web browsers with bookmark management [41], music player software [21] and interactive maps [8]. Respective scenarios can arguably be traced back to certain cognitive tasks [7], which are analogous to those in the Bolts Put-that-there setup [5]. In the Bolts experimental setup users were faced with an on-screen map on which they were able to create, delete and move geometrical shapes through speech and gesture inputs. They were then requested to perform certain operations.

These cognitive tasks of the user are in essence a visual search for a target (e.g. relevant piece of information, button to press or folder to open). In most cases, the target is placed somewhere between a set of distractors, with which it shares similarities in multiple properties, respectively [44]. This forces the user to apply a top-down search since there is no *pop out* effect. Compared to a feature search, where the target is clearly distinguishable from the distractors in one feature (e.g. a triangle among circles), such a *conjunction search* is inefficient with regard to the time it takes to find the target. Consequently, the task to be completed within the experimental setup we developed is a visual *conjunction search*.

12.1.2 Multimodal Interaction and Feedback

One major goal of research in HCI is the improvement of the interaction processes between users and technical systems such as a *Companion*-System. To this end, interfaces are enhanced with additional input and output modalities. The resulting adaptive multimodal user interfaces (MMUIs) are thus focused on mimicking interactions between humans.

One focal point in this field of study is adaptive multimodal inputs, which have already been investigated for some time [40]. Modern technologies enable more and more versatile ways for users to communicate with their devices, be it with the classic methods like mouse and keyboard or newer ways like touch, gestures, or speech. Combinations of these interaction modalities expand the range of possible interactions even further. Various reviews cover different aspects of multimodal interaction. They show that, besides a large portion of the research in this field being focused on technical aspects, like computer vision [16] and fusion engines [22], or frameworks and models [12], the majority of multimodal user interfaces combine speech inputs with a manual interaction form such as touch. Therefore the experimental setup we developed entails both speech and touch interaction, while the latter can be substituted by other interaction modalities (e.g. mouse, keyboard, gestures).

The other design factor involved is the feedback which is provided during an interaction. Basically every user interface is built with a screen to display information and most of them are accompanied by speakers for auditory backing of the information transfer to the user. Research has shown that users who are given acoustic, verbal feedback have a stronger tendency to interact through speech [3] and that multimodal feedback increases the performance of users in visual search tasks [19, 24]. Further aspects are the visual layout of the task, the instructions given to the user as well as the user's general expertise with multimodal systems [9]. Consequently, our experimental setup had to entail both visual and auditory feedback, which should be manipulable for research purposes.

12.1.3 Monitoring

One key attribute for successful *Companion*-Technology is flexibility. Given that such systems will most likely be used by multiple individuals in close succession and varying order, a *Companion*-System is required to adapt to, and accommodate, the widest range of users in the widest possible range of settings. In order to accomplish this goal, adaptive MMIs are expected to have a range of specific properties such as availability, cooperativeness and trustworthiness [4] (see Chap. 1). For example, in a possibly public interaction (e.g. ticket vending machine), the system should not force users to input necessary private information (e.g. bank details for payment) via speech [37].

In order to enable the *Companion*-System for such adaptation it is necessary for the system to collect data on the relevant context and user information through various sensory systems such as (infrared) cameras, microphones, pressure sensors and physiological sensors as well (e.g. pulse, respiration, heart rate, EEG). Correspondingly, an experimental setup is required to easily the system pair with various sensory systems in order to create multimodal data streams which are synchronized through time.

12.2 Relevant User Factors

In order for *Companion*-Systems to be capable of such adaptation it is necessary to, apart from monitoring immediate interaction factors such as the aforementioned publicity, predict the behavior of the current user as reliably as possible as well.

To this end the system needs to either collect or infer certain information about the user to be able to estimate the user's needs and demands in the current interaction. Corresponding research in the field of HCI is focused on the identification, discrimination and quantification of possible determinants of human behavior. Consequently, our testbed is required to enable the isolation, measurement, quantification and manipulation of a wide range of factors. These can be found on both sides of the human computer interaction.

Users themselves are, in essence, a cybernetic system on their own, resulting in the observation and interaction between two cybernetic systems, the human and the electronic. This dynamic process is called *meta* or *second-order* cybernetics. It acknowledges that the observer and the observed cannot be separated in terms of the resulting outcome of the interaction. Consequently, research in HCI needs a strong focus on users as well, in order to improve the system's models and predictions of their behavior.

There is still a lack of knowledge about which determinants affect multimodal interaction behavior to what extent, and how these determinants interact with each other. Furthermore, it is virtually impossible to find and test a set of participants with a homogeneous clearly defined and comparable multimodal interaction history. Depending on the social, financial and educational backgrounds, as well as gender and age, users' previous experiences with unimodal or multimodal interfaces are highly unique among them. Although a combination of speech and touch interfaces is getting more frequent in cars, these are still far from common or reliable, and therefore users of such systems are too scarce to constitute enough participants for studies of larger scale. Furthermore, there are various studies [13, 23, 32] identifying a wide range of factors affecting user behavior, which need to be addressed individually as well as in interaction with each other.

12.2.1 Modality Choice

Multimodal *Companion*-Systems enable the user to use multiple ways of communication with the system. With the resulting option of choice for the user, certain groups of users with distinct preferences are to be assumed. A corresponding adaptation of the system would thus improve user satisfaction and efficiency.

Current research on user preference shows that adaptive MMUIs are indeed preferred by users over unimodal interfaces and that they are able to improve user performance in both speed and efficiency, depending on their design [29]. This preference for MMUIs also holds true if the respective interface does not promote better performance, since users still experience multimodal interfaces as more enjoyable and effective [10, 18].

To investigate user preference, the experimental setup should support multiple combinations of interaction modalities. Furthermore, these should be either free of choice for the user, or limited by design, in order to investigate user performance and behavior under various conditions.

12.2.2 Proficiency and Context

However, the user's choice of interaction and behavior is not only influenced by the options given by the *Companion*-System. For instance, the context and proficiency of the user affect his interaction behavior. The higher the publicity of the interaction, the less are users willing to interact multimodally [17]. The proficiency of the user has the opposite effect on his multimodal interaction frequency; the more expertise the user has, the higher his use of multimodal inputs [9, 38]. Subsequently, a user's previous behavior is a strong indicator of current and future behavior with such systems [28].

Users themselves learn and change their behavior and preferences based on their own experience [11]. As previously mentioned, a biological adaptive multimodal system observes and interacts with an electronic one, and vice versa. Consequently, for an adaptive multimodal system to achieve more robust fusion decisions and to accustom the UI more efficiently to the user, it needs to monitor and predict the behavior of the current user constantly.

The domain and thus the kinds of the tasks to be solved is another of the contextual factors, if, on the one hand, users were to interact with a computer aided design (CAD) software they preferred to interact with through a combination of pen, speech and mouse, while, on the other hand, when interacting with map-based software, they omitted mouse interaction [39].

Consequently, the context and experience of the user are essential factors which are to be measured and quantified in the experimental paradigm. The testbed should therefore enable the induction and manipulation of the user's previous experience as well as the mobile application in different contexts.

12.2.3 Demographics

Given that the *Companion*-System will seldom be used by a single user, it will have to adapt to different demographic groups as well. Known factors discriminating different demographic groups are age, sex, gender, education, and cultural background. It is therefore necessary to assess the effects of these factors on human behavior in the context of HCI.

Studies have shown that the sex of the user does have an influence on the use of speech interaction [46], and that increasing cognitive load facilitates multimodal inputs [17, 31]. However, demographic factors are hard to discern from each other in effects, which is reinforced by the complexity of multimodal interfaces. Furthermore, known personal characteristics, such as gender or personal traits, usually explain less than 20% of the variance (e.g. [2, 42]), rendering them as such less useful for adaptive algorithms.

In order to remedy this issue, the experimental paradigm should bring about a strong control over the experimental situation and its factors. Through the rigorous control and manipulation of system and context factors as well as the dependant factors within the user, a clear analysis of the effects of demographic factors on user behavior in HCI is possible.

12.2.4 Workload

In designing a human computer interface one could assume that the system needs to entail and offer each and every way to communicate with the user. This might result in a multitude of suboptimal and redundant combinations which in turn could overwhelm or discomfort users. One way to solve this issue is to minimize the cognitive effort for the user by excluding the least favorable multimodal interaction patterns. Research in this area has tried to sharpen our knowledge of how cognitive load is generated and consolidated as well as how it can be quantified [6, 33, 34]. The cognitive load of HCI depends on what the user wants to achieve and the way the corresponding information is presented.

Working memory, however, is limited in its processing capacities. It can store up to seven items of new information for 20 s, after which the information needs to be refreshed or it is almost completely lost [45]. These restrictions in working memory are bypassed through schemata which are formed through repetition and stored in long-term memory. Cognitive schemata diminish the cognitive load of a complex task by combining the set of simple ideas it is comprised of into a complex and automated one. Consequently, with growing experience the cognitive load of a task is gradually diminished, which frees up space in working memory for additional tasks. An example for such automation of complex tasks is driving a car. At first one is overwhelmed by the complexity of controlling the car, but with sufficient experience enough automation takes place to enable us to do other tasks at the same time (e.g. compile or update to-do lists). Consequently, human behavior and preferences change over time with accumulating experience. This argues for individual HCI setups and opposes the previously mentioned general HCI setup.

12.2.5 User Engagement and Disposition

User behavior in general and while interacting with the *Companion*-System is influenced by his engagement and current dispositions. Given the example of a ticket vending machine, a user could approach it with different goals in mind. If, for example, the user is in a hurry to catch the next train, his focus is on speed rather than accuracy and detail. The corresponding behavior is significantly different from that of a user whose aim is to plan a trip far in the future. The former will ignore irrelevant additional information and his inputs are expected to be shorter and faster than those of the latter. Consequently, the user's engagement can be derived from his behavior. In order to quantify these changes in behavior, the experimental paradigm is to be extended by a factor which can manipulate user engagement.

Along those lines, the emotional dispositions a user can have while interacting with the *Companion*-System are important as well. Based on the experience the user previously collected with such a system or similar ones, he might have a positive or negative attitude towards the whole system or specific components thereof. Furthermore, other events which have no direct relation towards the *Companion*-System might affect the mood and therefore the behavior of the user. Lets assume a user who has a generally positive attitude towards the system, and who thus prefers speech input and audio output as interaction modalities. The system has, over multiple interaction instances, adapted to his preference. However, the current interaction was preceded by a strongly negative event for the user. This causes a change in the behavior of the user (e.g. the uses exclusively touch) which necessitates the system to adapt. Through respective sensory systems and models about the user, the *Companion*-System recognizes the respective shift in the user's mood and reacts faster and more efficiently.

However, for the system to form respective models, it is necessary to investigate the effects of the user's emotions, moods and dispositions. Consequently, our experimental setup should either be useful to effectively induce such changes in the user, or be easily extended by respective modules.

12.3 Test Bed

With the previously mentioned issues in mind we created an experimental design which enables researchers to control a vast range of factors, mix them in the intended proportions and measure their effects on user behavior. Because of its exemplary nature and its focus on the essential aspects of multimodal interaction, its results should be generalizable to more complex and adaptive MMIs. In the following, we present the experimental paradigm, demonstrate its use, report some first findings and carefully discuss its potential benefits and profits (Fig. 12.1).

As previously stated, current MMUIs in HCI virtually always require users to browse through a range of distractors to find and select a certain target object. Consequently, a visual conjunction search is the central component of our paradigm. We decided to use speech and touch inputs, which are common and frequent input modalities for MMI in HCI [8, 20, 27], and let the system respond through multimodal feedback (visual and auditory).

When it comes to multimodal interaction and user experience, it is still very difficult to acquire groups of users with comparable experience within groups and clearly distinct experiences between the experimental groups at the same time. This paradigm enables us to rectify this issue by the optional inclusion of an induction phase, and thus facilitate the investigation of previous user experience with regard to behavior with the current MMI. If, for example, users with exclusive speech experience are desired, participants can be instructed to go through a certain number of trials with the interaction modality being fixed respectively. The extent and particular properties of the past user experience can be manipulated by the

Fig. 12.1 Screen shot of a 3 × 3 matrix. In this example, the target (*red object* at position 7) had to be identified via two interactions. One interaction is used to select the color, the other is used to reference the position. Each interaction could be performed either via touch or speech. No temporal order was given in execution of these two interactions

increase or decrease of induction trials and their specific properties (e.g. difficulty). Therefore, this testbed enables the testing of clearly distinct but homogeneous user groups.

According to current user experience design rules [15], the cognitive load of the interface is kept as low as possible by only displaying the absolutely essential information and the application of gestalt laws in the design. Our testbed enables the manipulation and examination of the effects of cognitive load on user behavior and performance through task difficulty. This can be achieved and regulated by the number and complexity of distractors displayed as well as the restriction of the user's modality choice. Both measures manipulate the workload necessary to complete a task, which is supported by the notion that colors are best coded verbally while positional information is best coded spatially, which can be based on various psychological models, such as, to mention two, the multiple resources model by Wickens [47] or the dual code theory by Paivio [35]. Accordingly, if users are required to respond through a suboptimal modality, the cognitive load of the task increases.

Furthermore, the combination of clearly defined previous user experience with the concept of an ideal modality choice for said task facilitates the opportunity to explore the trade-off mechanics between those two factors. Questions such as to what extent previous user experience (e.g. exclusive speech interaction) shapes user behavior in light of the presence of an ideal modality choice (e.g. color via speech and position via touch) can be assessed.

The testbed enables the gradual manipulation and measurement of user engagement with the system. By displaying a timer the focus of the user's engagement can be directed towards speed. Gradual manipulation is then achieved through the time set for each trial as well as through the addition of acoustic feedback (e.g. a ticking sound after certain intervals of time have passed). It is possible to use the acoustic cues without a visible timer in order to investigate the effects of acoustic feedback alone. If, however, a focus on accuracy is to be induced in users to examine its effects on user behavior, a reward system can be displayed. Again, this can be fine-tuned by the number of points won with each trial, and the inclusion of punishment for wrong answers. Furthermore, the extent of that focus can be manipulated by whether the current score is depicted at all, and if so, at what points in time it is depicted (e.g. between or during each trial, or both). Consequently, very specific trade-offs between those two foci can be induced and measured in users by combining the aforementioned parameters.

In the test setting users are presented with a matrix of colored geometric shapes on a touch screen. We designed a conjunction search task along those two feature dimensions (i.e. color and shape) in order to avoid pop-up effects by preventing top-down advance cuing of the expected singleton [26]. The interactive system accepts touch and speech inputs while expecting two inputs to complete one interaction trial. These are simple geometric forms (e.g. circles, squares, triangles), which could be of red, blue or green color. Users are tasked to detect the one target object, which is a unique combination of color and shape (e.g. the single green triangle). Targets can appear at a random position in a matrix of distractors. Task difficulty can be varied by varying the number of distractors and thus the matrix size (e.g. 3×3, 4×4, 5×5). A correct answer consists of indicating the position and the color of the target irrespective of the user's applied input modality. All objects are labeled with increasing numbers for easier verbal reference.

The timer, indicating the remaining time left and the points to be won in that trial, can be shown on the side of the screen corresponding to the user's dominant hand. On the other side of the screen, three round buttons in the three aforementioned colors are depicted. These color-buttons are labeled with their corresponding color names. All objects are comfortably reachable. If not instructed otherwise, the task can be completed either exclusively by touch (touching the object and the corresponding color button), exclusively through speech (naming the number of the object and its color), or a combination of those modalities (touch object and name color or vice versa).

The modality choices, reaction times and errors of the users can be recorded. Furthermore, the order, duration and temporal relationships of individual inputs can be recorded as well. Additionally, the system records the user on video to document emotional reactions to the experiment.

12.4 User- and System-Centered Interaction History

Multimodal systems often ignore the previously described individual input behavior of users, and at the same time suffer from erroneous sensor inputs. Although many researchers have described user behavior in specific settings and tasks, little to nothing is known about the applicability of such information when it comes to increasing the robustness of a system for multimodal inputs. In the following, we describe the results of a study using the aforementioned testbed to investigate individual user behavior and the influence of previous experience on it. Furthermore, the major sensor errors that can occur in multimodal interactions are identified, and it is shown how interaction history can be applied to detect and recover from them. The remainder is a condensed version of [43].

12.4.1 Related Work and Research Questions

There are many works that have described multimodal user behavior in different applications. The most relevant are described in the following.

Oviatt et al. have initially shown that users reveal a consistent type of temporal interaction called multimodal integration patterns [30]. It is based on the temporal overlap of the on- and offsets of the two involved modalities which can be categorized as being either sequential or simultaneous in nature. Other findings suggest, that such categorization of user behavior is dependent on the involved modalities and the task at hand. In [23], speech and gesture were mostly performed in a simultaneous style. In [13] users almost always performed speech and single touch input in a sequential pattern. Other research focused on different measures of interaction behavior, e.g. in [10] onset distances of speech and touch input where investigated.

Summing up, users show different, but quite consistent behavior that highly depends on the involved modalities and tasks. It remains unclear to what extent the previous experiences of the user affect the behavior and how such user behavior should best be described in an interaction history, i.e. if input orders, onset distances, integration patterns or some other measures are the most adequate representation. Although suggested (e.g. in [32]), such information has not yet been used for making multimodal input systems more robust, other than prescribing a specific pattern for input.

There are four major questions to be answered in the presented study. (1) How does user-centered interaction history affect the behavior of the user? (2) Which representation of said user behavior should be used as interaction history, independently of the underlying task or modalities at hand? (3) What types of errors can occur during multimodal input? And finally (4), how can an interaction history be used to detect, or even resolve, error types in a multimodal input system?

12.4.2 Method

Using the described testbed, the data of 41 volunteers, 33 male and eight female, who were paid according to their success in the study, was analyzed. 88.1% of the participants were right-handed. The average age in the participant pool is 27.2 (SD = 9.18), with the youngest participant being 16 and the oldest being 63. 95.2% had prior experience with multi-touch interfaces like smartphones or tablets.

The first block of the experiment (*history* block) consisted of 90 trials, which in turn were segmented into three parts with increasing task difficulty. All tasks within the first block had to be solved with a specific modality combination (e.g. location-touch + color-speech) to build up the corresponding user-centered interaction history. The second block of the experiment (*free interaction* block) entailed 45 trials with randomized difficulty in which the participant had free choice of which modalities to use.

The following variables were extracted from each participant's data: On- and offsets of actual single inputs; event times of single input events; the modality combination of complete inputs. In addition, the data was semi-automatically labeled for the following information: system errors (i.e. wrong sensor input) and actual user intended input.

12.4.3 Results

A total of 5535 tasks was analyzed using IBM SPSS Statistics and Matlab from MathWork's. In the following, the results related to the three research questions are reported.

12.4.3.1 Representation of Typical User Behavior

Overall in the free choice block, a preference for *color speech-position touch* was observed with 51% of all choices, whereas the opposite, *color touch-position speech* was rarely chosen with 3.3% (see Fig. 12.2). This supports the previously mentioned notion that colors are best coded verbally while positional information is best coded spatially. These assumptions state that certain, mainly abstract, information is better (i.e. easier and faster) coded verbally, and other, rather concrete information, is better spatially coded. Deviations cause additional processing and thus increase cognitive load with corresponding consequences for performance. The exclusive interaction modes are comparable in frequency: *exclusive touch* 23.9% of cases, and *exclusive speech* 21.8% of cases.

Taking into account the induced user interaction history, the picture changes remarkably and varies between interaction history groups. *Exclusive touch* was used 99.2% more often when induced in advance, *color touch-position speech* 236% and *color speech-position touch* 45.5%. Exact portions are given in Fig. 12.2. That is, in all but the *exclusive speech* condition (−4.1%) users employed the induced interaction mode more frequently than the other groups of users.

Thus, the user-centered interaction history with a multimodal system has a large systematic effect on the user's choice of interaction modes and on performances. The unequal distribution of choices of interaction modes, irrespective of previous history alone, shows how important the acquisition of data about the user is. However, we did also observe a strong tendency to use *color speech-position touch*

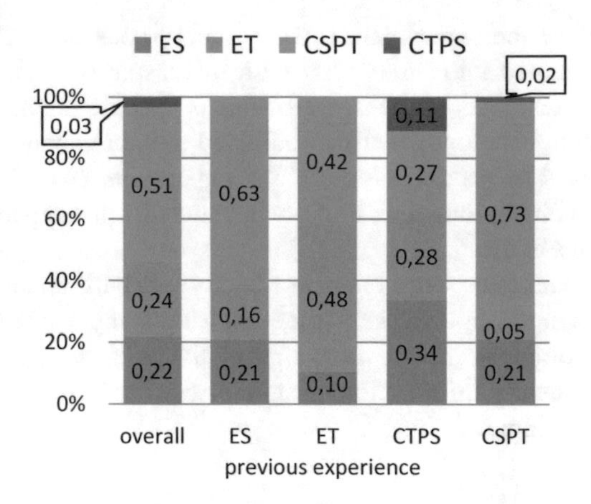

Fig. 12.2 The choice of interaction mode depending on previous experience with the system. The *first column* shows the overall distribution of choices. The *second* shows the choices of users who had experience with *exclusive speech* (ES). The *third* shows the choices of users who had experience with *exclusive touch* (ET). The *fourth* shows the choices of users who had experience with *color touch-position speech* (CTPS). The *fifth* shows the choices of users who had experience with *color speech-position touch* (CSPT)

throughout all conditions. This might be due to the gamification of the experiment which caused users to optimize effectivity and efficiency when choosing a certain interaction mode. Consequently, the inferences of these results on user behavior in non-competitive setups is limited. If users are not restricted in time or their gain is not their on it, other criteria might govern dependent behavior, which will be assessed in subsequent studies (Fig. 12.3).

One important point of interest is whether users not only enhance their effectivity and efficiency, but decrease their workload as well. In the current study, workload was investigated after the induction using the NASA TLX (Task Load Index) [14]. The workload ratings after the induction block were not significantly different among the four interaction mode groups ($F(3) < 1$). Consequently, we have to proceed with the assumption that the kinds of interaction modes used did not affect the subjective workload. Other research (e.g. [31]) has shown effects of interaction modes on subjective workloads for multimodal interactions. A reason for this might be the fact that both the induction and the free choice block contained the same ratio of the same difficulty settings.

Although related work suggests that multimodal integration patterns may be a distinctive measure for any user's multimodal interaction behavior, the experiment data shows too-low internal consistencies to be a distinctive measure. E.g. the consistency of users who provided location input via touch and color input via speech in a simultaneous or sequential manner was only about 83% opposed to well over 90% in [31, 32].

Regarding the order of inputs, data shows no clear distinctiveness either. Dividing participants into groups that predominantly start with color (60% of all inputs),

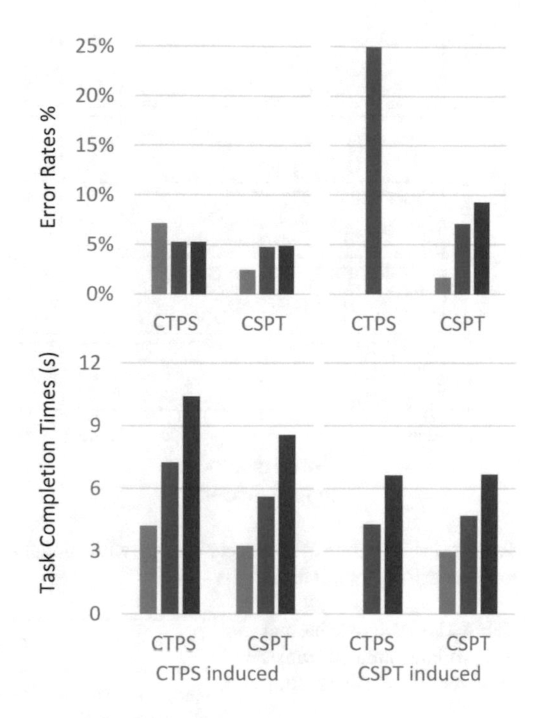

Fig. 12.3 The *first bar chart* shows the error rates of the *color touch-position speech* (CTPS) group. The *second bar chart* shows the error rates of the *color speech-position touch* (CSPT) group. The *third bar chart* shows the TCTs of the *color touch-position speech* (CTPS) group. The *fourth bar chart* shows the TCTs of the *color speech-position touch* (CSPT) group. All are separated by the three difficulty levels (easy = *green*, medium = *blue*, hard = *red*)

and those that predominantly start with location, there were 15 users predominantly starting with color and 22 with location input, while four users showed no dominant order. Overall, 24.4% of the 41 participants showed a consistency of less than 80%.

These results suggest that both concepts known so far (integration patterns and input orders) are not suited to represent typical user behavior and that a deeper analysis is required.

Individual Distributions of Different Metrics

In addition to *temporal gaps* that are used for integration patterns, there are other metrics that can be used as well. Figure 12.4 shows the histograms of the *onset distances* and the *temporal gaps* between the location and color inputs for some participants. As one can see, these distributions are highly different between users with varying means and standard deviations (not shown).

In total, 13 different variants of metrics for the interaction history data were calculated, as shown in Table 12.1. The metric called *Integration Index* is different from the other metrics, as it does not simply depend on the absolute distances between the time intervals of input modalities, but is a relative measure for the position (center) of one input related to the whole time interval of the other input. This way, it comprises more information than the other metrics.

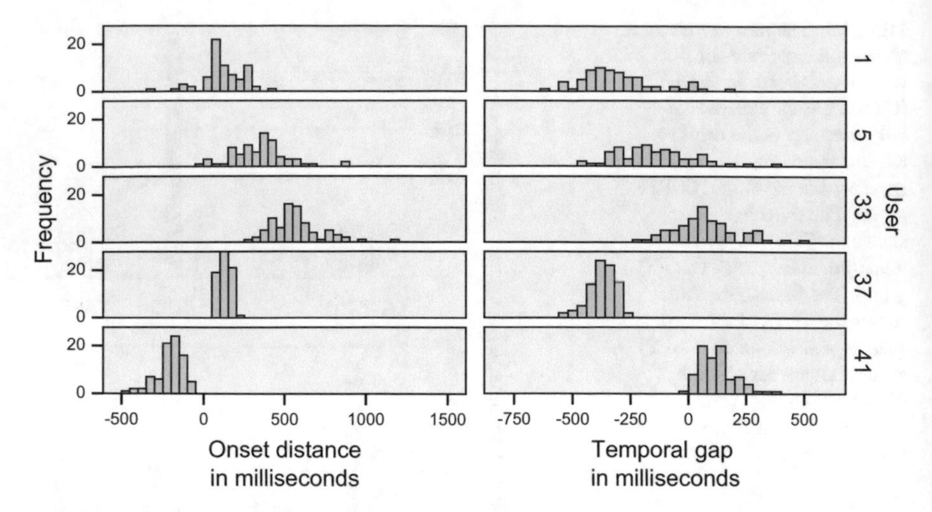

Fig. 12.4 Histograms for selected users with modality combination color-speech + location-touch reveal a user's individuality

Table 12.1 All metrics and their variants created from the onset and offset information

Metric	Illustration	13 variants, depending on
Temporal Gap (Offset -- Onset)		Input sequence Color relative Location relative
Onset distance		Input sequence Color relative
Offset distance		Input sequence Color relative
Total duration		Earliest onset and latest offset
Center distance		Input sequence Color relative
Integration index	−1 0 +1	Relative to longest input Color relative Location relative

Redundant variants are not listed (taken from [43])

In order to decide which of these metrics describes user behavior best, a metric must differentiate well between users, and at the same time stay as consistent as possible for a specific user. This information can be derived from the mean values between users and the standard deviation for each user. So a metric that shows a larger standard deviation of means and a lower mean of standard deviations should be chosen. Although there are in fact differences between the metrics, a final decision cannot be justified on the experiment data yet, as it also depends on how well a metric performs when used to detect and recover from errors of a multimodal input system (as described later).

12.4.3.2 Error Types During Multimodal Input

For the *free interaction* block, semi-automatic labelling revealed an average of 4.9% of inputs not handled as intended by the users. These errors, that all stem from some kind of erroneous sensor inputs (speech in our case), can be further categorized into the following types. The relative frequency of each error type is given in brackets.

False negatives (54.4%): A sensor did not react at all, or was too unconfident on the intended input.

False positives (10.0%): A sensor reacts upon an action of the user that was not intended as input.

Conflicts (16.7%): A special case of the false positive error type. A correctly recognized input of the user from one sensor is disturbed by a false positive from another sensor, in such a way that both sensors contradict each other.

Misunderstandings (8.9%): A sensor misinterprets an input.

Other (10.0%): This category contains unclear error types, like a combination of two or more errors from above, as well as user-initiated errors. These can be, for example, self corrections that are missed by the system or incomplete inputs by the user.

Although the frequency distribution certainly depends on the task at hand and the involved modalities, these types cover all relevant sensory errors that can occur in any multimodal interaction. The question remains about how an interaction history can be used to prevent or at least recover from such errors.

12.4.3.3 Using Interaction History for Error Detection and Recovery

When the typical user behavior in terms of temporal relation is known for a specific task and modality combination, it can be used to reveal sensor errors at different stages of input processing. Figure 12.5 depicts the complete process of detecting and recovering from errors for two input modalities that can easily be applied for an arbitrary number of inputs. The process can briefly be described as follows.

Whenever a potential *1st single input* (e.g. a location event) of a multimodal input occurs, the process starts by trying to detect a false negative. This is accomplished by calculating the maximum event distance of every possible multimodal input combination from the interaction history. When no further input is recognized up to that maximum time, a false negative is assumed. To recover from such an error, all sensors are requested to re-process their input data for the time in question with increased sensitivity. Using such a *back-channelling* approach, it may be possible to detect the missed input.

When the false negative is resolved or a *2nd single input* event occurs, the fusion of both single inputs may reveal a conflict error. So this error can be detected automatically by an input fusion system that is capable of doing so (e.g. the one

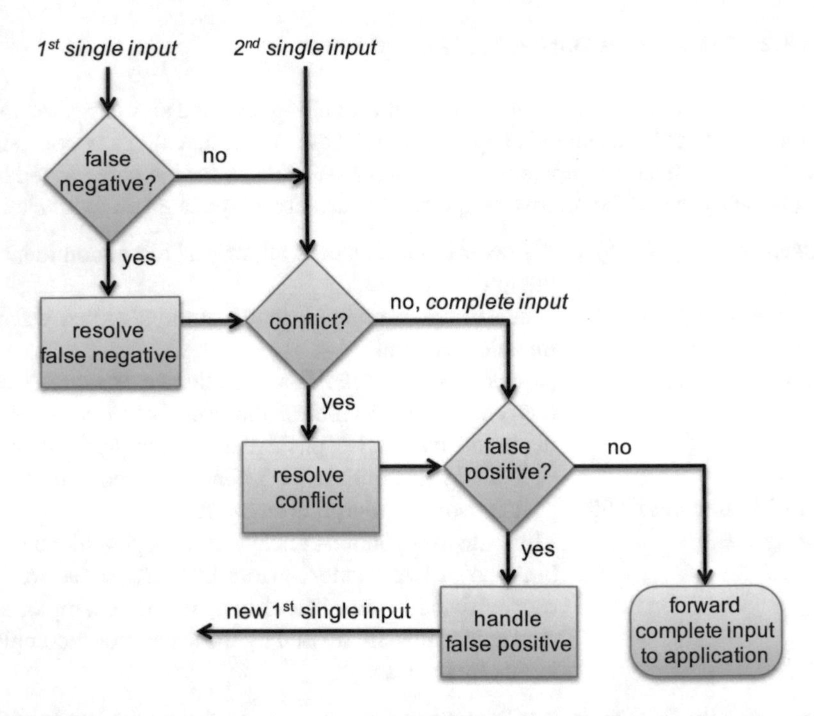

Fig. 12.5 Error detection and recovery conceptualized as a cascading process for two combined input modalities (taken from [43])

described in Chap. 10). To resolve a conflict, one needs to find out which of the conflicting inputs is the most likely one. This can be done by calculating a relative conformity with the interaction history as described in [43] and selecting the most appropriate one as *true* input while discarding the other.

When a solution is found, or if there was no conflict at all, the fused *complete input* (e.g. color: blue + location: 3) is tested for being a false positive. For this, *min* and *max* values of the metrics stored in the interaction history are compared to the current inputs. If a non-typical temporal relation that lies outside these boundaries is detected, the single input with the lower confidence value is assumed to be a false positive and is discarded. This results in restarting the complete process with the most confident input as the new 1st single input. When there seems to be no false positive, the complete input is assumed valid and forwarded to the application or dialog management.

Applying this process to the experiment data leads to a reduction of errors from initially 4.9% to a minimum of 1.2%, assuming complete success of the handling strategies for false negatives and false positives. Those could not be evaluated on the recorded data, which is why 7.2% of interactions cannot be regarded as false or correct with absolute certainty. Note that misunderstandings are the only error type that can be neither detected nor recovered using a temporal interaction history as described above.

12.4.4 Conclusion

The study shows that previous user experience does have a strong systematic effect on user behavior and performance. For example, the induced interaction experiences in this study and its effect on modality choice does interact with the tendency to use the optimal modality to solve the visual search tasks [25, 36].

The multimodal interaction data of the conducted experiment, as well as findings from other researchers (cf. Sect. 12.4.1), support the notion that integration patterns as defined by Oviatt et al. with their categorization into simultaneous and sequential users are oversimplified to be suitable as a representation of interaction history, at least for the purpose of detecting and recovering from sensor errors. The same holds true for other simple classifications based on the order of inputs.

As individual differences between users are so meaningful, we suggest using actual distributions of temporal metrics. While there is not enough data to commit to a single metric yet, the presented metrics clearly show individual differences between users (variances of means between participants are high), while retaining meaningful coherence per user (means of variances are low).

From the identified sensor error types of multimodal input systems, three out of four can be identified and even be recovered using the stepwise process described on p. 245. Applied to the experimental data, promising results in error rate reduction of up to 75% can be anticipated; but the presented recovering strategies still need further research. Nevertheless, the idea of adapting input recognition to the individual history of a user as described here also is not only valuable for *Companion*-Systems, but is also supposed to be applicable to any multimodal input system.

12.5 Outlook

As demonstrated, the testbed enables the fixing and gradual manipulation of a wide range of user- and system-centered determinants of the interaction. With its lean and flexible design, it is easily incorporated into different experimental setups, facilitating sound and easy comparison and combination of study results in various fields of study concerning *Companion*-Systems, such as the following.

12.5.1 Planing and Decision Making

A key attribute of *Companion*-Systems is the ability to construct and maintain declarative knowledge bases of the user and the situation. These are used by adaptive planning, reasoning and decision making processes in order to detect and prevent suboptimal trends in the course of the interaction. The corresponding databases, models and algorithms are then to be tested and evaluated within clearly defined laboratory conditions. The experimental paradigm in this chapter provides the necessary control over the conditions of a human computer interaction while maintaining the necessary ecological validity. With said control, specific and rare situations between the user and the *Companion*-System can be simulated in order to assess the performance of the models and algorithms.

12.5.2 Interaction and Accessibility

The use of multimodal input and output channels opens up new possibilities for *Companion*-Systems. Consequently, an effective dialogue management system is necessary which is focused on multimodal and iterative information seeking, given that the user is usually not able to clearly define the desired information. The conjunction search within our testbed with its flexible difficulty settings simulates a corresponding information search situation in which the user initially is not aware of what exactly he is looking for, and is thus a valuable testbed for corresponding dialogue managing modules.

With its lean and clean design, this testbed can be applied in the development and testing of new mobile user interfaces which require or induce certain multimodal user behavior which needs to be examined and accounted for in the development of corresponding fusion and fission algorithms.

12.5.3 Situation and Emotion

Emotion elicitation with this setup can be done in various ways and is in turn very useful for the recognition of dispositions through spoken language and paralinguistic features. The effective elicitation of emotions in this paradigm can facilitate research on vision-based emotion recognition if paired with a camera facing the user during his interaction with the system. Due to the easily manipulable extent of the interaction, this paradigm allows for the assessment and combination of both static and dynamic features. This modular extensibility of the experimental setup allows additional measurements of psycho-biological parameters, such as skin conductance rates (SCRs), heart rate variability (HRV), or respiration rates. This, combined with the emotion eliciting features of the paradigm, enables an in-depth assessment and recognition of emotions. By combining multiple sensory systems with our experimental paradigm it is possible to collect respective multimodal data on the user's emotions and dispositions. This spatio-temporally fused data can in turn be used to train corresponding classification algorithms in *Companion*-Systems.

Acknowledgements This work was done within the Transregional Collaborative Research Centre SFB/TRR 62 "*Companion*-Technology for Cognitive Technical Systems" funded by the German Research Foundation (DFG).

References

1. Ailomaa, M., Melichar, M., Rajman, M., Lisowska, A., Armstrong, S.: Archivus: a multimodal system for multimedia meeting browsing and retrieval. In: Proceedings of the COLING/ACL on Interactive Presentation Sessions, pp. 49–52. Association for Computational Linguistics, Stroudsburg, PA (2006)
2. Ajzen, I.: Attitudes, Personality, and Behavior. McGraw-Hill International, Maidenhead (2005)
3. Bellik, Y., Rebaï, I., Machrouh, E., Barzaj, Y., Jacquet, C., Pruvost, G., Sansonnet, J.P.: Multimodal interaction within ambient environments: an exploratory study. In: Human-Computer Interaction–INTERACT 2009, pp. 89–92. Springer, Berlin (2009)
4. Biundo, S., Wendemuth, A.: Companion-technology for cognitive technical systems. Künstl. Intell. **30**(1), 71–75 (2016). Special issue on companion technologies
5. Bolt, R.A.: "Put-That-There": Voice and Gesture at the Graphics Interface, vol. 14. ACM, New York (1980)
6. Camp, G., Paas, F., Rikers, R., van Merrienboer, J.: Dynamic problem selection in air traffic control training: a comparison between performance, mental effort and mental efficiency. Comput. Hum. Behav. **17**(5), 575–595 (2001)
7. Carter, S., Mankoff, J., Klemmer, S.R., Matthews, T.: Exiting the cleanroom: on ecological validity and ubiquitous computing. Hum. Comput. Interact. **23**(1), 47–99 (2008)
8. Cohen, P.R., Johnston, M., McGee, D., Oviatt, S., Pittman, J., Smith, I., Chen, L., Clow, J.: Quickset: multimodal interaction for distributed applications. In: Proceedings of the Fifth ACM International Conference on Multimedia, pp. 31–40. ACM, New York (1997)
9. De Angeli, A., Gerbino, W., Cassano, G., Petrelli, D.: Visual display, pointing, and natural language: the power of multimodal interaction. In: Proceedings of the Working Conference on Advanced Visual Interfaces, pp. 164–173. ACM, New York (1998)
10. Dey, P., Madhvanath, S., Ranjan, A., Das, S.: An exploration of gesture-speech multimodal patterns for touch interfaces. In: Proceedings of the 3rd International Conference on Human Computer Interaction, pp. 79–83. ACM, New York (2011)
11. Domjan, M.: The principles of learning and behavior. Cengage Learning, Stamford, CT (2014)
12. Dumas, B., Lalanne, D., Oviatt, S.: Multimodal interfaces: a survey of principles, models and frameworks. In: Lalanne, D., Kohlas, J. (eds.) Human Machine Interaction. Lecture Notes in Computer Science, vol. 5440, pp. 3–26. Springer, Berlin, Heidelberg (2009)
13. Haas, E.C., Pillalamarri, K.S., Stachowiak, C.C., McCullough, G.: Temporal binding of multimodal controls for dynamic map displays: a systems approach. In: Proceedings of the 13th International Conference on Multimodal Interfaces, pp. 409–416. ACM, New York (2011)
14. Hart, S.G., Staveland, L.E.: Development of NASA-TLX (task load index): results of empirical and theoretical research. Adv. Psychol. **52**, 139–183 (1988)
15. Hollender, N., Hofmann, C., Deneke, M., Schmitz, B.: Integrating cognitive load theory and concepts of human–computer interaction. Comput. Hum. Behav. **26**(6), 1278–1288 (2010)
16. Jaimes, A., Sebe, N.: Multimodal human–computer interaction: a survey. Comput. Vis. Image Underst. **108**(1), 116–134 (2007)
17. Jöst, M., Häußler, J., Merdes, M., Malaka, R.: Multimodal interaction for pedestrians: an evaluation study. In: Proceedings of the 10th International Conference on Intelligent User Interfaces, pp. 59–66. ACM, New York (2005)
18. Käster, T., Pfeiffer, M., Bauckhage, C.: Combining speech and haptics for intuitive and efficient navigation through image databases. In: Proceedings of the 5th International Conference on Multimodal Interfaces, pp. 180–187. ACM, New York (2003)
19. Kieffer, S., Carbonell, N.: How really effective are multimodal hints in enhancing visual target spotting? Some evidence from a usability study. J. Multimodal User Interfaces **1**(1), 1–5 (2007)
20. Koons, D.B., Sparrell, C.J., Thorisson, K.R.: Integrating Simultaneous Input from Speech, Gaze, and Hand Gestures. MIT, Menlo Park, CA, pp. 257–276 (1993)

21. Kruijff-Korbayová, I., Blaylock, N., Gerstenberger, C., Rieser, V., Becker, T., Kaisser, M., Poller, P., Schehl, J.: An experiment setup for collecting data for adaptive output planning in a multimodal dialogue system. In: Proceedings of ENLG (2005)
22. Lalanne, D., Nigay, L., Robinson, P., Vanderdonckt, J., Ladry, J.F., et al.: Fusion engines for multimodal input: a survey. In: Proceedings of the 2009 International Conference on Multimodal Interfaces, pp. 153–160. ACM, New York (2009)
23. Lee, M., Billinghurst, M.: A wizard of oz study for an ar multimodal interface. In: Proceedings of the 10th International Conference on Multimodal Interfaces, pp. 249–256. ACM, New York (2008)
24. Lee, J.H., Poliakoff, E., Spence, C.: The effect of multimodal feedback presented via a touch screen on the performance of older adults. In: Haptic and Audio Interaction Design, pp. 128–135. Springer, Berlin (2009)
25. Mignot, C., Valot, C., Carbonell, N.: An experimental study of future "natural" multimodal human-computer interaction. In: INTERACT'93 and CHI'93 Conference Companion on Human Factors in Computing Systems, pp. 67–68. ACM, New York (1993)
26. Müller, H.J., Krummenacher, J.: Visual search and selective attention. Vis. Cogn. **14**(4–8), 389–410 (2006)
27. Neal, J.G., Shapiro, S.C.: Intelligent multi-media interface technology. ACM SIGCHI Bull. **20**(1), 75–76 (1988)
28. Ouellette, J.A., Wood, W.: Habit and intention in everyday life: the multiple processes by which past behavior predicts future behavior. Psychol. Bull. **124**(1), 54 (1998)
29. Oviatt, S.: Multimodal interactive maps: designing for human performance. Hum. Comput. Interact. **12**(1), 93–129 (1997)
30. Oviatt, S., Cohen, P., Wu, L., Vergo, J., Duncan, L., Suhm, B., Bers, J., Holzman, T., Winograd, T., Landay, J., Larson, J., Ferro, D.: Designing the user interface for multimodal speech and pen-based gesture applications: state-of-the-art systems and future research directions. Hum. Comput. Interact. **15**(4), 263–322 (2000)
31. Oviatt, S., Coulston, R., Lunsford, R.: When do we interact multimodally?: Cognitive load and multimodal communication patterns. In: Proceedings of the 6th International Conference on Multimodal Interfaces, pp. 129–136. ACM, New York (2004)
32. Oviatt, S., Lunsford, R., Coulston, R.: Individual differences in multimodal integration patterns: what are they and why do they exist? In: Proceedings of the SIGCHI Conference on Human Factors in Computing Systems, pp. 241–249. ACM, New York (2005)
33. Oviatt, S., Arthur, A., Cohen, J.: Quiet interfaces that help students think. In: Proceedings of the 19th Annual ACM Symposium on User Interface Software and Technology, pp. 191–200. ACM, New York (2006)
34. Paas, F., Tuovinen, J.E., Tabbers, H., Van Gerven, P.W.: Cognitive load measurement as a means to advance cognitive load theory. Educ. Psychol. **38**(1), 63–71 (2003)
35. Paivio, A.: Mental Representations: A Dual Coding Approach. Oxford University Press, Oxford (1990)
36. Ratzka, A.: Explorative studies on multimodal interaction in a PDA- and desktop-based scenario. In: Proceedings of the 10th International Conference on Multimodal Interfaces, pp. 121–128. ACM, New York (2008)
37. Reeves, L.M., Lai, J., Larson, J.A., Oviatt, S., Balaji, T.S., Buisine, S., Collings, P., Cohen, P., Kraal, B., Martin, J.C., McTear, M., Raman, T., Stanney, K.M., Su, H., Wang, Q.Y.: Guidelines for multimodal user interface design. Commun. ACM **47**(1), 57–59 (2004)
38. Reis, T., de Sá, M., Carriço, L.: Multimodal interaction: real context studies on mobile digital artefacts. In: Haptic and Audio Interaction Design, pp. 60–69. Springer, Berlin (2008)
39. Ren, X., Zhang, G., Dai, G.: An experimental study of input modes for multimodal human-computer interaction. In: Advances in Multimodal Interfaces-ICMI 2000, pp. 49–56. Springer, Berlin (2000)
40. Ruiz, N., Chen, F., Oviatt, S.: Multimodal input. In: Multimodal Signal Processing: Theory and Applications for Human-Computer Interaction, p. 231. Academic, Boston (2009)

41. Savidis, A., Stephanidis, C.: Unified user interface design: designing universally accessible interactions. Interact. Comput. **16**(2), 243–270 (2004)
42. Schüssel, F., Honold, F., Weber, M.: Influencing factors on multimodal interaction during selection tasks. J. Multimodal User Interfaces **7**(4), 299–310 (2013)
43. Schüssel, F., Honold, F., Weber, M., Schmidt, M., Bubalo, N., Huckauf, A.: Multimodal interaction history and its use in error detection and recovery. In: Proceedings of the 16th ACM International Conference on Multimodal Interaction, ICMI '14, pp. 164–171. ACM, New York (2014). doi:10.1145/2663204.2663255
44. Treisman, A.M., Gelade, G.: A feature-integration theory of attention. Cogn. Psychol. **12**(1), 97–136 (1980)
45. Van Merrienboer, J.J., Sweller, J.: Cognitive load theory and complex learning: recent developments and future directions. Educ. Psychol. Rev. **17**(2), 147–177 (2005)
46. Wasinger, R., Krüger, A.: Modality preferences in mobile and instrumented environments. In: Proceedings of the 11th International Conference on Intelligent User Interfaces, pp. 336–338. ACM, New York (2006)
47. Wickens, C.D.: Multiple resources and mental workload. Hum. Factors J. Hum. Factors Ergon. Soc. **50**(3), 449–455 (2008)

Chapter 13
LAST MINUTE: An Empirical Experiment in User-*Companion* Interaction and Its Evaluation

Jörg Frommer, Dietmar Rösner, Rico Andrich, Rafael Friesen, Stephan Günther, Matthias Haase, and Julia Krüger

Abstract The LAST MINUTE Corpus (LMC) is a unique resource for research on issues of *Companion*-technology. LMC not only comprises 57.5 h of multimodal recordings (audio, video, psycho-biological data) from interactions between users—133 subjects in sum, balanced in age and gender—and a WoZ-simulated speech-based interactive dialogue system. LMC also includes full verbatim transcripts of all these dialogues, sociodemographic and psychometric data of all subjects as well as material from 73 in-depth user interviews focusing the user's individual experience of the interaction. In this chapter the experimental design and data collection of the LMC are shortly introduced. On this basis, exemplifying results from semantic analyses of the dialogue transcripts as well as from qualitative analyses of the interview material are presented. These illustrate LMC's potential for investigations from numerous research perspectives.

13.1 Introduction

In order to examine the system side as well as the user side of user-*Companion* interaction (UCI), naturalistic data of such interactions is indispensable. The LAST MINUTE corpus (LMC) includes approximately 57.5 h of such data. It is based on a widely standardized Wizard of Oz (WoZ) experiment in which the participants interacted with a simulated speech-based interactive dialog system [5, 29]. This system represented the main characteristics of future *Companion*-Systems, e.g. by pretending to be an individualized assistant in everyday life. The experimental modules simulated the following interaction situations: the collection

J. Frommer (✉) • M. Haase • J. Krüger
Universitätsklinik für Psychosomatische Medizin und Psychotherapie, Otto-von-Guericke-Universität, Magdeburg, Germany
e-mail: joerg.frommer@med.ovgu.de; matthias.haase@med.ovgu.de; julia.krueger@med.ovgu.de

D. Rösner • R. Andrich • R. Friesen • S. Günther
Institut für Wissens- und Sprachverarbeitung, Otto-von-Guericke Universität, Magdeburg, Germany
e-mail: roesner@ovgu.de; andrich@ovgu.de; friesen@ovgu.de; stguenth@ovgu.de

S. Biundo, A. Wendemuth (eds.), *Companion Technology*, Cognitive Technologies,
DOI 10.1007/978-3-319-43665-4_13

of personal as well as private user information for the purpose of personalization and individualization, and a mundane planning situation with the need for re-planning because of the accumulation of various planning constraints—a situation in which half of all participants got an emphatic system-initiated intervention. The following research questions led the development of the experiment and the design of the simulated system:

- What are the semantic markers in user utterances that enable us to identify negative dialog courses and the risk of a decrease in cooperativeness or even a communication break-up?
- How do the participants individually experience the interaction with the system, what do they ascribe to the system in order to develop their individual view on it and what emotions occur during the interaction?
- Which types of users can be differentiated by the semantic markers in participants' utterances, the sociodemographic as well as psychological characteristics of the participants and the subjective experiences of the system and the interaction with it?

The LMC provides multimodal recordings of all interactions including audio, video and psycho-biological data. These recordings are supplemented by transcripts of all interactions, data from well-established psychological instruments and material from in-depth user interviews focusing on the subjective experience of the experiment. The variety of recordings and supplementary material marks one of the ways the LMC goes beyond other naturalistic corpora [31]. This chapter is structured as follows: Initially, the LMC is introduced by a short overview on experimental design and data collection. On this basis, exemplifying insights from analyses of the collected data are presented—on the one hand with regard to semantic analyses of the interactions, and on the other hand with regard to qualitative analysis of the in-depth user interviews.

13.2 LAST MINUTE Experiment and Its Realization

The LAST MINUTE corpus contains multimodal recordings from a WoZ experiment that allows us to investigate how users interact with a *Companion*-System in a mundane situation with the need for planning, re-planning and strategy change. The design of this experiment is described in [27].

Before the interaction the subjects are instructed that they will interact with a speech-driven system and that they can begin the interaction by saying that they want to start. All system output is pronounced via a text-to-speech system (TTS).

The overall structure of an experiment is divided into (1) a personalization module, followed by (2) the LAST MINUTE module. These modules serve quite different purposes and are further substructured in a different manner (for more details, cf. [29]).

In the personalization module the system welcomes the subject, gives a short self-description and prompts the subject to tell and spell his name and (further) introduce himself, prompts for missing information and then gives a summary of the collected information. The system then stipulates user narratives on recent positive and negative events, his hobbies and his experiences with technical devices.

In the bulk of LAST MINUTE—the problem solving phase—the subject is expected to pack a suitcase for a 2-week holiday trip by choosing items from an online catalogue with 12 different categories that are presented in a fixed order. The options of each category are given as a menu with icons on the subject's screen.

In a normal packing subdialog a user requests articles from the current selection menu (e.g. 'ten t-shirts') followed by a confirmation of the system (e.g. 'ten t-shirts have been added'). For the whole packing a total of 15 min are allocated.

There are two ways to finish the current category and to proceed to the next one: (1) the user explicitly asks for a change of the category (in the following *SICC* for subject-induced category change) or (2) the system changes the category because a time limit is reached (*WICC* for wizard-induced category change). In the latter case the system informs the user that the selection from the current category has to end and that the following category is now available.

The normal course of a sequence of repetitive subdialogs is modified for all subjects at specific time points. These modifications or barriers are:

- after the sixth category, the current contents of the suitcase are listed verbally (*listing* barrier),
- during the eighth category, the weight limit for the suitcase is reached (*weight limit* barrier, WLB).
- at the end of the tenth category, the subject is informed that he has to pack the suitcase for cold weather in Waiuku instead of summer weather (*weather info* barrier, WIB).

Additional difficulties for the subjects may occur depending on the course of the dialog. These are typically caused by user errors or limitations of the system or a combination of both.

Nearly half of the subjects—randomly chosen—get an empathic intervention designed according to the principles of Rogerian psychotherapy [26] after the weather info barrier.

After the optional intervention or, when no intervention was given, immediately after the weather info barrier, the subjects finish packing the suitcase. We distinguish between two types of endings of the selection phase: (1) the user ends the selection on his own (*early enders*) or (2) the system ends the selection due to the global time limit reached. From a total of $N = 133$ subjects, $n_1 = 41$ subjects (with $n_2 = 20$ from the elderly and $n_3 = 21$ from the young group) ended the selection on their own; in the other $n_4 = 92$ cases the system closed the session and blocked additional user input.

After the end of the selection phase the system prompts the user to rate the outcome of the packing and reveal his plans for the holiday trip.

The system closes the session, thanks the user for his cooperation and says goodbye. Many users answer with (variants of) goodbye as well.

13.3 LAST MINUTE Corpus

13.3.1 Sample

User characteristics have an important influence on human computer interaction. Currently, age and gender are most debated sociodemographic data in HCI (e.g. [14, 23, 35]). A lot of studies evaluate the effect of gender on performance with technical devices, usage behavior, and experience of self-competence as well as the amount of anxiety while dealing with those devices. Furthermore, there is a vast number of studies regarding age. Headwords are the effect of social media usage on child development or possibilities to improve and integrate the handling of technical devices for elder users. It is noticeable that age cohorts are analyzed mostly exclusively from each other, which considerably complicates an actual comparison regarding usage and interaction behavior or performance while using technical devices. In order to take this into account we differentiated between participants at the age of 18–28 years and participants over 60 years. We minded equal distribution of gender and the level of education. We distinguished between higher educational level (general matriculation standard, studies at a university or a university of applied sciences) and lower educational level (secondary school or secondary modern school certificate, apprenticeship as highest educational/occupational qualification). We additionally evaluated sociodemographic data like profession, experience and skills using technical devices (especially computer systems). After finishing the 'LAST MINUTE' module, we employed the AttrakDiff [8] questionnaire to evaluate the hedonic and pragmatic quality of the simulated computer system. Participants filled out further psychometric questionnaires on personality factors [20], interpersonal difficulties [10], coping strategies [18], technology affinity [12] and attributional style [24] during a separate appointment. On average, participants needed 90 min for this process. Participants were recruited via advertisement in local newspapers and bulletins in vocational schools and universities as well as through sporting or recreational associations. Altogether we recruited 135 participants. In two experiments, there were technical problems during recording. In addition, three subjects did not complete the psychometric questionnaires.

Thus, the total sample consists of $N_{total} = 130$ subjects of which one could not be assigned properly to one of the educational levels (see Table 13.1).

Table 13.1 Sample distribution with regard to age, gender and educational level

	Male	Female	Total
Age-group 18–29 years			
Higher educational level	22	23	45
Lower educational level	13	12	25
Age-group over 60 years			
Higher educational level	14	13	27
Lower educational level	14	18	32
Total	63	66	129

13.3.2 Recorded Data

The LAST MINUTE corpus comprises multimodal recordings (audio, high-resolution video from multiple perspectives, psychobiological indicators such as skin reductance, heartbeat, and respiration) and verbatim transcripts of all completed experiments (N = 133). A complete WoZ session takes approx. 30 min. The total lengths of sessions vary from 19 to 39 min. The technical setup and used hardware (cf. Figs. 13.1, 13.2, and 13.3) are described in detail in [28].

GAT 2 Minimal Standard The recordings of the experiments and of the interviews were transcribed according to the GAT 2 minimal standard [33]. Besides the textualization of utterances, the standard also considers pauses, breathing, and overlaps (simultaneously spoken utterances) and allows for descriptive annotations.

Fig. 13.1 Operator room. The functionality of the WoZ system was controlled by the *wizards* using the *right screens*. The recordings were coordinated by the operators using the *left screens*

Fig. 13.2 The hardware setting in the subject room. C=High resolution camera, H=Heart beat clip, M=Microphone, R=Respiration belt, S=Skin conductance clip, T=Stereo camera, W=Observation webcam. Not in the picture: Headwear microphone

Fig. 13.3 Subject room total view. The subjects sat in a comfortable chair in front of the subject screen in a room furnished like a living room

In the transcription process, we tried to come as close as possible to the actual wording and pronunciation of subjects. The transcripts, therefore, include typical phenomena of spontaneous spoken speech (e.g. discontinuities, repairs, restarts, incongruencies) as well as non-standard spellings (e.g. for dialect).

Quality Assurance The transcription was executed by trained personnel using the transcription software FOLKER [32]. Our own software was employed to support the transcribers in detecting and correcting possible misspellings and other defects. In addition, each transcript was examined multiple times by different transcribers.

13.3.3 Annotations

Dialog Act Representation (DAR) As reported in [30] the packing phase of the LAST MINUTE corpus was annotated with dialog acts. The utterances of the subjects show a large variance with respect to many linguistic features, so all dialog acts from the packing phase of each subject were annotated by a human rater. The annotation process was computer-assisted: raters verified automatic annotation suggestions and annotated remaining non-annotated segments. Most of the Wizard contributions were preformulated and could be annotated automatically by using regular expressions.

The DAR annotations are hierarchical. The first capital letter indicates the speaker (S for subject, W for wizard) while the second capital letter denotes the dialog act (x acts as a placeholder variable):

Sx: R = request, O = offtalk, Np = non-phonological and pauses, A = answer
Wx: A = accept, Rj = reject, I = information, Q = question/request

A third capital letter may refine a dialog act subtype:

SRx: P = packing, U = unpacking, E = exchange, F = finalization, L = listing, C = category change
SOx: T = general offtalk, Q = question
WAx: P = packing, U = unpacking, F = finalization
WRjx: P = packing, U = unpacking, Np = non-processable
WIx: C = category change, W = weather, F = finalization, L = listing
WQx: F = finalization, I = intervention, C = comment, E = elaboration

The lowercase letters (as in 'Rj') are inserted for readability.

Dialog Success Measures (DSMs) User requests may be either accepted and confirmed by the system or they may be rejected. The relation between subject requests and their acceptance or rejection allows us to define measures for the global dialog success (or failure) in the problem solving phase of LAST MINUTE [30]:

- first approach: ratio between the accepted subject requests and the total number of subject requests (termed **DSM1**)
- refinement: ratio between the accepted subject requests and the total number of turns (i.e. not only subject requests) in problem solving (termed **DSM2**)

Thus for all subjects the following must hold: $0 \leq DSM2 \leq DSM1$.
These measures are further addressed in Sect. 13.4.1 and Chap. 14.

LAST MINUTE Workbench The DAR is part of the LAST MINUTE workbench—a collection of tools for the exploration of the LAST MINUTE corpus. There are several visual representations of the DAR optimized for different tasks such as overview, search for patterns or calculation of statistics and other measures. The DAR and the LAST MINUTE workbench are described in more detail in [30].

An Example The DAR example in Table 13.2 is taken from a dialog segment where a subject (20110401adh) tries to pack a (winter) coat but the packing attempt is rejected several times (SRP WRjP pairs) and therefore the subject has to unpack several other items (SRU WAU) in order to create sufficient space. Please note the emotional expression of relief ('gott sei dank', Eng. 'thank god') when the subject finally succeeds.

Wizard Problems: Errors and Inconsistencies On the one hand the LMC experiment is a carefully designed and highly standardized experiment, based on a detailed manual [5], performed by intensively trained personnel (wizards) with elaborate computer support [27, 29].

Table 13.2 Transcript example

Tag	German text	English gloss
	...	...
SRP	ein mantel	A coat
WRjP	der artikel mantel kann nicht hinzugefügt werden (.) anderenfalls würde die maximale gewichtsgrenze des koffers überschritten werden	The item coat cannot be added (.) otherwise the weight limit of your suitcase will be exceeded
SNp	((raschelt)) ((schmatzt))	Rustles, smacks
SNp	(-)	(-)
SRU	ein buch raus	One book out
WAU	ein buch wurde entfernt	A book has been removed
SRP	ein mantel	A coat
WRjP	der artikel mantel kann nicht hinzugefügt werden (.) anderenfalls würde die maximale gewichtsgrenze des koffers überschritten werden	The item coat cannot be added (.) otherwise the weight limit of your suitcase will be exceeded
SRU	badelatschen raus	Beach slippers out
WAU	ein paar badelatschen wurden entfernt	A pair of beach slippers has been removed
SRP	ein mantel	A coat
WRjP	der artikel mantel kann nicht hinzugefügt werden (.) anderenfalls würde die maximale gewichtsgrenze des koffers überschritten werden	The item coat cannot be added (.) otherwise the weight limit of your suitcase will be exceeded
SOT	tja	Well
SNp	(1.77)	(1.77)
SOQ	was kann man denn noch rausnehmen	Well what else can be removed
SNp	(1.48)	(1.48)
SNp	pf pf pf pf pf pf pf	pf pf pf pf pf pf pf
SNp	(4.8)	(4.8)
SRU	zwei bh raus	Two bras out
WAU	zwei bhs wurden entfernt	Two bras have been removed
SRP	ein mantel	A coat
WAP	ein mantel wurde hinzugefügt	A coat has been added
SOT	gott sei dank	Thank god
	...	...

SNPs stand for nonphonological utterances. Following the GAT-2 minimal standard [33], short pauses are noted as (.) and (-), longer pauses with their duration in brackets, e.g. (1.77)

In spite of intensive training and a detailed manual [5] the wizards did not always operate consistently and accurately. We found, for example, inconsistent wizard behavior by analyzing the subject-initiated category changes. It turned out that some rejected wordings would have been accepted by different wizards or even by the same wizard in other situations.

Table 13.3 Tau statistics for sociodemographic features and successful actions

Kendall's tau	Age	d_{Cohen}	Gender	d_{Cohen}
Successes (abs.)			0.209***	0.51
Successes (rel.)	−0.117*	0.39	0.166**	0.41

*$p < 0.05$; **$p < 0.01$; ***$p < 0.001$ (cf. [30])

We also found some wizard errors, meaning situations where a wizard did not operate according to the guidelines of the manual. One type of such a wizard error is the rejection of a subject request with 'your input could not be processed' (WRjNp) when indeed the intention of the subject was clearly recognizable and the intended action was performable.

In sum, a manual like [5] is *necessary*, but by *no means sufficient* for successful experiments. A manual gives overall structure of experiments, but for non-trivial interactions nearly necessarily many questions will remain open.

In other words, spontaneous improvisation by wizards seems unavoidable, but it has a price: unreflected and unsupervised improvisation may—very likely—result in inter-session inconsistencies in wizard behavior. Indeed, the LAST MINUTE corpus contains many occurrences of inter-session wizard inconsistencies.

An example of such W inconsistencies is accepting or rejecting synonyms. In some cases wizards accepted synonyms of packed items, in other cases they did not.

Recommendations Thorough analysis of wizard errors and inconsistencies across the full LAST MINUTE corpus gives strong support for the following recommendations for future Wizard of Oz experiments:

- Invest in wizards. They need not only training, they need continuous supervision and monitoring.
- Expect the unexpected; therefore define appropriate meta rules for wizards.
- Invest in transcription and annotation, i.e. at least three independent transcribers with majority voting.

DAR-Based Measures The DAR characterizes dialog courses with sequences of dialog act labels. This allows the definition of a variety of measures for dialog success or failure.

One approach is to calculate the relative frequency of successful subject commands (SRx). As already reported in [30], Table 13.3 shows the correlation of absolute and relative frequency of successful actions with age and gender calculated using Kendall's tau. Missing values were not significant on a $p < 0.05$ level.

Successful Packing and Unpacking Actions The variety of actions changes during the experiment. At the beginning the subjects (normally) only pack items and change the category—which is usually accepted by the system (SRP WAP).[1] After the weight limit barrier (WLB) other actions become necessary. The subjects have to successfully unpack items first in order to be able to pack others.

[1] Some subjects try to unpack items before it is necessary, but these are only single cases.

A more refined dialog success measure is given by the number of successful packing actions after the WLB. This quantitative measure uncovers a subgroup of eight subjects (of $N_{total} = 133$) that did not have a single successful packing action after the WLB at all. This is in sharp contrast to the top performer group in the fourth quartile with at least nine and up to 14 successful packings after the WLB. Why users fail in such a drastic way was then a matter of qualitative in-depth analyses of the transcripts. It turned out that virtually in all eight cases with zero successful packing actions after the WLB, wizard errors and inconsistencies caused or heavily contributed to the negative dialog courses that ended without any more progress after the WLB.

A closer look at those top performers with 10 or more packing successes after WLB reveals that from a total of 24 subjects in this subgroup we find 13 that are young and female (cf. Table 13.4), far more than would be expected in a random sample.

These observations motivate us to have a closer look at the differences between the age group- and gender-based subgroups of subjects with respect to the number of successful packing actions after the WLB (cf. Figs. 13.4 and 13.5). We exclude the eight outliers with zero successes and work with $N_{nozero} = 125$ subjects. A Kruskal-Wallis test with the four age group and gender pairings yields a significant result (Kruskal-Wallis chi-squared = 9.4945, df = 3, $p = 0.02339$*). The mean for young women is larger than the means of the other three groups each (the mean values in decreasing order are mean(yw) = 8.182, mean(ym) = 6.706, mean(ew) = 6.333, and mean(em) = 5.857). Wilcoxon tests give the following results when we test young women against the three other pairings: yw against ym yields W = 413, p = 0.06408; yw against ew W = 319, $p = 0.01499$*, $d_{Cohen} = 0.59$; yw against

Table 13.4 Top performers after weight limit (for details see text)

Age/Sex	Male	Female
Young	5	13
Elderly	4	2

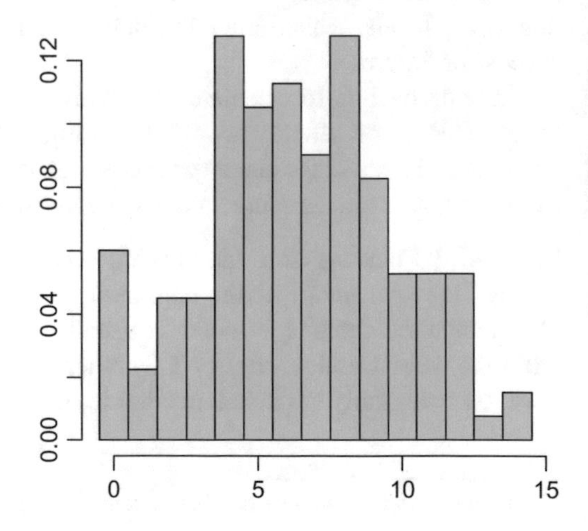

Fig. 13.4 Distribution of number of successful packing actions after WLB per transcript ($N = 133$). Please note the amount of zero values

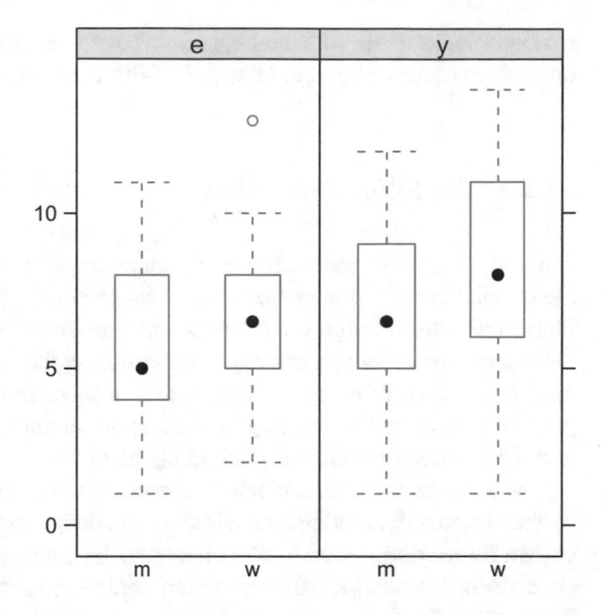

Fig. 13.5 Distributions of number of successful packing actions after WLB conditioned by age group and gender; eight outliers with zero actions removed (see text)

em W = 275, $p = 0.006328^{**}$, $d_{Cohen} = 0.73$). This finding replicates other results about age group- and gender-based differences in dialog success in the LMC. But as a dialog success measure, 'successful packing actions after WLB' has a major weakness, it is too task-dependent. For more general and task-independent success measures and the resp. results see Sect. 13.4.1 and Chap. 14.

Initiative in LAST MINUTE Dialogs An important aspect of dialogs is the initiative: Which participant has the initiative and is thus driving the dialog? While the system generally has the initiative during the personalization phase, the problem solving phase is primarily characterized by user initiative. The subject expresses a request (R) for a system action and, as a response, the system either confirms (A) or rejects (Rj) the request. An action may be rejected based on aspects of the user's utterance ('your request could not be processed') or—although the utterance was 'understood' and accepted—because the action cannot be performed for task-related reasons. In sum, the pairs of successive dialog acts are SRx WAx, SRx WRjx or SRx WRjNp, but there are exceptions to this general rule.

One example for system initiative is category change. If the allocated processing time of approximately 1 min for a selection category is over, the system takes initiative, informs the user about the timeout (WIF) and performs the change to the next category (WIC).

Self-initiated changes of the category for selection are an indicator for the degree of control of the dialog flow and for the taken initiative that a subject exhibits. As already reported in [30]: An in-depth analysis of self- vs. system-initiated category changes reveals a highly significant correlation between age and number of successful self-initiated category changes after the weight limit barrier; calculating Kendall's tau for these two quantities reveals that, with a tau statistic of −0.27, a

p-value smaller than 10^{-5} and $d_{Cohen} = 0.69$ as effect size, younger subjects have on average a higher number of such self-initiated category changes than the elderly.

13.3.4 Post-Hoc Interviews

Currently, there are research investigations regarding methods and techniques for the technical realization of *Companion*-Systems (e.g. [9]). Comparatively, only few theoretical considerations and hardly any empirical investigations about the users' inner processes, feelings and experiences during the interaction with such a system exist (e.g. [37]). In order to contribute to a more in-depth understanding of what goes on in users' minds during user-*Companion* interactions (UCI), it was decided to realize interviews with approximately half of all participants [17].

The interviews were conducted subsequent to the experiment. They were guided by the idea that the participants will make up their individual views on the simulated system by ascribing human-like characteristics, motives, wishes, beliefs, etc. to it (Intentional Stance, [4]). Besides intentional ascriptions to the system the interview focused on: emotions occurring during the interaction, the subjective experience of the speech-based interaction and the intervention (if given), the role of technical devices in autobiography and the general evaluation of the system.

The interviews were semi-structured to enable the participants to express their individual views and experiences freely and non-restrictedly (e.g. [7]). A preluding non-specific narration stimulus evoked a so-called initial narration regarding the experience of the experiment ("You have just done an experiment. Please try to put yourself back in this experiment. Please tell me how you did during the experiment. Please tell me in detail what you thought and experienced!"). If necessary, questions for clarifying and specifying vague narration sequences were given (immanent questions). Afterwards, pre-determined open questions (exmanent questions) from an interview guide were used, which dealt with the topics mentioned above and were handled flexibly with regard to formulation and chronological order. These questions were extended if unexpected relevant aspects were reported.

The interviewees were recruited according to a qualitative sampling scheme. Therein, age, educational level, sex and assignment to experimental or control group were considered to get a heterogeneous sample in order to maximize variance of subjective experience. Finally the sample included 73 participants of the total sample of the LMC, nearly balanced in regard to the mentioned variables. In total, 96 h and 22 min of interview material were recorded and transcribed in large parts according to the GAT-2 minimal standard [33].

13.4 Insights from Analyses

13.4.1 Discourse Analysis of Transcripts

Turn-Based Measures For the analysis of (two party) dialogs, so-called *'adjacency pairs'*—i.e. pairs of consecutive dialog acts or turns of the two participants—are a fundamental unit [11].

In LAST MINUTE typical adjacency pairs have two different structures [30]:

- In the personalization phase, the typical adjacency pair is a wizard question or prompt followed by the subject's answer or narrative.
- In the problem solving phase, a typical adjacency pair is made up of a user request for some action (packing some items, unpacking, changing category, etc.) followed by either a confirmation or a rejection by the wizard.

Given this dialogic structure, the number of turns of the participants is a measure of the total extent of the complete interaction or of its subparts. Taking the total number of turns in the problem solving phase (without intervention) we find a high variance within the whole cohort of $N = 133$ (cf. Figs. 13.6 and 13.7).[2,3]

The primary result about differences in total number of turns in problem solving (without intervention) is that gender matters (cf. Fig. 13.7 and Table 13.5). Men (fewer turns) significantly differ from women; young men significantly differ from

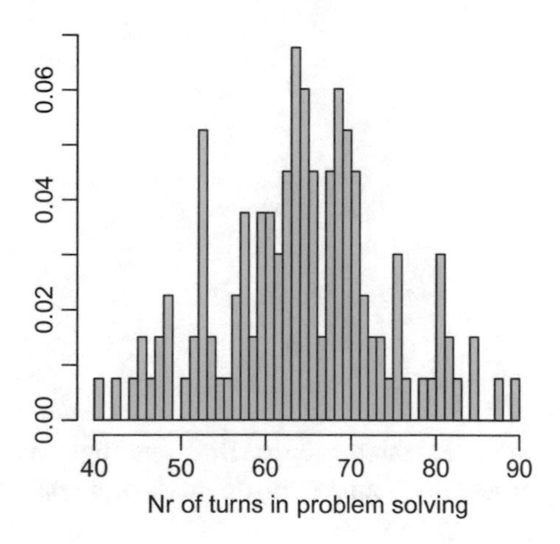

Fig. 13.6 Distribution of total number of user turns in problem solving (without intervention) per transcript (mean = 64.75, median = 65.00, sd = 9.79, N = 133)

[2]The distribution is visualized here—and in other figures—as a trellis boxplot: the rectangles represent the interquartile range (i.e. the range of 25% of the values above and below the median resp.); the filled dot gives the median; the whiskers extending the rectangle extend to the range of values, but maximally to 1.5 of the interquartile range; outlier values beyond the maximal whisker range are given as unfilled dots (cf. [1]).

[3]Unless noted otherwise, all statistical tests and calculations have been performed with the R, language [1, 25].

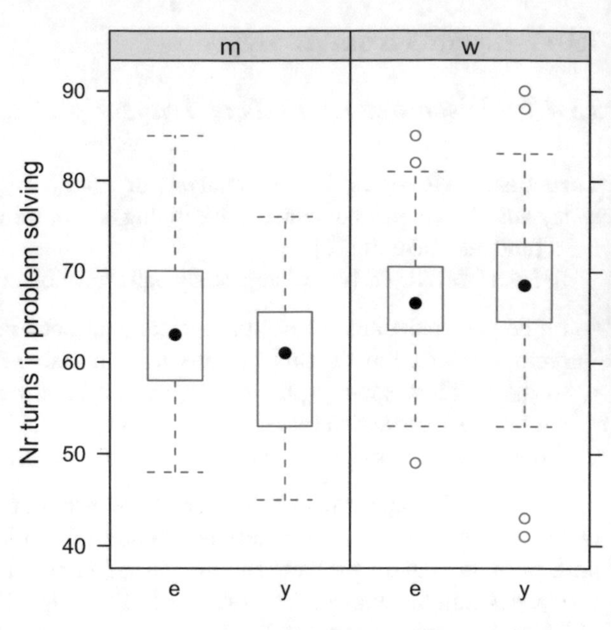

Fig. 13.7 Total nr of turns in problem solving (without intervention) per transcript conditioned by age group (e = elder, y = young subjects) and gender (m = men, w = women; please note outliers in the subgroup of young and elder women)

Table 13.5 Results from t tests for differences in *number of turns in problem solving* for subgroups of all subjects

g1	rel	g2	*p* value	d_{Cohen}
m	<	w	0.001219**	0.57
e	<	y	n.s	n.a.
ym	<	em	0.04275*	0.53
ym	<	ew	0.0007628***	0.86
ym	<	yw	0.000746***	0.83
em	<	yw	n.s.	n.a.
em	<	ew	n.s.	n.a.
ew	<	yw	n.s.	n.a.

Please note: Shapiro normality tests allow normality assumptions for all investigated subgroups
Significance levels: $*p < 0.05$; $**p < 0.01$; $***p < 0.001$

the other three groups. Both findings have a domain-dependent reason: men, and especially young men, do *on average* pack fewer items and therefore employ fewer packing requests.

Dialog Success Measures Measures based on simply counting turns have a major fault: They do not differentiate between successful and unsuccessful adjacency pairs. This is remedied by measures taking (local) success and failure into account. In the following, we therefore work with the dialog success measures as defined in Sect. 13.3.3.

For differences between sociodemographic groups with respect to both dialog success measures we find that age group matters (cf. Table 13.6), whereas gender

Table 13.6 Results from Wilcoxon tests for differences between sociodemographic groups in dialog success measures *DSM1* and *DSM2*

g1	rel	g2	p DSM1	d_{Cohen}	p DSM2	d_{Cohen}
y	>	e	0.007459**	0.32	0.001061**	0.51
w	>	m	n.s	n.a	n.s.	n.a
yw	>	ym	0.01281*	0.54	0.02627*	0.43
yw	>	em	0.01589*	0.635	0.002896**	0.69
yw	>	ew	0.001619**	0.77	0.0005762***	0.79
ym	>	em	n.s.	n.a	n.s.	n.a
ym	>	ew	n.s.	n.a	n.s.	n.a
em	>	ew	n.s.	n.a	n.s.	n.a

Significance levels: $^{*}p < 0.05$; $^{**}p < 0.01$; $^{***}p < 0.001$

alone does not give a significant difference. In detail: The difference between young and elderly subjects is significant. The subgroup of *young women* shows significantly **higher** values (of DSM1 and DSM2) than the other three groups.

Phasewise Dialog Success Values The three major subphases of problem solving in the LAST MINUTE experiment are demarcated by the weight limit barrier (WLB) and the weather info barrier (WIB). Up to the weight limit barrier there is no need for unpacking (although some subjects do some unpacking, for example as part of exchange actions). After the weight limit barrier, unpacking becomes crucial. Without at least one successful unpacking there is no further progress possible. Finally, the weather info demands a strategy change. Now items for cold and rainy weather are needed in exchange for bathing suits and other summer items.

The three phases reach

- P2s: from the start of problem solving to the weight limit barrier (WLB),
- P2b: from the weight limit barrier to the weather info (WIB), and finally
- P2w: from the weather info to the end of the experiment.

Global success values differ remarkably within the different phases. The distributions for the successful requests to all request ratios (DSM1) are presented in Fig. 13.8. Note that in the P2s phase (before the weight limit barrier) more than 75% of the subjects have DSM1 values over 0.95, whereas in the subsequent phases, P2b and P2w, the means drop to 0.6877 and 0.7096 resp. The differences between the mean values for DSM1 before the WLB and those of P2b and P2w are highly significant (Wilcoxon tests: P2s vs. P2b, W = 2932.5, $p < 2.2 \times 10^{-16}$, $d_{Cohen} = 1.35$; P2s vs. P2w, W = 490.5, $p < 2.2 \times 10^{-16}$, $d_{Cohen} = 2.06$), whereas the slight differences between P2b and P2w are insignificant ($W = 9451$, $p = 0.3337$, $d_{Cohen} = 0.10$).

We get a similar picture for the phasewise values of the successful requests to all turn ratios (DSM2) (cf. Fig. 13.9). The differences between the mean values for DSM2 before the WLB and those of P2b and P2w are highly significant (Wilcoxon tests: P2s vs. P2b, W = 444.5, $p < 2.2 \times 10^{-16}$, $d_{Cohen} = 2.595$; P2s vs. P2w, W

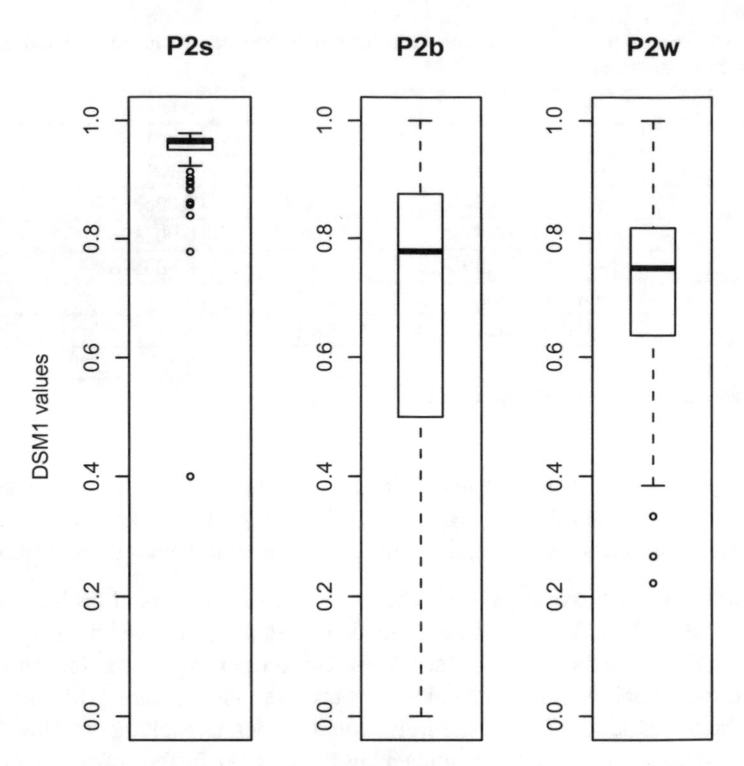

Fig. 13.8 Distributions of DSM1 values in the three subphases of problem solving ($N = 133$)

$= 356.5, p < 2.2 \times 10^{-16}, d_{Cohen} = 3.395$), whereas the slight differences between P2b and P2w are insignificant ($W = 8303, p = 0.3883, d_{Cohen} = 0.15$).

Evaluations with Phasewise DSMs Dialog success measures for subphases of the complete LAST MINUTE dialogs allow us to investigate questions that are raised by findings with the global measures. A point in case for such more fine-grained analyses is questions about differences between user success with respect to the barriers or the intervention.

When we investigate differences in global dialog success between the subgroups of subjects with and without intervention, we find that the differences between the two groups with respect to global DSM2 values are close to significant (Wilcoxon test: $W = 1794, p = 0.06258, d_{Cohen} = 0.40$; the resp. differences for DSM1 are insignificant).

This raises the question of whether the higher global DSM2 values for the group with intervention can be attributed to the intervention. We therefore perform a fine-grained investigation of the phasewise DSM values. We first compare the subgroups of subjects with and without intervention in the dialog phase P2w after the weather information (and thus after the intervention). We get differences in the means for this phase for both DSM values but these are not significant (Wilcoxon rank sum

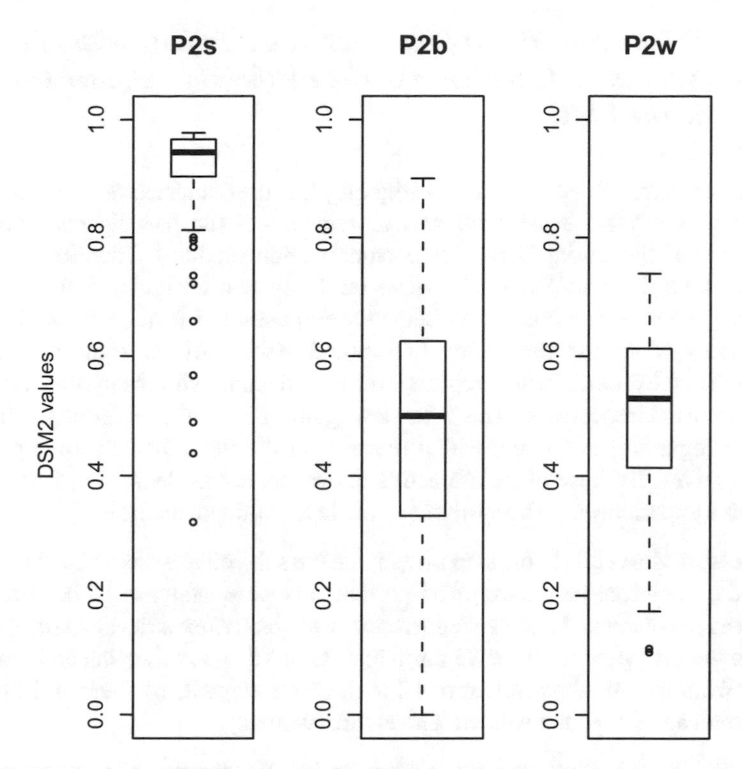

Fig. 13.9 Distributions of DSM2 values in the three subphases of problem solving ($N = 133$)

test, DSM1: $W = 2240.5$, $p = 0.8854$, $d_{Cohen} = 0.03$; DSM2: $W = 1946.5$, $p = 0.2397$, $d_{Cohen} = 0.22$).

When we then compare phasewise DSM values in the phases from the beginning of problem solving to the weight limit (P2s) and from there to the weather info (P2b), we see additionally that already in those phases (long before the intervention) the subgroup with intervention has higher DSM values than the subgroup without. Some differences between the subgroups are even close to significant (e.g. Wilcoxon for DSM2 in P2s: W = 1780.5, $p = 0.0544$, $d_{Cohen} = 0.42$).

In other words, the dialog behavior of the groups with and without intervention already differed long before the intervention took place with the former group being more successful on average than the latter. This analysis result is in line with a general observation: As soon as user–*Companion* dialogs show enough degrees of freedom and approximate realistic dialogs—as is the case with the LAST MINUTE dialogs—their course under varying constraints is hard to predict in advance and fine-grained analyses are needed.

13.4.2 *"Okay Now It Is a Computer Voice Talking with Me": Subjective Experience of the Artificial Computer Voice in the LMC*

One of the main objectives in investigating the user interviews conducted subsequent to the WoZ experiment was to reconstruct the participants' subjective experience of the Initial Dialogue (personalization module). Therefore, analyses focused on participants' individual views on the system, including their ascriptions regarding the system's interaction behavior and essential features as well as their experiences of themselves during this module (see [16] for some of the main results). Thereby, participants' reports explicitly dealing with the system's artificial voice occurred repeatedly. The interview guide did not contain any exmanent question regarding this issue, which means that the participants came up with it by themselves. Because of its subjective relevance, it was decided to focus on the subjective experiences of the system's voice in an in-depth analysis.

Sample and Material In order to investigate the subjective experience of the Initial Dialogue, the initial narratives of 31 participants were analyzed.[4] The aim of this study was to collect and analyze reports on the subjective experience of the system's voice in the interviews of these 31 participants. Up to now, collecting and analyzing reports from the initial narratives of 12 of these participants has been finished (five young men and six young women, one elderly woman).

Analyses The interview sequences were broken up into so-called meaning units and then were condensed, summarized and interpreted using methods of qualitative content analysis [19]. Throughout this abstraction process categories were developed which arrange the summarized meaning units.

Results Three categories were developed out of the material. These represent the variance of participants' experiences regarding the artificial computer voice.

1. *Characterization of the voice—from weird to agreeable:* The system's voice is mainly characterized as *"monotone" (e.g. KK[5]), "non-human" (e.g. UK), "choppy" (e.g. CT), "weird" (e.g. AK)* and *"mechanical" (e.g. SS)*. It is clearly associated with a technical device and differentiated from a human voice *("a very technical voice which has so little of a human face", SR)*. There are participants who ascribe to it that it is unable to convey emotions *("affectionately cold", SR)*. Taken together with the absence of dialect and prosody, this leads to criticism regarding the voice's lack of individuality as well as regarding the system's strange- and differentness experienced by the participants *("well it was like*

[4] In order to maximize variance, interviews were chosen randomly one by one considering a balanced allocation of groups according to the qualitative sample plan (cf. Sect. 13.3.4). When no more increase of variance could be detected, no further material was added (criterion of theoretical saturation).

[5] Participant's initials.

talking to someone who is not German and just started to learn the language and pronounces the words differently (.) it's really like yeah almost like an accent so to say", UK). There are only few participants who generally perceive the voice as *"agreeable" (AK)* or experience variability in the voice during the interaction *(MS).*

2. *The voice's impact on the participant—from adaption to distancing:* The voice contributes to experiencing the interaction as a man-machine interaction. It is perceived as unusual, and it is strange to talk to an artificial voice. Especially at the beginning, the voice evokes feelings of scariness, irritation or anxiety *("I felt uneasy about it", MS).* Furthermore, it can inhibit the development of trust in the system (*"if it had been a nice and lovely voice one would of course have had more trust even though it is a computer", KK).* Many participants refrain from using familiar communication principles from human-human interaction (HHI) *("one don't even try to speak with it like with a human being instead one switches to the mode okay now it is a computer voice speaking with me", UK).* Based on experiencing the voice as monotone and non-human, the participants ascribe to the system that it is demanding, does not tolerate contradictions or allows breaking its flow of words. These issues create a feeling of losing control. Furthermore, participants' initiative reduces, because they feel hampered in influencing the interaction *("because of this strange voice it was like it asks me something and I react", FW).* The voice is experienced as keeping the participant at a distance *("it appeared like rejecting", SS)* or otherwise evoking the desire for distancing in the participant. Some participants continuously struggle with accepting the voice, whereas others make lots of substantial adjustments to adapt to it. On the one hand, they make cognitive efforts *("really strain oneself", MH).* On the other hand, they adapt their behavior to the system's abilities, which they anticipate as being deficient because of the system's voice quality *("that I have to speak clearly and can't use slang because maybe it will not understand that", UK).*
3. *Wishes regarding the voice—from somehow different to more human-like:* Besides participants who would simply appreciate a change of the voice, there are participants who express more specific ideas regarding this change, e.g. the wish for a more modulated voice *("more fluently (...) like a wave while talking", CT)* or the preference for a female voice. Furthermore, some would appreciate a more individual voice using, e.g. idioms or proverbs, to be enabled to develop a *"personal counterpart" (FW)*, which appears less strange and more likeable. This all culminates in the wish for a more human-like voice, which features all of these characteristics and is imagined as much more pleasant. All in all, the idea of individualization of the voice runs through the participants' reports, e.g. the possibility to choose out of a set of predefined voices.

Discussion Participants' descriptions of the computer voice are comparable to those reported in other studies (e.g. [36]). Its artificial sound triggers associations with a technical counterpart. The participants require a lot of cognitive effort to understand it and adapt their behaviour to anticipated system's abilities. They want to understand the system and want to be understood by it. Regarding the emotional level, they are frightened, scared and alienated, feel like they are losing control and wish to distance themselves from the voice—feelings which can be discussed regarding the uncanny valley phenomenon [22]. Other studies [13, 36] underline the dissatisfaction with artificial systems' voices in HCI. In our analysis, participants consequently withdraw from the interaction by reducing their initiative. This is a critical finding when considering the aim of *Companion*-Systems to be experienced as trustworthy attendants and relational partners [3, 37]. Besides the desire to generate mutual understanding, the need to relate to the system could be conveyed by the wish for a more pleasant and human-like voice, which is in line with users' ideas for improving HCI by referring to experiences from HHI [36]. Maybe participants imagine that they could enter a deeper and more trustworthy relationship with such a voice. According to [15] this could be explained by the humans' need to belong [2], which appears even in interactions with simple artifacts showing only basic social cues like speech.

It has to be considered that the presented results are preliminary, because they are based on material of mainly young participants, which is currently being expanded regarding sample size and considered interview sequences. Nevertheless, first meaning structures occurred, which will be verified and extended in further analyses.

13.5 Summary

The LMC stands out from other available naturalistic corpora considering sample size, sample heterogeneity, total length of each interaction, number and quality of the recordings as well as further user information from questionnaires and interviews (cf. [21, 29]). It marks a cornerstone for work in the SFB/TRR 62 [3]. Working groups in the SFB consortium focus on research questions regarding the automatic detection and classification of markers for users' emotions during the interaction in different modalities (e.g. [6, 34], see also Chap. 14). In this chapter, exemplary results were presented regarding the users' observable dialog behavior and dialog success as well as results regarding their unobservable inner processes, feelings and evaluations during the interaction. All in all, the LMC provides potential for investigations from various perspectives, and researchers are invited to explore it according to their particular research interests.

Contributions This paper reports on joint work—i.e. the design and execution of the LAST MINUTE experiments—as well as on distinct contributions of the two involved groups. The responsibility for the dialog act annotation of the LMC and for the discourse analysis of transcripts (cf. Sects. 13.3.3 and 13.4.1) lies with D. Rösner, R. Andrich, R. Friesen, and S. Günther. The responsibility for sample and post-hoc interviews (cf. Sects. 13.3.1, 13.3.4, 13.4.2) lies with J. Frommer, M. Haase and J. Krüger. For the discussion and the conclusions in Sect. 13.5 the responsibility is shared by all authors.

Availability The LAST MINUTE corpus is available for research purposes upon written request (via email or mail) from project A3 of SFB TRR 62 (heads: Prof. Frommer and Prof. Rösner).

Acknowledgements This work was done within the Transregional Collaborative Research Centre SFB/TRR 62 "*Companion*-Technology for Cognitive Technical Systems" funded by the German Research Foundation (DFG).

References

1. Baayen, R.H.: Analyzing Linguistic Data – A Practical Introduction to Statistics Using R. Cambridge University Press, Cambridge (2008)
2. Baumeister, R.F., Leary, M.R.: The need to belong: desire for interpersonal attachments as a fundamental human motivation. Psychol. Bull. **117**(3), 497 (1995)
3. Biundo, S., Wendemuth, A.: *Companion*-technology for cognitive technical systems. Künstl. Intell. (2016). doi:10.1007/s13218-015-0414-8
4. Dennett, D.C.: The Intentional Stance. MIT, Cambridge (1987)
5. Frommer, J., Rösner, D., Haase, M., Lange, J., Friesen, R., Otto, M.: Project A3 - Detection and Avoidance of Failures in Dialogs. Pabst Science Publisher, Lengerich (2012)
6. Frommer, J., Michaelis, B., Rösner, D., Wendemuth, A., Friesen, R., Haase, M., Kunze, M., Andrich, R., Lange, J., Panning, A., Siegert, I.: Towards emotion and affect detection in the multimodal last minute corpus. In: Calzolari, N., Choukri, K., Declerck, T., Doğan, M.U., Maegaard, B., Mariani, J., Moreno, A., Odijk, J., Piperidis, S. (eds.) Proceedings of the Eighth International Conference on Language Resources and Evaluation (LREC'12). European Language Resources Association (ELRA), Paris (2012)
7. Galletta, A.: Mastering the Semi-Structured Interview and Beyond: From Research Design to Analysis and Publication. NYU, New York (2013)
8. Hassenzahl, M., Burmester, M., Koller, F.: AttrakDiff: ein Fragebogen zur Messung wahrgenommener hedonischer und pragmatischer Qualität. In: Ziegler, J., Szwillus, G. (eds.) Mensch & Computer 2003. Berichte des German Chapter of the ACM, vol. 57, pp. 187–196. Vieweg+Teubner Verlag, Stuttgart (2003)
9. Honold, F., Schüssel, F., Weber, M., Nothdurft, F., Bertrand, G., Minker, W.: Context models for adaptive dialogs and multimodal interaction. In: 2013 9th International Conference on Intelligent Environments (IE), pp. 57–64. IEEE, New York (2013)
10. Horowitz, L.M., Alden, L.E., Wiggins, J.S., Pincus, A.L.: Inventory of Interpersonal Problems Manual. Psychological Cooperation, San Antonio (2000)
11. Jurafsky, D., Martin, J.H.: Speech and Language Processing: An Introduction to Natural Language Processing, Computational Linguistics, and Speech Recognition, 2nd edn. Prentice Hall, Englewood Cliffs (2008)
12. Karrer, K., Glaser, C., Clemens, C., Bruder, C.: Technikaffinität erfassen – der Fragebogen TA-EG. In: Lichtenstein, A., Stößel, C., Clemens, C. (eds.) Der Mensch im Mittelpunkt technischer Systeme, ZMMS Spektrum, Reihe 22, vol. 29, pp. 196–201. VDI Verlag GmbH, Düsseldorf (2008)
13. Kastner, M., Stangl, B.: Exploring a text-to-speech feature by describing learning experience, enjoyment, learning styles, and values–a basis for future studies. In: 2013 46th Hawaii International Conference on System Sciences (HICSS), pp. 3–12. IEEE, New York (2013)
14. King, J., Bond, T., Blandford, S.: An investigation of computer anxiety by gender and grade. Comput. Hum. Behav. **18**(1), 69–84 (2002)
15. Krämer, N.C., Eimler, S., von der Pütten, A., Payr, S.: Theory of companions: what can theoretical models contribute to applications and understanding of human-robot interaction? Appl. Artif. Intell. **25**(6), 474–502 (2011)

16. Krüger, J., Wahl, M., Frommer, J.: Making the system a relational partner: users' ascriptions in individualization-focused interactions with companion-systems. In: Proceedings of the 8th International Conference on Advances in Human-oriented and Personalized Mechanisms, Technologies, and Services (CENTRIC 2015), pp. 48–54. IARIA XPS Press (2015). http://www.thinkmind.org/index.php?view=article&articleid=centric_2015_3_20_30079
17. Lange, J., Frommer, J.: Subjektives Erleben und intentionale Einstellung in Interviews zur Nutzer-Companion-Interaktion. In: Informatik 2011. Lecture Notes in Informatics, vol. 192, p. 240. Köllen, Bonn (2011). http://www.user.tu-berlin.de/komm/CD/paper/060332.pdf
18. Levenstein, S., Prantera, C., Varvo, V., Scribano, M.L., Berto, E., Luzi, C., Andreoli, A.: Development of the Perceived Stress Questionnaire: a new tool for psychosomatic research. J. Psychosom. Res. **37**(1), 19–32 (1993)
19. Mayring, P.: Qualitative content analysis. Theoretical foundations, basic procedures and software solution. n.p., Klagenfurth (2014). http://nbn-resolving.de/urn:nbn:de:0168-ssoar-395173[retrieved:09,2015]
20. McCrae, R.R., Costa, P.T.: A contemplated revision of the NEO Five-Factor Inventory. Personal. Individ. Differ. **36**(3), 587–596 (2004)
21. McKeown, G., Valstar, M.F., Cowie, R., Pantic, M.: The SEMAINE corpus of emotionally coloured character interactions. In: 2010 IEEE International Conference on Multimedia and Expo (ICME), pp. 1079–1084. IEEE, New York (2010)
22. Mori, M., MacDorman, K.F., Kageki, N.: The uncanny valley [from the field]. IEEE Robot. Autom. Mag. **19**(2), 98–100 (2012)
23. Naumann, A., Hermann, F., Peissner, M., Henke, K.: Interaktion mit Informations-und Kommunikationstechnologie: Eine Klassifikation von Benutzertypen. In: Herczeg, M., Kindsmüller, M.C. (eds.) Mensch & Computer 2008: Viel Mehr Interaktion, pp. 37–45. Oldenbourg Verlag, München (2008)
24. Peterson, C., Semmel, A., Baeyer, C.v., Abramson, L.Y., Metalsky, G.I., Seligman, M.E.P.: The attributional Style Questionnaire. Cogn. Ther. Res. **6**(3), 287–299 (1982)
25. R Development Core Team: R: A Language and Environment for Statistical Computing. R Foundation for Statistical Computing, Vienna (2010)
26. Rogers, C.: A theory of therapy, personality and interpersonal relationships as developed in the client-centered framework. In: Koch, S. (ed.) Psychology: A Study of a Science. Formulations of the Person and the Social Context, vol. 3. McGraw Hill, New York (1959)
27. Rösner, D., Friesen, R., Otto, M., Lange, J., Haase, M., Frommer, J.: Intentionality in interacting with companion systems – an empirical approach. In: Jacko, J. (ed.) Human-Computer Interaction. Towards Mobile and Intelligent Interaction Environments. Lecture Notes in Computer Science, vol. 6763, pp. 593–602. Springer, Berlin/Heidelberg (2011)
28. Rösner, D., Frommer, J., Andrich, R., Friesen, R., Haase, M., Kunze, M., Lange, J., Otto, M.: LAST MINUTE: a novel corpus to support emotion, sentiment and social signal processing. In: Devillers, L., Schuller, B., Batliner, A., Rosso, P., Douglas-Cowie, E., Cowie, R., Pelachaud, C. (eds.) Proceedings of LREC'12 - Workshop Abstracts, Istanbul, p. 171 (2012)
29. Rösner, D., Frommer, J., Friesen, R., Haase, M., Lange, J., Otto, M.: LAST MINUTE: a multimodal corpus of speech-based user-companion interactions. In: LREC, pp. 2559–2566 (2012)
30. Rösner, D., Friesen, R., Günther, S., Andrich, R.: Modeling and evaluating dialog success in the LAST MINUTE corpus. In: Chair, N.C.C., Choukri, K., Declerck, T., Loftsson, H., Maegaard, B., Mariani, J., Moreno, A., Odijk, J., Piperidis, S. (eds.) Proceedings of LREC'14, Reykjavik (2014)
31. Rösner, D., Haase, M., Bauer, T., Günther, S., Krüger, J., Frommer, J.: Desiderata for the design of companion systems – insights from a large scale wizard of Oz experiment. Künstl. Intell. **30**(1), 53–61 (2016). Online first: Oct 28, 2015; doi:10.1007/s13218-015-0410-z
32. Schmidt, T., Schütte, W.: Folker: an annotation tool for efficient transcription of natural, multi-party interaction. In: Chair, N.C.C, Choukri, K., Maegaard, B., Mariani, J., Odijk, J., Piperidis, S., Rosner, M., Tapias, D. (eds.) Proceedings of the Seventh conference on International Language Resources and Evaluation (LREC'10). European Language Resources Association (ELRA), Valletta (2010)

33. Selting, M., Auer, P., Barth-Weingarten, D., Bergmann, J.R., Bergmann, P., Birkner, K., Couper-Kuhlen, E., Deppermann, A., Gilles, P., Günthner, S., et al.: Gesprächsanalytisches Transkriptionssystem 2 (GAT 2). Gesprächsforschung-Online-Zeitschrift zur verbalen Interaktion **10** (2009)
34. Siegert, I., Philippou-Hübner, D., Hartmann, K., Böck, R., Wendemuth, A.: Investigation of speaker group-dependent modelling for recognition of affective states from speech. Cogn. Comput. **6**(4), 892–913 (2014). doi:10.1007/s12559-014-9296-6
35. Sieverding, M., Koch, S.C.: (Self-) evaluation of computer competence: how gender matters. Comput. Educ. **52**(3), 696–701 (2009)
36. Veletsianos, G.: How do learners respond to pedagogical agents that deliver social-oriented non-task messages? Impact on student learning, perceptions, and experiences. Comput. Hum. Behav. **28**(1), 275–283 (2012)
37. Wilks, Y.: Close Engagements with Artificial Companions: Key Social, Psychological, Ethical and Design Issues, vol. 8. John Benjamins Publishing, Amsterdam (2010)

Chapter 14
The LAST MINUTE Corpus as a Research Resource: From Signal Processing to Behavioral Analyses in User-*Companion* Interactions

Dietmar Rösner, Jörg Frommer, Andreas Wendemuth, Thomas Bauer, Stephan Günther, Matthias Haase, and Ingo Siegert

Abstract The LAST MINUTE Corpus (LMC) is one of the rare examples of a corpus with naturalistic human-computer interactions. It offers richly annotated data from $N_{total} = 130$ experiments in a number of modalities. In this paper we present results from various investigations with data from the LMC using several primary modalities, e.g. transcripts, audio, questionnaire data.

We showed that sociodemographics (age, gender) have an influence on the global dialog success. Furthermore, distinct behavior during the initial phase of the experiment can be used to predict global dialog success during problem solving. Also, the influence of interventions on the dialog course was evaluated.

D. Rösner (✉)
Institut für Wissens- und Sprachverarbeitung (IWS), Otto-von-Guericke Universität, 39016 Magdeburg, Germany

Center for Behavioral Brain Sciences, 39118 Magdeburg, Germany
e-mail: roesner@ovgu.de

J. Frommer • M. Haase
Universitätsklinik für Psychosomatische Medizin und Psychotherapie, Otto-von-Guericke Universität, 39120 Magdeburg, Germany
e-mail: joerg.frommer@med.ovgu.de; matthias.haase@med.ovgu.de

A. Wendemuth
Institut für Informations- und Kommunikationstechnik (IIKT), Otto-von-Guericke Universität, 39016 Magdeburg, Germany

Center for Behavioral Brain Sciences, 39118 Magdeburg, Germany
e-mail: andreas.wendemuth@ovgu.de

T. Bauer • S. Günther
Institut für Wissens- und Sprachverarbeitung (IWS), Otto-von-Guericke Universität, 39016 Magdeburg, Germany
e-mail: tbauer@iws.cs.uni-magdeburg.de; stguenth@iws.cs.uni-magdeburg.de

I. Siegert
Institut für Informations- und Kommunikationstechnik (IIKT), Otto-von-Guericke Universität, 39016 Magdeburg, Germany
e-mail: ingo.siegert@ovgu.de

S. Biundo, A. Wendemuth (eds.), *Companion Technology*, Cognitive Technologies,
DOI 10.1007/978-3-319-43665-4_14

Additionally, the importance of discourse particles as prosodic markers could be shown. Especially during critical dialog situations, the use of these markers is increasing. These markers are furthermore influenced by user characteristics.

Thus, to enable future *Companion*-Systems to react appropriately to the user, these systems have to observe and monitor acoustic and dialogic markers and have to take into account the user's characteristics, such as age, gender and personality traits.

14.1 Introduction

Wizard of Oz (WoZ) experiments are a well-established approach in research about *Companion* technology (cf. [24, 43]) and human-computer interaction (HCI) in general. WoZ experiments allow us to investigate many of the open issues in user-*Companion* interaction (UCI) without the need to actually have to implement the functionalities of the envisaged *Companion*-System [44]. Examples of such questions that demand empirical answers include the following: How do 'naive' users spontaneously interact with a *Companion*-System if it allows them to converse in spoken natural language? Can distinct user groups be detected based on observed behavior? How do observed linguistic markers correlate with sociodemographic or psychometric data of the users?

Multimodal recordings from such WoZ experiments are valuable assets, but their impact remains limited if they are not prepared for and not made available for third-party usage within the research community in the form of a corpus. To convert raw data records from an experiment into a corpus usable as research resource is by no means a trivial and easy task. On the contrary, this task is both challenging conceptually and expensive with respect to time and effort needed. This is one of the reasons why publicly accessible corpora with naturalistic human-computer interactions are still rare exceptions (cf. [10]).

Converting raw data recorded in experiments into a corpus demands at least two major steps: *transcription* and *annotation*. In transcription audio records from the interactions need to be converted into written records that are then amenable to analysis by methods of computational linguistics and corpus linguistics. Annotation is the process of adding interpretative labels to the recorded data. Annotations may serve multiple purposes. They may be used in further analyses of the data or they may serve as input to machine learning procedures that are, for example, employed in training and testing of respective classifiers.

Research Questions The LAST MINUTE Corpus (LMC) is one of the rare examples of a corpus with naturalistic human-computer interactions. It offers richly annotated data from $N_{total} = 133$ experiments in a number of modalities (cf. Chap. 13). The LMC thus allows researchers from many disciplines to investigate research questions from a multitude of perspectives and with a plethora of approaches and methods. In this paper we exemplify these options with example investigations and their results from three independent, yet cooperating, groups. The following research questions will be addressed:

- How do user groups based on sociodemographics (age, gender) differ with respect to linguistic aspects of the interaction and especially in global dialog success (cf. Sect. 14.4.1)?
- How do user groups that are defined based on distinct behavior during the experiment differ in global dialog success (cf. Sect. 14.4.2)?
- Does the intervention show effects in the course of the dialogs (cf. Sect. 14.4.3)?
- How do human raters annotate the emotional content of selected audio and video excerpts from the corpus (cf. Sect. 14.3.3)?
- What is the extent of improvement of classifier performance when classifiers are trained separately for the four age- and gender-based subgroups (cf. Sect. 14.4.4)?
- How do the age- and gender-based subgroups differ with respect to the use of discourse particles before and after the weight limit barrier (cf. Sect. 14.4.4)?
- Do personality traits of subjects influence the use or non-use of discourse particles before and after the weight limit barrier (cf Sect. 14.4.5)?

The investigations differ not only in the methods and perspectives but also in the primary modalities that are employed (e.g. transcripts, audio, questionnaire data) and in the size of the subcohorts of subjects included. The latter ranges from a small sample of just 13 sets of excerpts in an investigative study with human raters (cf. Sect. 14.3.3) to the full set of $N_{total} = 133$ verbatim transcripts (cf. Sects. 14.4.1 and 14.4.2).

14.2 Material: LAST MINUTE Corpus

The experiment that is underlying the multimodal recordings in the LAST MINUTE Corpus (LMC) was designed in such a way that the dialogs between the simulated system and the users were on the one hand restricted enough but on the other hand still offered enough room for individual variation [8, 31]. The domain chosen was mundane enough not to demand any specialist knowledge as a prerequisite. On the other hand an inherent need for re-planning (unpacking after weight limit) and for strategy change (from summer to winter items after the weather information barrier WIB) was built into the scenario.

After having completed about two thirds of the experiment, participants received additional information from the computer system, calling into question the way they handled the task so far. This so-called "weather information barrier" (`WIB`) represented a complex set of problems [9], because participants had to consider a large number of interacting variables and had no insights regarding the dynamics of the course of the experiment. As a result of undergoing the `WIB`, participants had to adapt their strategy to the new circumstances. Subsequent to the `WIB`, a randomly selected part of the participants received an affect-oriented intervention, the design of which was based on general factors of psychotherapy (resource activation, problem actualization, accomplishment and clarification) [11]. Prior studies have already shown that empathic interventions initialized by computer systems can alter affective states that interfere with processes of communication (cf. [4, 16, 21]).

The WoZ scenario of LAST MINUTE is described in Chap. 13 in detail and with transcripts of example interactions.

The investment into the careful design of the scenario of the LAST MINUTE experiments pays off now. The resulting LMC is a valuable resource based on a large number of highly formalized, yet still variable, experiments with subjects balanced with respect to gender and age group.

As a resource the LMC is 'middle ground' between data (or a corpus) from a (small)scale experiment with a single hypothesis only and a corpus based on recordings from virtually unrestricted real-life interactions (e.g. Vera am Mittag [13], with records from a German TV talk show).

14.3 Methods

14.3.1 Analysis of Transcripts

Discourse Analysis The LAST MINUTE corpus comprises transcripts of all N = 133 experiments performed. On average, each experiment takes approximately 30 min real time. In order to be able to quantitatively compare and contrast different dialog courses, an adequate representation is needed [32, 33].

We employ a dialog representation based on the series of subsequent dialog acts of user and system, the so-called dialog act representation (DAR, [32]). This level of representation is independent of the domain of discourse, i.e. it is by no means restricted to the task in LAST MINUTE but is applicable to all types of task-oriented user-*Companion* dialogs.

In the following we use the dialog success measures (DSMs) as defined in Chap. 13. They allow the following types of investigations: How do user groups based on sociodemographics differ in global dialog success (cf. Sect. 14.4.1)? How do user groups that are defined based on distinct behavior during the experiment differ in global dialog success (cf. Sect. 14.4.2)?

The methods employed in discourse analysis of the LMC are as follows: The transcripts are available as an XML-based data structure in the FOLKER format [36]. This highly structured format contains not only the transcription of all user and wizard contributions during each experiment plus their relative temporal order, it comprises also additional annotations ranging from recorded nonphonological events (e.g. sighing, coughing) up to annotations on the discourse level (e.g. dialog act labels). For further details, cf. Chap. 13 or [32].

Starting from the FOLKER-encoded transcripts we determine features (or markers) either for complete transcripts or for their subparts (e.g. personalization

Table 14.1 Examples of empirical distributions of features calculated for complete transcripts (N = 133)

Marker	Min.	First Qu.	Median	Mean	Third Qu.	Max.	SD	Total sum
Tokens	266.0	444.0	545.0	602.7	699.0	1601.0	247.34	80,160
Turns	62	81	86	86.08	91	111	9.95	11,448
Tokens per turn	2.804	5.143	6.282	7.060	8.109	19.290	2.95	n.a.

vs. problem solving or their resp. subphases). Such features are calculated on all levels of the linguistic system, i.e. from the lexical level (e.g. occurrence counts for classes of lexical items) via syntax (e.g. preferred syntactic style in user commands) to semantic classifications (e.g. local meaning of user utterances) and pragmatic concerns (e.g. can the user's current intention be detected?).

The feature sets derived in this way then undergo a thorough data analysis in which we combine quantitative and qualitative approaches from corpus linguistics [12]. The quantitative methods start with the empirical distributions of the feature values. These are visualized appropriately and tested with respect to normality vs. skewness. Transcripts of (extreme) outliers are additionally checked qualitatively in order to detect possible reasons for the deviations.

A repeating finding for virtually all investigated features is that the distributions of feature values show a large variance. This even holds for features that quantify aspects of the overall extent of the highly standardized experiments (cf. Table 14.1).

Analyzing the reasons for the observed variance is a major issue in the work reported here. The different user groups based on sociodemographic features—i.e. age group (young subjects vs. elderly subjects) and the four combinations of age group with gender—are a primary potential source for the observed variance. Indeed, for many features the differences between the age groups and for subgroups based on subconditioning with gender prove to be significant (cf. Sect. 14.4.1).

When significant differences in the distribution of feature values have been found between sociodemographic groups, then the additional question arises about whether these differences correlate with significant differences in dialog success (as measured with DSM1 and DSM2).

Behavioral Analysis In behavioral analyses, errors that users make and problems they run into are valuable assets. This holds especially when early occurrences of problematic user behavior prove to be predictive for later global dialog success or failure. As will be elaborated in Sect. 14.4.2, early errors in the personalization phase can be detected that have this predictive power. The data analysis methods employed in evaluating observed differences in user behavior are the same as presented above. The only difference is that user groups are now defined on *observed differences in behavior in the course of the dialogs* and no longer on a priori differences between subjects like age group or gender.

14.3.2 Analysis of Psychometric Data

The authors examined the effectiveness of an affect-oriented intervention, which was given to participants after a confrontation with a complex set of problems, and in addition investigated its influence on participants' interaction behavior. In contrast to the rest of the experiment, where the interaction was guided by the ideational metafunction [14], the intervention was designed to address the participant on a conversant interactional level (interpersonal and textual metafunction). Therefore, the authors analyzed the influence of interpersonal problems on the effectiveness of the intervention using questionnaires and a self-developed criterion ('the dialog exchange'). Psychological questionnaires are broadly used research instruments for data acquisition. Generally, traits are measured via single questions (items) and potential answers are differentiated via Likert scales. Single items are summarized in subscales.

The **Inventory of Interpersonal Problems (IIP-C)** [17] measures problems which occur within interpersonal relationships. Applying the interpersonal circumplex model helps to assess behavior that is problematic for the test person as well as behavior he or she tends to show excessively. Eight scales are used for evaluation: domineering/controlling, vindictive/self-centered, cold/distant, socially inhibited, nonassertive, overly accommodating, self-sacrificing, intrusive/needy. These eight scales are in accordance with the octants of an circumplex model of interpersonal behavior, traits, and motives. The IIP-C was used for analyses presented in Sects. 14.4.3 and 14.4.5.

The **dialog exchange**. Analysis of the linguistic interactions in the WoZ experiment is possible by focusing more closely either on content- or on conversation-related aspects. Considering the research questions, an analysis of the conversation dynamics as such seems reasonable. Thus, the dialog exchange criterion was conceptualized in reference to the dialogism of interpersonal interactions. In conversation analysis, dialog is characterized by the 'boundaries of utterances', which are determined by aspects like 'change of speakers' (which is a fundamental characteristic of spoken language) [3] as well as the internal closure of single speaker contributions. In the WoZ experiment, the number of verbal contributions (so-called 'logs') given by the simulated system was recorded automatically [31]. The system was designed to respond to the participant. With the help of recorded logs we were able to determine the number of changes in the dialog between the system and the participant (dialog exchange criterion). Neither the content nor the length of utterances was considered.

The authors investigated the **following questions**. What is the impact of an affect-oriented intervention on participants' interaction behavior (dialog exchange) after a complex problem situation (barrier)? How does the extent of interpersonal problems influence the effectiveness of the affect-oriented intervention?

To answer these questions, the authors applied a range of different **methods**. According to the standardized experimental scenario, all subjects had to pass the identical procedure, which allowed for an exact definition of the course of events before and after the barrier. Initially, the experiment was divided into four parts. In the baseline condition (BSL), the participants who accomplished the task proceeded without any further limitations. The first momentous limitation was the weight limit barrier (WLB), which the participants did not expect to occur. Later in the experiment, the system provided weather information (WIB) prompting participants to change their strategy. In the revision stage (RES), subjects got the opportunity to repack their suitcase under increasing time pressure [8]. In order to gather further information on the effect of the intervention on the dialog exchange, the authors compared the intervention and control group before and after the barriers. For statistical analysis, repeated measures ANOVA's has been used to test the effects of different independent variables on the dialog exchange. We conducted one within subjects ANOVA to test only the effects of the different conditions over time (BSL, WLB, WIB and RES) and used the Greenhouse Geisser correction of degrees of freedom when a significant Mauchly Test indicated lack of sphericity.

14.3.3 Analysis of Audio Records

For a realistic scenario, the development of the interaction is important and the users' reaction within critical events has to be assessed. Therefore, it has to be confirmed that the users show emotional reactions after the experimental barriers and that this reaction is different for the different kinds of barriers. This is later used to assess the type of barrier a user is allocated by using his acoustics (cf. Sect. 14.4.4).

To **evaluate the emotional content** right after the barriers, we created short excerpts for all four events containing video and audio utilizing a subset of $N_{labeling} = 13$ speakers from the LMC.

These clips are given to the labelers, who should rate each clip. The used labels are inspired by a previous experiment, as described in [38]: *surprise*, *interest*, *relief*, *joy*, *contempt*, *confusion*, *sadness*, *hope*, and *helplessness*. The labelers can choose between one of these predefined labels, but are also allowed to not give any label to a clip or to give a self-defined label. Six labelers, all not familiar with the corpus, conducted that labeling task. This results in the distribution of labels given in Fig. 14.1. It reveals that the dialog phase after each barrier has its own distribution of several emotional states. This shows that the experimental barriers evoke different reactions by the users. The distribution of emotional states after each barrier confirms the expected reaction. To select the barriers worth for later automatic analyzes, the amount of the user's speech data has to be taken into account. As for CLB and WIB, the user is hardly involved as only information is presented; further experiments are conducted between BSL and WLB. Further details can be found in [37, 39].

Furthermore, we analyze **discourse particles (DPs) as an interaction pattern**. During Human–Human interaction (HHI) several semantic and prosodic cues are exchanged between the interaction partners and used to signalize the progress of

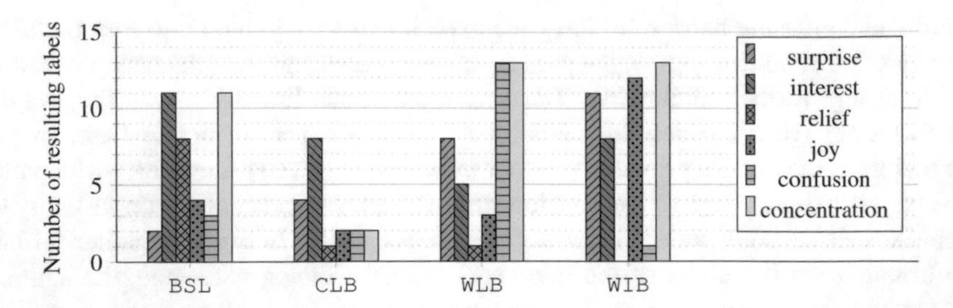

Fig. 14.1 Distribution of emotions over the dialog barrier phases of LMC gathered by manual labeling on a subset of 13 subjects (cf. [39]). `BSL` denotes the baseline, `CLB` the listing barrier, `WLB` the weight limit barrier, and `WIB` the weather information barrier

the dialog [1]. Especially, the intonation of utterances transmits the communicative relation of the speakers and also their attitude towards the current dialog. Thus, DPs can be seen as pattern exposing information about the current interaction.

Furthermore, it is assumed that these short feedback signals are uttered in situations of a higher cognitive load [5] where a more articulated answer cannot be given. As, for instance, stated in [23, 35], specific monosyllabic verbalizations, the DPs, have a specific intonation. In [35] it is stated that DPs like "hm" or "uhm" cannot be inflected but can be emphasized and are occurring at crucial communicative points. The DP "hm" is seen as a "neutral consonant" whereas "uh" and "uhm" can be seen as "neutral vocals". The intonation of these particles is largely free of lexical and grammatical influences. Schmidt called that a "pure intonation" [35].

Additionally, an empirical study of German is presented in [35] determining seven form-function relations of the DP "hm" due to listening experiments. Several studies confirmed the form-function relation for HHI; cf. [20, 27]. In Sects. 14.4.4 and 14.4.5 it is investigated whether these cues are also used within HCI and can serve an indicator for critical parts of the dialog. Furthermore, influences on the usage of DPs are analyzed.

14.4 Results

14.4.1 Discourse Analysis: Age and Gender Matters

Differences in Verbosity The ratio of tokens per turn (TpT) is an adequate verbosity measure for dialogs. Given the different nature of the different phases in the experiment, the measure varies between more narrative-oriented phases and phases with a preference for usually shorter commands (Fig. 14.2).

As a major result for problem solving (without intervention) we get that age group matters and that young subjects are significantly less verbose than elderly one

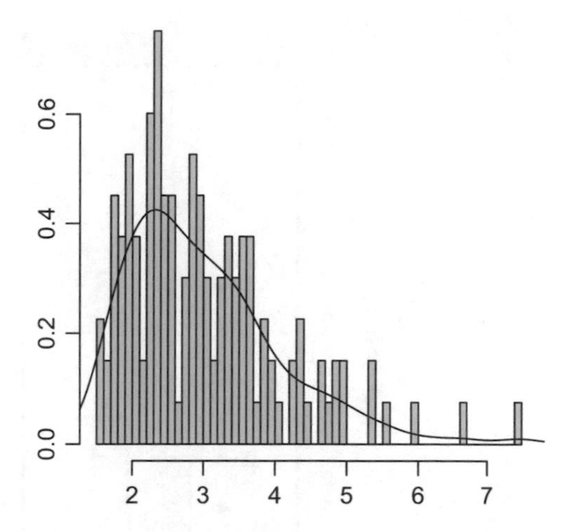

Fig. 14.2 Distribution of tokens per turn (TpT) ratios for problem solving without intervention ($N = 133$)

(cf. Fig. 14.3,[1] Table 14.2; Wilcoxon: $W = 1722, p = 0.03251^*$, $d_{Cohen} = 0.24$). In contrast, gender gives insignificant differences only. In addition, the pairings of age group and gender result in significant differences as well (Kruskal-Wallis chi-squared = 8.375, $df = 3, p = 0.03886^*$).[2] Similar results hold for TpT values for other parts of the experiment. A case in point is, for example, the narratives phase in personalization (cf. Table 14.2).

Politeness Particles as Indicators for CASA When humans conversing with a computer system do employ politeness particles when they address the system, this can be seen as an indicator for (mindlessly) treating Computers as Social Actors (CASA, [26]).

Counting the number of occurrences of politeness particles *'bitte'* (Engl. *'please'*) and *'danke'* (Engl. *'thank you'*) in user utterances per transcript, we get distributions for all N = 133 subjects as depicted in Figs. 14.4 and 14.5. Note: 55 subjects have *not a single occurrence* of one of these politeness particles and the median for all subjects lies at one occurrence.

Again, age matters. The subgroup above the median of counts of used politeness particles is clearly dominated by elderly subjects, whereas the subgroups at the median and below the median are dominated by young subjects (cf. Table 14.3).

Tests show significant differences for the two age groups (Wilcoxon W = 1138, $p = 7.078e - 07^{***}$, $d_{Cohen} = 0.51$) and the four pairings of gender and age group

[1]The distributions are visualized—here and in other figures—as trellis box plots: the rectangles represent the interquartile range (i.e. the range of 25% of the values above and below the median resp.); the filled dot gives the median; the whiskers extending the rectangle extend to the range of values, but maximally to 1.5 of the interquartile range; outlier values beyond the maximal whisker range are given as unfilled dots (cf. [2]).

[2]Unless noted otherwise, all statistical tests and calculations have been performed with the R language [30] [2]. Significance levels are denoted by $^*p < 0.05$; $^{**}p < 0.01$; $^{***}p < 0.001$.

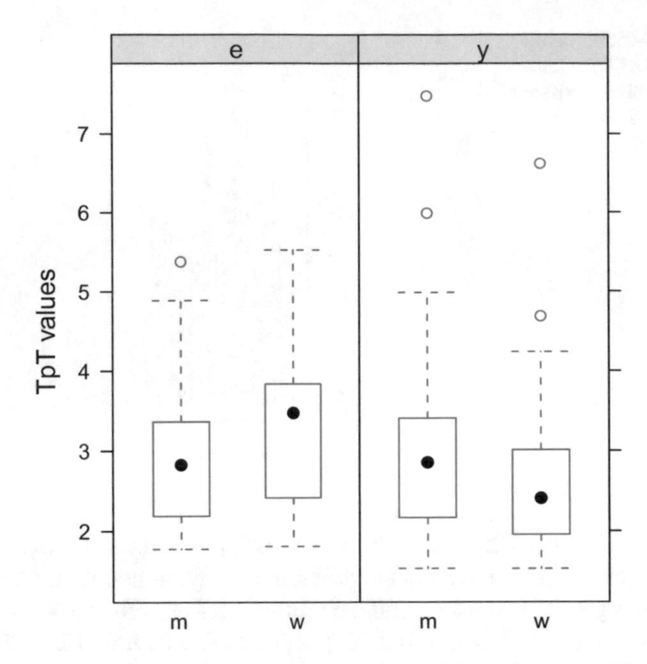

Fig. 14.3 Distributions of tokens per turn (TpT) ratios for problem solving conditioned by gender and age group ($N = 133$; m = men, w = women; e = elder, y = young)

Table 14.2 Differences in mean TpT values for sociodemographic groups

Marker	g1	Rel	g2	*p* value	Test	d_{Cohen}
TpT problem solving	y	lt	e	0.03251*	Wilcoxon	0.24
TpT problem solving	m	lt	w	n.s.	Wilcoxon	n.a.
TpT pers. narratives	y	lt	e	0.00369**	Wilcoxon	0.47
TpT pers. narratives	w	lt	m	n.s.	Wilcoxon	n.a.

(Kruskal chi-squared = 26.0632, df = 3, $p = 9.251e-06^{***}$), but for gender we get insignificant differences only. The most significant and largest pairwise difference between subgroups with respect to the use or non-use of politeness particles is between young women and elderly women (Wilcoxon W = 204, $p = 1.688e{-}06^{***}$, $d_{Cohen} = 0.72$; cf. Fig. 14.5).

The Impact of Age and Gender Young subjects on average use none or significantly fewer politeness particles, they do employ significantly fewer tokens per turn (TpT) in personalization narratives and in problem solving than elderly subjects. In addition, young subjects on average are significantly more successful than elderly subjects with respect to the different dialog success measures (cf. Chap. 13).

When all women vs. all men are contrasted, gender makes no global significant differences with respect to use of politeness particles, tokens per turn (TpT) and differences in control (i.e. wizard-induced category changes; cf. Sect. 14.4.2). The

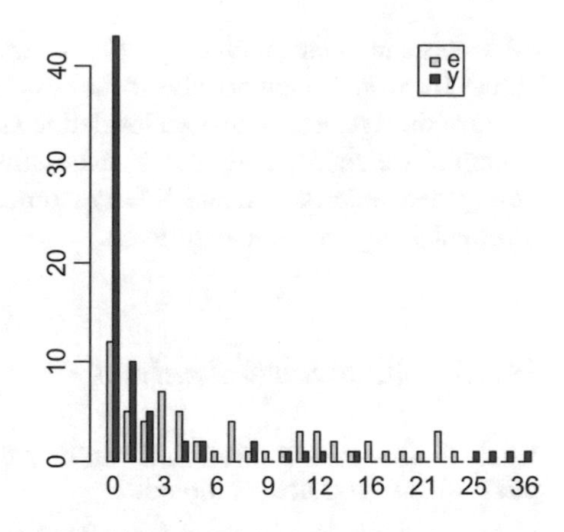

Fig. 14.4 Subgroup comparison for young (*dark bars*; N = 72) vs. elderly subjects (*light bars*; N = 61): Distributions of number of occurrences of politeness particles in user utterances per transcript (please note the number of zero occurrences)

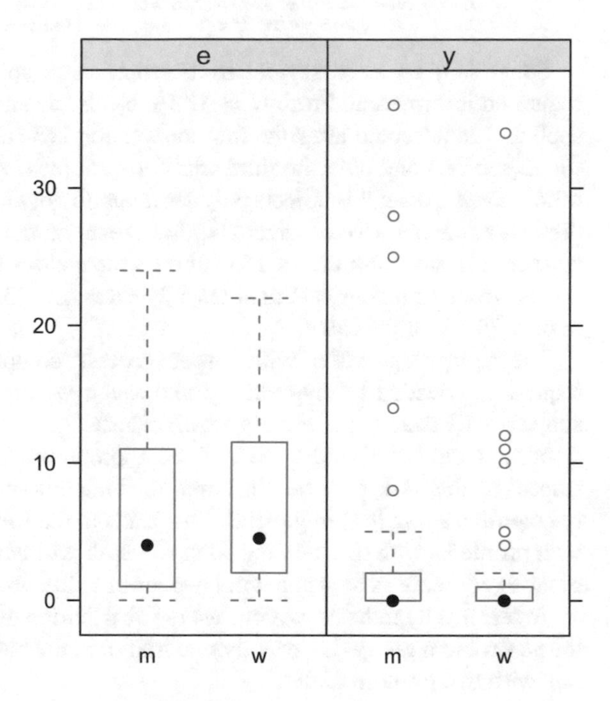

Fig. 14.5 Distributions of number of occurrences of politeness particles in user utterances per transcript, conditioned by age group and gender ($N = 133$; m = men, w = women; e = elder, y = young. Please note the zero medians for the young subgroups and the outliers)

Table 14.3 Distributions of gender and age group pairings for subgroups of subjects in relation to median of used politeness particles

Subgroup	*em*	*ew*	*ym*	*yw*	Sum
Above median	**19**	**25**	11	8	63
AT median	3	2	**6**	**4**	15
Below median	7	5	**19**	**24**	55

Boldface is used to emphasize the two largest values in each row of this table differences in means with respect to the dialog success measures (with larger mean values for women) are not significant. But there are significant differences between some of the age group and gender-defined subgroups of subjects. Young women, for example, are significantly more successful (in both dialog success measures) than young men, elderly men and elderly women. Pairwise differences between the latter three subgroups are not significant.

14.4.2 *Behavioral Analyses*

Early Problems with 'Tell and Spell' At the very beginning of the personalization phase all subjects are prompted:

```
Bitte nennen und buchstabieren Sie zunächst Ihren Vor- und Zunamen!
Please tell and spell your first name and surname!
```

Some subjects need several trials; some even completely fail to provide the requested information. From $N = 133$ subjects the answer to the prompt 'Tell and spell ...' is accepted after the first answer for 113 subjects, after the second trial for 12 subjects and after the third trial for 8 subjects. Actually the task completion ratio is even worse: 20 subjects only *spell* but do not tell their name; two more leave the first name out. (Note: wizards did not react on these latter types of incomplete answers.) In sum: from $N = 133$ subjects the answer to the prompt 'Tell and spell ...' is wrong or incomplete in at least 34 cases (i.e. 25.6%); full task completion is in only 74.4% of the cases.

The age groups differ with respect to task completion: exactly two spelling request are needed by five elderly and seven young subjects, whereas the eight subjects with exactly three trials are all elderly.

Why should 'tell and spell ...' be a problem? The failure of subjects with respect to this task may be attributed to 'inattentional deafness' [7] or to effects of cognitive aging [44] in general. This leads to the following **hypothesis**: Subjects with problems with the 'tell and spell ...' task will have problems with other parts of the experiment as well and will have lower values in the dialog success measures.

To test this hypothesis we contrast the distribution of dialog success measures for the no problem group (i.e. exactly one trial) and the complementary problem group (i.e. with two or more trials).

The difference in mean for DSM2 (no problem: 0.7075; problem: 0.6612) is significant as a Wilcoxon test reveals (W = 773, $p = 0.02482^*$, $d_{Cohen} = 0.56$; the distribution of the no problem group clearly differs from a normal distribution).

Similar results hold for DSM1: the problem group has poorer dialog success values and—again—these differences between the no problem and the problem group are significant (Wilcoxon: W = 770.5, $p = 0.02382^*$, $d_{Cohen} = 0.4993$).

In sum, problems with the very first task in personalization are an early predictor of later problems in the problem solving dialog of LAST MINUTE proper.

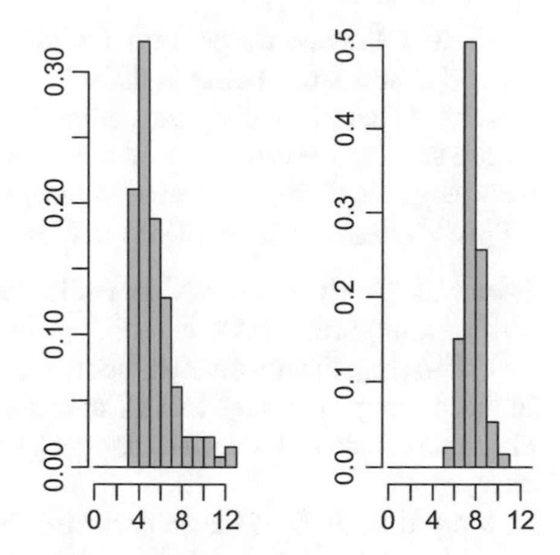

Fig. 14.6 Distributions of total number of user turns in subphases of personalization per transcript ($N = 133$): data acquisition (*left*), narratives (*right*)

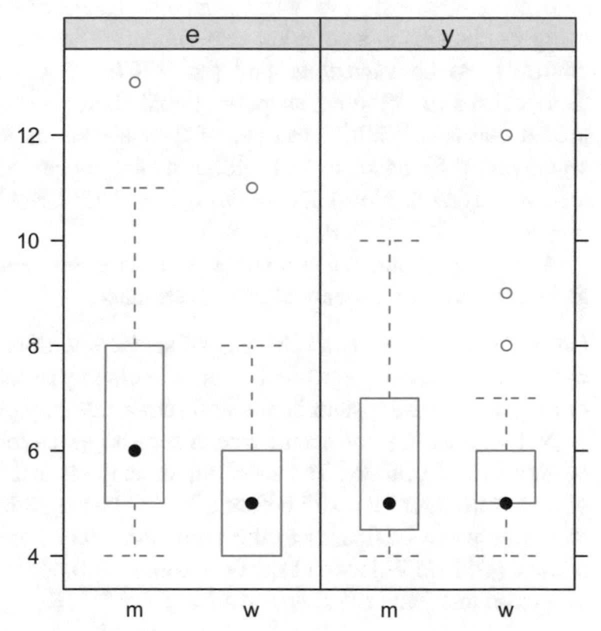

Fig. 14.7 Distributions of total number of user turns in data acquisition per transcript, conditioned by age group and gender ($N = 133$; m = men, w = women; e = elder, y = young)

Early Predictor: Data Acquisition in Personalization In the personalization phase, initiative lies primarily with the system. Here a typical adjacency pair is made up of a wizard prompt or question followed by a user narrative or answer.

In cases of a normal dialog course the sources of variation are reprompts (e.g. 'tell and spell'), the number of questions of the 'bitte ergänzen sie angaben zu ...' type and the number of prompts for 'more detail'. Sources of variation in unforeseen courses are user questions, e.g. caused by understanding problems (Fig. 14.6).

Note: more adjacency pairs in personalization, thus in general, indicate problems. The empirical distributions of the total number of user turns in data acquisition, conditioned by age group and gender, are depicted in Fig. 14.7.

In the following we perform a median split with respect to the total number of turns (i.e. adjacency pairs) in the subphase 'data'. The overall result: the subgroup of subjects below (and at) the median (of 5) has significantly better values for both dialog success measures in problem solving. For both dialog measures, Wilcoxon tests judge the differences between the groups as significant (DSM1: W = 1746, $p = 0.04035^{*}$, $d_{Cohen} = 0.265$; DSM2: W = 1604.5, $p = 0.00718^{**}$, $d_{Cohen} = 0.435$).

Issues of Control: Pauses as Indicators of Helplessness Being in control or not is an important issue in a dialog. In the LM experiments the issue of control is underlying the distinction between two types of category change: subject-induced category change (SICC, the subject explicitly utters a request for category change) vs. wizard-induced category change (WICC, the wizard enforces a category change).

More than 50% of the subjects are 'in control' in this sense. They have either zero or only one or two wizard-induced category changes (from a total of 14 category changes in a complete experiment). The complement of this group ('poor control') has between three and ten WICCs. Poor control of category changes (i.e. `WICCs > 2`) predicts poor global dialog success. The two subgroups—at and below the WICC median of 2 or above the WICC median, resp.—show significant differences in both global dialog success measures (DSM1: Wilcoxon test, W = 1667.5, $p = 0.02614^{*}$, $d_{Cohen} = 0.45$; DSM2: Wilcoxon test, W = 1139, $p = 3.610 \times 10^{-06***}$, $d_{Cohen} = 0.96$).

Again: age group makes a major difference between the two subgroups whereas gender differences are only of minor relevance.

Long Pauses There is a subgroup of subjects with poor control that—after some choices in a category—passively wait without any further action, sometimes for 40 s or longer, until the system finally enforces a category change (WICC).

Not surprisingly, the occurrence of such a type of long pause is again a predictor of global dialog failure. The subgroup of subjects that have at least one occurrence of a pause longer than 10 s before a WICC has significantly poorer dialog success measures when compared to the complementary group of subjects without such pauses. (DSM1: Wilcoxon test, W = 268, $p = 0.003109^{**}$, $d_{Cohen} = 1.71$; DSM2: Wilcoxon test, W = 138.5, $p = 4.87 \times 10^{-05***}$, $d_{Cohen} = 2.36$).

14.4.3 Results from Analysis of Psychometric Data

First of all, the influence of the intervention on the general experimental course was examined. The authors used repeated-measures ANOVA for data analysis to analyze the effects of interpersonal problems (IIP) on dialog exchange. One ANOVA was conducted with intervention (intervention $N = 62$ vs. control group $N = 68$) as

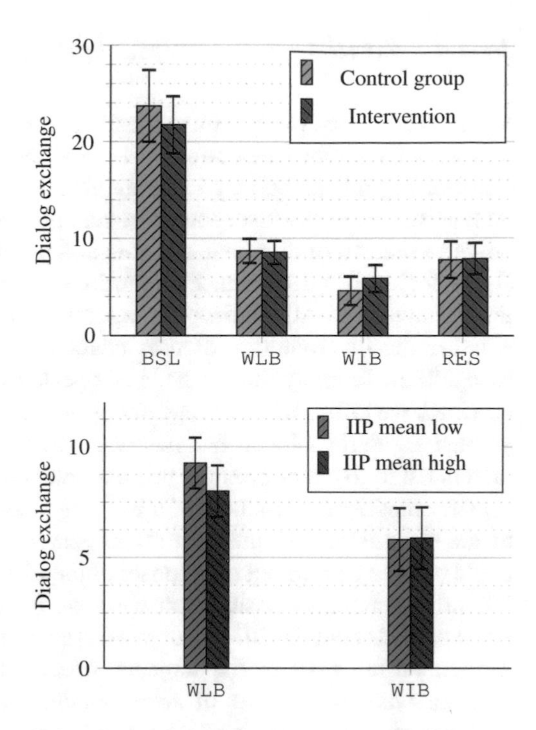

Fig. 14.8 Dialog exchange (control and intervention group) during the experimental course

Fig. 14.9 Dialog exchange considering the level of interpersonal problems (high vs. low) between WLB and WIB

between subject factor. This revealed a significant interaction of intervention on dialog exchange over time ($F(1.89, 242.36) = 3.39$, $p < 0.038$), indicating that subjects who had received an intervention had a higher level of dialog exchange. An observation of dialog exchange over time revealed the difference between intervention and control group at the BSL already ($d_{Cohen} = 0,29$); see Fig. 14.8. At this point, the course did not differ in both groups yet.

The next step was to examine the impact of interpersonal problems on the effectiveness of a system-initiated intervention. Averaged over all test intervals, this revealed no statistical significant difference ($F(1, 60) = 2.02$, $p < 0.160$) between the extent of interpersonal problems (dichotomization by means of a median split led to groups with high vs. low levels of interpersonal problems) and the response behavior of a system-initiated intervention. However, the descriptive account indicates that participants with pronounced interpersonal problems show a lower dialog exchange prior to WIB and intervention ($d_{Cohen} = 0,55$). This difference can't be identified after the intervention anymore ($d_{Cohen} = 0,03$). Both groups (IIP level high vs. low) show no statistically significant difference regarding the amount of dialog exchange. Figure 14.9 shows how the intervention seems to be more effective for participants with a high degree of interpersonal problems.

14.4.4 Results from Analysis of Audio Records

In this section, we present our acoustic analysis to identify whether a user has no problems within the interaction, or is experiencing a barrier. Therefore, we define a two-class problem and try to distinguish the dialog phases after the `BSL` and the `WLB` events, where the user should be set into a certain clearly defined condition. As we have seen from discourse analysis and psychometric data analysis and already shown for acted and spontaneous emotions, we can identify different speaker groups that behave differently during this naturalistic interaction. We therefore investigate whether the incorporation of user characteristics can improve the speech-based recognition. To apply our methods of Speaker Group Dependent (SGD)-modeling on LMC, we utilize the same age and gender groupings as in the previous sections: young vs. elderly subjects and men vs. women. The combination of both grouping factors led to four sub-groups: (`ym`, `em`, `yw`, and `ew`).

The emotional assessment of the dialog phases is described in Sect. 14.3.3. Due to the quite time-consuming work to generate the transcripts, these experiments could only be carried out on a subset of the LMC, containing just $N_{acoustic-HSDP2} =$ 79 participants. As classification baseline, the Speaker Group Independent (SGI) set is used. It contains all 79 subjects. Age and gender of the speakers are known a priori on the basis of the subjects' transcripts. Different age-gender groupings together with the number of corresponding subjects are depicted in Fig. 14.10. Training and testing is based on the subjects' utterances of the two dialog phases after the `BSL` and `WLB`. These utterances are extracted automatically on the basis of the transcripts. This results in 2301 utterances with a total length of 31 min. Furthermore, the following acoustic characteristics are utilized as features: 12 Mel-Frequency Cepstral Coefficients, Zeroth cepstral coefficient, Fundamental frequency, and Energy. The Δ and $\Delta\Delta$ regression coefficients of all features are used to include contextual information. As channel normalization technique, RelAtive SpecTrAl (RASTA)-filtering is applied. Gaussian Mixture Models (GMMs) with 120 mixture components utilizing four iteration steps are used as classifiers. For validation we use a Leave-One-Speaker-Out (LOSO) strategy. As performance measure, the unweighted average recall (UAR) is applied.

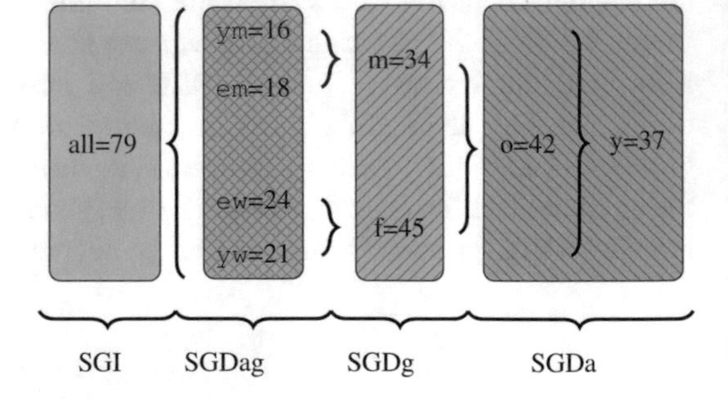

Fig. 14.10 Subjects' distribution into speaker groups (SG) on LMC. Abbreviations: I = independent set, D = dependent on a = age, g = gender

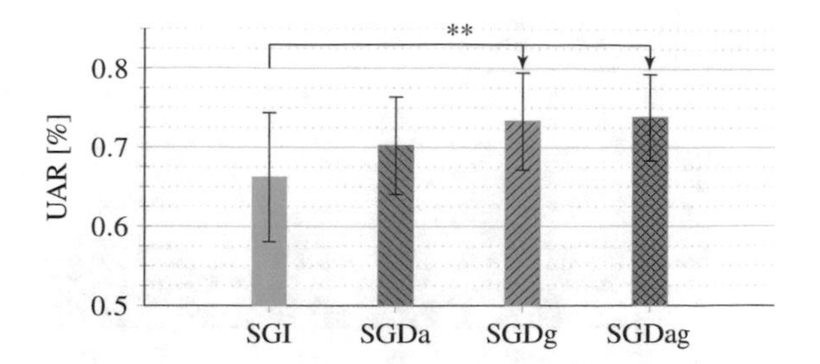

Fig. 14.11 Recognition results for SGI and different SGD configurations. The *stars* denote the significance level: **($p < 0.01$) using ANOVA

Afterwards, we performed the experiments on the SGI set as well as in sets grouped according to age (a) and gender (g), training the corresponding classifiers in an LOSO manner. To allow a comparison, we combined the different results of all speaker groupings. For instance, the results for each male and female speaker are put together to get the overall result for the SGDg set. This result can then be directly compared with results gained on the SGI set. The outcome is shown in Fig. 14.11.

The classification achieved with LMC shows that SGDag grouping can significantly outperform the SGI results with a rate of 73.3%. The improvement is significant for SGI to SGDg ($F = 8.706, p = 0.0032, d_{Cohen} = 0.492$) and SGI to SGDag ($F = 10.358, p = 0.0013$ $d_{Cohen} = 0.526$). Both comparisons are within the zone of desired effects, after Hattie [15]. When comparing the achieved UARs utilizing either age or gender groups, it can be seen that the gender grouping outperforms the age grouping. Further details can be found in [37, 41].

Discourse Particles as Interaction-Patterns in HCI In the following, we analyze whether discourse particles (DPs) can be seen as interaction patterns occurring at critical situations within an HCI. We start by using the whole session and analyzing global differences in DP usage. Afterwards, the local usage within significant situations is analyzed, referring to the `WLB` barrier. All investigations are performed utilizing $N_{acoustic-SH66} = 90$ subjects of LMC. Based on the transcripts, all DPs are automatically aligned and extracted, utilizing a manual correction phase. The extraction results in a total number of 2063 DPs, with a mean of 23.18 DP per conversation and a standard deviation of 21.58. This result shows that DPs are used in our HCI experiments, although the conversational partner, the technical system, was not enabled to express them or react to them. The average DP length is approx. 1 s$\pm$0.4 s. Only 2600 tokens from all 82,000 tokens represent DPs, illustrating the small number of uttered DPs. As a statistical test, an one-way non-parametric ANOVA is used to compare the means of our two median-split samples [22].

To provide valid statements on the DP usage in a naturalistic HCI within the different SGD groups, two aspects have to be taken into account. The first aspect is the verbosity, denoting the number of tokens a speaker has made during the experiment. We analyze both verbosity and DP usage over the dialog-phases starting after specified barrier-events. As a second aspect, the usage of DP depending on age and gender of the subjects is analyzed, analogously to our previous approach in affect recognition. We again use the same speaker grouping; see Fig. 14.10. From the

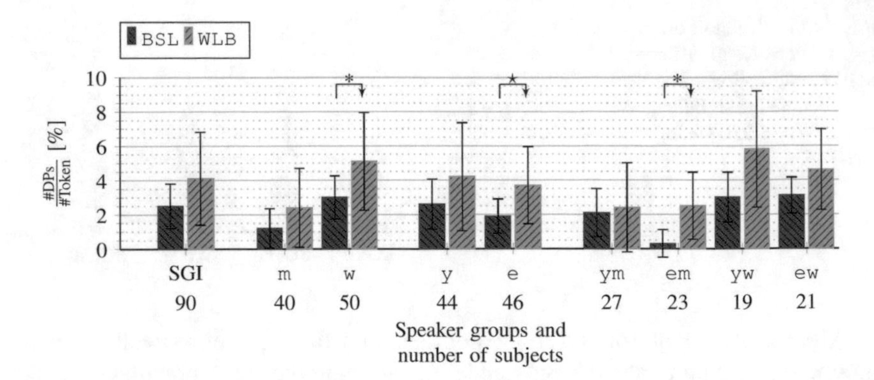

Fig. 14.12 Mean and standard deviation for the DP usage distinguishing the dialog phases after BSL and WLB regarding different speaker groups in LMC. For comparison, the group-independent frequency (SGI) is given. The significance level with ($p < 0.05$) is denoted with *, *star* denotes close proximity to significance level. The effect size is within the zone of desired effect according to [15], using d_{Cohen}

barriers' description, we assume that a higher cognitive load due to the re-planning task WLB increases the DP usage, since DPs are known to indicate a high cognitive load (cf. [5]). For the analysis of this assumption, we calculated the relation of uttered DP and verbosity within the dialog phases after both dialog barriers and distinguished this from the previously used speaker groupings. Both comparisons are within the zone of desired effects, reaching from 0.4 to 1.0, after Hattie [15]. The results are depicted in Fig. 14.12.

Regarding the DP usage between the two dialog barriers BSL and WLB, it is apparent that for all speaker groupings the average number of DPs for WLB is higher than for BSL. This is significant for the speaker group w, and near the significance level in the speaker group e. This observation supports the statement from [29] that male users and young users tend to have less problems to overcome the experimental barriers, confirming the findings of Sect. 14.4.1 that the age and the gender of the subjects matter. Considering the combined age-gender grouping, only for the em grouping can a significant difference between BSL and WLB be observed. Thus, it can be summarized that DPs are capable of serving as interaction patterns indicating situations where the user is confronted with a critical situation in the dialog [40]. This investigation reveals the need to detect and interpret these signals (cf. [25, 28]).

14.4.5 *Combined Analyses of the Impact of Personality Traits and Discourse Particles*

As one can see from the previous section, particularly for the two groups yw and ew, the standard deviation for the DP usage after the WLB is quite high. This also indicates that other factors influence the individual DP usage. Therefore, we analyze

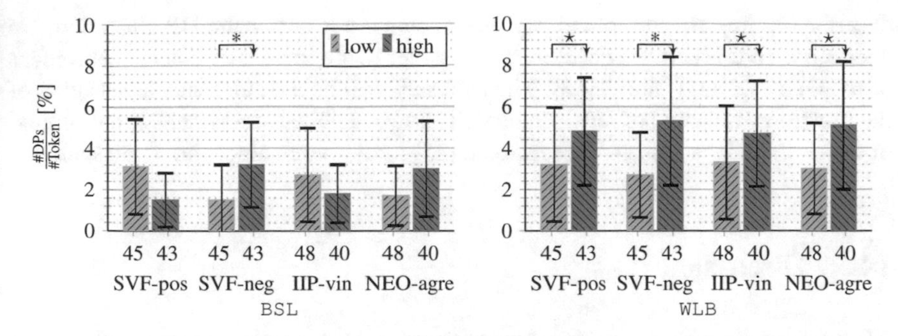

Fig. 14.13 Mean and standard deviation for the frequency of the DPs of the two barriers regarding different groups of user characteristics. The significance level with ($p < 0.05$) is denoted with *, *star* denotes close proximity to significance level using ANOVA

personality traits as additional kinds of factors. Among the gathered personality traits (cf. Chap. 13), we chose those which are in connection with stress coping. To analyze whether a specific personality trait influences DP usage, we utilized the same sample of $N_{acoustic-SH66} = 88$ and used a median split to distinguish between users with low traits (below median) and those with high traits (above median). The resulting numbers of subjects for each group are given in Fig. 14.13. We tested all personality traits available for the LMC, but report only those factors close to the significance level. These factors are determined by the following personality questionnaires: NEO Five-Factor Inventory (NEO-FFI) [6], Inventory of Interpersonal Problems (IIP-C) [18], and Stress-coping questionnaire (SVF) [19].

Considering each psychological trait, no significant differences are noticeable between the two dialog styles "personalization" and "problem solving". In addition to the analysis based on the two experimental phases, we also investigated the different usage of DPs between the dialog phases following the experimental events `BSL` and `WLB` regarding the personality trait factors `SVF-pos`, `SVF-neg`, `IIP-vin`, and `NEO-agre` (cf. Fig. 14.13). In this case, the `SVF-neg` factor shows significant results to distinguish between the low and high groups for both `BSL` ($H = 6.340$, $p = 0.012$, $d_{Cohen} = 0.452$) and `WLB` ($H = 4.617$, $p = 0.032$, $d_{Cohen} = 0.497$), whereas for `SVF-pos`, `IIP-vin` and `NEO-agre` users belonging to the high group show an increased DP usage after the `WLB` barrier that is close to the significance level (cf. [42]).

Thus, it can be assumed that "negative" psychological characteristics stimulate the usage of DPs. A person having a bad stress regulation capability will be more likely to use DPs in a situation of higher cognitive load [5] than a person having a good stress regulation capability.

The investigations presented in this section reveal that the occurrences of DPs can provide hints about specific situations of the interaction. We show that not just the mere occurrence of the DPs is essential, but also the context in which they are used. DPs are occurring more frequently in situations of a higher cognitive load and thus are an important interaction pattern. For the automatic usage of this phenomenon, described in this chapter, obviously further steps, e.g. automatic DP allocation, are necessary. To this end, we developed a classifier to automatically detect occurrences of the DPs "äh" and "ähm" [28]. The next step will be the automatic assessment of the functional meaning of DPs, which will allow a detailed assessment of the context in which the DPs are used. Therefore, in [25] a rule-based algorithm is presented.

14.5 Discussion

We have reported on a number of investigations of three different research groups into data from the LMC and on their respective results. Some results have cross-fertilized other work. For example, the findings about the significant differences between age group- and gender-based subgroups of the whole cohort of subjects (cf. Sect. 14.4.1) have inspired the experiments with training of subgroup-specific classifiers (cf. Sect. 14.4.4) and the investigation of differences between these subgroups with respect to use of discourse particles (cf. Sect. 14.4.4). Some investigations yielded negative results. For example, no significant effect of the intervention has been found in the data of the LMC (cf. Sect. 14.4.3 and Chap. 13). The strongest results—both with respect to significance levels as well as effect sizes—were achieved from the in-depth behavioral analyses of the transcripts (cf. Sect. 14.4.2).

Major Insights from Analyses User groups based on sociodemographics matter in the LMC. This holds especially for the differences between young and elderly subjects, with the former being more successful *on average*. On the other hand, gender matters only when taken into account as a further subcondition after an age group-based primary grouping.

One reason for communication problems in the LMC is that some subjects have difficulties comprehending and memorizing information that was given as spoken language utterances by the system. Such problems occur significantly more often with elderly subjects. Early occurrences of such problems in speech understanding are a strong predictor of global failure of the (independent) later problem solving dialog (cf. Sect. 14.4.2). Another strong indicator of a potential user problem is an overly long pause when the user actually has the turn, i.e. the right for the next utterance (cf. Sect. 14.4.2).

Design Considerations for *Companion*-Systems The findings from the analyses of the dialogs in the LAST MINUTE corpus may contribute to design considerations for *Companion*-Systems that are based on speech interaction with their users [34]. On the one hand, differences between sociodemographic groups—especially differences between age groups—have to be taken into account by the dialog management of *Companion*-Systems. On the other hand, the broad variance between individuals [44] demands personalized calibration of dialog management strategies. Tests that are easy to perform and evaluate and that have large predictive power for potential problems in the subsequent global dialog course—cf. Sect. 14.4.2—may be employed for this purpose. In addition the dialog history of the user-*Companion* interactions needs to be monitored continuously. Special emphasis shall be given to situations where the user has the turn but does not take it within a certain time span. As discussed in Sect. 14.4.2, such overly long pauses are strong indicators of problems and possibly helplessness on the user's side, and demand an adequate response by the system.

Contributions: The results reported in this paper have been contributed by three different groups. The responsibility for discourse analysis and behavioral analysis of transcripts (e.g. Sects. 14.4.1 and 14.4.2) lies with D. Rösner, Th. Bauer and St. Günther; for analysis of psychometric data (e.g. Sect. 14.4.3) responsibility lies with J. Frommer and M. Haase; and for analysis of audio records (e.g. Sect. 14.4.4) responsibility lies with A. Wendemuth and I. Siegert. For the discussion and the conclusions in Sect. 14.5, the responsibility is shared by all authors.

Acknowledgements This work was done within the Transregional Collaborative Research Centre SFB/TRR 62 "*Companion*-Technology for Cognitive Technical Systems" funded by the German Research Foundation (DFG).

References

1. Allwood, J., Nivre, J., Ahlsén, E.: On the semantics and pragmatics of linguistic feedback. J. Semant. **9**, 1–26 (1992)
2. Baayen, R.: Analyzing Linguistic Data – A Practical Introduction to Statistics Using R. Cambridge University Press, Cambridge (2008)
3. Bakhtin, M.M., Holquist, M., McGee, V., Emerson, C.: Speech Genres and other Late Essays, vol. 8. University of Texas Press, Austin (1986)
4. Bickmore, T., Gruber, A., Picard, R.: Establishing the computer–patient working alliance in automated health behavior change interventions. Patient Educ. Couns. **59**(1), 21–30 (2005)
5. Corley, M., Stewart, O.W.: Hesitation Disfluencies in Spontaneous Speech: the Meaning of *um*. Lang. Ling. Compass. **2**, 589–602 (2008)
6. Costa, P.T., McCrae, R.R.: Domains and facets: hierarchical personality assessment using the revised NEO personality inventory. J. Pers. Assess. **64**, 21–50 (1995)
7. Dalton, P., Fraenkel, N.: Gorillas we have missed: sustained inattentional deafness for dynamic events. Cognition **124**(3), 367–372 (2012)
8. Frommer, J., Rösner, D., Haase, M., Lange, J., Friesen, R., Otto, M.: Project A3 - Detection and Avoidance of Failures in Dialogs. Pabst Science Publisher, Lengerich (2012)
9. Funke, J.: Complex problem solving: a case for complex cognition? Cogn. Process. **11**(2), 133–142 (2010)
10. Georgila, K., Wolters, M., Moore, J., Logie, R.: The MATCH corpus: a corpus of older and younger users' interactions with spoken dialogue systems. Lang. Resour. Eval. **44**(3), 221–261 (2010)
11. Grawe, K.: Grundriß einer Allgemeinen Psychotherapie. Psychotherapeut **40**, 130–145 (1995)
12. Gries, S.T.: Quantitative Corpus Linguistics with R: A Practical Introduction. Routledge, Abingdon (2009)
13. Grimm, M., Kroschel, K., Narayanan, S.: The Vera am Mittag German audio-visual emotional speech database. In: Proceedings of the 2008 IEEE ICME, pp. 865–868 (2008)
14. Halliday, M.A.K: Language as a Social Semiotic: The Social Interpretation of Language and Meaning. Hodder & Stoughton Educational, London (1976)
15. Hattie, J.: Visible Learning. A Bradford Book. Routledge, London (2009)

16. Hone, K.: Empathic agents to reduce user frustration: the effects of varying agent characteristics. Interacting Comput. **18**(2), 227–245 (2006)
17. Horowitz, L.M., Alden, L.E., Wiggins, J.S., Pincus, A.L.: Inventory of interpersonal problems manual. Psychological Cooperation, San Antonio (2000)
18. Horowitz, L.M., Strauß, B., Kordy, H.: Inventar zur Erfassung interpersonaler Probleme (IIPD), 2nd edn. Beltz, Weinheim (2000)
19. Jahnke, W., Erdmann, G., Kallus, K.: Stressverarbeitungsfragebogen mit SVF 120 und SVF 78, 3rd edn. Hogrefe, Göttingen (2002)
20. Kehrein, R., Rabanus, S.: Ein Modell zur funktionalen Beschreibung von Diskurspartikeln. In: Neue Wege der Intonationsforschung, Germanistische Linguistik, vol. 157–158, pp. 33–50. Georg Olms, Hildesheim (2001)
21. Klein, J., Moon, Y., Picard, R.W.: This computer responds to user frustration: theory, design and results. Interacting Comput. **14**, 119–140 (2002)
22. Kruskal, W., Wallis, W.A.: Use of ranks in one-criterion variance analysis. J. Am. Stat. Assoc. **47**, 583–621 (1952)
23. Ladd, R.D.: Intonational Phonology. In: Studies in Linguistics, vol. 79. Cambridge University Press, Cambridge (1996)
24. Legát, M., Grůber, M., Ircing, P.: Wizard of oz data collection for the czech senior companion dialogue system. In: Fourth International Workshop on Human-Computer Conversation, pp. 1–4. University of Sheffield (2008)
25. Lotz, A.F., Siegert, I., Wendemuth, A.: Automatic differentiation of form-function-relations of the discourse particle "hm" in a naturalistic human-computer interaction. In: Proceedings of the 26th ESSV, Eichstätt (2015)
26. Nass, C., Moon, Y.: Machines and Mindlessness: Social Responses to Computers. J. Soc. Issues **56**(1), 81–103 (2000)
27. Paschen, H.: Die Funktion der Diskurspartikel HM. Master's thesis, University Mainz (1995)
28. Prylipko, D., Egorow, O., Siegert, I., Wendemuth, A.: Application of image processing methods to filled pauses detection. In: Proceedings of INTERSPEECH'14, Singapore (2014)
29. Prylipko, D., Rösner, D., Siegert, I., Günther, S., Friesen, R., Haase, M., Vlasenko, B., Wendemuth, A.: Analysis of significant dialog events in realistic human-computer interaction. J. Multimodal User Interfaces **8**(1), 75–86 (2014)
30. R Development Core Team: R: A Language and Environment for Statistical Computing. R Foundation for Statistical Computing, Vienna (2010). http://www.R-project.org
31. Rösner, D., Frommer, J., Friesen, R., Haase, M., Lange, J., Otto, M.: LAST MINUTE: a multimodal corpus of speech-based user-companion interactions. In: Proceedings of the 8th LREC, Istanbul, pp. 96–103 (2012)
32. Rösner, D., Friesen, R., Günther, S., Andrich, R.: Modeling and evaluating dialog success in the LAST MINUTE corpus. In: Proceedings of the 9th LREC, Reykjavik (2014)
33. Rösner, D., Andrich, R., Bauer, T., Friesen, R., Günther, S.: Annotation and Analysis of the LAST MINUTE corpus. In: Proceedings of the International Conference of the German Society for Computational Linguistics and Language Technology, pp. 112–121 (2015)
34. Rösner, D., Haase, M., Bauer, T., Günther, S., Krüger, J., Frommer, J.: Desiderata for the Design of Companion Systems – Insights from a Large Scale Wizard of Oz Experiment. Künstliche Intelligenz **30**(1), 53–61 (2016). Online first: Oct 28 (2015). doi:10.1007/s13218-015-0410-z
35. Schmidt, J.E.: Bausteine der Intonation. In: Neue Wege der Intonationsforschung, Germanistische Linguistik, vol. 157–158, pp. 9–32. Georg Olms, Hildesheim, Germany (2001)
36. Schmidt, T., Schütte, W.: Folker: An annotation tool for efficient transcription of natural, multiparty interaction. In: Proceedings of the 7th LREC, Valletta (2010)
37. Siegert, I.: Emotional and user-specific cues for improved analysis of naturalistic interactions. Ph.D. thesis, Otto von Guericke University Magdeburg (2015)
38. Siegert, I., Böck, R., Philippou-Hübner, D., Vlasenko, B., Wendemuth, A.: Appropriate Emotional Labeling of Non-acted Speech Using Basic Emotions, Geneva Emotion Wheel and Self Assessment Manikins. In: Proceedings of the 2011 IEEE ICME, Barcelona (2011)

39. Siegert, I., Böck, R., Wendemuth, A.: The influence of context knowledge for multimodal annotation on natural material. In: Joint Proceedings of the IVA 2012 Workshops, pp. 25–32. Santa Cruz (2012)
40. Siegert, I., Hartmann, K., Philippou-Hübner, D., Wendemuth, A.: Human behaviour in HCI: complex emotion detection through sparse speech features. In: Salah, A., Hung, H., Aran, O., Gunes, H. (eds.) Human Behavior Understanding. Lecture Notes on Computer Science, vol. 8212, pp. 246–257. Springer, Berlin (2013)
41. Siegert, I., Böck, R., Wendemuth, A.: Inter-Rater Reliability for Emotion Annotation in Human-Computer Interaction – Comparison and Methodological Improvements. J. Multimodal User Interfaces **8**, 17–28 (2014)
42. Siegert, I., Haase, M., Prylipko, D., Wendemuth, A.: Discourse particles and user characteristics in naturalistic human-computer interaction. In: Kurosu, M. (ed.) Human-Computer Interaction. Advanced Interaction Modalities and Techniques. Lecture Notes on Computer Science, vol. 8511, pp. 492–501. Springer, Berlin (2014)
43. Webb, N., Benyon, D., Bradley, J., Hansen, P., Mival, O.: Wizard of Oz experiments for a companion dialogue system: eliciting companionable conversation. In: Proceedings of the Seventh Conference on International Language Resources and Evaluation (LREC'10). European Language Resources Association (ELRA), Paris (2010)
44. Wolters, M., Georgila, K., Moore, J., MacPherson, S.: Being old doesn't mean acting old: how older users interact with spoken dialog systems. ACM Trans. Access. Comput. **2**(1), 2:1–2:39 (2009)

Chapter 15
Environment Adaption for *Companion*-Systems

Stephan Reuter, Alexander Scheel, Thomas Geier, and Klaus Dietmayer

Abstract One of the key characteristics of a Companion-System is the adaptation of its functionality to the user's preferences and the environment. On the one hand, a dynamic environment model facilitates the adaption of output modalities in human computer interaction (HCI) to the current situation. On the other hand, continuous tracking of users in the proximity of the system allows for resuming a previously interrupted interaction. Thus, an environment perception system based on a robust multi-object tracking algorithm is required to provide these functionalities. In typical Companion-System applications, persons in the proximity are closely spaced, which leads to statistical dependencies in their behavior. The multi-object Bayes filter allows for modeling these statistical dependencies by representing the multi-object state using random finite sets. Based on the social force model and the knowledge base of the companion system, an approach to modeling object interactions is presented. In this work, the interaction model is incorporated into the prediction step of the sequential Monte Carlo (SMC) of the multi-object Bayes filter. Further, an alternative implementation of the multi-object Bayes filter based on labeled random finite sets is outlined.

15.1 Introduction

In addition to adapting the system behavior to the user's preferences and its cognitive state, a *Companion*-System is expected to adapt to the current environment. An intuitive example is the adaptation of the system's input and output modalities. For example, audio output should not be used for providing confidential information if other persons are in the proximity of the *Companion*-System. Further, the presence of other persons typically increases the uncertainty of speech and gesture input.

S. Reuter (✉) • A. Scheel • K. Dietmayer
Institute of Measurement, Control, and Microtechnology, Ulm University, Albert-Einstein-Allee 41, 89081 Ulm, Germany
e-mail: stephan.reuter@uni-ulm.de; alexander.scheel@uni-ulm.de; klaus.dietmayer@uni-ulm.de

T. Geier
Institute of Artificial Intelligence, Ulm University, James-Franck-Ring, 89081 Ulm, Germany
e-mail: thomas.geier@alumni.uni-ulm.de

S. Biundo, A. Wendemuth (eds.), *Companion Technology*, Cognitive Technologies,
DOI 10.1007/978-3-319-43665-4_15

To realize these capabilities, the *Companion*-System requires an exact model of its environment including all persons in its proximity.

The environment model is realized using a multi-object tracking system which jointly estimates the number of persons as well as the current state of the individual persons. The continuous tracking of the persons in the proximity of the *Companion*-System additionally facilitates the resumption of previously interrupted interactions. Standard multi-object tracking approaches like the Joint Probabilistic Data Association (JPDA) [1] filter, the Joint Integrated Probabilistic Data Association (JIPDA) filter [16], and Multiple Hypotheses Tracking (MHT) [18] are bottom-up approaches which extend the Kalman filter [7] to facilitate the tracking of multiple objects. During the last decade, approximations of the Multi-Object Bayes filter [11] became very popular in multi-object tracking applications. The representation of the multi-object state using random finite sets (RFSs) naturally represents the uncertainty in the number of objects as well as in their individual states. Hence, a realization of an RFS delivers an estimate for the number of persons in the proximity of the *Companion*-System. Since the number of objects in each realization is fixed, an RFS allows for the incorporation of dependencies between the objects, which is not possible in standard multi-object tracking approaches. Especially in crowded environments, the modeling of statistical dependencies between the objects is required since the presence of other objects physically restricts the possible movements of the considered object. A popular approach to modeling the interactions of persons is the Social Force Model [5], which is widely used for the simulation of evacuation scenarios.

This chapter is outlined as follows: First, the basics of random finite sets and multi-object tracking are introduced. In Sect. 15.3, a sequential Monte Carlo implementation of the multi-object Bayes filter as well as possibilities to integrate object interactions in the filtering algorithm are presented. Finally, an accurate and efficient approximation of the multi-object Bayes filter, the labeled multi-Bernoulli filter, is introduced and the differences with the sequential Monte Carlo implementation of the multi-object Bayes filter are illustrated.

15.2 Random Finite Sets and the Multi-object Bayes Filter

Random vectors are typically used to represent the state of an object in single-object tracking. A commonly used approach to applying random vectors for multi-object tracking is to stack the vectors of the individual objects. However, a drawback of this approach is the missing representation of the uncertainty about the number of objects and the ordering of the stacked vectors. In contrast, a random finite set (RFS)

$$\mathrm{X} = \{x^{(1)}, \ldots, x^{(n)}\} \tag{15.1}$$

comprises a random number $n \geq 0$ of unordered points whose states are represented by random vectors $x^{(1)}, \ldots, x^{(n)}$. Hence, an RFS X implicitly captures the uncertainty in the number of objects of the multi-object state. Similarly to single-object tracking, the state of the individual objects is represented using random vectors. Due to the varying number of objects in the sensor's field of view and the possibility of missed detections and false alarms, the measurement process typically returns a random number of measurements. Further, the values of the measurements are also random. Consequently, an RFS

$$\mathrm{Z} = \{z^{(1)}, \ldots, z^{(m)}\} \tag{15.2}$$

is well suited to represent the uncertainty of the measurement process, where $z^{(i)}$ denotes a single measurement. Finite set statistics (FISST) facilitates calculations with RFSs using the notion of integration and density in a way which is consistent with point process theory. Hence, FISST provides a mathematically well-founded way to extend the well-known single-object Bayes filter to multi-object tracking applications using RFSs—the multi-object Bayes Filter [11]. By filtering a finite set-valued random variable over time, the estimate obtained by the multi-object Bayes filter captures the uncertainty in number of objects in addition to the state uncertainty of the individual objects. Similarly to the single-object Bayes filter, the multi-object Bayes filter comprises a prediction and an update step, which are outlined in the following. For additional details as well as the derivations, the reader is referred to [11].

In the prediction or time update step, the multi-object posterior density at time k is predicted to the time of the next measurement. In contrast to single-object tracking, where a prediction of the object's state x to the time of the next measurement using a Markov density $f_+(x_+|x)$ is sufficient, the motion models required for multi-object tracking are far more complex. In addition to the state transition of the individual objects, the multi-object motion model is required to handle object appearance and disappearance. In some applications, it may even be necessary to incorporate object spawning, i.e. that an already existing object originates a new object. Since spawning is not relevant for the environment perception of a *Companion*-System, it is neglected in the following.

The standard multi-object motion model introduced by Mahler [11] comprises the following assumptions:

- an object survives during the transition to the time of the next measurement with probability $p_S(x)$,
- each object is assumed to move independently of other objects in the scene based on a Markov transition density $f_+(x_+|x)$,
- new-born objects follow a Poisson distributed birth density $\pi_B(\mathrm{X})$ which is statistically independent of the persisting objects.

Based on these assumptions, the multi-object Markov density is given by

$$f_+(\mathrm{X}_+|\mathrm{X}) = \pi_B(\mathrm{X}_+)\pi_+(\emptyset|\mathrm{X}) \sum_{\theta} \prod_{i:\theta(i)>0} \frac{p_S(x^{(i)}) \cdot f_+(x_+^{(\theta(i))}|x^{(i)})}{(1 - p_S(x^{(i)})) \cdot \lambda_B p_B(x_+^{(\theta(i))})}. \tag{15.3}$$

Here, the state-dependent survival probability is denoted by $p_S(\cdot)$ and the expected number of new-born objects λ_B as well as the probability density $p_B(\cdot)$ are parameters of the birth model. The sum in (15.3) includes all possible associations $\theta : \{1, \ldots, n'\} \to \{0, 1, \ldots, n\}$; the association $\theta(i) = 0$ represents the disappearance of object i and $\theta(i) > 0$ implies the persistence of object i. The probabilities of all objects being new-born and of the disappearance of all objects are given by

$$\pi_B(\mathrm{X}_+) = e^{-\lambda_B} \prod_{i=1}^{n} \lambda_B p_B(x_+^{(i)}) \tag{15.4}$$

and

$$\pi_+(\emptyset|\mathrm{X}) = \prod_{i=1}^{n'} \left(1 - p_S(x^{(i)})\right). \tag{15.5}$$

Observe that the contribution of a state vector $x^{(i)}$ to (15.4) and (15.5) is canceled out for all associations $\theta(i) > 0$ by the denominator of the product in (15.3).

Using the Chapman-Kolmogorov equation and the multi-object Markov density (15.3), the prediction of the multi-object Bayes filter to the time of the next measurement is given by

$$\pi_+(X_+) = \int f(X_+|X)\pi(X)\delta X, \tag{15.6}$$

where $\pi(X)$ is the prior multi-object density. Observe that the Markov assumption used in (15.6) implies that the multi-object posterior density $\pi(\mathrm{X})$ captures the entire information about the multi-object state at a time k. The integral in (15.6) is a set integral which integrates over all possible cardinalities.

The update step of the multi-object Bayes filter is based on a multi-object likelihood function $g(\mathrm{Z}|\mathrm{X}_+)$ incorporating the single-object likelihood function $g(z|x_+)$, the field of view (FOV) of the sensor, the detection probability, and the false alarm rate. Here, the single-object likelihood function $g(z|x_+)$ provides the likelihood that a measurement z has been generated by object x_+ based on the spatial distance and the corresponding uncertainties. Further, the state-dependent detection probability incorporates the handling of the sensor's FOV. The standard multi-object measurement model [11] is illustrated by Fig. 15.1 and uses the following assumptions:

- A measurement is generated by at most one object and each object is observed by the sensor according to a single-object spatial likelihood function $g(z|x_+)$,
- Each object gives rise to a measurement according to the state dependent detection probability $p_D(x_+)$ and it is not detected with probability $1 - p_D(x_+)$,
- The sensor delivers Poisson-distributed false alarms with mean number of λ_c measurements. The false alarms follow the spatial distribution $c(z)$ which is

Fig. 15.1 Illustration of the events represented by the multi-object likelihood function: object detections (*red rectangles*), missed detections (no measurement for partially occluded person on the *right side*), false alarms (*red dashed rectangles*)

usually modeled by a uniform distribution over the sensor's FOV. Further, the object detection process and the false alarm process are assumed to be statistically independent and the measurements have to be conditionally independent of the objects' states.

In multi-object tracking, the track-to-measurement association, i.e. which measurement belongs to which target, is ambiguous in most scenarios due to the spatial uncertainty of the objects' states and the measurements. Further, the possibility of missed detections and false alarms additionally increases the ambiguity. To handle these ambiguities, the multi-object likelihood function averages over all possible association hypotheses, which is the best one can do if no prior knowledge about the track-to-measurement association is available. Similarly to the multi-object Markov density, the association hypotheses for n objects and m measurements are represented by $\theta : \{1, \ldots, n\} \rightarrow \{0, 1, \ldots, m\}$, where the measurement '0' covers possible missed detections of some of the objects. The assumption that a measurement belongs to at most one of the objects is ensured by $\theta(i) = \theta(j) > 0$ if and only if $i = j$, which uniquely assigns a measurement $z_{\theta(i)}$ to an object i. The missed detection of an object is represented by $\theta(i) = 0$.

Using the associations θ, the multi-object likelihood function covering missed detections and false alarms is given by

$$g(\mathrm{Z}|\mathrm{X}_+) = \pi_C(\mathrm{Z})\pi(\emptyset|\mathrm{X}_+)\sum_{\theta}\prod_{i:\theta(i)>0}\frac{p_D(x_+^{(i)})\cdot g(z_{\theta(i)}|x_+^{(i)})}{(1-p_D(x_+^{(i)}))\cdot\lambda_c c(z_{\theta(i)})}, \tag{15.7}$$

where $p_D(\cdot)$ denotes the state-dependent detection probability and $g(\cdot|\cdot)$ is the single-object likelihood function representing the likelihood of a measurement $z_{\theta(i)}$ given an object with predicted state $x_+^{(i)}$. The expected number of false alarms λ_c and the spatial distribution $c(\cdot)$ are the parameters of the Poisson clutter process. The factor

$$\pi(\emptyset|\mathrm{X}_+) = \prod_{i=1}^{n}(1-p_D(x_+^{(i)})) \tag{15.8}$$

denotes the probability that none of the objects has been detected by the sensor at the current time step and

$$\pi_C(\mathrm{Z}) = e^{-\lambda_c}\prod_{z\in\mathrm{Z}}\lambda_c c(z) \tag{15.9}$$

is the probability that all measurements $z \in \mathrm{Z}$ are originated by the clutter process.

Using (15.7), the multi-object posterior density after integrating the current set of measurements is calculated using the multi-object Bayes filter update,

$$\pi(X|Z) = \frac{g(Z|X)\pi(X)}{\int g(Z|X)\pi(X)\delta X}. \tag{15.10}$$

The recursive update of the multi-object posterior density is consequently realized by applying (15.6) and (15.10) each time a new measurement is obtained. Similarly to the single-object Bayes filter, an analytical implementation of the multi-object Bayes filter is not possible in general. However, the multi-object Bayes filter facilitates an approximation using sequential Monte-Carlo (SMC) methods as well as a closed-form implementation using δ-generalized labeled multi-Bernoulli (δ-GLMB) RFSs which are presented in detail in Sects. 15.3 and 15.4.

Further approximations of the multi-object Bayes filter, which will not be discussed in in this chapter, are the probability hypothesis density (PHD) filter [9], the cardinalized probability hypothesis density (CPHD) filter [10], and the cardinality balanced multi-target multi-Bernoulli (CB-MeMBer) filter [32]. While the PHD and CPHD filters approximate the multi-object posterior by the first statistical moment (and the cardinality distribution in case of the CPHD filter), the CB-MeMBer filter approximates the posterior using a Multi-Bernoulli distribution. Further details as well as implementations of these filters using Gaussian mixture (GM) and SMC methods are given in [10, 28, 30–33].

15.3 SMC Implementation of the Multi-object Bayes Filter and Modeling of Object Interactions

In typical applications of *Companion*-Systems, a high number of humans in the proximity is expected. Obviously, the movement of the individual persons is restricted in these scenarios by the presence of other persons, leading to statistical dependencies in their movements. In the following, the sequential Monte-Carlo (SMC) implementation of the multi-object Bayes filter incorporating object interactions is presented. For further details, the reader is referred to [19, 20, 22] as well as [8, 11, 26, 30].

In the SMC implementation of the single-object Bayer filter, vector-valued particles $x^{(i)} \in \mathbb{R}^n$ are typically used to approximate the spatial distribution $p(x)$. For the sequential Monte-Carlo multi-object Bayes (SMC-MOB) filter, each multi-object particle has to be a sample of a random finite set and is consequently given by a set of state vectors

$$\mathrm{X}^{(i)} \triangleq \left\{x^{(1)}, \ldots, x^{(n)}\right\}, \tag{15.11}$$

where the number of objects n as well as the state vectors $x^{(j)}$ are random. In the following, the state vectors $x^{(j)}$ of the multi-object particle $\mathrm{X}^{(i)}$ are conveniently called "particles". Using the ν multi-object particles, the multi-object probability density is approximated by

$$\pi(\mathrm{X}) \cong \sum_{i=1}^{\nu} w^{(i)} \cdot \delta_{\mathrm{X}^{(i)}}(\mathrm{X}). \tag{15.12}$$

15.3.1 *Prediction*

The prediction step of the SMC-MOB filter has to predict each multi-object particle according to the multi-object Markov density (15.3), incorporating the motion of persisting objects as well as object appearance and disappearance. Consequently, the prediction of a multi-object particle is obtained by the union of the set of surviving particles $\mathrm{X}^{(i)}_{+,S}$ and the set of new-born particles $\mathrm{X}^{(i)}_{B}$:

$$\mathrm{X}^{(i)}_{+} = \mathrm{X}^{(i)}_{+,S} \cup \mathrm{X}^{(i)}_{B}. \tag{15.13}$$

The set of persisting particles of a multi-object particle $\mathrm{X}^{(i)} = \{x^{(1)}, \ldots, x^{(n)}\}$ is obtained by a multi-Bernoulli distribution using the persistence probability $p_S(x^{(j)})$ as a parameter. Since a multi-Bernoulli distribution is the union of several independent Bernoulli distributions, the persistence of each object is assumed to be statistically independent of other objects. Thus, the persistence of a subset

$\{x^{(1)}, \ldots, x^{(n')}\}$ has a probability of

$$\pi\left(\{x^{(1)}, \ldots, x^{(n')}\}|\mathrm{X}^{(i)}\right) = \prod_{x \in \mathrm{X}^{(i)}} (1 - p_S(x)) \cdot \prod_{\tilde{x} \in \{x^{(1)}, \ldots, x^{(n')}\}} \frac{p_S(\tilde{x})}{1 - p_S(\tilde{x})}. \tag{15.14}$$

Instead of drawing the persisting particles directly using (15.14), the independence of the M Bernoulli distributions within the multi-Bernoulli distribution facilitates sampling the persistence of each particle independently. This can be realized by drawing a uniformly distributed random number $\zeta^{(j)}$ for each particle in $\mathrm{X}^{(i)}$. Consequently, the set of persisting particles follows

$$\mathrm{X}_S^{(i)} = \{x : \zeta^{(j)} < p_S(x^{(j)}) \; \forall \, j = 1, \ldots, |\mathrm{X}^{(i)}|\}. \tag{15.15}$$

Hence, a particle only persists if its state-dependent survival probability is greater than the drawn random number. Finally, the $j = 1, \ldots, n'$ persisting particles have to be predicted to the time of the next measurement using a single-object Markov transition density,

$$x_+^{(j)} \sim f_+\left(\cdot|x^{(j)}\right), \tag{15.16}$$

in order to obtain the predicted set of surviving particles:

$$\mathrm{X}_{+,S}^{(i)} = \left\{x_+^{(1)}, \ldots, x_+^{(n')}\right\}. \tag{15.17}$$

The birth process is utilized to obtain the set of new-born particles $\mathrm{X}_B^{(i)}$. Therefore, the number of appearing objects n_B is sampled from a Poisson-distributed cardinality distribution $\rho_B(n)$ with an expectation value of λ_B. The state of the new-born particles is obtained by sampling from the spatial distribution p_B of new-born objects:

$$x_+^{(j)} \sim p_B(\cdot) \; \forall \, j = 1, \ldots, n_B. \tag{15.18}$$

15.3.2 Update

In the update step of the SMC-MOB, the weight of each multi-object particle has to be updated using the multi-object likelihood function (15.7). The usage of a hypotheses tree [12–14] facilitates an intuitive representation of all valid association hypotheses. An example of a hypotheses tree is illustrated by Fig. 15.2 for a scenario with two objects and two measurements. A complete association hypothesis for a multi-object particle corresponds to the path from the root of the tree to a leaf. Since different cardinalities are represented by additional hypotheses trees and the assignment of measurements to the clutter source is realized by the

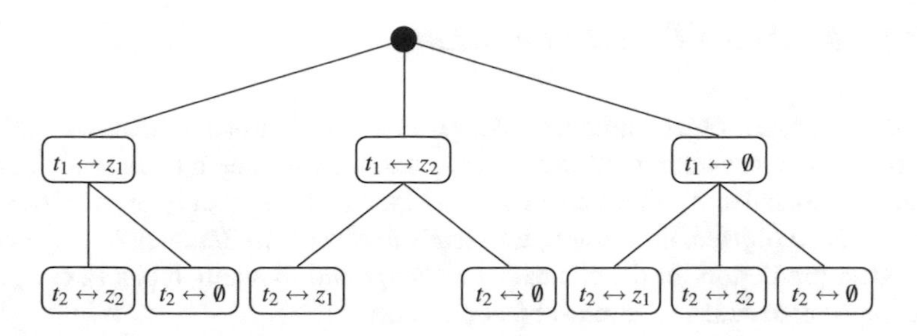

Fig. 15.2 Hypotheses tree for a scenario with two objects, t_1 and t_2, and two measurements, z_1 and z_2. Each node represents an association of the object t_i to measurement z_j (i.e., $\theta(i) = j$) or the missed detection $\emptyset$ (i.e., $\theta(i) = 0$)

factor $\pi_C(\mathrm{Z})$, the hypotheses tree for the SMC-MOB filter is less complex than the one for the joint integrated probabilistic data association (JIPDA) algorithm in [12, 14]. The likelihood of an association $\theta(i)$ corresponds to an edge of the hypotheses tree. The value of each summand of (15.7) is calculated by multiplying the edge likelihoods from the root of the tree to the corresponding leaf. The likelihood of the measurement Z for the multi-object particle $\mathrm{X}_+^{(i)}$ is obtained by accumulating the likelihoods for all paths and a subsequent multiplication with the clutter factor $\pi_C(\mathrm{Z})$ and the missed detection factor $\pi\,(\emptyset|\mathrm{X}_+^{(i)})$. Similarly to [12, 14], the likelihoods for all track-to-measurement assignments are calculated a priori and stored in a lookup table. The implementation of the hypotheses tree using recursion is straightforward.

The update step of the SMC-MOB filter does not affect the state of the multi-object particles, i.e. the posterior multi-object particles are identical to the predicted ones:

$$\mathrm{X}^{(i)} \triangleq \mathrm{X}_+^{(i)}. \tag{15.19}$$

However, the weights of the multi-object particles are updated using the multi-object likelihood function:

$$w^{(i)} \triangleq \frac{\pi\left(\mathrm{Z}|\mathrm{X}_+^{(i)}\right)}{\sum_{e=1}^{\nu} \pi\left(\mathrm{Z}|\mathrm{X}_+^{(e)}\right)}. \tag{15.20}$$

Here, the denominator is a normalizing constant which ensures that the weights still sum up to 1 after the update.

After several measurement updates, the weights typically tend to concentrate on one or only a few multi-object particles since the prediction step increases the variance of the particles and the update does not decrease the variance. Hence, standard resampling approaches used for SMC implementation of the single-object Bayes have to be applied [25, 27].

15.3.3 Modeling Object Interactions

The standard multi-object motion model given by (15.7) assumes that the motion of each object depends only on its current state and the assumed motion model, i.e. the objects are considered to be statistically independent. Especially in scenarios with closely spaced objects, this assumption leads to physically impossible multi-object states after prediction. In the context of a *Companion*-System, typical examples of these impossible states are multi-object particles

In order to avoid invalid multi-object states, an appropriate model for human motion is required. In [6], Henderson observed correlations between fluid dynamics and human motion. However, this approach facilitates only a macroscopic formulation which, e.g., delivers the mean velocity of a group of people. In contrast, the Social Force Model proposed by Helbing and Molnar [5] uses a microscopic model to represent human motion where changes of the individual behaviors due to the current environment are modeled by attractive and repellent force vectors. Repellent forces are typically used to avoid collisions with other persons as well as static obstacles. Attractive forces are used to model the destination of a person. Further, the model is based on the knowledge that each person tries to reach its destination on the shortest path while moving with its desired velocity.

In addition to scenarios with closely spaced objects, the incorporation of object interaction is also recommended for scenarios with occlusions or in the case of low measurement rates. In Sect. 15.3.3.1, an approach based on the incorporation of physical constraints is proposed which avoids collisions of the persons and may be realized without any additional information. Further, Sect. 15.3.3.2 outlines the possibilities to improve the tracking results by using the information available in the *Companion*-Systems knowledge base.

15.3.3.1 Set-Based Weight Adaption

In order to obtain only valid predicted multi-object states, an incorporation of object inter-dependencies in the transition densities is required. Since the computation of such transition densities is computationally demanding, the proposed method predicts all objects within a multi-object particle independently, and a subsequent weight adaption of the multi-object particles is applied to remove invalid ones.

The weight adaption is based on the repellent forces used in the social force model, which are modeled using exponential functions [17]. In the case of circular objects with radius r_p, the likelihood that a multi-object particle comprises two objects s and t follows

$$\Lambda_d(x^{(s)}, x^{(t)}) = \begin{cases} 0 & \text{if } d(x^{(s)}, x^{(t)}) < 2r_p \\ 1 - \exp\left(-\frac{(d(x^{(s)},x^{(t)})-2r_p)^2}{2\sigma_d^2}\right) & \text{otherwise,} \end{cases} \tag{15.21}$$

where $d(x^{(s)}, x^{(t)})$ denotes their Euclidean distance. Obviously, a likelihood of 0 is assigned if two objects are overlapping and the exponential function facilitates a smooth transition of the likelihood function for all distances up to the preferred inter-object distance. Afterwards, the weight of the multi-object particle $\mathrm{X}^{(i)}$ is adapted using the minimum likelihood of all possible pairings (s, t):

$$\widetilde{w}_{+}^{(i)} = \min_{s=1,\ldots,|\mathrm{X}^{(i)}|} \left(\min_{t=1,\ldots,|\mathrm{X}^{(i)}|, t \neq s} \left(\Lambda_d(x^{(s)}, x^{(t)}) \right) \right) \cdot w_{+}^{(i)} . \qquad (15.22)$$

The weight of a multi-object particle is set to 0 if any two of its objects are colliding. In contrast, the weight of the multi-object particle is unchanged if the distances between all of its objects are higher than the preferred distance.

15.3.3.2 Integration of Destinations Using Knowledge Base

A *Companion*-System comprises a multitude of different components, each of which can potentially produce and/or consume information (see Fig. 15.3). As decision making and inference across different modules has to be kept consistent, a central probabilistic *knowledge base* (KB) is tasked with maintaining a global filtered probabilistic belief state X_{KB}. Consistency requires that the local belief state X_{C} of a component C correspond to the marginalization of X_{KB} over the variables not included in C. To achieve this global synchronization, probabilistic state information may flow bidirectionally between the KB and the interfacing modules. By maintaining a globally consistent belief state, the knowledge base provides mutual abstraction between all interfacing components, such that every module has to deal only with its local view of the global state. Such joint treatment of belief across components fosters synergistic effects, which may also improve the state prediction of multi-object trackers. Information originating from high-level components can be used to improve track continuity—in particular in situations

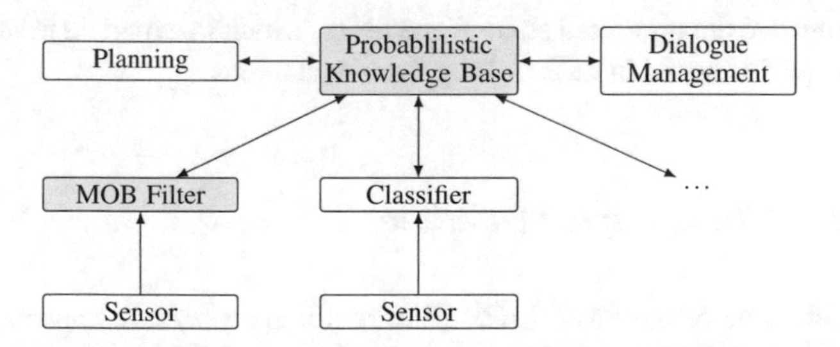

Fig. 15.3 Architecture of a prototypical *Companion*-System. The central knowledge base maintains a filtered belief state. It integrates between lower-level sensor processing modules, like the multi-object tracker and further classifier, and high-level functionality, including decision making/planning and routines of the user interface

featuring occlusions or low measurement rates, where the associations of the tracks between the individual time steps are ambiguous.

Basically every correlation between the global belief state X_{KB} and the true location of a certain object/user can be used to improve the association quality. We can identify several sources of potentially useful information, although a large part of the model is application-dependent. As human users are supposed to interact with the system, we can harvest this interaction to gain hints about their true location. A registered touch event at a stationary device gives a strong indication that a user instead of a non-user is standing in front of the device. Further, knowledge about screen content increases the chance that a user will be moving toward the screen to read the information, even if no touch contact takes place. Besides information originating from the dialog management, we can also use knowledge obtained from planning. The planning component maintains a future course of action for the user [2] to follow. If some actions are known to be connected to a certain location, we can exploit this knowledge to improve the tracking accuracy. An exemplary situation happens when the system issues a job to a printer, and the user is supposed to fetch the produced document. Then this knowledge, in combination with knowledge of the location of the printer, can be used to disambiguate which observed object corresponds to the current user.

To exploit such hints on the true user location, one can follow two approaches. One approach consists of the multi-object tracking algorithm maintaining the track labels only in cases of high confidence, trying to avoid any wrong associations. This results in many spawned tracks belonging to the same object in challenging situations. Then a track to person association can be maintained by the knowledge base using available background information. The feasibility of this approach has been demonstrated successfully in an experimental setting [3, 4] using a probabilistic model formulated with Markov Logic [24]. One major disadvantage could be identified in the requirement of discretizing the user position, because of the limitations of the modeling language. The second approach is to improve the performance of the multi-object tracking algorithm itself by integrating the hints about current and future positions of human users. The (typically imprecise) knowledge about future destinations of the user may be used within the Social Force Model to improve the predicted state of the users, which is expected to significantly improve the performance in case of long term occlusions.

15.3.4 Real-Time Implementation

An implementation of the SMC-MOB filter requires a very large number of multi-object particles to obtain a sufficient approximation of the multi-object posterior. The reason for this is that the dimension of the space of the multi-object particles is given by the dimension of the state vector times the number of objects in the scene. The prediction and update steps of the SMC-MOB filter facilitate a massively parallel implementation since the calculations for each multi-object particle do not

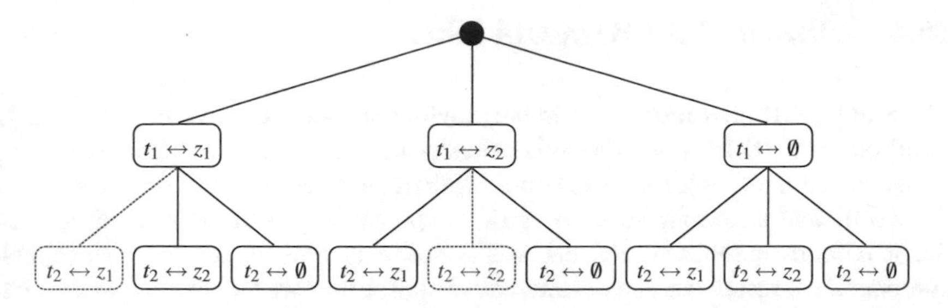

Fig. 15.4 Approximate multi-object likelihood function: compared to Fig. 15.2, two additional nodes (marked by *red dashed lines*) have been added which facilitate the assignment of one measurement to multiple tracks

depend on any other multi-object particle. Consequently, graphics processing units (GPUs) are well suited for the implementation of the SMC-MOB filter.

Due to the combinatorial complexity and the restrictions concerning recursive functions on GPUs, an exact computation of the multi-object likelihood function is only feasible for a limited number of tracks and measurements. The reason for the complexity is the assumption that a measurement is created by at most one object. Neglecting this assumption, the multi-object likelihood function simplifies to [19]

$$\widetilde{g}(\mathrm{Z}|\mathrm{X}_+) = \pi_C(\mathrm{Z})\pi(\emptyset|\mathrm{X}_+) \cdot \prod_{i=1}^{n} \left(1 + \sum_{j=1}^{m} \frac{p_D\left(x_+^{(i)}\right) \cdot g\left(z_j|x_+^{(i)}\right)}{\left(1 - p_D(x_+^{(i)})\right) \lambda_c c(z_j)} \right), \tag{15.23}$$

i.e. the multi-object likelihood may be calculated using two for loops and the computational complexity reduces to $\mathcal{O}(mn)$, where n is the number of objects and m is the number of measurements. The corresponding hypotheses tree for two measurements and two tracks is depicted by Fig. 15.4. Obviously, the approximation leads to two additional nodes in the tree (marked by dashed lines) and the approximation error is negligible if each measurement has a significant likelihood for at most one object, i.e. the association hypotheses due to the two additional nodes have an insignificant contribution to the multi-object likelihood function. Modeling object interactions as presented in Sect. 15.3.3; the approximation errors are negligible if the extent of the objects is significantly larger than the standard deviation of the measurement noise.

In [21, 22], it is shown that the proposed approximation of the multi-object likelihood function facilitates a real-time capable implementation of the SMC-MOB filter using a GPU. With a total number of 25,000 multi-object particles, an Nvidia Tesla C2075 GPU processes the prediction step, the update step and the track extraction in less than 40 ms for a scenario with up to seven objects.

15.4 Labeled Multi-Bernoulli Filter

The SMC-MOB filter introduced in the previous section requires a huge number of multi-object particles since the dimension of the sample space increases linearly in the number of objects. Since the required number of multi-object particles for a sufficient state representation grows exponentially with the state dimension, the maximum number of objects in the scene is limited due to the available computational resources. Hence, alternative approaches are required to handle large numbers of objects.

In [29], Vo and Vo showed that the class of δ-generalized labeled multi-Bernoulli (δ-GLMB) RFSs is closed under the prediction and update equations of the multi-object Bayes filter[1] for the standard multi-object motion model as well as the standard multi-object likelihood. Hence, the δ-GLMB filter [29] facilitates an analytical implementation of the multi-object Bayes filter. Similarly to the SMC-MOB filter, the number of components required within the δ-GLMB filter is combinatorial. The labeled multi-Bernoulli (LMB) filter [19, 23] approximates the δ-GLMB distribution by an LMB distribution which facilitates the tracking of a huge number of objects due to the application of principled approximations.

15.4.1 Labeled Random Finite Sets

The SMC-MOB filter as well as the PHD, CPHD, and CB-MeMBer filters require a (typically heuristic) post-processing to extract object tracks out of the estimated multi-object probability density. The underlying idea of the class of labeled RFSs is to augment the state by track labels. Thus, filtering a labeled RFS over time delivers a joint estimate of the number of tracks, their individual positions as well as their trajectories.

In a labeled RFS, each object's state $x \in \mathbb{X}$ is augmented by a label $\ell \in \mathbb{L}$, where $\mathbb{L}$ is a discrete label space (e.g., the set of positive integers $\mathbb{N}$). Consequently, a labeled multi-object state is represented by the set $\mathbf{X} = \{\mathbf{x}^{(1)}, \ldots, \mathbf{x}^{(n)}\}$ on the space $\mathbb{X} \times \mathbb{L}$, where the labeled state vectors are abbreviated using $\mathbf{x} = (x, \ell)$. In multi-object tracking applications, it is required that the object labels are distinct, i.e. a label ℓ may be assigned to at most one object in each realization. In order to ensure distinct labels within each realization of a labeled RFS $\mathbf{X}$, the distinct label indicator [29]

$$\Delta(\mathbf{X}) = \delta_{|\mathbf{X}|}(|\mathscr{L}(\mathbf{X})|) \tag{15.24}$$

[1] Observe that the number of components of the δ-GLMB distribution increases during these steps.

requires the cardinality of a realization $\mathbf{X}$ to be equal to the number of distinct track labels $|\mathscr{L}(\mathbf{X})|$, where the set of track labels is given by the projection

$$\mathscr{L}(\mathbf{X}) = \{\mathscr{L}(\mathbf{x}) : \mathbf{x} \in \mathbf{X}\} \tag{15.25}$$

with $\mathscr{L}(\mathbf{x}) = \mathscr{L}((x,\ell)) = \ell$.

15.4.1.1 Labeled Multi-Bernoulli Random Finite Set

The representation of the uncertainty about object existence is intuitively realized using a Bernoulli RFS X. With the existence probability r, the Bernoulli RFS is given by a singleton. Consequently, the RFS X corresponds to the empty set with probability $1 - r$. The probability density of a Bernoulli RFS follows [11, pp. 368]

$$\pi(\mathrm{X}) = \begin{cases} 1 - r, & \text{if } \mathrm{X} = \emptyset, \\ r \cdot p(x), & \text{if } \mathrm{X} = \{x\}, \end{cases} \tag{15.26}$$

where $p(x)$ is the spatial distribution of the object on the space $\mathbb{X}$. Obviously, the cardinality distribution follows a Bernoulli distribution with parameter r. A multi-Bernoulli distribution X [11] is the union of M independent Bernoulli RFSs $\mathrm{X}^{(i)}$, i.e. $\mathrm{X} = \cup_{i=1}^{M} \mathrm{X}^{(i)}$.

By interpreting the component indices of the multi-Bernoulli distribution as track labels, the LMB RFS [29] is obtained which is completely defined by the parameter set

$$\boldsymbol{\pi}(\mathbf{X}) = \{(r^{(\ell)}, p^{(\ell)})\}_{\ell \in \mathbb{L}}. \tag{15.27}$$

Using the multi-object exponential notation, an LMB RFS is expressed by

$$\boldsymbol{\pi}(\mathbf{X}) = \Delta(\mathbf{X}) w(\mathscr{L}(\mathbf{X})) p^{\mathbf{X}}, \tag{15.28}$$

where $h^{\mathrm{X}} = \prod_{x \in \mathrm{X}} h(x)$ and $h^{\emptyset} = 1$. The weights of the realizations are given by the multi-Bernoulli distribution

$$w(L) = \prod_{i \in \mathbb{L}} \left(1 - r^{(i)}\right) \prod_{\ell \in L} \frac{1_{\mathbb{L}}(\ell) r^{(\ell)}}{1 - r^{(\ell)}}, \tag{15.29}$$

and the spatial distributions are $p(x,\ell) = p^{(\ell)}(x)$. An example for an LMB RFS is illustrated by the upper part of Fig. 15.5.

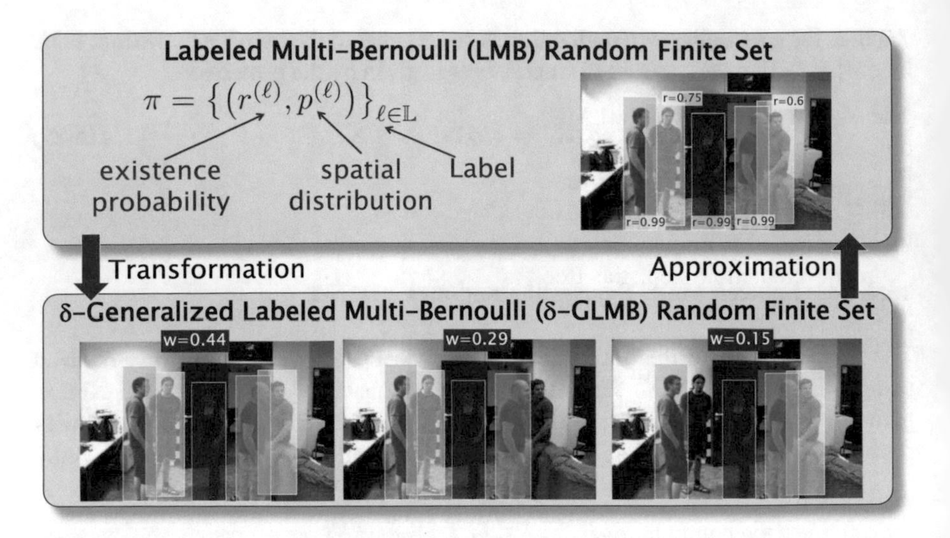

Fig. 15.5 Representation of the multi-object state using LMB and δ-GLMB RFSs. An LMB RFS can be equivalently rewritten in δ-GLMB form. In contrast, a δ-GLMB RFS can only be approximated using an LMB RFS

15.4.1.2 δ-Generalized Labeled Multi-Bernoulli Random Finite Set

An LMB RFS facilitates exactly one realization for a given set of track labels due to the assumption of statistical independence of the tracks. In contrast, a δ-GLMB RFS provides the possibility of several realizations for each set I of track labels. The distribution of a δ-GLMB RFS is given by

$$\pi(\mathbf{X}) = \Delta(\mathbf{X}) \sum_{(I,\xi)\in\mathscr{F}(\mathbb{L})\times\Xi} w^{(I,\xi)} \delta_I(\mathscr{L}(\mathbf{X})) \left[p^{(I,\xi)}\right]^{\mathbf{X}}, \tag{15.30}$$

where ξ denotes the history of track-to-measurement associations. Thus, the δ-GLMB RFS is able to represent the ambiguity in the track-to-measurement association during the filter update using several components or hypotheses for each set of track labels.

The difference between an LMB RFS and a δ-GLMB RFS is depicted by Fig. 15.5 (observe that only a subset of all hypotheses of the δ-GLMB RFS is shown). While the LMB RFS requires the tracks to be statistically independent, the δ-GLMB RFS facilitates the representation of statistical dependencies. Since an LMB RFS is a special case of a δ-GLMB RFS, it can be transformed into the corresponding δ-GLMB representation. In contrast, a δ-GLMB RFS can only be approximated by an LMB RFS.

15.4.2 Implementation of the Labeled Multi-Bernoulli Filter

The labeled multi-Bernoulli (LMB) filter is based on the representation of the multi-object state using an LMB RFS. A complete cycle of the LMB filter is conceptually illustrated by Fig. 15.6. In the following, the main ideas behind the LMB filter are outlined for an implementation using GMs. For additional details, the derivation of the filter, and SMC implementations, refer to [19, 23].

In the prediction step, the Bernoulli distribution of each track ℓ is predicted independently. First, the spatial distribution of the track is predicted using the well-known Kalman filter equations. In the case of slightly non-linear motion models, the corresponding equations of the extended Kalman filter (EKF) or unscented Kalman filter (UKF) have to be applied. The prediction of the track's existence probability is realized by multiplying the posterior existence probability with the survival probability p_S. Finally, the tracks of the birth distribution have to be appended to the predicted LMB RFS.

To reduce the computational complexity of the filter update, the predicted LMB density is partitioned using a grouping procedure. The grouping procedure returns groups of closely spaced objects and their associated measurements where the groups can be assumed to be statistically independent in case of sufficiently large gating values. Thus, the filter update can be applied to each group independently, which significantly reduces the computational load [23].

The grouping procedure enables parallel processing of each group during filter update. The update of each group is performed as follows: After transforming the LMB RFS of each group to δ-GLMB form, the full δ-GLMB update [29] is applied, which results in several hypotheses for each set of track labels due to the ambiguity of the track-to-measurement association. The hypotheses are again given by the tree in Fig. 15.2, where each path from the root to a leaf represents a single association hypothesis. In order to reduce computational load, only the k best association hypotheses are evaluated for large groups using Murty's algorithm [15]. After calculating the updated hypotheses, the posterior δ-GLMB density of

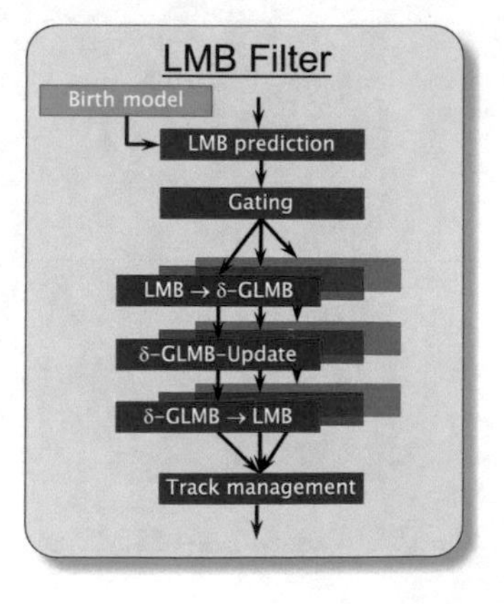

Fig. 15.6 LMB filter schematic

each group is approximated by an LMB RFS. The approximation matches the first moment of the δ-GLMB density, i.e. the spatial distribution and the mean value of the cardinality distribution of the approximation are equivalent while the cardinality distribution itself differs. Finally, the LMB RFSs of the groups are merged and the subsequent track management module is extracting track estimates and pruning tracks with very small existence probabilities.

15.5 Conclusion

This chapter presented two multi-object tracking algorithms based on random finite sets, which are suitable to track all humans in the proximity of a *Companion*-System. The SMC-MOB filter facilitates the integration of object-interactions as well as the information of a knowledge-base in the filtering algorithm. In contrast, the LMB filter requires significantly smaller computational resources and is capable of tracking even huge numbers of objects. The table in Fig. 15.7 summarizes the differences of the two filters and illustrates that the choice for the most convenient tracking algorithm strongly depends on the scenarios that should be handled.

The presented multi-object tracking algorithms facilitate an adaption of the *Companion*-System's behavior to the current environment. Examples for the adaption are given in the context of the demonstration scenario 3 (see Chap. 25), e.g., activation

	LMB	SMC-MOB (approx.)
Approximation Type	- parameters of labeled multi-Bernoulli distribution	- multi-object particles
Approximations	- approximation of posterior using LMB RFS - truncation of update hypotheses using Murty's algorithm	- exact representation requires huge amount of particles - likelihood approximation
Complexity (prediction/update)	- linear / cubic	- linear (one per multi-object particle) / linear
Pros	- integrated track labeling - reduced complexity due to grouping - tracking huge number of objects	- modeling of object interactions
Cons	- track coalescence possible	- number of particles grows exponentially in number of objects

Fig. 15.7 Comparison of LMB and SMC-MOB filters

of the system or including the group size into the purchase process. Further, the continuous tracking of the user provides the possibility to resume previously interrupted interactions. Other possibilities include the adaption of the input and output modalities to the current situation.

Acknowledgements This work was done within the Transregional Collaborative Research Centre SFB/TRR 62 "*Companion*-Technology for Cognitive Technical Systems" funded by the German Research Foundation (DFG).

References

1. Bar-Shalom, Y., Fortmann, T.: Tracking and Data Association. Academic, New York (1988)
2. Biundo, S., Bercher, P., Geier, T., Müller, F., Schattenberg, B.: Advanced user assistance based on AI planning. Cogn. Syst. Res. **12**(3-4), 219–236 (2011). doi:10.1016/j.cogsys.2010.12.005. http://www.sciencedirect.com/science/article/B6W6C-51TGG9K-1/2/4f8891da44f7fbe776d553396911a589. Special Issue on Complex Cognition
3. Geier, T., Biundo, S., Reuter, S., Dietmayer, K.: Track-person association using a first-order probabilistic model. In: IEEE 24th International Conference on Tools with Artificial Intelligence, vol. 1, pp. 844–851 (2012). doi:10.1109/ICTAI.2012.118
4. Geier, T., Reuter, S., Dietmayer, K., Biundo, S.: Goal-based person tracking using a first-order probabilistic model. In: Proceedings of the Ninth UAI Bayesian Modeling Applications Workshop (UAI-AW 2012) (2012)
5. Helbing, D., Molnar, P.: Social force model for pedestrian dynamics. Phys. Rev. E **51**, 4282–4286 (1995). doi:10.1103/PhysRevE.51.4282
6. Henderson, L.F.: The statistics of crowd fluids. Nature **229**, 381–383 (1971)
7. Kalman, R.E.: A new approach to linear filtering and prediction problems. Trans. ASME J. Basic Eng. **82**(Series D), 35–45 (1960)
8. Ma, W.K., Vo, B.N., Singh, S., Baddeley, A.: Tracking an unknown time-varying number of speakers using TDOA measurements: a random finite set approach. IEEE Trans. Signal Process. **54**, 3291–3304 (2006)
9. Mahler, R.: Multitarget Bayes filtering via first-order multitarget moments. IEEE Trans. Aerosp. Electron. Syst. **39**(4), 1152–1178 (2003). doi:10.1109/TAES.2003.1261119
10. Mahler, R.: PHD filters of higher order in target number. IEEE Trans. Aerosp. Electron. Syst. **43**(4), 1523–1543 (2007). doi:10.1109/TAES.2007.4441756
11. Mahler, R.: Statistical Multisource-Multitarget Information Fusion. Artech House Inc, Norwood (2007)
12. Mählisch, M.: Filtersynthese zur simultanen Minimierung von Existenz-, Assoziations- und Zustandsunsicherheiten in der Fahrzeugumfelderfassung mit heterogenen Sensordaten. Ph.D. thesis, Ulm University (2009)
13. Maskell, S.: Sequentially structured Bayesian solutions. Ph.D. thesis, Cambridge University (2004)
14. Munz, M., Mählisch, M., Dietmayer, K.: Generic centralized multi sensor data fusion based on probabilistic sensor and environment models for driver assistance systems. IEEE Intell. Transp. Syst. Mag. **2**(1), 6–17 (2010). doi:10.1109/MITS.2010.937293
15. Murty, K.: An algorithm for ranking all the assignments in order of increasing cost. Oper. Res. **16**, 682–687 (1968)
16. Musicki, D., Evans, R.: Joint integrated probabilistic data association: JIPDA. IEEE Trans. Aerosp. Electron. Syst. **40**(3), 1093–1099 (2004). doi:10.1109/TAES.2004.1337482
17. Pellegrini, S., Ess, A., Schindler, K., van Gool, L.: You'll never walk alone: Modeling social behavior for multi-target tracking. In: Proceedings of the 12th IEEE International Conference on Computer Vision, pp. 261–268 (2009). doi:10.1109/ICCV.2009.5459260

18. Reid, D.: An algorithm for tracking multiple targets. IEEE Trans. Autom. Control **24**(6), 843–854 (1979). doi:10.1109/TAC.1979.1102177
19. Reuter, S.: Multi-object tracking using random finite sets. Ph.D. thesis, Ulm University (2014)
20. Reuter, S., Dietmayer, K.: Pedestrian tracking using random finite sets. In: Proceedings of the 14th International Conference on Information Fusion, pp. 1–8 (2011)
21. Reuter, S., Dietmayer, K., Handrich, S.: Real-time implementation of a random finite set particle filter. In: Proceedings of the 6th Workshop Sensor Data Fusion: Trends, Solutions, Applications (2011). http://www.user.tu-berlin.de/komm/CD/paper/100149.pdf
22. Reuter, S., Wilking, B., Wiest, J., Munz, M., Dietmayer, K.: Real-time multi-object tracking using random finite sets. IEEE Trans. Aerosp. Electron. Syst. **49**(4), 2666–2678 (2013). doi:10.1109/TAES.2014.6619956
23. Reuter, S., Vo, B.T., Vo, B.N., Dietmayer, K.: The labeled multi-Bernoulli filter. IEEE Trans. Signal Process. **62**(12), 3246–3260 (2014)
24. Richardson, M., Domingos, P.: Markov logic networks. Mach. Learn. **62**(1–2), 107–136 (2006). doi:10.1007/s10994-006-5833-1. http://www.springerlink.com/index/10.1007/s10994-006-5833-1
25. Ristic, B., Arulampalam, S., Gordon, N.: Beyond the Kalman Filter: Particle Filters for Tracking Applications. Artech House Inc., Norwood (2004)
26. Sidenbladh, H., Wirkander, S.L.: Tracking random sets of vehicles in terrain. In: Conference on Computer Vision and Pattern Recognition Workshop, p. 98 (2003). doi:10.1109/CVPRW.2003.10097
27. Thrun, S., Burgard, W., Fox, D.: Probabilistic Robotics (Intelligent Robotics and Autonomous Agents). The MIT Press, Cambridge (2005)
28. Vo, B.N., Ma, W.K.: The Gaussian mixture probability hypothesis density filter. IEEE Trans. Signal Process. **54**(11), 4091–4104 (2006). doi:10.1109/TSP.2006.881190
29. Vo, B.T., Vo, B.N.: Labeled random finite sets and multi-object conjugate priors. IEEE Trans. Signal Process. **61**(13), 3460–3475 (2013)
30. Vo, B.N., Singh, S., Doucet, A.: Sequential Monte Carlo methods for multitarget filtering with random finite sets. IEEE Trans. Aerosp. Electron. Syst. **41**(4), 1224–1245 (2005). doi:10.1109/TAES.2005.1561884
31. Vo, B.T., Vo, B.N., Cantoni, A.: Analytic implementations of the cardinalized probability hypothesis density filter. IEEE Trans. Signal Process. **55**(7), 3553–3567 (2007). doi:10.1109/TSP.2007.894241
32. Vo, B.T., Vo, B.N., Cantoni, A.: The cardinality balanced multi-target multi-Bernoulli filter and its implementations. IEEE Trans. Signal Process. **57**(2), 409–423 (2009). doi:10.1109/TSP.2008.2007924
33. Zajic, T., Mahler, R.: A particle-systems implementation of the PHD multitarget-tracking filter. In: Signal Processing, Sensor Fusion, and Target Recognition XII, SPIE, vol. 5096, pp. 291–299, Bellingham, WA (2003)

Chapter 16
Non-intrusive Gesture Recognition in Real *Companion* Environments

Sebastian Handrich, Omer Rashid, and Ayoub Al-Hamadi

Abstract Automatic gesture recognition pushes Human-Computer Interaction (HCI) closer to human-human interaction. Although gesture recognition technologies have been successfully applied to real-world applications, there are still several problems that need to be addressed for wider application of HCI systems: Firstly, gesture-recognition systems require a robust tracking of relevant body parts, which is challenging, since the human body is capable of an enormous range of poses. Therefore, a pose estimation approach that identifies body parts based on geodetic distances is proposed. Further, the generation of synthetic data, which is essential for training and evaluation purposes, is presented. A second problem is that gestures are spatio-temporal patterns that can vary in shape, trajectory or duration, even for the same person. Static patterns are recognized using geometrical and statistical features which are invariant to translation, rotation and scaling. Moreover, stochastical models like Hidden Markov Models and Conditional Random Fields applied to quantized trajectories are employed to classify dynamic patterns. Lastly, a non-gesture model-based spotting approach is proposed that separates meaningful gestures from random hand movements (spotting).

16.1 Introduction

Multimodal human behavior analysis in modern human computer interaction (HCI) systems is becoming increasingly important, and it is supposed to outperform the single modality analysis. Like other modalities, for instance facial expression and prosody, gestures play an important role, since they are very intuitive and close to natural human-human interaction. Gestures are spatio-temporal patterns which can be categorized into dynamic patterns (hand movements), static morphs of the hands (postures), or a combination of both. Although gesture recognition technologies have been successfully applied to real-world applications, there are still three major problems that need to be addressed for wider applications of HCI systems: Firstly,

S. Handrich (✉) • O. Rashid • A. Al-Hamadi
Otto-von-Guericke University Magdeburg, Magdeburg, Germany
e-mail: sebastian.handrich@ovgu.de; omer.ahmad@ovgu.de; ayoub.al-hamadi@ovgu.de

S. Biundo, A. Wendemuth (eds.), *Companion Technology*, Cognitive Technologies,
DOI 10.1007/978-3-319-43665-4_16

a gesture-recognition system requires a robust detection and tracking of the relevant body parts, in particular the hands and arms. Since the human body is capable of an enormous range of poses and due to further complexities like self-occlusions and appearance changes, this remains a challenging task. The second major problem is caused by the fact that gesture may vary in shape, trajectory or duration for different persons, or even for the same person when a gesture is performed multiple times. Therefore, features and classifiers must be employed that are invariant to these changes. Furthermore, in a real-world scenario, there are often a lot of random hand movements. The number of these non-gestures is practically infinite and therefore they cannot all be learned in order to be separated from meaningful gestures. In this chapter, the proposed approach is fragmented into two parts. In the first part, the human body parts are detected, which utilizes torso fitting, geodesic distance and path assignment algorithms and builds the basis for the human pose estimation. From the extracted human body parts, the detected hands are utilized in the second part to extract meaningful information for gesture recognition. Keeping in view these goals, Sect. 16.2 presents the literature review of human pose estimation and hand gesture recognition. Section 16.3 is dedicated for the extraction of human poses whereas hand gesture recognition is presented in Sect. 16.4. The experimental results are presented in Sect. 16.5 and conclusion is presented in Sect. 16.6.

16.2 Related Work

16.2.1 Human Pose Estimation

The estimation of the human pose and the robust detection of gesture relevant body parts like hands and arms builds the basis for most gesture recogntion systems. Over the last decade, much research has been done in the area of hand-detection and tracking or even recovering the complete human pose from color or depth cameras. There are several criteria by which these methods can be classified. One of these criteria is whether 3D-data is available or not. Pure image-based methods are usually based on features like skin color [5, 20], contours[14], or silhouettes [4], but often lack the ability to resolve ambiguities, in particular when body parts occlude each other. Sometimes, this problem was addressed by the use of markers [27] (e.g. gloves). Since the aim of the gesture recognition is to push the HMI-systems closer to natural human-human-interactions, the wearing of markers appears to be too cumbersome, but still has its use, for example in the generation of ground-truth data. With the emergence of 3D-sensors like the Microsoft Kinect which are capable of providing dense depth maps in real time, these ambiguities did not disappear but could be more easily resolved.

Human pose estimation methods can be further divided into generative, discriminative and hybrid ones. Discriminative approaches [10, 17, 24] try to match several extracted features to a set of previously learned poses. For this, machine-learning methods like neural networks, support-vector-machines and decision-trees are used. The authors of [24], for example, have used decision forests to determine for each pixel of a depth image the respective body part using a Kinect sensor. This requires a very large training data set (up to $500K$ images) and the training of the classifiers is computationally very expensive, but attempts to reduce the complexity have been made [3]. Multiple authors have examined the accuracy and robustness of the Kinect's pose estimation and skeleton joint localization. An overview can be found in [11]. In realistic experimental setups, containing general poses, the typical error of the Kinect skeletal tracking was about 10 cm [16]. Another problem of discriminative models is that the pose estimation is often limited to poses that have previously been learned, e.g. [25], where the authors match synthesized depth images of a pose database to observe depth images, where detected hand and head candidates are employed to reduce the size of the search space.

In contrast, generative approaches [18, 21, 23] try to fit an existing model of the human body, in particular a skeleton, to the observed data. In general these approaches are capable of detecting arbitrary poses. However, an initial calibration for the body model is often required (usually a T-pose) and problems can occur when motions are abrupt and fast. For example, the authors of [23] proposed an approach that combines geodesic distances and optical flow in order to segment and track several body parts. The final pose is obtained by adapting a skeleton model consisting of 16 joints to the body part locations. Reported joint errors were in the range of 7–20 cm. A hybrid approach was proposed in [9], where the human pose is estimated from sequences of depth images. In the generative part, a local model-based search is performed that tries to update the configuration of a skeleton model. If this fails, discriminatively trained patch classifiers are used to localize body parts and re-initialize the skeleton configuration. The authors reported average pose errors between 5 and 15 cm.

16.2.2 Gesture Recognition

Gesture recognition is one of the important research domains of computer vision, where the detection, feature modeling and classification of patterns (i.e. gestures drawn by hand) are the major building blocks of the recognition system. Briefly speaking, in gesture recognition, meaningful patterns are identified in the continuous video streams defining the gestural actions. Practically, this is a difficult problem due to the interferences caused by hand-hand or hand-face occlusion, lighting changes or background noise, which cause distortion in the detected patterns and lead to the misclassification of gestural actions. The researchers have utilized various types of image features such as hand color, silhouette, motion and their derivatives. However, the literature presented here is confined to the appearance-based approaches as follows:

Yeo et al. [30] proposed an approach for the hand tracking and gesture recognition in a gaming application. In their approach, after the skin segmentation, the features (i.e., position, direction) are measured for the hand gesture recognition, where the fingertips are utilized to compute the orientation in consecutive frames. Afterwards, a Kalman filter is employed to estimate the optimal hand positions, which results in smoother hand trajectories. The experimental results are reported for the hand gesture dataset, which includes open and closed palm, claw, pinching and pointing. Similarly, Elmezain et al. [7] presented an approach for the isolated and meaningful gestures with the hand location as its feature using HMM. In their approach, the isolated gestures are tested on the number from 0 to 9, which are then combined to recognize the meaningful gesture based on zero-codeword with constant velocity motion. Finally, the experimental results are reported with the average recognition rate of 98.6% and 94.29% respectively for isolated and meaningful gestures.

Wen et al. [28] proposed an approach to detect the hand and recognize the gestures by utilizing image and depth information. In their approach, first the skin color is employed along with K-means clustering to segment the hands and this is followed by extraction of contour, convex hull and fingertip features to represent the gestures. Similarly, Nanda et al. [15] proposed an approach for hand tracking in cluttered environments using 3D depth data captured by Time of Flight (TOF) sensors. The hand and face are detected by applying the distance transform and k-component-based potential fields by calculating weights. The experiments are conducted on a dataset that comprises ten people in which the hands are tracked to recognize the gestures under occlusion. However, the suitability of the proposed approach is affected due to hand shape variation in real scenarios.

Using TOF cameras paired with an RGB camera, Van den Bergh et al. [26] proposed a method for gesture recognition. In their approach, the background is eliminated and the hand is detected by applying a skin segmentation approach. Further, the average 15 Neighborhood Margin Maximization transformation is measured to build the classifier for gesture recognition, where the Haarlet coefficients are calculated to match hand gestures stored in a database. The experiments are presented for both cameras, where the accuracy of 99.2% is acquired with depth-based (i.e., from the TOF cameras) hand detection whereas the accuracy of color-based detection is 92%. The accuracy of gesture recognition is 99.54% and 99.07% for RGB and depth images respectively. Similarly, Yang et al. [29] recognized gestures for the application of a media player. In their approach, eight gestures are recognized based on the hand trajectory features through HMM. Moreover, the tracking of hand is carried out using a continuously adaptive Mean-shift algorithm by taking the depth probability and updates its state using depth histogram. Raheja et al. [19] proposed a Kinect-based recognition system for the hand palm and fingertips tracking in which the depth defines the thresholding criterion for hand segmentation. Finally, the palm and fingertips are determined using the image differencing methods. Beh et al. [2] proposed a gesture recognition system by modeling spatio-temporal data in a unit-hypersphere space approach; a Mixture of von Mises-Fisher (MvMF) Probability Density Function (PDF) is incorporated into HMM. Further, the modeling of trajectories on a unit-hypersphere addresses the constraints of the user's arm length or distance between the user and camera. The results are presented with public datasets, InteractPlay and UCF Kinect datasets, with superior recognition performance compared to relevant state-of-the-art techniques.

To summarize, the problem of gesture recognition is addressed using a wide range of approaches which is impossible to cover here. In fact, the gesture recognition problem is highly context-sensitive, depending on indoor or outdoor environment, types of gesture (i.e. simple to complex gestural commands), single or multiple interaction mediums and different types of sensor. Due to this, it is difficult to generalize the methodologies suggested for different scenarios like gaming applications, air drawn gestures, and pointing gestures. In this chapter, we have investigated in particular the approaches utilizing appearance characteristics and contributes to the feature extraction phase by modeling the features for hand gesture classification.

16.3 Human Pose Estimation

In this work, we propose a generative method for estimating the human pose from depth images. Our approach is based on the measurement of geodesic distances along the surface of the human body and the extraction of geodesic paths to several key feature points. Finally, a hierarchical skeleton is adapted in order to match these paths (Fig. 16.1). The sequences of depth images $\{\boldsymbol{D}_t\}$ are either captured by a depth sensor or synthetically created, depending on the experimental setup. We assume that in each image the user is present and that he is the only object in the scene; so segmentation of the person is beyond the scope of this work. Based on a pinhole camera model, we further compute a 3D point-cloud $\mathbf{W} = \{\mathbf{w}_i \in \mathbb{R}^3\}_{i=1}^{n_x \times n_y}$ from each depth image with resolution $n_x \times n_y$ as follows:

$$\mathbf{w}_i = \begin{pmatrix} -d_i(x_i - c_x)/c_k \\ -d_i(y_i - c_y)/c_k \\ d_i \end{pmatrix}, \tag{16.1}$$

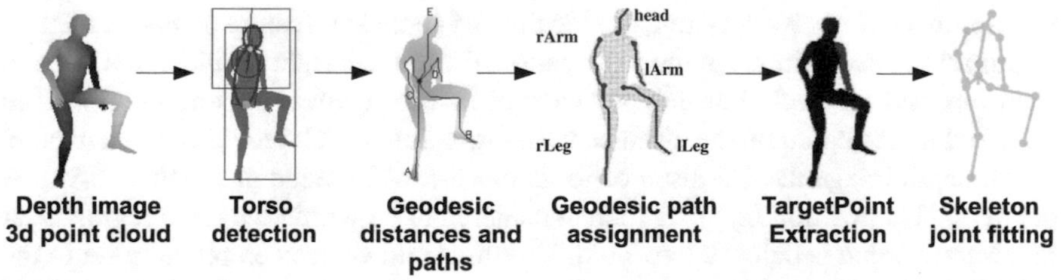

Fig. 16.1 Flow chart of the proposed approach for human pose estimation

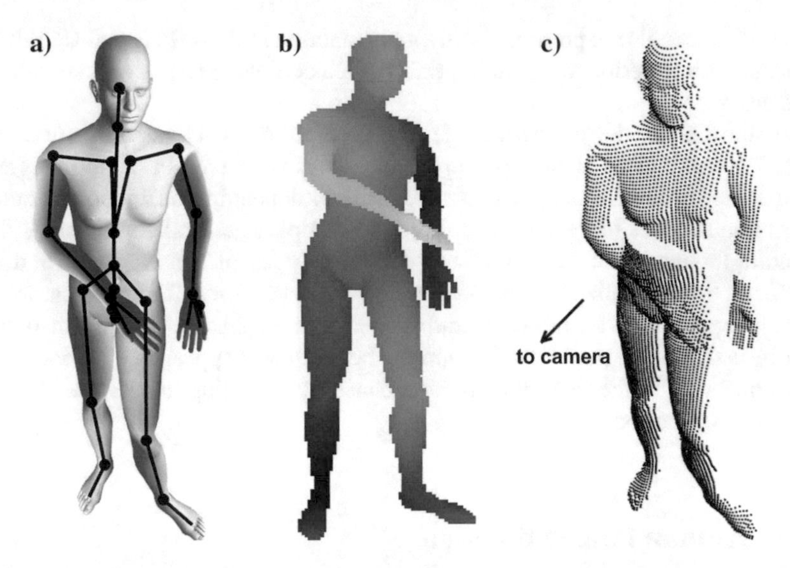

Fig. 16.2 (**a**) Recorded skeleton used to animate a 3D-character. (**b**) Rendered synthetic depth image. (**c**) 3D-point cloud computed from the depth image assuming an ideal pinhole camera model

where $(x_i, y_i)^T$ and d_i are the position and depth of each depth image pixel, respectively, and c_k denotes the camera constant. The image distortion is neglected and the principal point $(c_x, c_y)^T$ is set to the center of the depth image. From here on, the projected 2D position of a 3D point $\mathbf{w}_i$ will be denoted by $\mathbf{w}'_i$.

In order to create synthetic data, we have recorded the skeleton configurations of a person using a Kinect sensor. Using a linear skinning technique, the mesh of a 3D human character model is animated according to the skeleton configuration in each frame and subsequently into a depth buffer (OpenGL). This is shown in Fig. 16.2.

16.3.1 Torso Fitting

Our approach starts by detecting the location of some key-feature points ℓ_j, namely the approximate positions of the torso center, the left and right shoulder, the left and right hip, and the neck. For this, we need at first to segment the torso region. The segmentation is based on the distance transformation, which provides for each pixel of the depth image the 2D-distance to its closest background pixel (Fig. 16.3a). A depth pixel at position $(x, y)^T$ is then considered to be within the torso region $\mathscr{T}$ if its distance value is below a threshold. The threshold value is empirically set to the half of the maximum distance value (Eq. (16.2)). We further compute the centroid of the segmented torso region $\overline{\mathscr{T}}$ and its orientation (Fig. 16.3b). The latter is given by the direction of the largest eigenvector U of the covariance matrix Σ of the torso

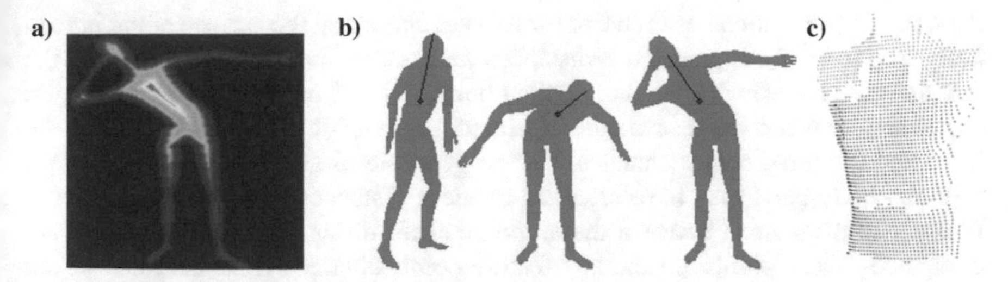

Fig. 16.3 (**a**) Distance transformation for the torso segmentation. (**b**) Initial centroid $\overline{\mathcal{T}}$ (*red dot*) and orientation U (*red line*) of the segmented torso region for several poses. (**c**) 3D torso model

region (Eq. (16.3)).

$$\mathcal{T} = \{(x, y)^T | \, I_{DT}(x, y) < 0.5 \max(I_{DT})\}. \tag{16.2}$$

$$\Sigma = \frac{1}{|\mathcal{T}|} \cdot (\mathcal{T} - \bar{\mathcal{T}}) \cdot (\mathcal{T} - \bar{\mathcal{T}})^T. \tag{16.3}$$

Our next goal is to fit a static 3D-torso model (Fig. 16.3c) to the observed 3D-points within the segmented torso region. Here, static means that only rigid body transformation, i.e. translation, rotation and scaling, are applied. The torso model is initially translated and rotated to match with the centroid $\overline{\mathcal{T}}$ of the torso region and its orientation (given by the eigenvector U). Using the point cloud registration algorithm described in [13], the transformation of the torso model is further refined. This requires the determination of a corresponding point set $C = \{(\mathbf{x}_i, \mathbf{y}_i)^T, \mathbf{x}_i \in \mathbf{X}, \mathbf{y}_i \in \mathbf{Y}\}$ between the torso model 3D points $\mathbf{X}$ and the observed 3D points $\mathbf{Y}$ located in the segmented torso region. For this purpose, we determine for each model point its closest point in the torso region and vice versa. Only points that match each other and are less then 0.1 m apart, are selected as corresponding points. In each frame, the transformation of the torso model in the previous frame is re-used. However, if the residual error $r = \sum_{(\mathbf{x}_i, \mathbf{y}_i)^T \in C} ||\mathbf{x}_i - \mathbf{y}_i||_2$ is too large, the torso fitting is re-initialized using the aforementioned distance transform-based method. Let $\mathbf{T}_{tor}$ denote the determined transformation matrix of the torso model; the positions of the key-feature points are computed by

$$\ell_j = \mathbf{T}_{tor} \mathbf{T}_j \, (0\;0\;0\;1)^T, \tag{16.4}$$

where T_j are transformation matrices specifying the relative transformations of the shoulder, neck, and hip joints in the skeleton model.

16.3.2 Geodesic Distance Measurement

A geodetic distance is the length of a geodesic path, which in turn represents the shortest connection between two nodes of a graph. As described below, such a graph can be constructed from the point-cloud vertices W. Thus, a geodesic path denotes the shortest connection between two arbitrary points along the surface of the human body. The idea of using geodesic distances for pose estimation is motivated by the fact that the geodesic distances of distinct body parts are invariant to pose changes. For example, when the user moves his hand, the Euclidean distance between the hand and the torso center changes, but the geodesic distance does not. However, different body parts may have identical geodetic distances to a certain root point. In order to distinguish between them, the geodesic distances to several points are computed. These points are the key-feature points obtained from the torso fitting step.

A graph consists of nodes and edges $\mathscr{G} = (\mathscr{N}, \mathscr{E})$. The nodes are the 3D-points of the point cloud W. Two criteria are introduced that determine whether two nodes are connected by an edge or not. The first criterion connects two nodes (3D-points) if the Euclidean distance between them is below a threshold. The threshold ϵ_{c1} depends on the resolution of the depth sensor. Typically, we have used 2 cm. Instead of comparing each 3D-point to all others, we make use of the fact that W is an organized point-cloud, i.e. for each 3D-point there is a unique 2D-point in the depth image. We then only compare 3D-points that correspond to adjacent depth pixels. A 3×3 neighborhood is used, so each node can be connected to at most eight others. Using only this criterion, a complete graph can only be generated when there are no self-occlusions (like in a T-pose). Otherwise, the occluding body parts separate individual graph regions from each other. Therefore, a second criterion is introduced that connects two nodes if they are only separated by foreground points, i.e. points that are closer to the camera and as such have a lower depth value. This criterion is only applied to nodes that have not already been fully connected by the first criterion. Furthermore, an edge is only created if neither of the two nodes has already been connected in the direction of the other. For example, a node missing a connection in the left direction (2D) can only be connected to a node that has no connection in the right 2D-direction.

The first edge criterion is given by Eq. (16.5). Here, $\mathbf{w}_i, \mathbf{w}_j \in W$ are two 3D points of the point cloud, $\mathbf{w}'_i, \mathbf{w}'_j$ their corresponding positions in the depth image and $d() = ||\mathbf{w}'_i - \mathbf{w}'_j||_2$ the spatial distance between them:

$$\mathscr{C}_1(i,j) = ||\mathbf{w}_i - \mathbf{w}_j||_2 \leq \epsilon_{c1} \wedge d(\mathbf{w}'_i, \mathbf{w}'_j) \leq 1. \tag{16.5}$$

The second criterion is defined by Eq. (16.6), where $\overline{\boldsymbol{w}'_i\boldsymbol{w}'_j}$ is the set of all depth pixels located at the straight line between $\boldsymbol{w}'_i$ and $\boldsymbol{w}'_j$ and $D_t(\boldsymbol{w}'_k)$ denotes their depth values.

$$\mathscr{C}_2(i,j) = D_t(\mathbf{w}'_k) < \min(D_t(\mathbf{w}'_i), D_t(\mathbf{w}'_j)) \ \forall \mathbf{w}'_k \in \overline{\mathbf{w}'_i\mathbf{w}'_j} \wedge d(\mathbf{w}'_i, \mathbf{w}'_j) > 1. \tag{16.6}$$

Two nodes $[\mathscr{N}_i, \mathscr{N}_j] = [w_i, w_j]$ are then connected by an edge if either criterion $\mathscr{C}_1(i,j)$ or $\mathscr{C}_2(i,j)$ evaluates to true (16.7). For each edge we store its length $|\mathscr{E}_{i,j}|$, which is the Euclidean distance between the two connected nodes.

$$\mathscr{E} = \{[\mathscr{N}_i, \mathscr{N}_j] \in \mathscr{N} \times \mathscr{N} \mid \mathscr{C}_1(i,j) \vee \mathscr{C}_2(i,j)\}. \tag{16.7}$$

Let $g_{\mathbf{r}}(\mathbf{a})$ denote the geodesic distance of an arbitrary point $\mathbf{a} \in \mathbf{W}$ to a given root point $\mathbf{r}$. Using the graph $\mathscr{G}$, we compute the geodesic distances of all point-cloud points using the Dijkstra algorithm [6]:

1. Assign to each node of $\mathscr{G}$ a geodesic distance $g_{\mathbf{r}}(\mathscr{N}_i)$. Set the distance of the root point to 0 and all others to ∞. Label the root point as *open*.
2. From all *open* nodes, find the one with the smallest geodesic distance to the root point, $\mathscr{N}_j = \text{argmin } g_{\mathbf{r}}(\mathscr{N}_i)$, and perform the following steps:
   a. Label $\mathscr{N}_j$ as *closed*.
   b. Compute for all *open* nodes $\mathscr{N}_k$ that are connected to $\mathscr{N}_j$ by an edge $\mathscr{E}_{j,k}$ the new geodesic distance $\hat{g}_{\mathbf{r}}(\mathscr{N}_k) = g_{\mathbf{r}}(\mathscr{N}_j) + |\mathscr{E}_{j,k}|$.
   c. If $\hat{g}_{\mathbf{r}}(\mathscr{N}_k) < g_{\mathbf{r}}(\mathscr{N}_k)$, set $g_{\mathbf{r}}(\mathscr{N}_k)$ to $\hat{g}_{\mathbf{r}}(\mathscr{N}_k)$ and mark the current node $\mathscr{N}_j$ as a predecessor of $\mathscr{N}_k$.
   d. If $\mathscr{N}_k$ is not labeled as *closed*, label it as *open*.

The first node to which the geodesic distances are calculated is the node that is closest to the center of the torso. The reason for that is that hands and feet—as the endpoints of the limbs—have high geodesic distances to the torso center and can therefore be found by searching for points of geodesic maxima. This is shown for some selected poses in Fig. 16.4. A very robust way to determine the geodesic maxima positions is to detect at first the node with the highest geodesic distance and connect it to the torso center by a zero weight edge, $|\mathscr{E}_{j,k}| = 0$. After that, the geodesic distances to all nodes are recomputed and the next absolute maximum node

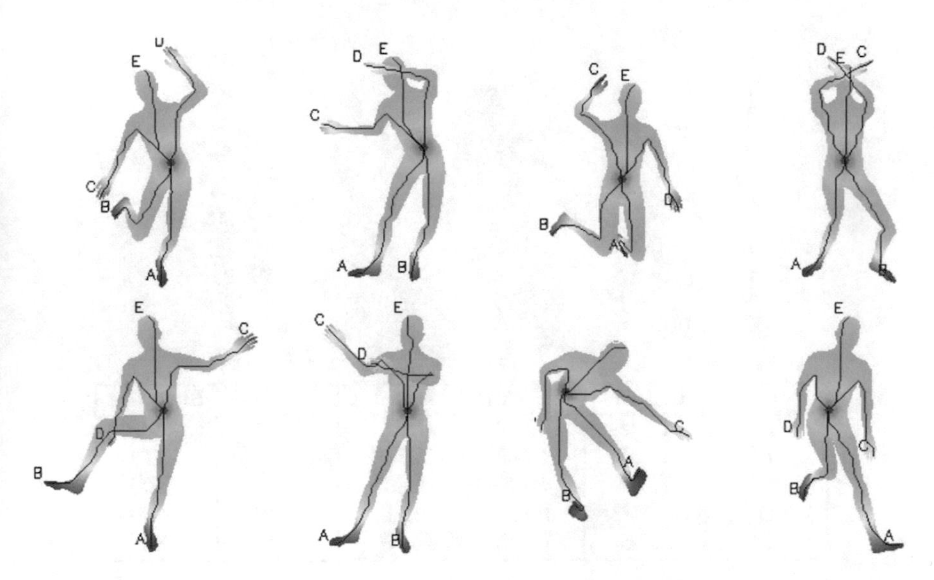

Fig. 16.4 Geodesic distances of all points to the torso center, drawn as colors from *blue* (low distances) to *red* (high distances). *Black lines* are the geodesic paths from the geodesic maxima (labeled as 'A'–'E') to the torso center

is determined. This is repeated five times in order to find the positions for both hands and feet and the head. In Fig. 16.4 the maxima positions are labeled as 'A'–'E'.

16.3.3 Geodesic Path Assignment

Let $m_i \in M$ denote the detected geodesic maximum points. After they have been determined, the next step is to decide to which limb (extremity) each maximum belongs. There are five extremities, denoted by $\mathscr{L}_1 \ldots \mathscr{L}_5$: the left and right arm, the head, and the two legs. Each extremity has a start point $\{\ell_j\}_{1\ldots5}$, which are the left and right shoulder, the neck and the leg and right hip position, respectively. Having determined the torso model transformation in Sect. 16.3.1, their positions are approximated (see Eq. (16.4)). Geodesic maximum points are assigned to limb start points ℓ_j based on their geodesic distance to ℓ_j. We therefore proceed by computing the geodesic paths between all limb start points ℓ_j and all geodesic maximum points m_i. This is shown for an exemplary pose in Fig. 16.5b. The assignment problem can be considered as a combinatorial optimization problem. A cost matrix $\mathbf{C}$ is created that stores for each geodesic maximum point the path lengths to ℓ_j:

$$\mathbf{C}(i,j) = g_{\ell_j}(m_i). \tag{16.8}$$

In total, there are $K = |M| \cdot 5$ possible assignments, where 5 is the number of the extremities and $|M|$ the number of detected geodesic maximum points. From all

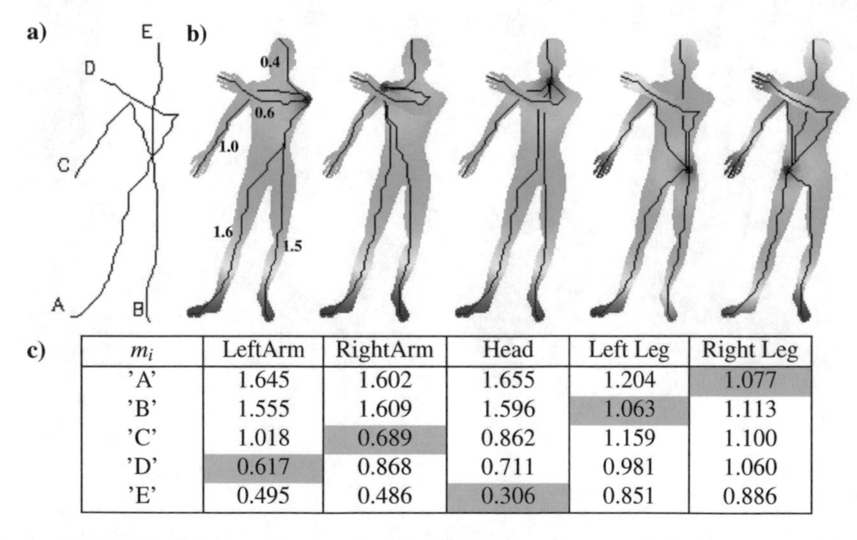

m_i	LeftArm	RightArm	Head	Left Leg	Right Leg
'A'	1.645	1.602	1.655	1.204	1.077
'B'	1.555	1.609	1.596	1.063	1.113
'C'	1.018	0.689	0.862	1.159	1.100
'D'	0.617	0.868	0.711	0.981	1.060
'E'	0.495	0.486	0.306	0.851	0.886

Fig. 16.5 (**a**) Geodesic maximum points and their paths to the torso center. (**b**) Geodesic paths between maximum points and limb start points ℓ_j (*dark blue dots*). The numbers in the *very left image* are the path lengths. (**c**) Cost matrix used to assign geodesic maximum points to limb start points. The numbers denote the path lengths in meters. Assigned pairs of geodesic maximum points and limb start points are *highlighted*

possible assignments $\Lambda = \{\Lambda_k\}_{1...K}$ between geodesic maximum points and limb start points, we determine the one $\hat{\Lambda}$ with minimal cost, i.e. the one that minimizes the cumulative geodesic distance (Eq. (16.9)).

$$\hat{\Lambda} = \underset{\Lambda_k \in \Lambda}{\operatorname{argmin}} \sum_{\lambda \in \Lambda_k} C(\lambda) \tag{16.9}$$

In Fig. 16.5c the cost matrix **C** for the shown geodesic paths in Fig. 16.5b is depicted. Assigned pairs $\hat{\Lambda}$ of geodesic maximum points and limb start points are highlighted. It can be seen that the geodesic maximum labeled as 'C' is correctly assigned to the right arm, 'D' to the left arm and so on.

16.3.4 Skeleton Fitting

Having assigned the geodesic maximum points to their respective limb points, the final step is to adapt a hierarchical skeleton model in order to determine the final pose. In our approach, each limb is adapted separately. A set of target points U_i to which each limb is adapted needs to be extracted as follows: Let $\{\mathscr{P}_i\}$ denote the set of geodesic paths between the geodesic maximum points and their assigned limb start points $\ell_{\hat{\Lambda}(i)}$ as shown in Fig. 16.6a. A set of target points $\mathbf{U}_i$ is then given by:

$$\mathbf{U}_i = \{\mathbf{w} \in \mathbf{W} \mid \; ||\mathbf{w} - \mathbf{p}_j||_2 < d_e \wedge |g(\mathbf{w}) - g(\mathbf{p}_j)| < d_e, \; \forall \mathbf{p}_j \in \mathscr{P}_i\}. \tag{16.10}$$

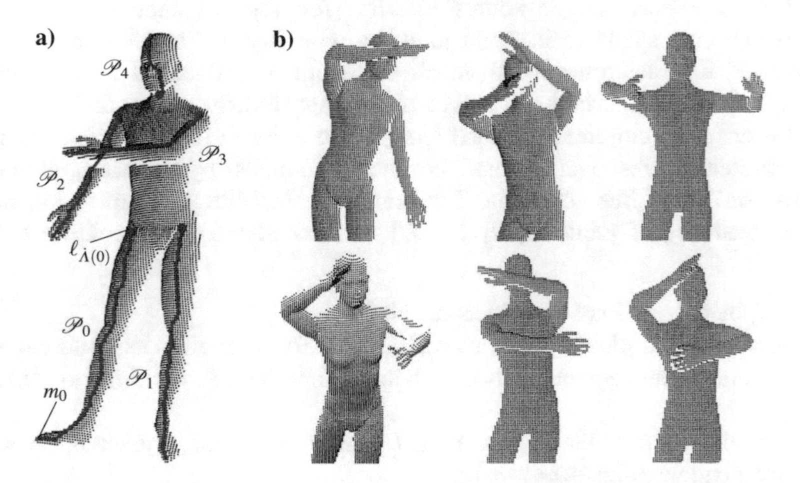

Fig. 16.6 (**a**) Geodesic paths $\mathscr{P}_i$ between the geodesic maximum points m_i and their assigned limb start points $\ell_{\hat{\Lambda}(i)}$. The paths are used to find a set of target points U_i to which a kinematic chain of a hierarchical skeleton model is adapted. (**b**) Extracted sets of target points in various poses to which the left arm of a kinematic skeleton model is adapted (*red dots*)

According to Eq. (16.10), a 3D point $\mathbf{w}$ is considered to be a target point for limb $\mathscr{L}_i$ if its Euclidean distance to a point that is an element of path $\mathscr{P}_i$ is below a threshold d_e and if it has a similar geodesic distance from the torso center. The latter condition prevents points that are close to a path point but belong to a different body part from being added to the set of target points. The threshold value d_e represents the diameter of a limb. We set $d_e = 5\,\text{cm}$ for the arms and $d_e = 10\,\text{cm}$ for the legs and the head. In Fig. 16.6b, a few examples of extracted sets of target points are shown.

In our skeleton model, each limb is represented by a kinematic chain consisting of two joints. The first joint describes the transformation of the upper limb segment, e.g. the upper arm or upper leg, respectively. Accordingly, the second joint describes the transformation of the lower limb segment (fore-arm or lower leg) relative to the first joint. The aim of the model adaptation is to minimize the distance between the target points and the bones of the skeleton model. One way to achieve this is to first determine for each target point whether it belongs to the upper or lower joint and then adapt the joint transformations successively and separately from each other, i.e. to fit, for example, the upper arm joint first and subsequently the lower arm. There are, however, some major problems with this approach:

1. If there are no or there is only an insufficient number of target points for the upper joint, the lower one cannot be adapted correctly, even if target points for it exist.
2. It is not guaranteed that the total error is minimized.
3. The joint rotation cannot be determined in all three axes.

These problems can be addressed when for a given joint also the target points that correspond to its child joints are considered. Therefore, a weight is introduced that determines how much each joint is influenced by each target point. This weight is high for target points that directly correspond to a joint and lower for its children. For each joint, a line segment (bone) to its child joint is computed. Corresponding points are then determined by projecting the target points onto these line segments. The complete algorithm to adapt a kinematic chain consisting of joints $\{J_0, J_1, \ldots, J_n\}$ to a set of target points $\mathbf{U}_i$ is as follows:

1. Let J_j be the first joint of a kinematic chain.
2. Recompute the global positions $\mathbf{b}_j, \mathbf{b}_{j+1}, \ldots$ for J_j and all its children. Also compute the line segments between all joints: $h_j = \mathbf{b}_j + t\mathbf{d}_j = b_j + t(\mathbf{b}_{j+1} - \mathbf{b}_j)$, $t \in [0, 1]$.
3. Determine for each target point $\mathbf{u}_k \in \mathbf{U}_i$ and for each line segment h_j the set of corresponding points $\mathbf{A}_j = \{a_k\}_j$:

   a. Project each target point u_k onto each line segment h_j.
   b. If the projected point is an element of the line segment, i.e. $0 < s = (\mathbf{u}_k - \mathbf{b}_j) \cdot \mathbf{d}_j < 1$, add the projected point to the set of corresponding points

$\mathbf{A}_j \leftarrow \{\mathbf{A}_j, \mathbf{b}_j + s * \mathbf{d}_j\}$. Otherwise add the endpoint of the line segment $\mathbf{A}_j \leftarrow \{\mathbf{A}_j, \mathbf{b}_{j+1}\}$.

4. Compute the weights. Determine for each target point $\mathbf{u}_k$ the index of the joint with the smallest distance:

$$\hat{j} = \underset{j}{\operatorname{argmin}} \, ||\mathbf{A}_j(k) - \mathbf{u}_k||_2. \tag{16.11}$$

If $J_{\hat{j}}$ is the current joint J_j, set the weight $\omega_{j,k}$ to 1. If it is a child joint ($\hat{j} > j$), set the weight to a small value ω_{suc}. Otherwise, set $\omega_{j,k} = 0$.

5. Given the weights and the set of corresponding points $\mathbf{A}_j$, solve for the new transformation $\mathbf{T}_j$ of the joint J_j. This is done by minimizing Eq. (16.12) using the quaternion-based closed-form solution provided in [13].

$$\sum_k \omega_{j,k} ||\mathbf{T}_j(\mathbf{A}_j(k) - \mathbf{b}_j) - (\mathbf{u}_k - \mathbf{b}_j)||^2. \tag{16.12}$$

6. Apply the computed transformation $\mathbf{T}_j$ to J_j. Proceed with the next joint $J_j \leftarrow J_{j+1}$ and go to (2).
7. Repeat steps (1)–(6) until the total error (the cumulative distances between the target points and their closest joints) falls below a threshold or a maximum number of iterations is reached.

The actual value of ω_{suc} is not a crucial parameter, but should be much smaller than 1, so that a joint is more influenced by target points that correspond directly to it. We set $\omega_{suc} = 0.05$. In Fig. 16.7 an example of kinematic chain fitting is shown.

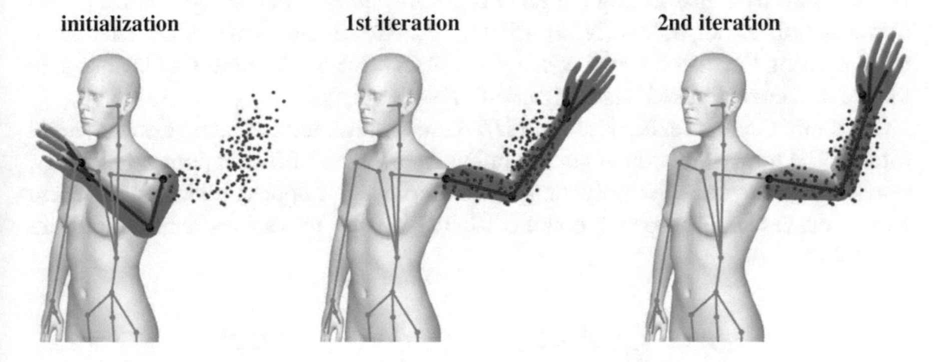

Fig. 16.7 A kinematic chain (*red lines*) that represents the transformation of left arm of a human model is adapted to a set of target points (*red dots*). After two iterations the algorithm has converged

16.4 Gesture Recognition

The gesture recognition process takes as input hand blob to detect features. The features describe an object's underlying characteristics, which should retain distinctly for every specific action. Specifically, the hand features are inherently a static feature set, but for the meaningful gestures these features are measured over a series of frames (*Fr*) to recognize complete gestural patterns (i.e., determined over a period of 1 s) utilizing the Bezier descriptor as follows.

16.4.1 Bezier Descriptors

In the proposed approach, the drawn gestures are recognized by modeling the computed hand centroid points (i.e., blob's center of detected skin segments) using Bezier curves to form Bezier points. The extracted Bezier points are used to compute the features and are then concatenated to form the Bezier descriptors. The proposed approach differs from other approaches as it is not relying on input hand centroid points but on fitted curves to form smoother trajectories for the classification process. Utilizing Bezier curves, the Bezier descriptors contain reliable features even for lower frame rates (i.e., captured frames per second, which is not possible because of its entire dependence on hand features data). The utilization of Bezier descriptors by fitting N (i.e., 5, 10, 15, 20, 30) points makes our approach independent of the HMM state model to be trained (i.e., in HMM, the features length has to be same as the number of states inside HMM, but this features length is hard to handle with varying frame rates). The Bezier descriptor is computed by transforming the hand centroid points which act as control points into a set of Bezier points ($N = 15$) [1, 12]. The Bezier features are calculated by measuring the difference between two consecutive Bezier points which are then binned and concatenated, resulting in a Bezier descriptor.

Mathematically, Bezier descriptors (B_d) are represented for each gesture symbol through the transformation of control points in the form of Bezier points. Normally, polynomials or piecewise polynomials are employed to approximate and represent Bezier curves, B. These polynomials [8, 12] can be of various degrees and are defined as:

$$B(t) = \sum_{i=0}^{d} C_i t^i = C_0 + C_1 t + C_2 t^2 + \cdots + C_d t^d, \tag{16.13}$$

$$C_i = \frac{n!}{(n-i)!} \sum_{j=0}^{i} \frac{(-1)^{(i+j)} P_j}{j!(i-j)!}. \tag{16.14}$$

The representation of these polynomials results in the approximation of Bezier curves [1, 12]. The Bezier curves can be generalized to higher dimensions but it is difficult to exercise control in higher-dimensional space, so we have designed the Bezier points to represent our gestures. A linear Bezier curve [22] which represents a line segment has the following form:

$$B(t) = (1-t)P_0 + tP_1; \quad t \in [0,1], \tag{16.15}$$

where P_0 and P_1 are input points. This mathematical representation [8, 22] is extended to higher-dimensional space as follows:

$$B(t) = (1-t)^d P_0 + dt(1-t)^{(d-1)} P_1 + \cdots + t^d P_d \tag{16.16}$$

$$= \sum_{i=0}^{d} \frac{d!}{i!(d-i)!} t^i (1-t)^{(d-i)} P_i \tag{16.17}$$

$$= \sum_{i=0}^{d} P_i b_{(i,d)}(t), \tag{16.18}$$

where the $b_{(i,d)}$ are Bernstein polynomials and form the basis of space for all the polynomials of $degree \leq d$. The motivation of using Bezier curves is to exploit its convex hull property, which ensures that the curve is always confined and controlled with its control points.

Utilizing these Bezier points, the orientation ϑ_i is computed between the two consecutive points P to extract the feature vector. The orientation between the two consecutive Bezier points ($Bz^i_{x,y}$, $Bz^{i+1}_{x,y}$) is presented as:

$$\vartheta_i = arctan\left(\frac{Bz^{i+1}_y - Bz^i_y}{Bz^{i+1}_x - Bz^i_x}\right); \; i = 1, 2, \ldots, T-1, \tag{16.19}$$

where T represents and length of the gesture drawing path, the $Bz^{i+1}_{x,y}$ and $Bz^i_{x,y}$ are two consecutive Bezier points.

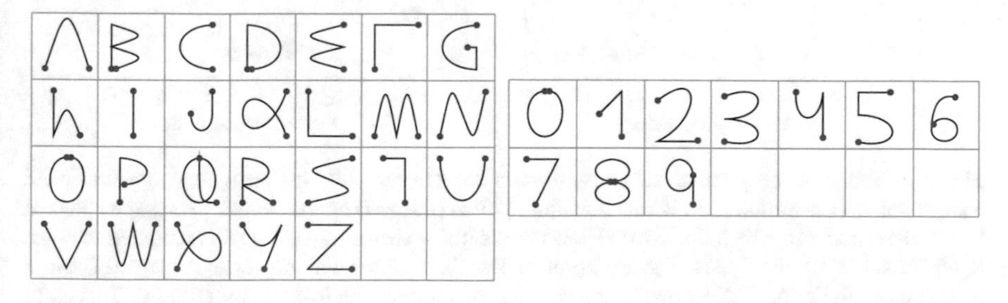

Fig. 16.8 Gesture dataset symbols

16.4.2 Features Binning

In the features binning process, the measured orientation ϑ is binned down into different indexes compatible for the classification process. In the proposed approach, orientation ϑ is scaled down eight bins with the factor of 45° to get quantized features (Qf), and by concatenating these features, Bezier descriptors ($B_d = \{\vartheta_1, \cdots, \vartheta_{T-1}\}$) are formed. Figure 16.9 presents a sequence performed by a subject along with its features in graphs. The images are presented at various instances with hand centroid points (control points in yellow) along with Bezier points (red). The Bezier features ϑ are computed from the consecutive Bezier points, which are binned to generate the feature vector. Here, the gesture stream (1 second data $\approx$ 25 fps) is represented by 14 Bezier points (i.e. shown by green curve in the tracked trajectory). The left graph shows the quantized features generated from

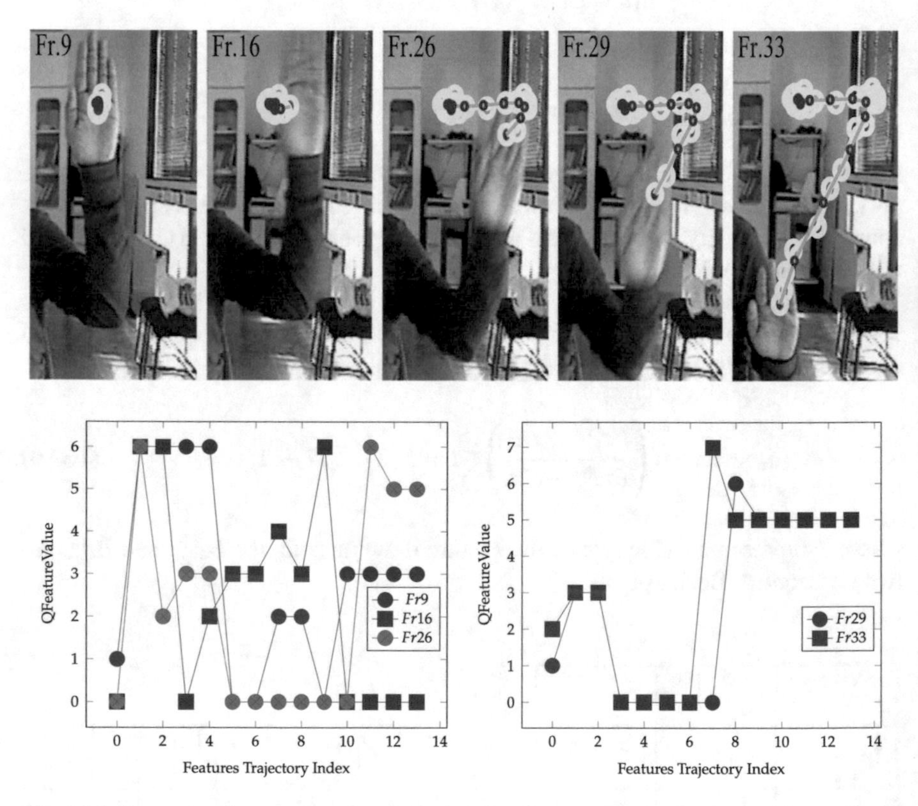

Fig. 16.9 Example of gestural action formation for *Gesture* '7'. In the figure, detected hand control points (i.e., *yellow*) and Bezier signature points (i.e., *red*) are presented. The graphs present Bezier descriptors in which the *X*-axis shows the features vector trajectory and quantized features (Qfs) are shown on the *Y*-axis. The *left graph* shows the features with no gesture detected *Gesture* '−1' (i.e., Fr 9, Fr 16, Fr 26) whereas the *right graph* presents the features for *Gesture* '7' (i.e., Fr 29, Fr 33)

Bezier points with no gestural pattern detected at *Fr*. 9, *Fr*. 16 and *Fr*. 26, whereas right graph shows Bezier points for detected *Gesture* '7' at *Fr*. 29 and *Fr*. 33.

16.4.3 Gesture Classification

Classification is the final phase, where the input class is assigned to one of predefined classes. The features extracted are the key elements of the classifier because the variability in the features affects the recognition in the same class and between different classes. In gesture classification, the Baum-Welch algorithm is employed to train HMM parameters by discrete vectors (i.e., Bezier descriptors). Inside the HMM, Left-Right banded topology is used with 14 states for classification of gestural symbols, and this is done by selecting the maximal observation probability of gesture model using the Viterbi algorithm.

16.5 Experimental Results

16.5.1 Human Pose Estimation

The proposed method has been validated with a set of synthetic test sequences. Each sequence consists of a series of depth maps that were generated by rendering an animated 3D-character into an OpenGL depth buffer. The character mesh was designed with the MakeHuman application and consists of a 3D mesh and a skeleton (hierarchy of joints). To obtain animation data (sequences of skeleton joint orientations) a human actor performed several motions that were recorded by a Kinect sensor. Even, if the Kinect does not detect the actors' pose correctly, the rendered depth image will match the pose that the Kinect has recognized. In this way, the exact ground truth joint positions are known for each depth image, without having to manually label them. In total, $N = 11{,}250$ frames, separated into ten sequences of variable length between 35 and 45 s and showing various poses, have been created. To each depth image all steps of the processing chain, i.e. measuring geodesic distances, extracting and classifying geodesic paths and adapting a skeleton model to the sets of target points, are applied.

The skeleton model in this approach consists of 16 joints (three for each leg and arm, two within the torso, one for each the neck and the head). Their positions in frame t are denoted by $x_i(t) \in \mathbb{R}^3$ and the ground truth joint positions are denoted by $y_i(t) \in \mathbb{R}^3$. As an error metric, we compute in each frame both the mean Euclidean

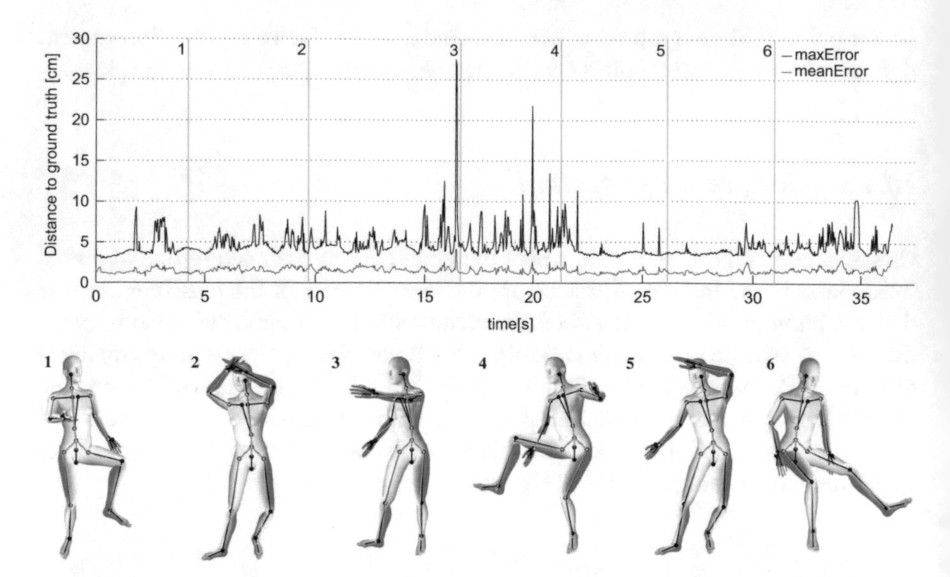

Fig. 16.10 *Top*: Mean joint error (*blue*) and maximum joint error (*black*) per frame in centimeters for a single sequence. *Bottom*: The estimated pose for some selected frames is depicted by the *blue skeleton*. The *black skeleton* (ground truth) was used to animate a 3D-character in order to create synthetic depth images

distance (Eq. (16.20)) and the maximum Euclidean distance (Eq.(16.21)).

$$e_m(t) = \frac{1}{16}\sum_{i=1}^{16} |x_i(t) - y_i(t)|_2 \tag{16.20}$$

$$e_h(t) = \max|x_i(t) - y_i(t)|_2, \ \forall i = 1 \dots 16 \tag{16.21}$$

$$e_j(i) = \frac{1}{N}\sum_{t=1}^{N} |x_i(t) - y_i(t)|_2 \tag{16.22}$$

Both errors are shown for a single sequence in Fig. 16.10 (top). It can be seen that the mean Euclidean error is about $\overline{e_m}(t) = 2$ cm and therefore very low. The maximum error rises up to 27 cm in single frames. This occurs when the arm is pointing directly toward the camera. In this case, there are too few 3D points to which the skeleton model can be adapted and the relative joint rotations remain unchanged. For several selected frames in Fig. 16.10, the animated characters are shown. It can be seen that also in cases of self-occlusions the estimated pose (blue skeleton) matches closely with the ground truth pose (black skeleton). In Fig. 16.11, we further computed the per joint errors (Eq. (16.22)) averaged over all N generated depth images. The error is between 1 cm for the neck and 4.1 cm for the feet. This can be compared to the also graph-based approach provided in [21], whose authors

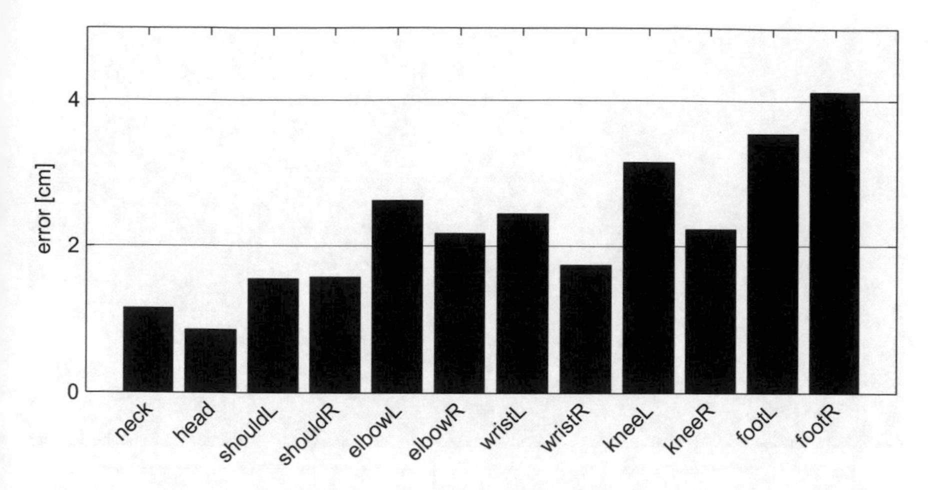

Fig. 16.11 Euclidean distances between estimated and true joint positions averaged over all created depth images

report per-joint errors between 3 and 6.8 cm. The pose is estimated with approx. 40 frames per second. Thus, our method is suitable for the subsequent online gesture recognition system.

16.5.2 Gesture Recognition

The experimental setup involves the tasks of data acquisition, feature extraction and classification. We have demonstrated the applicability of proposed system on real situations where the gestures are recognized as satisfying flexibility and naturalness (i.e., bare hand, complex background). The proposed framework runs with real-time processing at 25 fps on a 2.83 GHz, 4-core Intel processor with 480×640 pixels resolution. The experiments are conducted on 15 video observations per gesture of six subjects performing various hand gestures wearing short to long sleeves, and the gesture dataset contains the symbols from $A \rightarrow Z$ and $0 \rightarrow 9$ (see Fig. 16.8).

Figure 16.12 presents the sequence where the subject is drawing the *Gesture* '8' along with the feature trajectories in graphs. In this sequence, the features are detected from the hand centroid points and transformed to Bezier points $N = 15$ (i.e., marked with red color). The consecutive Bezier points are then utilized to extract Bezier features ϑ, which are then binned to build Bezier descriptors. Bezier descriptors are used inside HMM using LRB topology with 14 states to recognize gestural actions. The left graph shows the Bezier descriptors with gestural action *Gesture* '8' detected at Fr 20, Fr 25, Fr 30, and Fr 33, whereas the right graph shows Bezier descriptors where no gestural action (i.e., Fr 1, Fr 8, Fr 12, Fr 16) is detected.

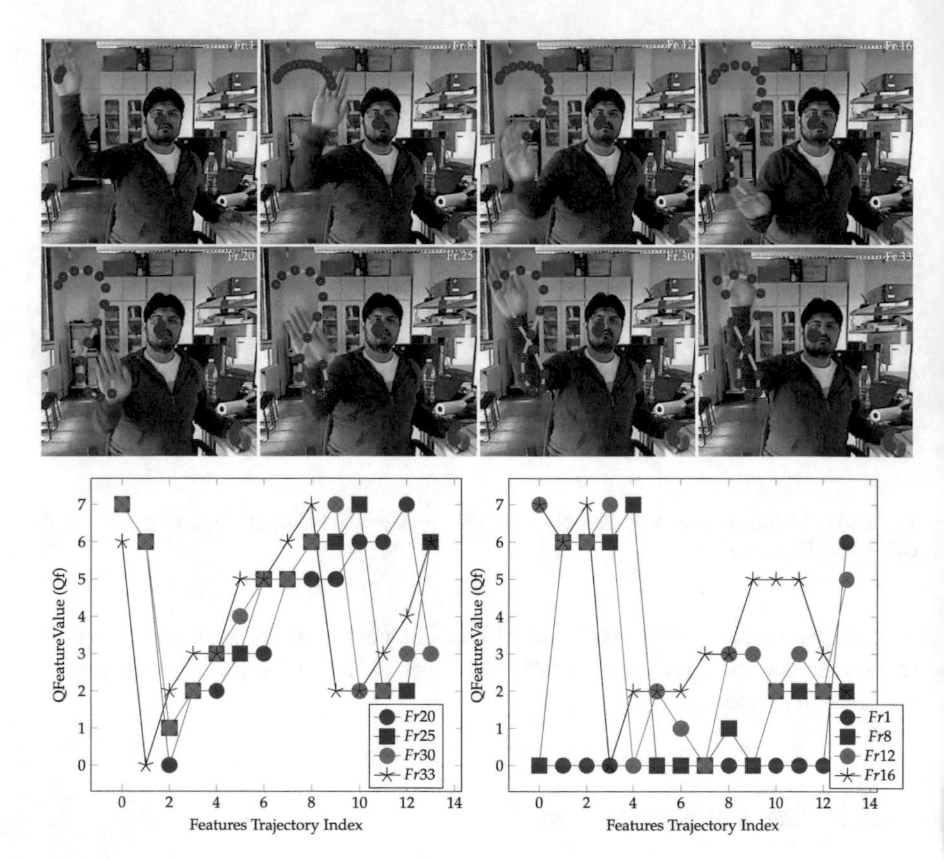

Fig. 16.12 Gestural action of a subject drawing *Gesture* '8' using a Bezier descriptor ($N = 15$). In the sequence, Bezier points (marked as *red points*) are presented whereas the graphs present extracted quantized features with trajectories. The *left graph* shows the features of detected *Gesture* '8' whereas in right graph no gesture *Gesture* '−1' is detected

Figure 16.13 presents the classification results of different Bezier descriptors (i.e., with different HMM states) for the sequence in Fig. 16.12 along with ground truth data. It can be seen that Bezier descriptor $N = 15$ results in the highest recognition rate for this sequence.

The confusion matrix of gestural symbols is presented for $N = 15$ (see Fig. 16.14a), where the diagonal elements show the gestural symbol classification. The higher rate of classification indicates that the Bezier descriptor is capable of recognizing the gestural symbols. Moreover, the performance of Bezier descriptor $N = 15$ is measured against various descriptors like $N = 5$, $N = 10$, $N = 20$ and $N = 30$. Figure 16.14b provides a comparative analysis by adjusting the N parameter (i.e., N is the number of points) for fitting Bezier curves in which the performance of the Bezier descriptor $N = 15$ is superior amongst all with an overall accuracy of 98.3%.

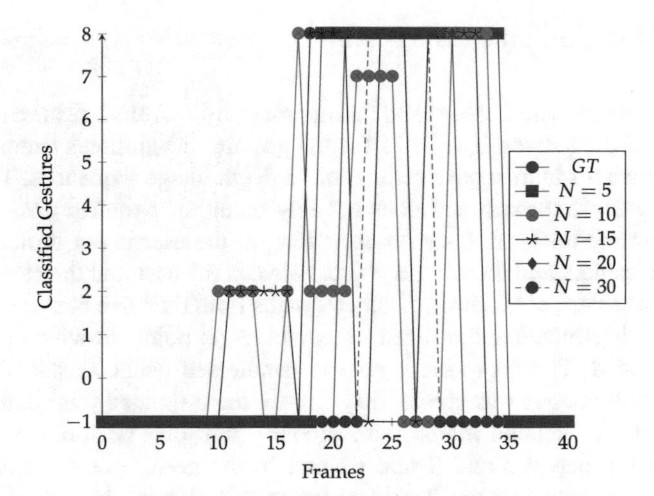

Fig. 16.13 Classification of Bezier descriptors (i.e., $N = \{5, 10, 15, 20, 30\}$) for the sequence using HMM with recognized *Gesture* '8' along with ground truth data *GT*

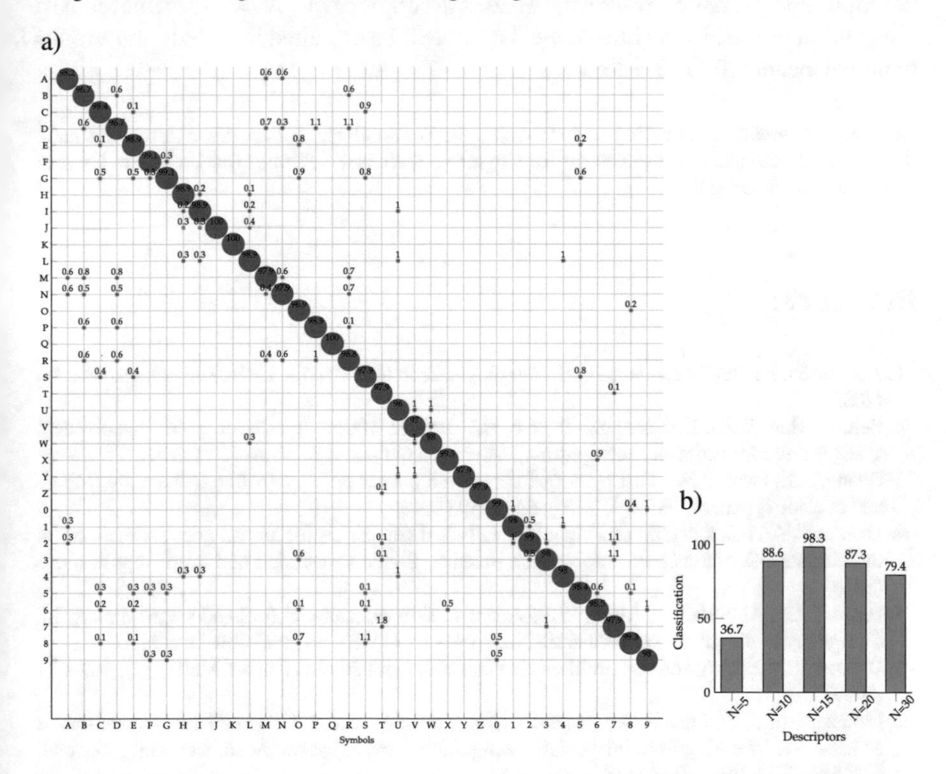

Fig. 16.14 (**a**) Confusion matrix of gesture symbols for Bezier descriptor $N = 15$. (**b**) Classification rate of Bezier descriptors ($N = 5, 10, 15, 20, 30$)

16.6 Discussion and Conclusion

In this work, we present a robust and fast approach for real-time gesture recognition suitable for HCI systems. The basis for the gesture recognition system is formed by an approach for human pose estimation in depth image sequences. The method is a learning-free approach and does not rely on any pre-trained pose classifiers. We can therefore track arbitrary poses as long as the user is not turned away too far from the camera and there is no object between the user and the camera. Based on the measurement of geodesic distances, paths from the torso center to the hands and feet are determined and utilized to extract target points to which a kinematic skeleton is fitted. The approach is able to handle self-occlusions. A database of synthetic depth images was created to compare the estimated joint configurations of a hierarchical skeleton model with ground truth joint positions. On average, the errors were between and 3 and 6.8 cm. In the gesture recognition, features are extracted by modeling the Bezier curves to build Bezier descriptors which are then classified using HMM. Moreover, a comparative analysis of different Bezier descriptors is presented where the Bezier descriptor with $N = 15$ achieves best recognition results. In the future, the dataset will be extended for words and actions to derive meaningful inferences.

Acknowledgements This work was done within the Transregional Collaborative Research Centre SFB/TRR 62 "*Companion*-Technology for Cognitive Technical Systems" funded by the German Research Foundation (DFG).

References

1. Andersson, F.: Bezier and B-spline Technology. Technical Report, Umea Universitet, Sweden (2003)
2. Beh, J., Han, D.K., Durasiwami, R., Ko, H.: Hidden Markov model on a unit hypersphere space for gesture trajectory recognition. Pattern Recogn. Lett. **36**, 144–153 (2014)
3. Chang, J.Y., Nam, S.W.: Fast random-forest-based human pose estimation using a multi-scale and cascade approach. ETRI J. **35**(6), 949–959 (2013)
4. Daniel, C.Y. Chen and Clinton B. Fookes. Labelled silhouettes for human pose estimation. In: International Conference on Information Science, Signal Processing and Their Applications (2010)
5. Dawod, AY., Abdullah, J., Alam, M.J.: Adaptive skin color model for hand segmentation. In: Computer Applications and Industrial Electronics (ICCAIE), pp. 486–489, Dec 2010
6. Dijkstra, E.W.: A note on two problems in connexion with graphs. Numer. Math. **1**(1), 269–271 (1959)
7. Elmezain, M., Al-Hamadi, A., Appenrodt, J., Michaelis, B.: A hidden Markov model-based isolated and meaningful hand gesture recognition. Proc. World Acad. Sci. Eng. Technol. (PWASET) **31**, 394–401 (2008)
8. Farouki, R.T.: The Bernstein polynomial basis: a centennial retrospective. Comput. Aided Geom. Des. **29**(6), 379–419 (2012)
9. Ganapathi, V., Plagemann, C., Koller, D., Thrun, S.: Real time motion capture using a single time-of-flight camera. In: CVPR, pp. 755–762 (2010)

10. Girshick, R., Shotton, J., Kohli, P., Criminisi, A., Fitzgibbon, A.: Efficient regression of general-activity human poses from depth images. In: ICCV, pp. 415–422 (2011)
11. Han, J., Shao, L., Xu, D., Shotton, J.: Enhanced computer vision with microsoft kinect sensor: a review. IEEE Trans. Cybern. **43**(5), 1318–1334 (2013)
12. Hansford, D.: Bezier techniques. In: Farin, G., Hoschek, J., Kim, M. (eds.) Handbook of Computer Aided Geometric Design, pp. 75–109. North-Holland, Amsterdam (2002)
13. Horn, B.K.P.: Closed-form solution of absolute orientation using unit quaternions. J. Opt. Soc. Am. **4**, 629–642 (1987)
14. Liang, Q., MiaoMiao, Z.: Markerless human pose estimation using image features and extremal contour. In: ISPACS, pp. 1–4 (2010)
15. Nanda, H., Fujimura, K.: Visual tracking using depth data. US Patent 7590262, Sept 2009
16. Obdrzalek, S., Kurillo, G., Ofli, F., Bajcsy, R., Seto, E., Jimison, H., Pavel, M.: Accuracy and robustness of kinect pose estimation in the context of coaching of elderly population. In: Engineering in Medicine and Biology Society (EMBC), pp. 1188–1193 (2012)
17. Plagemann, C., Ganapathi, V., Koller, D., Thrun, S.: Real-time identification and localization of body parts from depth images. In: ICRA, pp. 3108–3113 (2010)
18. Qiao, M., Cheng, J., Zhao, W.: Model-based human pose estimation with hierarchical ICP from single depth images. In: Advances in Automation and Robotics. Lecture Notes in Electrical Engineering, vol. 2, pp. 27–35. Springer, Berlin (2012)
19. Raheja, J.L., Chaudhary, A., Singal, K.: Tracking of fingertips and centers of palm using kinect. In: Computational Intelligence, Modelling and Simulation, 248–252 (2011)
20. Rasim, A., Alexander, T.: Hand detection based on skin color segmentation and classification of image local features. Tem J. **2**(2), 150–155 (2013)
21. Rüther, M., Straka, M., Hauswiesner, S., Bischof, H.: Skeletal graph based human pose estimation in real-time. In: Proceedings of the British Machine Vision Conference, pp. 69.1–69.12. BMVA Press, Guildford (2011). http://dx.doi.org/10.5244/C.25.69
22. Salomon, D.: Curves and Surfaces for Computer Graphics. Springer, New York (2006)
23. Schwarz, L.A., Mkhitaryan, A., Mateus, D., Navab, N.: Human skeleton tracking from depth data using geodesic distances and optical flow. Image Vis. Comput. **30**(3), 217–226 (2012)
24. Shotton, J., Girshick, R., Fitzgibbon, A., Sharp, T., Cook, M., Finocchio, M., Moore, R., Kohli, P., Criminisi, A., Kipman, A., Blake, A.: Efficient human pose estimation from single depth images. IEEE Trans. Pattern Anal. Mach. Intell. **35**(12), 2821–2840 (2013)
25. Siddiqui, M., Medioni, G.: Human pose estimation from a single view point, real-time range sensor. In: CVPRW, pp. 1–8, June 2010
26. Van den Bergh, M., Van Gool, L.J.: Combining RGB and ToF cameras for real-time 3d hand gesture interaction. In: WACV, pp. 66–72. IEEE Computer Society, New York (2011)
27. Wang, R.Y., Popović, J.: Real-time hand-tracking with a color glove. ACM Trans. Graph. **28**(3), 63 (2009)
28. Wen, Y., Hu, C., Yu, G., Wang, C.: A robust method of detecting hand gestures using depth sensors. In: IEEE International Workshop on Haptic Audio Visual Environments and Games, pp. 72–77 (2012)
29. Yang, C., Jang, Y., Beh, J., Han, D., Ko, H.: Gesture recognition using depth-based hand tracking for contactless controller application. In: ICCE, pp. 297 –298 (2012)
30. Yeo, H., Lee, B., Lim, H.: Hand tracking and gesture recognition system for human-computer interaction using low-cost hardware. Multimed. Tools Appl. **74**, 1–29 (2013)

Chapter 17
Analysis of Articulated Motion for Social Signal Processing

Georg Layher, Michael Glodek, and Heiko Neumann

Abstract *Companion* technologies aim at developing sustained long-term relationships by employing non-verbal communication (NVC) skills. Visual NVC signals can be conveyed over a variety of non-verbal channels, such as facial expressions, gestures, or spatio-temporal behavior. It remains a challenge to equip technical systems with human-like abilities to reliably and effortlessly detect and analyze such social signals. In this proposal, we focus our investigation on the modeling of visual mechanisms for the processing and analysis of human-articulated motion and posture information from spatially intermediate to remote distances. From a modeling perspective, we investigate how visual features and their integration over several stages in a processing hierarchy take part in the establishment of articulated motion representations. We build upon known structures and mechanisms in cortical networks of primates and emphasize how generic processing principles might realize the building blocks for such network-based distributed processing through learning. We demonstrate how feature representations in segregated pathways and their convergence lead to integrated form and motion representations using artificially generated articulated motion sequences.

17.1 Introduction and Motivation

The analysis of verbal and non-verbal communication (NVC) plays a key role in generating elaborate social behavior in humans [1, 8, 19]. We focus here on the selection of visual features and their compositions, to be processed for the support of such non-verbal channels and signals in communication. Such investigations aim at gaining more detailed insights regarding human-like communication and transforming them into capabilities of future intelligent technologies. The design of *Companion* technologies necessitates creating agents that are personalized in terms of building relationships with their (human) interlocutors on the basis of intelligent

G. Layher (✉) • M. Glodek • H. Neumann
Institute of Neural Information Processing, Ulm University, James-Franck-Ring, 89069 Ulm, Germany
e-mail: georg.layher@uni-ulm.de; michael.glodek@uni-ulm.de; heiko.neumann@uni-ulm.de

S. Biundo, A. Wendemuth (eds.), *Companion Technology*, Cognitive Technologies,
DOI 10.1007/978-3-319-43665-4_17

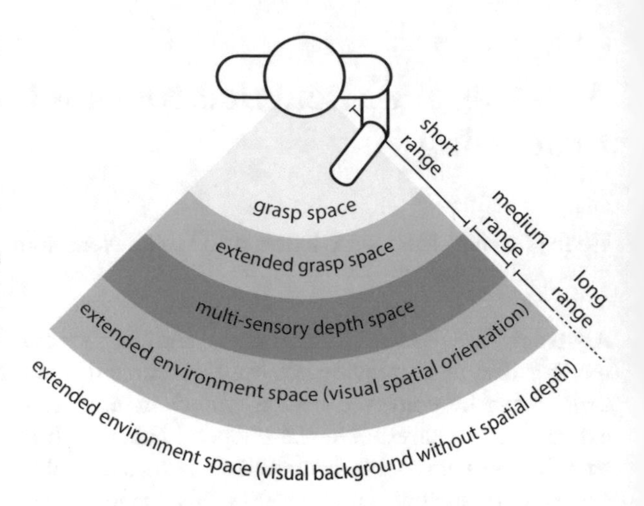

Fig. 17.1 The extra-personal interaction space of an agent is shown (adapted from [22]). From the center to the periphery, the extra-personal space can be partitioned into four regions: the immediate space of reaching, an extended grasp space, a depth space of multi-sensory orientation and an extended environment space. Each of the regions entails different requirements and constraints for the sensory processing

multi-modal and affective interaction. A key functionality of such artifacts, which function as true companions, is their ability to build human-computer relationships which develop over a long-term scale [6–8] (Fig. 17.1), a property which renders them different from simple assistive systems. Socially perceptive *Companion*-Systems should be affect-sensitive at the encoding as well as the decoding side of the communication channel. In order to detect such social signals the features and signal properties must be defined, and then proper mechanisms need to be developed [35]. Within this chapter, we particularly focus on the analysis of bodily movements and postures of humans and the underlying mechanisms of visual information processing. In order to define and structure relevant non-verbal visual signals to be detected, we suggest a multi-level approach to account for the analysis of different types of cues. We subdivide the different regions of the extra-personal (interaction) space in accordance to the categories proposed in [22]. From a perspective of the available affect-related visual information, the three distance ranges defined in Fig. 17.1, namely short-, medium-, and long-range interaction, can be defined as metric distance categories (compare [15]). In *short-range* interaction face-to-face and multi-modal analysis of non-verbal cues should be enabled, such as facial expressions, head gestures and orientation, eye gaze, posture, or body expressiveness. For an extended distance between a user and a *Companion*-System, *medium-range* interaction allows us to analyze states and events of the human user, e.g., to infer his intentions or detect the initiation of an interaction. Indicators derived from full-body movements need to be analyzed based on motion detection and motion pattern interpretation (e.g., contraction and expansion to detect approach, analyze waving gestures, etc.) or to recognize simple actions. If the user is somewhere in the same environment as the *Companion*-system but not in its immediate range, *long-range* affective interaction capabilities should be triggered to build hypotheses about affect-related states. Examples are the detection of a predisposition to interact, the generic detection of the presence of people in the surrounding, and accounting for the frequency at which they move in proximity, or to analyze group effects, i.e. whether people move together, temporally meet and chat in a group, etc.

Such capabilities directly reflect the tendency of humans to infer other people's intentions from their actions [9], which determines a basis for human affect analysis at a distance. Inferring the intention of others requires the robust detection and analysis of motion information. In this contribution, we focus on the analysis of articulated and biological motion.[1] We develop a model framework that builds upon previous investigations of modeling the main computational mechanisms of biological motion [20] but extends this along several directions. We adopt a perspective that utilizes knowledge about the biological mechanisms in the visual cortex of primates that contribute to the analysis of articulated motion. The primary motivation of such a biologically inspired approach is that human-like capabilities are extremely sophisticated in analyzing social signals of a broad spectrum and with subtle details. We believe that understanding the underlying functionality is a great source of inspiration for building advanced technical systems. The proposed model suggests how motion and form representations contribute to building sequence-selective representations at different stages in the cortical hierarchy [42]. The framework makes predictions about how mutual interaction between the segregated motion and form pathways helps to learn representations in which key poses are automatically captured from articulated events out of a stream of continuous articulated motions. It will be demonstrated that such a generic framework of learning and mutually coupled visual representations defines a core repertoire of inferring intentions from biological or articulated motion sequences and, in addition, explains how articulated postures in still images lead to implied motion activation that is registered in motion representations as well. Section 17.2 starts with a motivation for mechanisms of articulated motion processing, namely the usefulness of motion as well as form information and their combination. Here, an overview of technical approaches in computer vision is presented, which is augmented by a discussion of the evidence that has been acquired from perceptual and neuroscientific investigations. Section 17.3 contains the core part, with an overview description of a neural model architecture for the analysis of articulated, or biological, motion. Apart from the generic model description, model simulations are demonstrated to show that such an approach yields several robustness and invariance properties. The chapter ends with a final summary and brief discussion of a future perspective on such modeling developments.

[1] We use the terms *articulated motion* and *biological motion* in a somewhat loose sense. In order to be more precise, articulated motion refers to the movement of parts, or limbs, which are connected by joints. These are themselves composed of elementary movements and concerted into a sequence of actions. The term biological motion is used in the social and cognitive neuroscience community to refer to moving animate objects, which can be attributed as being locomotive. Efforts have been devoted to impoverishing the stimuli depicting such animate movements in order to reveal the key features underlying the perception of such locomotions, e.g., the point-light motion sequences proposed by Johansson [27].

17.2 Processing and Representation of Articulated Motion

The term *articulated motion* could be described in a general sense as the motion of an animated object consisting of two or more interconnected subcomponents (limbs connected through joints), which combine into a skeleton structure. Consequently, an articulated motion sequence can either be regarded as a deformation of such structure over time or as a temporal succession of skeletal configurations and the transitions in between. The analysis and recognition of articulated motion sequences thus can be based on two different kinds of input representation: dynamic *motion information*, reflecting the translation, rotation and deformation of an object over time, and *form information*, using distinctive skeletal configurations (key poses) to derive indications on the underlying articulated motion, or a combination of *both*. Figure 17.2 intuitively illustrates how motion and form might contribute to the recognition of articulated motion. Neurophysiological experiments indicate that the learning of articulated motion sequences in the primate brain might indeed make use of *both* input modalities and that combined form and motion representations are established within areas located higher up in the visual processing hierarchy [24]. Likewise, technical approaches in computer vision try to recognize articulated motions, or actions, in image sequences using form and motion information either in isolation or in combination (e.g. [39]). In the following, we will first discuss the role of form- and motion-based image features in technical action recognition systems, before we give some more details on how articulated and biological motion are assumed to be represented in the brain.

Fig. 17.2 Contributions of form and motion information to the recognition of articulated motion. Six static snapshots of an articulated motion sequence are shown (*top row*) together with corresponding optical flow fields (*bottom*). The direction of flow vectors is color-encoded, speed is indicated by opacity. The form information contained in the *second*, *fourth* and *sixth* image intuitively implies the kind of motion performed by the subject ("jumping jack"). The *first* and the *fifth* snapshots, on the other hand, do not convey the impression that the subject is moving at all. The motion information displayed in the second row reveals the difference behind these two static images. In the first image the subject has not yet started to perform the action, whereas in the fifth image, the subject is moving upwards in between two repetitions of the action

17.2.1 Technical Approaches

The analysis of articulated motion for action recognition has been a topic of intense investigation in computer vision in recent years. Due to space limitations we will only briefly summarize some of the key developments and proposals (see a recent discussion in [43]). Approaches of image-based action recognition can be subdivided into global and local approaches. *Global* space-time representations have been proposed in the form of motion-energy and motion-history images to generate templates of movement-related spatio-temporal changes over a temporal window [10]. Other authors use slices through the spatio-temporal cube of movement-related activity and calculate discriminative global statistical parameters [38]. More recently, [21] suggested analyzing the 3D shape of the space-time shapes generated by the articulation of the animated character movement.

Alternatively, approaches of *local* space-time representation focus on indicators of the presence of localized spatio-temporal keypoints. The computational approaches can be typically decomposed into (1) a detection stage that is followed by (2) a stage at which a local description of the spatio-temporal structure is extracted. The most prominent approach for detecting spatio-temporal interest points is the generalization of the structure tensor using the covariances of the spatial as well as the temporal derivatives of the video input [31]. Descriptors are then extracted from local cuboids in the space-time input located around the detected keypoints for subsequent matching purposes. Examples are gradient histograms (with dimension reduction), combined image-gradient and motion histograms, and spatio-temporal Gabor filters [16, 32]. Also, the detection of keyframes in videos relies on local features in order to cope with the variabilities of the events that demarcate the distinguishing events.

Several computer vision approaches to action recognition have been proposed which make use of knowledge about the neural processing of form and motion information and combining them into an action classifier. These approaches have been influenced by the biologically inspired model of Giese and Poggio [20], who proposed a hierarchical feedforward network architecture that aims at explaining the computational mechanisms underlying the perception of body motion including recognition from impoverished stimuli, such as point-light walkers. The model builds upon knowledge about the primary stages of processing along the dorsal and the ventral pathways in the primate visual cortex. In order to derive a functional model for computer vision applications, the core mechanisms of the model architecture have been combined with principles of a computational framework for hierarchical form processing [37] and extended to analyze motion information in addition, like the Giese-Poggio model (see [26, 39]). Here, the relative contributions of form and motion features to the classification of actions have been investigated. A different scheme emphasizes the initial stages of visual processing using local motion features as well as contrasts in the local motion representation [17, 18]. In most of these proposed models, the mechanisms for hierarchical motion (and form) processing are predefined and learning occurs at the level of a final classifier to distinguish different action categories.

17.2.2 Evidence from Biology

Animated movements in actions, like walking, turning, etc., can be robustly detected from video sequence input, and predictions about future occurrences can be derived from such spatio-temporal patterns. As mentioned above, Giese and Poggio [20] compiled an overview of the current state of knowledge of the different processing stages in the cortex along the motion and form pathways which are involved in the processing of articulated or biological motion. The proposed feedforward model architecture aims at explaining the computational mechanisms underlying the perception and recognition of bodily motion including impoverished stimuli, such as point-light walkers. The use of such stimuli that are mainly reduced of any static compositional features has been popularized by the Swedish psychologist G. Johansson, who demonstrated vivid perceptions of moving forms from animated compositions of formless dot arrangements [27]. Since then, such point-light stimuli have been utilized to reveal the key features underlying the perception of biological motion. In a nutshell, two prominent ideas have been developed. One emphasizes the dominant role of motion features, or more complex patterns thereof, that enable us to perceive articulated motions from real videos as well as animated point-light (PL) sequences. Such features can be extracted from the sensory input as long as the PL elements are in motion, but diminish when PL arrangements stop moving [14]. The other emphasizes the role configurational information plays in the recognition of biological motion that is predominantly reflected by form features. It is suggested that the human figure can be considered as a deformable template that is animated by the motion of the limbs of the moving agent [29, 30]. We propose a new learning-based hierarchical model for analyzing animated motion sequences that depend on both motion as well as form features. Prototype representations in the form and motion pathways are established using a modified Hebbian learning scheme. We suggest how snapshot prototypes might be automatically selected from continuous input video streams combining features from the motion pathway and the form pathway. Minima in the motion energy are indicative of the occurrence of specific snapshots with strongly articulated configurations, that serve as key postures. Such snapshot representations for keyposes can be considered as templates, as suggested by Lappe [30]. However, within the proposed model they are not deliberately deformed to match a continuation of changing configurations. On the other hand, the complementary pattern motion prototypes are activated by input motions such that they contribute the proper temporal changes in the configurations. Sequence-selective representations of articulated motions in the superior temporal sulcus (STS) are driven *jointly* by input activations from both motion and form prototypes (see Fig. 17.3). In addition, feedback connections are learned to enable STS neurons predicting expected input from the form selective inferior temporal cortex (IT) and the motion-sensitive medial superior temporal area (MST). We argue that for inputs presenting articulated postures without continuing motion, STS representations are fed by the corresponding snapshot prototype activations [24]. In turn, STS will send feedback to stages in the segregated pathways for

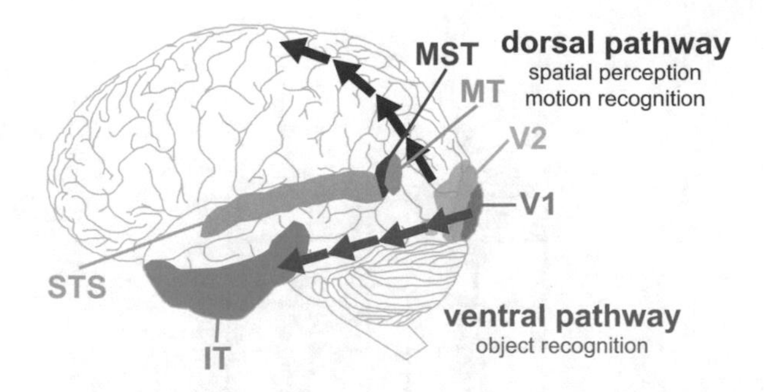

Fig. 17.3 Form and motion information both contribute to the neural processing of biological motion. The ventral (form) and the dorsal (motion) pathway both converge to the superior temporal sulcus (STS) complex, where there is evidence that biological motion-specific representations are built

form as well as motion processing. Stationary images which depict articulated postures consequently generate effects of implied motion, which have been shown in functional magnetic resonance imaging (fMRI) studies [28]. Within the proposed model, this behavior is achieved by the combined form and motion representations in model area STS, which can be driven by static form (snapshot) information only. Once activated, a sequence representation sends feedback to *both* form and motion representations, thus amplifying motion representations even if no direct motion input is present.

17.3 A Neural Model of Learning Representations of Articulated Motion Sequences

We developed a biologically inspired model for the learning of articulated motion representations. The model extends the work of Giese and Poggio [20] by combining it with mechanisms for the hierarchical feedforward and feedback processing of motion and form information along the dorsal and ventral pathways [4, 46]. The overall structure of the model is depicted in Fig. 17.4. Characteristic prototypes of intermediate-level form representations (model area IT) and motion patterns (model area MST) are learned using a Hebbian learning scheme. In addition, a motion-dependent reinforcement mechanism is used to regulate the learning of prototypical form representations. This prototypical learning scheme is inspired by empirical findings. Psychophysical and physiological experiments suggest that body poses which show a high degree of articulation allow a higher performance in the recognition of human actions ([3, 41]; compare the static snapshots in Fig. 17.2) and evoke the percept of implied motion as well [28]. Here, we suggest a simplistic but generic approach that selects such characteristic poses out of continuous streams of video sequences in an unsupervised fashion. In the case of such selections, learning

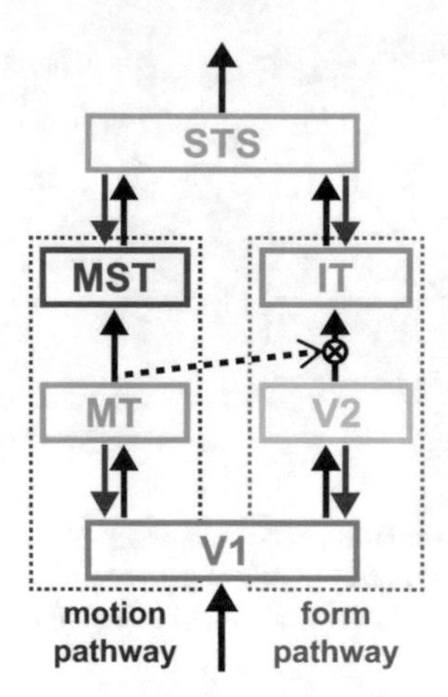

Fig. 17.4 Model overview. The model consists of two separate processing pathways, which both converge in model complex STS. In the (dorsal) motion pathway local motion features (model area V1) are combined in optical flow features (MT) and finally optical flow patterns are learned (MST). In the (ventral) form pathway, local oriented contrasts (V1) are grouped into corners and junctions (V2) used to build static form representations (IT). Feedback connections in the early processing stages increase the robustness to noise and resolve ambiguities within both pathways. At the level of STS, combined representations of MST and IT activities are learned via an instar/outstar learning scheme in combination with a trace rule. The weights of afferent connections are used to recognize the closest matching form and motion patterns. The weights of efferent connections are fed back to preceding stages, realizing a prediction mechanism. In addition, we suggest an interaction between the pathways at the level of MST/IT which enables the selective learning of key poses in the form pathway

is enabled to build representations which preferentially focus on poses with a high degree of articulation. Both pathways converge in model complex STS,[2] where sequence-selective representations are built using a combined Hebbian bottom-up and top-down learning (instar/outstar; [13]) alongside a short-term memory trace, realizing a temporal associative memory. Details about the core model components and simulation results are outlined in the following (for a complete description of the model, refer to [33]).

17.3.1 Form and Motion Processing

Processing the raw input data utilizes an initial stage of orientation and direction-selective filtering (in model area V1; see Fig. 17.4). These responses are fed

[2]The superior temporal sulcus (STS) is anatomically not an area, but a region that contains several areas and subcomponents thereof. We use the term "complex" for the model in order to highlight its specific functionality within the model as a convergent zone of information fusion.

into separated pathways which are selective to static form representations (areas V2 and IT) and characteristic optical flow patterns (areas MT and MST). The membrane potential of individual model neurons is calculated by conductance-based mechanisms of feed-forward integration of excitatory and inhibitory feeding input and a passive leakage. The potential can be enhanced by a gating mechanism to amplify the efficacy of the current potential by a matching top-down feedback signal. The membrane potential is finally regulated by a gain control mechanism that leads to activity normalization for a pool of neurons through mutual divisive inhibition (shunting inhibition; compare [12]). These mechanisms are summarized in a three-stage hierarchy of processing that includes input filtering, modulatory feedback, and pool normalization. The output of a cell is defined by a signal function which converts the membrane potential into a firing rate, or activity. Such model cells are grouped into layers which form abstract models of cortical areas. The computational properties of the model neurons and the three-stage cascade have been reported in, e.g., [4, 11, 23, 36]. We will not go into further detail to explain the computational mechanisms here. The representations that are learned in the segregated form and motion pathways define the input activities for model STS cells. Based on [25] we suggest that form and motion activities converge to drive cells in categorial representations of temporal movement selectivity.

17.3.2 Learning of Form and Motion Prototypes

In order to generate feature representations of complex form (model area IT) and motion patterns (model area MST) we employ an unsupervised learning mechanism based on a modified Hebbian learning scheme. The modification stabilizes the learning such that the growing of weight efficacies is constrained to approaching (bounded) activity levels in the input or the output activation [34]. We combined the modified Hebbian learning mechanism with a short-term memory trace of prolonged activity of the pre- or the post-synaptic cells (*trace rule*; see, e.g., [45]). The adaptation of weights is controlled by post-synaptic cells, which, in turn, mutually compete for their ability to adjust their incoming connection weights. The post-synaptic cells which gate the learning of their respective input weights are arranged in a layer of neurons competing for the best matching response and their subsequent ability to adapt their kernel that denotes the pattern of spatial input weights. In a nutshell, the layer of post-synaptic neurons competes to select a winning node for a given input presentation, which, in turn, is allowed to automatically adapt their incoming synaptic weights. The temporal trace (or short-term memory) establishes representations that encode the average input over a short temporal interval and thus gain more robustness by allowing small perturbations for the changing input signals.

17.3.3 Automatic Selection of Key Pose Frames

The Giese-Poggio model [20] suggests that sequence-selectivity for biological motion recognition is driven by sequences of static snapshots. While the original model relies on snapshots that were regularly sampled over time, we suggest a mechanism that automatically selects snapshots that correspond to strongly articulated poses. Such snapshot representations are learned in the form pathway (model area IT) by utilizing a gating reinforcement signal which is driven by the complementary representation of motion in the motion pathway (model areas MT/MST; see the dashed arrow in Fig. 17.4). For the whole body motion considered here, we simply integrated the motion energy over the entire optical flow field. The motion energy signal itself is a function of time, which is used to steer the instar learning in the form pathway. We suggest that different subpopulations of static form, or snapshot, representations can be learned that correspond to either weakly or strongly articulated postures. In the proposed learning scheme, we focus on snapshot poses corresponding to *highly* articulated postures with signatures of maximum limb spreading. The overall motion energy at the limbs drops during phases of high articulation when their apparent direction of motion reverses. This, in turn, leads to local minima in the overall motion energy of an articulating actor which coincide with the above-mentioned events of articulated key poses. In Fig. 17.5 we show the test case of a walker with two snapshots selectively read out at points of local minima and local maxima in the motion energy. Using the motion energy to drive a reinforcement signal that controls the learning of key poses demonstrates the usefulness of the extended learning for establishing key pose frame representations in the form pathway.

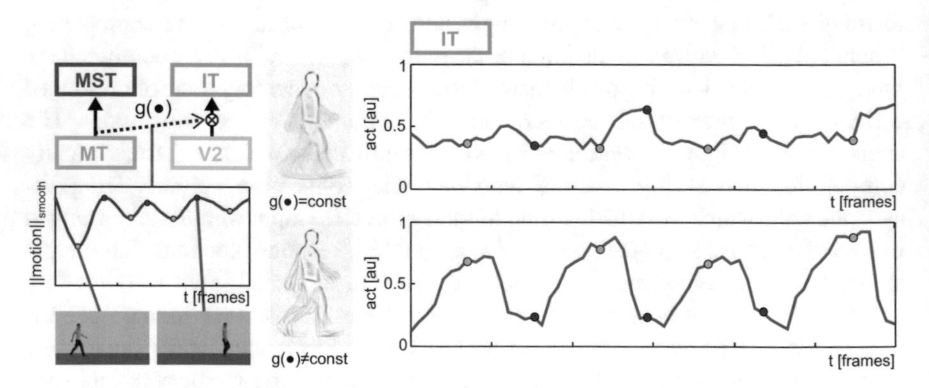

Fig. 17.5 Unsupervised learning of key body postures. The motion energy derived within the motion pathway is used to gate the learning of the form prototypes. Local minima in the motion energy signal correspond to body postures containing a high degree of articulation (see plot on the *left*). The weight of those postures is amplified during the learning process. Without incorporating the gating signal $g(\bullet)$, the statistical mean of the whole sequence is learned (on the *top right*). If $g(\bullet)$ is applied on the learning as a weighting function, a prototype representation selectively responding to highly articulated poses is established (*bottom right*). Adapted from [33]

17.3.4 Learning of Sequence-Selective Representations

The categorial representations in the form and motion pathways, namely in model areas IT and MST, which were learned at the previous stage, propagate their activations to the stage of STS to generate hierarchical representations of moving form. In order to stabilize the representations and activity distributions, even in the case of partial loss of input signals, the STS sequence-selective representations send top-down signals to their respective input stages. Prototypical representations with spatio-temporal sequence-selectivity are suggested to exist in the cortical STS complex where both form and motion pathways converge [44]. The selectivities of model STS neurons are learned again by using a modified Hebbian learning mechanism similar to the separate learning of form and motion prototypes described above (Sect. 17.3.1). In a conceptional format the representations involved in the learning process can be denoted by $\{\text{IT, MST}\} \rightarrow \text{STS}$. An important component is that sequence-selective prototypes activated in STS in turn learn the output weights back to the segregated form and motion prototype representations, namely STS $\rightarrow \{\text{IT, MST}\}$. Unlike the bottom-up, or feedforward, learning mechanisms, the learning of the top-down, or feedback, weights in STS is gated by the pre-synaptic cell. The learning of the feedback weights is combined with a temporal trace, which leads to a temporal delay in the activities of the STS cells. This establishes a representation in which each STS sequence-selective prototype encodes and memorizes in its weight pattern the following driving input activity pattern from the form and motion pathways. Such a top-down weighting pattern can then be used to generate predictions concerning the expected future input given the current maximally activated prototype at the STS level. The unsupervised learning in combination with the trace rule thus realizes the memorization and association of temporally correlated input patterns.

17.4 Model Simulations

We experimentally analyzed the capabilities of the proposed network in a number of computational experiments using artificially generated input sequences. All input sequences were generated by animating motion capturing data using a 3D animation software (SmithMicro Poser Pro 2014). In the following, we summarize such experiments, which aim at demonstrating the key functionalities of the architecture and its components.

Experiment I: Form and Motion In the first experiment we highlight the contributions of the individual components in the complex distributed architecture. Motion-captured data of a person walking was used to generate five different input configurations. The input configurations included a sequence containing both form and motion data (original), a point light sequence of the walker (PLW), an articulated and a not articulated static posture, as well as a sequence showing a

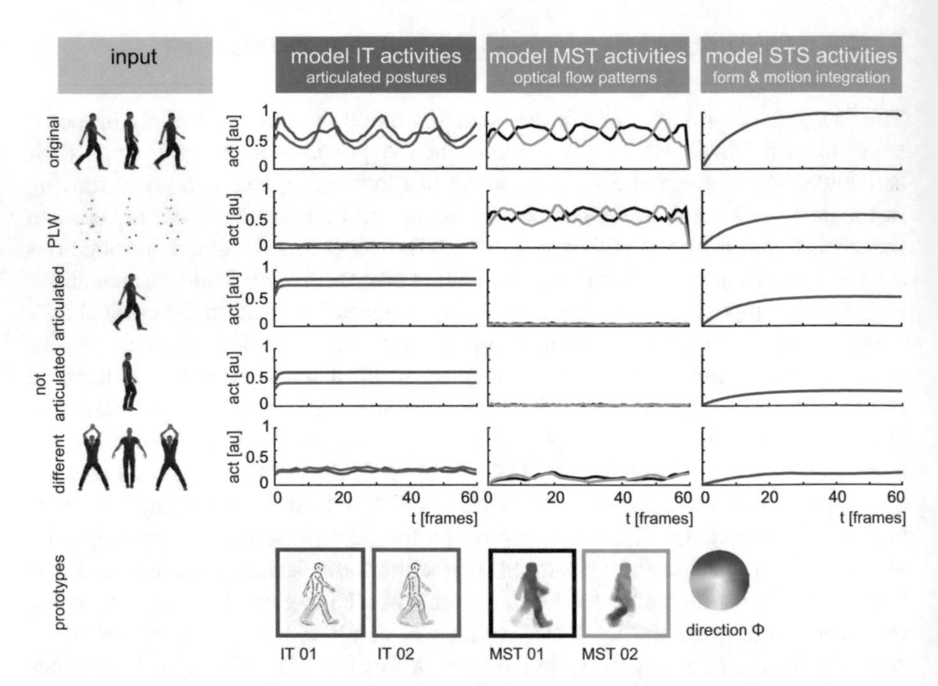

Fig. 17.6 Activations of form, motion and sequence-selective prototype cells. We used five artificially generated image sequences to demonstrate the network responses to different input configurations (*first column*). The *last row* shows two form and two motion representations (at model areas IT and MST) which were established during the training. First, the data the network was trained on was used as input, resulting in activities at their maximum in all model areas (original). Note that the activity of model IT (*second column*) selectively increases when a high degree of articulation is present. Second, we probed the network with a point-light-walker sequence, generated on the basis of the same articulated motion as in the original case (PLW). As expected, the activities of model IT cells almost drop to zero. The activities of model MST (*third column*) cells, on the contrary, stay at a level comparable to the original case, and, in combination with the feedback, trigger a medium level excitation of the model STS cell. The *third* and the *fourth columns* show the results of single pictures being continuously presented to the network as a static input. Two different static images were used, one containing a highly articulated pose and the other barely showing any articulation. The activity of the model IT, as well as STS cells, is clearly increased if the sequence shows an articulated posture. The *fifth row* shows how the activities drop throughout all model areas when a different motion sequence is used as input and, thus, the model's capability of distinguishing between different articulated motion sequences

person performing a jumping jack (different). Figure 17.6 shows the activations of the form and motion prototypes at model areas MST and IT, as well as the sequence-selective cells (model area STS) after learning. The Sequence selective representations at STS respond at a maximum level to the input they were trained on (original). In contrast, they respond at their minimum when probed with a different motion sequence (different), showing that the model is capable of learning and distinguishing different articulated motions. In line with experimental evidence, the sequence-selective cells at model STS show increased activities either if just motion (point-light-walker; see, e.g., [5]) or solely form (articulated pose) is presented to the network (compare [3, 24]). In case of static input, model STS cells respond at a higher level for input pictures showing an articulated posture compared to input images containing a not articulated pose.

Experiment II: Occlusion In a second experiment, we simulated how network activities are affected if a walker passes behind an occluder while performing an articulated motion. The network was trained using same sequence of a walker as in the first experiment. Figure 17.7 shows the result of the simulation. The sequence-selective cell in model STS shows an increase in activity until the walker starts to step behind the occluder. During the phase of occlusion, the activity only decreases slowly until the walker reappears and the activity increases again (compare [2]). In addition, our model makes the prediction that the level of activation during the occlusion—and thus the temporal duration the network "remembers" an observed motion—is dependent on the degree of articulation on stepping behind the occluder.

Experiment III: Direction Invariance Invariance to changes in perspective is one of the major challenges for action recognition systems. In the third presented experiment, we investigated the tuning of the sequence-selective prototypes to inputs showing a sequence from different perspectives. Again, input sequences were generated using the same motion capture of a walking gait as in the first experiment. Here, we configured different walkers with varying movement directions and speeds with reference to a previously learned representation of a rightward moving walker. Walking directions in the test cases were rotated by $\pm\{5°, 10°, 20°, 40°\}$ around the vertical axis. Model simulations result in a direction sensitivity of STS cells with

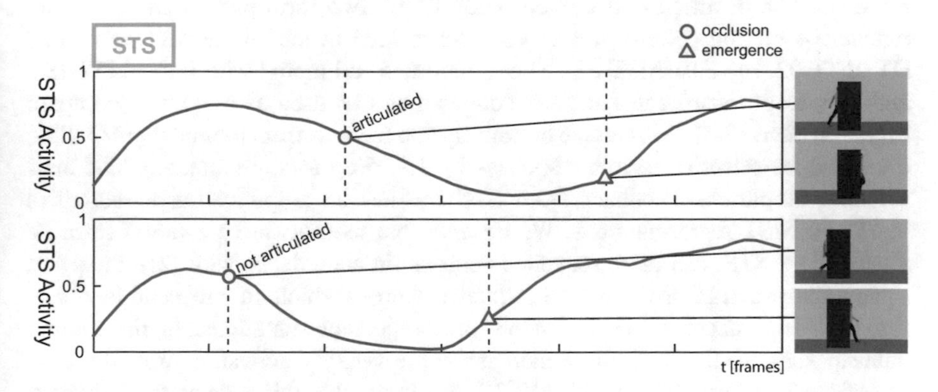

Fig. 17.7 Occlusion of an observed articulated motion. After training, a sequence containing an occluder was presented to the network. After the subject disappears behind the occluder, the activities of the sequence-selective model STS cell don't drop instantaneously, but slowly decrease until they increase again when the walker reappears (*first row*). The model predicts that the speed at which the activities decrease during occlusion is dependent on the degree of articulation of the walker at the moment he steps behind the occluder (*first row* versus *second row*)

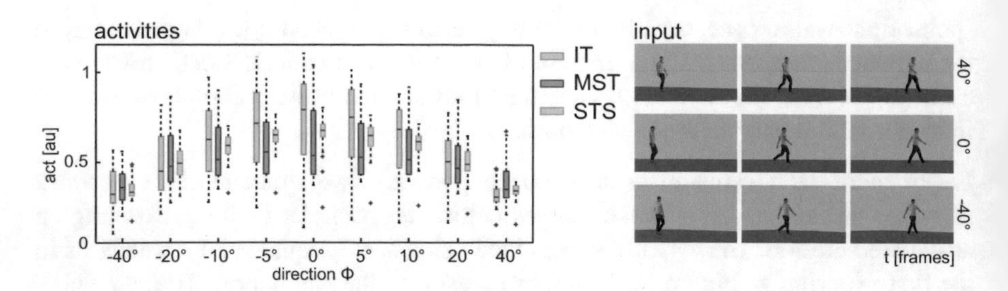

Fig. 17.8 Direction selectivity of model area IT, MST and STS cells. We trained a walker moving in a horizontal direction for $\phi = 0°$. Activities of prototypical cells are shown (*left*) which were probed by different inputs with varying movement directions, i.e. walkers approaching or receding at different angles with respect to the horizontal reference axis (*right*). Data has been summarized into box plots showing the response variabilities of models cells as well as the monotonic decline in response to deviations from the target view tuning. The tuning width at half the maximum response is around $\pm 40°$. Adapted from [33]

half amplitude of approximately $\pm 40°$ (see Fig. 17.8). IT and MST cells, on the other hand, also show a decrease in activities but have a much larger variability. This latter observation renders the selective form or motion features unreliable as an indicator for estimating the spatial walking direction. The fused signals that trigger sequence-selective activations allow us to differentiate between changes in pose due to their increased angle relative to the learned reference direction.

Experiment IV: Lesioning Finally, we show the result of selectively extinguishing connections between model areas or complexes that participate in the distributed representation of articulated motions (Fig. 17.9). Two form pattern and the same number of motion pattern prototypes were trained at model areas IT and MST (IT 01/IT 02, MST 01/MST 02). The unmodified full model with trained IT/MST and STS feedforward and feedback connections was used as reference (compare Fig. 17.4, Sect. 17.3). Without the bottom-up connections from motion input (MST), the sequence-selective neuron responses in STS drop to approximately half their response amplitude. Feedback from STS invokes an amplification of activities in IT and MST representations. We observe that feedforward activation from IT alone, IT $\rightarrow$ STS, can drive sequence neurons (in accordance with [2]). Snapshot representations in IT drive the STS sequence neurons which, in turn, send feedback signals to the stages of IT and MST prototype representations. In the motion pathway such feedback elicits an increase in pre-synaptic activation of model cells in MT/MST (IT $\rightarrow$ STS $\rightarrow$ {IT, MST}). We argue that this reflects the induction of increased fMRI BOLD response in human MT+ following the presentation of static implied motion stimuli [28, 40]. We observed the same effect when cutting bottom-up connections from form input (IT $\rightarrow$ STS), but with reversed roles of the MST and IT representations.

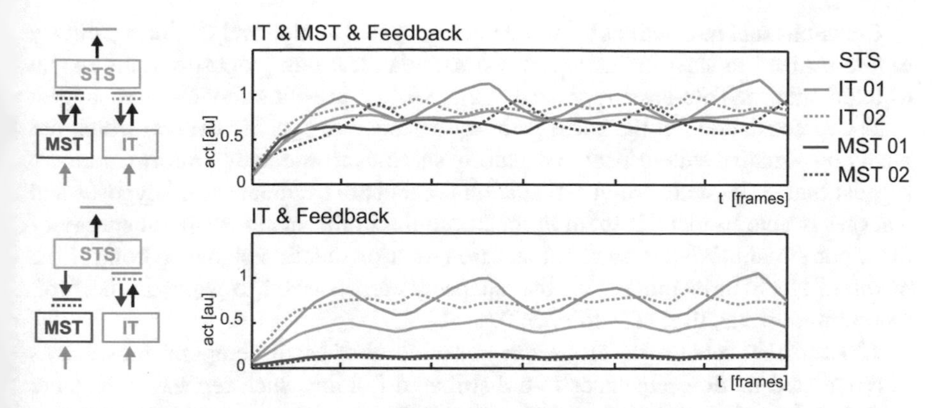

Fig. 17.9 Selective removal of interconnections. The model was left untouched to provide a reference (*top*). Bottom-up (feedforward) connections between areas MST and STS were removed, preventing any motion-related signals being propagated to STS (*bottom*). The amplitude of the IT prototype activities remains almost the same, whereas the sequence-selective STS cell responds only at about half-magnitude (because of the missing support from the motion pathway). Note the feedback activities propagated from STS to MST optical flow pattern prototypes. We argue that this reflects the induction of increased fMRI BOLD response in human MT+ following the presentation of static implied motion stimuli. Adapted from [33]

17.5 Discussion and Conclusion

Companion-Systems need to be affect-sensitive in order to detect and analyze social signals that are generated by human subjects operating in the system's surround region of the extra-personal space. Here, we have focused on the analysis of articulated motion of subjects based on features that can be detected in a medium- to long-range interaction region of the extra-personal space (Fig. 17.1). A model architecture has been proposed that is inspired by the architecture and function of the primate visual system. This biologically inspired visual architecture is equipped with segregated processing pathways which are specialized to motion and shape, or form, processing. Representations generated in these pathways converge to generate a fused sequence-selective representation which, together with the motion and form representations, can be considered a distributed representation of articulated, or biological, motion. The signal flows in this model architecture are bidirectional to define bottom-up, or feedforward, and top-down, or feedback, counter stream networks which enable us to build rich representations from the sensory input that is combined by context information at different stages of the processing. The effective weights of such feedforward as well as feedback connections are learned in an unsupervised fashion, which enables the system to self-adapt its internal representations to the statistics of the environment. The proposed model architecture has been tested by running selected simulations of the computational stages. The main observations and outcomes of the computational demonstrations are summarized below.

Cross-channel interactions between the segregated motion and the form pathway can be utilized to steer the unsupervised Hebbian learning proposed in the model to build intermediate-level representations. Unlike previous models, we suggest static representations in the form pathway selective to specific events, which are highly informative with respect to capturing significant articulation information. We suggest that poses with strong articulation content are candidates for keyposes and that one is able to identify them at local minima in the spatio-temporal energy of the input stream. We use such information as a modulation signal to control the modified Hebbian learning such that snapshots are selected to generate the form representation adaptively (Sect. 17.3.3).

The learning scheme develops a representation of spatio-temporal movements of articulated motion sequences in a distributed fashion. Such representations are established in the MT/MST and the V2/V4/IT stage, respectively. Here, at an intermediate level of detail, prototypical motion patterns as well as prototypical snapshots are learned in the response of an input stream that contains articulated motion of an animated character. In addition, the prototypical representations also activate sequence-selective representations in the STS complex, where inputs from the motion and form pathways converge to build a combined coding of moving form. It has been demonstrated that such a distributed but interacting representation is selective to different sequences, e.g., segregating walking from jumping. Gradual manipulations of such sequences, e.g., by reversing the order of presentation of the components of a sequence or scrambling the phase of the temporal compositions, leads to gradual changes in the responsiveness of such a representation. This replicates findings from experimental investigations. In addition, gradual changes in the view direction of an articulated sequence leads to a gradual decline in the responsiveness of the prototypical representation that has been trained for a reference view. The results show that such representations at the sequence-selective stage of model STS show a broad tuning function, but that the selective motion and form representations show a less prominent tuning (Sect. 17.4, Experiment II).

The representations in the model are bidirectionally connected such that feed-forward and feedback processing are combined according to the experimental literature. In combination with the initial learning of the feedforward as well as the feedback connections, a network functionality emerges that contains convergent feature representations from the fusion of motion and form. It is demonstrated that the selective extinction of one input channel, e.g., missing form input in the case of impoverished sequence input using point-light walkers, can be partially compensated for in the fused sequence-selective representations (Sect. 17.4, Experiment I). In addition, since the fused representations makes reciprocal connections into the motion and form pathways, respectively, the activated sequence representations generate expectations for the most probable input features in the segregated representations in these pathways. This leads to a transfer of information from one pathway into the complementary one. For example, articulated still images perceptually lead to the impression of an implied motion that has been frozen into a selective event of high dynamics (Sect. 17.4, Experiment I/Experiment IV). Even in experiments, an activation has been observed in the motion pathway in response to such still images. Our model makes the prediction that articulated motions activate prototypical pose representations in the form pathway which are fed forward into the convergent representation in the STS. Such representations, in turn, send feedback to *both* expected inputs and thus pre-activate the corresponding, most probable, inputs of prototypical motions.

Related to the top-down predictions that lead to an information transfer from one into the complementary pathway are any consequences generated by the feedback activation to generate predictions while continuous input is generated via the sensory input channels. In humans, it has been demonstrated that differences in activation strength occur when a walking person disappears behind an occluder. When the occlusion occurs in the case of an articulated posture, then the activation generated during the temporal period of occlusion is significantly higher than in the case of occlusion in a neural pose. We predict that such anticipation by the visual system is generated via the signals propagated along feedback paths that are increased due to the coactivated keyposes in the form pathway to generate a more prolonged activity trace for the expected input (Sect. 17.4, Experiment II).

These computational competencies, though motivated by principles of biological processing of articulated motion sequences of observed actors, have a high potential to build future technologies in the observation and analysis of human actors. The segregated motion and form representation as well as their fusion into a combined sequence-selective representation is indicative for building efficient and robust mechanisms and their associated representations. It is demonstrated how a selective loss of partial information can be compensated for by such a systems architecture that is composed of such distributed but mutually coupled representations. In addition, the unsupervised learning of such representations serves as a potential candidate for building technical systems which are capable of self-adapting and autonomously learning specific internal representations necessary to build long-term relationships for human-*Companion* interaction. The employed learning mechanisms are generic in the sense that a core principle of correlation learning via Hebbian mechanisms has been utilized and specified to achieve certain functionalities. The model architecture builds upon recurrent feedforward and feedback processing, like in primate visual cortex. While feedforward hierarchical processing is also utilized in many machine vision approaches, the incorporation of a feedback signal pathway is much less prominent so far. Here, such top-down signal streams are hypothesized to use the activations in a representation that already fused partial information from segregated pathways to gather more context in the interpretation of articulated motion and form. Such information, in turn, is re-entered at earlier stages and as such can help to disambiguate locally noisy and error-prone sensory input to yield more robust computational results.

The proposed model is not complete as it does not cover all details of a fully elaborate functional set of competencies. The representations reported here are all holistic in their nature. Bodies and their articulations, especially when they are observed from different view directions, can be naturally decomposed into body parts and also key features, such as joint and connected limbs. Several key mechanisms discussed in this chapter can be applied to the acquisition of representations through learning and the specific selection of prominent events in the spatio-temporal sequence. What is still a challenge is how such different levels of the representation can be combined into a network of distributed representations that is dedicated to the multi-scale analysis of articulated motion. In addition, we have not addressed here the issue of how multiple motions that occur in a crowded scene might be robustly segregated prior to the analysis of the articulated motion sequence. While these topics determine an outline for future investigation they remain focused on the complex interplay of processing information in the visual sensory stream. One direction of further investigation is how such visual processes might be combined with proprioceptive and motor information. Consider the problem of how articulated motion observed from a moving person (or animal) can be distinguished from the complex movement of an artificial object that can change the view perspective of the observation. Standard computer vision or vision-dominated approaches investigate the extraction of a rich set of features and their combinations and how these can be learned to build a representation that serves as an input for a subsequent classifier. A different idea is guided by the observation that articulated motions can be accomplished by one's own movements such that an observer can mimic those limb motions and virtually recapitulate such motions through simulation, at least in principle. This offers a new road for investigating articulated and biological motion perception by considering a combined vision and proprioceptive/motion account in which a large-scale network is involved in analysis and recognition processes that sense the input data and, at the same time, mentally simulate such observed movements to arrive at a consensus interpretation of the action in the scene.

Acknowledgements This work was done within the Transregional Collaborative Research Centre SFB/TRR 62 "*Companion*-Technology for Cognitive Technical Systems" funded by the German Research Foundation (DFG).

References

1. Argyle, M.: Bodily Communication. Methuen & Co Ltd, London (1988)
2. Baker, C., Keysers, C., Jellema, T., Wicker, B., Perrett, D.: Neuronal representation of disappearing and hidden objects in temporal cortex of the macaque. Exp. Brain Res. **140**(3), 375–381 (2001)
3. Barraclough, N.E., Xiao, D., Oram, M.W., Perrett, D.: The sensitivity of primate STS neurons to walking sequences and to the degree of articulation in static images. Prog. Brain Res. **154**, 135–148 (2006)
4. Bayerl, P., Neumann, H.: Disambiguating visual motion through contextual feedback modulation. Neural Comput. **16**(10), 2041–2066 (2004)
5. Beauchamp, M.S., Lee, K.E., Haxby, J.V., Martin, A.: FMRI responses to video and point-light displays of moving humans and manipulable objects. J. Cogn. Neurosci. **15**(7), 991–1001 (2003)
6. Benyon, D., Mival, O.: Landscaping personification technologies: from interactions to relationships. In: Proceedings of the CHI '08, Extended Abstracts on Human Factors in Computing Systems, CHI EA '08, pp. 3657–3662. ACM, New York (2008)

7. Benyon, D., Mival, O.: Scenarios for companions. In: Your Virtual Butler. Lecture Notes in Computer Science, vol. 7407, pp. 79–96. Springer, Berlin (2013)
8. Bickmore, T.W., Picard, R.W.: Establishing and maintaining long-term human-computer relationships. ACM Trans. Comput.-Hum. Interaction **12**, 293–327 (2005)
9. Blakemore, S.J., Decety, J.: From the perception of action to the understanding of intention. Nat. Rev. Neurosci. **2**(8), 561–567 (2001)
10. Bobick, A.F., Davis, J.W.: The recognition of human movement using temporal templates. IEEE Trans. Pattern Anal. Mach. Intell. **23**(3), 257–267 (2001)
11. Bouecke, J.D., Tlapale, E., Kornprobst, P., Neumann, H.: Neural mechanisms of motion detection, integration, and segregation: from biology to artificial image processing systems. EURASIP J. Adv. Signal Process. **2011**(1), 781561 (2010)
12. Carandini, M., Heeger, D.J., Movshon, J.A.: Linearity and gain control in V1 simple cells. Cereb. Cortex (13), 401–444 (1999)
13. Carpenter, G.A.: Neural network models for pattern recognition and associative memory. Neural Netw. **2**(4), 243–257 (1989)
14. Casile, A., Giese, M.A.: Critical features for the recognition of biological motion. J. Vis. **5**(4), 6 (2005)
15. Castellano, G., McOwan, P.W.: Towards affect sensitive and socially perceptive companions. In: Your Virtual Butler. Lecture Notes in Computer Science, vol. 7407, pp. 42–53. Springer, Berlin (2013)
16. Dollár, P., Rabaud, V., Cottrell, G., Belongie, S.: Behavior recognition via sparse spatio-temporal features. In: 2nd Joint IEEE International Workshop on Visual Surveillance and Performance Evaluation of Tracking and Surveillance, 2005, pp. 65–72. IEEE, New York (2005)
17. Escobar, M.J., Kornprobst, P.: Action recognition via bio-inspired features: the richness of center–surround interaction. Comput. Vis. Image Underst. **116**(5), 593–605 (2012)
18. Escobar, M.J., Masson, G.S., Vieville, T., Kornprobst, P.: Action recognition using a bio-inspired feedforward spiking network. Int. J. Comput. Vis. **82**(3), 284–301 (2009)
19. Frith, C.D., Wolpert, D.M.: The Neuroscience of Social Interaction: Decoding, Imitating, and Influencing the Actions of Others. Oxford University Press, Oxford (2004)
20. Giese, M.A., Poggio, T.: Neural mechanisms for the recognition of biological movements. Nat. Rev. Neurosci. **4**(3), 179–192 (2003)
21. Gorelick, L., Blank, M., Shechtman, E., Irani, M., Basri, R.: Actions as space-time shapes. IEEE Trans. Pattern Anal. Mach. Intell. **29**(12), 2247–2253 (2007)
22. Grüsser, O.J.: Grundlagen der neuronalen Informationsverarbeitung in den Sinnesorganen und im Gehirn. In: GI - 8. Jahrestagung, pp. 234–273. Springer, Berlin (1978)
23. Hansen, T., Neumann, H.: A recurrent model of contour integration in primary visual cortex. J. Vis. **8**(8), 1–25 (2008)
24. Jellema, T., Perrett, D.I.: Cells in monkey STS responsive to articulated body motions and consequent static posture: a case of implied motion? Neuropsychologia **41**(13), 1728–1737 (2003)
25. Jellema, T., Maassen, G., Perrett, D.I.: Single cell integration of animate form, motion and location in the superior temporal cortex of the macaque monkey. Cereb. Cortex **14**(7), 781–790 (2004)
26. Jhuang, H., Serre, T., Wolf, L., Poggio, T.: A biologically inspired system for action recognition. In: Proceedings of the 11th IEEE International Conference on Computer Vision, pp. 1–8 (2007)
27. Johansson, G.: Visual perception of biological motion and a model for its analysis. Percept. Psychophys. **14**(2), 201–211 (1973)
28. Kourtzi, Z., Kanwisher, N.: Activation in human MT/MST by static images with implied motion. J. Cogn. Neurosci. **12**(1), 48–55 (2000)
29. Lange, J., Lappe, M.: A model of biological motion perception from configural form cues. J. Neurosci. **26**(11), 2894–2906 (2006)

30. Lappe, M.: Perception of biological motion as motion-from-form. e-Neuroforum **3**(3), 67–73 (2012)
31. Laptev, I.: On space-time interest points. Int. J. Comput. Vis. **64**(2-3), 107–123 (2005)
32. Laptev, I., Caputo, B., Schüldt, C., Lindeberg, T.: Local velocity-adapted motion events for spatio-temporal recognition. Comput. Vis. Image Underst. **108**(3), 207–229 (2007)
33. Layher, G., Giese, M.A., Neumann, H.: Learning representations of animated motion sequences - a neural model. Top. Cogn. Sci. **6**(1), 170–182 (2014)
34. Oja, E.: Simplified neuron model as a principal component analyzer. J. Math. Biol. **15**(3), 267–273 (1982)
35. Pentland, A.: Social Signal Processing. IEEE Signal Process. Mag. **24**(4), 108–111 (2007)
36. Raudies, F., Mingolla, E., Neumann, H.: A model of motion transparency processing with local center-surround interactions and feedback. Neural Comput. 1–45 (2011)
37. Riesenhuber, M., Poggio, T.: Hierarchical models of object recognition in cortex. Nat. Neurosci. **2**, 1019–1025 (1999)
38. Rittscher, J., Blake, A., Hoogs, A., Stein, G.: Mathematical modelling of animate and intentional motion. Philos. Trans. R. Soc. Lond. Ser. B Biol. Sci. **358**(1431), 475–490 (2003)
39. Schindler, K., Van Gool, L.: Action snippets: how many frames does human action recognition require? In: Proceedings of the 2008 IEEE Conference on Computer Vision and Pattern Recognition, pp. 1–8. IEEE Computer Society, New York (2008)
40. Senior, C., Barnes, J., Giampietroc, V., Simmons, A., Bullmore, E.T., Brammer, M., David, A.S.: The functional neuroanatomy of implicit-motion perception or 'representational momentum'. Curr. Biol. **10**(1), 16–22 (2000)
41. Thirkettle, M., Benton, C.P., Scott-Samuel, N.E.: Contributions of form, motion and task to biological motion perception. J. Vis. **9**(3), 28 (2009)
42. Thompson, J.C., Clarke, M., Stewart, T., Puce, A.: Configural processing of biological motion in human superior temporal sulcus. J. Neurosci. **25**(39), 9059–9066 (2005)
43. Turaga, P., Chellappa, R., Subrahmanian, V.S., Udrea, O.: Machine recognition of human activities: a survey. IEEE Trans. Circuits Syst. Video Technol. **18**(11), 1473–1488 (2008)
44. Ungerleider, L.G., Pasternak, T.: Ventral and dorsal cortical processing streams. Vis. Neurosci. **1**(34), 541–562 (2004)
45. Wallis, G., Rolls, E.: Invariant face and object recognition in the visual system. Prog. Neurobiol. **51**(2), 167–194 (1997)
46. Weidenbacher, U., Neumann, H.: Extraction of surface-related features in a recurrent model of V1-V2 interactions. PloS ONE **4**(6), e5909 (2009)

Chapter 18
Automated Analysis of Head Pose, Facial Expression and Affect

Robert Niese, Ayoub Al-Hamadi, and Heiko Neumann

Abstract Automated analysis of facial expressions is a well-investigated research area in the field of computer vision, with impending applications such as human-computer interaction (HCI). The conducted work proposes new methods for the automated evaluation of facial expression in image sequences of color and depth data. In particular, we present the main components of our system, i.e. accurate estimation of the observed person's head pose, followed by facial feature extraction and, third, by classification. Through the application of dimensional affect models, we overcome the use of strict categories, i.e. basic emotions, which are focused on by most state-of-the-art facial expression recognition techniques. This is of importance as in most HCI applications classical basic emotions are only occurring sparsely, and hence are often inadequate to guide the dialog with the user. To resolve this issue we suggest the mapping to the so-called "Circumplex model of affect", which enables us to determine the current affective state of the user, which can then be used in the interaction. Especially, the output of the proposed machine vision-based recognition method gives insight to the observed person's arousal and valence states. In this chapter, we give comprehensive information on the approach and experimental evaluation.

18.1 Introduction

In contemporary human-computer interaction (HCI), machine-based vision increasingly gains pace, whereas, besides gesture control, analysis of faces is an important application area. In that field, not only person identification is focused on, but also the deciphering of non-verbal communication through facial expression, as this can provide feedback about user behavior [17]. In automated camera-based

R. Niese (✉) • A. Al-Hamadi
Institute for Information Technology and Communications (IIKT), University of Magdeburg, Magdeburg, Germany
e-mail: Robert.Niese@ovgu.de; Ayoub.Al-Hamadi@ovgu.de

H. Neumann
Institute for Neural Information Processing, University of Ulm, Ulm, Germany
e-mail: Heiko.Neumann@uni-ulm.de

S. Biundo, A. Wendemuth (eds.), *Companion Technology*, Cognitive Technologies,
DOI 10.1007/978-3-319-43665-4_18

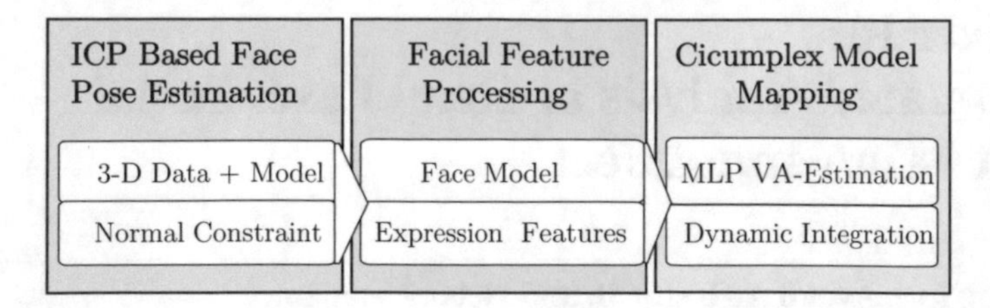

Fig. 18.1 Processing chain of the three main modules

facial expression analysis mostly fixed emotion categories have been utilized in the past, as described by Ekman [6]. However, due to the rare occurrence of classic basic emotions in HCI applications, usually this strategy is of limited use. For that reason, several groups have attempted to map visual and audio-visual utterances, predominantly facial expressions to dimensional emotion-models, like the Circumplex Model of Affect [19], by inferring Valence-Arousal (V-A) parameters. Having these affective user state parameters, the course of the interaction can generally be guided more intuitively, i.e. the machine can provide help to a puzzled user. In the dimensional emotion model, the parameters represent states from negative to positive for valence and calm to aroused for arousal. It has also been shown that the V-A transformation can be disturbed by inaccuracies in the image-based processing [14]. In the presented concept, we focus on this problem through hierarchical analysis. Further, we derive the intensity of a particular expression state, which provides a useful parameter for interaction, i.e. a user with high arousal can be given a different response.

In the following three sections, we present the components of our system as depicted in Fig. 18.1. The first one is used to determine the observed person's head pose, followed by facial feature processing and, third, by classification. In our concept, these are successive modules, which can also be substituted by alternative approaches, i.e. a different pose estimator, feature set, or classification strategy, in order to adapt to a particular application. The evaluation of the presented methods is based on a tilt sensor for the pose, analysis of a 3-D database for facial expressions of emotion, as well as online examples, as shown at the end each section.

18.2 ICP-Based Face Pose Estimation

Automated analysis of image content requires precise knowledge about the arrangement in the captured scene. This does not only include detection and recognition of the interesting objects, but also the determination of their orientation. This general principle also holds for automated analysis of human subjects observed by a camera system, and of course, it is easy to see that the evaluation of a rotated face differs much from a frontal one. Driven by the availability of depth sensors at affordable prices, e.g. Microsoft Kinect, ASUS WAVI Xtion or SoftKinetic, in the recent few years numerous market-feasible applications have been developed for human-machine interaction, mostly for gesture recognition in real time. It has been shown that by using active depth sensors, many problems can be tackled, e.g. illumination changes, strong rotation, occlusion and difficult background. This gives motivation for the presented face pose estimation approach. Fanelli et al. [8] have presented an efficient, but training-intensive depth-based head pose estimation that uses the Discriminative Regression Forest technique. In contrast, our approach achieves a high accuracy and robustness and can handle strong rotations even without excessive training [15]. It is based on an extended variant of the Iterative Closest Point (ICP) registration algorithm, which was originally introduced by Besl and McKay [2]. In the applied ICP approach a user-adapted face model is registered with measured point cloud data. The adaptation is carried out only once in an initialization step. The important processing steps and the used parametrization are given in the following.

18.2.1 Acquisition of 3-D Scene Data

In the first processing step the camera's depth and color data is captured using the software frameworks OpenCV/OpenNI [3]. Under the assumption of a pinhole camera model and a given camera constant, we compute point cloud **W** (18.1) of the scene from the depth image [15]. Further, we define a box as the operating volume **V** (18.2) that limits the amount of point cloud data and excludes the background (Fig. 18.2). The parametrization depends upon the experimental setup, e.g. we have mounted the camera on top of the monitor.

$$\mathbf{W} = \{\mathbf{p}_1, \ldots, \mathbf{p}_n\}, \mathbf{p}_i \in \mathbb{R}^3 \tag{18.1}$$

$$\mathbf{V} = \{\mathbf{p}_{min}, \mathbf{p}_{max}\}, \tag{18.2}$$

with $\mathbf{p}_{min} = (-0.5, -0.5, 0.5)$, $\mathbf{p}_{max} = (0.5, 0.5, 1.0)$ in meters.

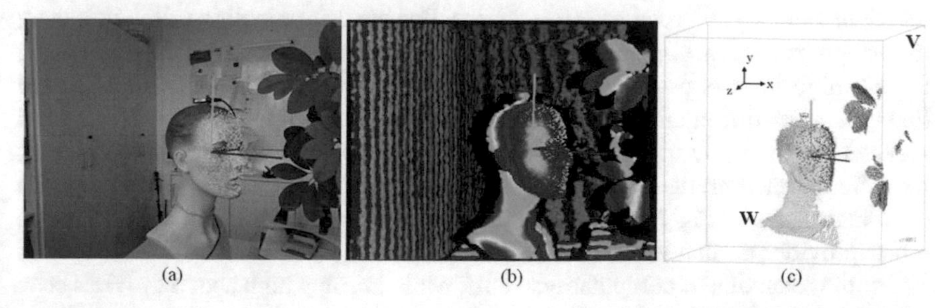

Fig. 18.2 Captured scene. (**a**) Image and pose encoded in a coordinate system (RGB $\simeq$ XYZ) in the centroid of the head, (**b**) depth map, (**c**) point cloud **W** and volume **V**

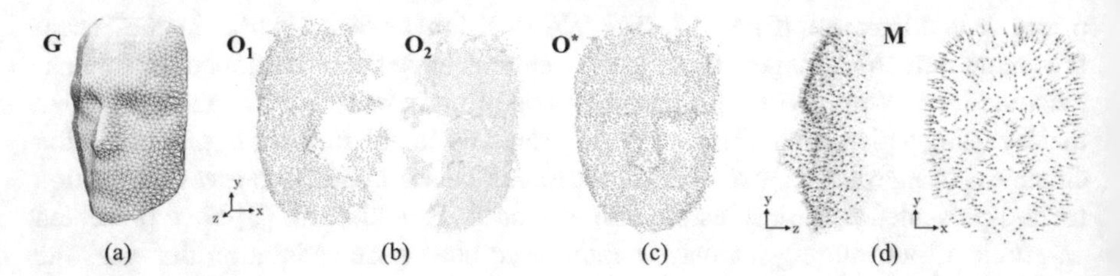

Fig. 18.3 Creation of ICP model. (**a**) Generic model **G**, (**b**) Point clouds $\mathbf{O}_1$ and $\mathbf{O}_2$ of *left* and *right* sides, (**c**) fused point cloud $\mathbf{O}^*$, (**d**) ICP fitting model **M** containing $n = 500$ vertices $\mathbf{a}_j$ and normals $\mathbf{b}_j$

18.2.2 Creation of User-Specific ICP Fitting Model

In this work, we apply a generic geometrical face model **G** (18.3) for the creation of user-specific ICP fitting models, which present the basis for pose estimation. There are several techniques for the creation of 3-D face models; the easiest way is a face scanner [10]. During processing, model **G** represents a smoothed coarse geometric average of ten evaluated subjects [15] (Fig. 18.3a). Thus, model **G** represents a general face shape and consists of a set of vertices $\mathbf{a}_i$, normal vectors $\mathbf{b}_i$ and triangle indices w_j. That enables the model to serve for rough pose estimation of up to $\pm 40°$ rotation angles when dealing with unknown faces.

$$\mathbf{G} = (\{\mathbf{a}_1, \ldots, \mathbf{a}_n\}, \{\mathbf{b}_1, \ldots, \mathbf{b}_n\}, \{w_1, \ldots, w_m\}),\ \mathbf{a}_j, \mathbf{b}_j \in \mathbb{R}^3, w_k \in \mathbb{N} \tag{18.3}$$

However, in order to achieve accurate results, especially for large rotation angles, it is beneficial to adapt the model shape and size as close as possible to the actual face. It is to be noted that transient face shape changes that may occur due to facial expression can be neglected at this point, as it has been shown in experiments that these changes do not have relevant influence on the pose estimation with the presented approach [15].

For the creation of the person-specific accurate ICP model, several point clouds $\mathbf{O}_i$ of the respective person are combined from different views, ideally from the left and right sides ($\pm 25°$ rotation) (Fig. 18.3b). For the determination of the points $\mathbf{O}_i$ the generic model **G** is approximated to the captured point cloud **W** (18.1) by using the ICP algorithm as presented in this chapter. Next, all measuring points are used that have a Euclidean minimum distance to model **G**. The combination of all point clouds $\mathbf{O}_i$ is done by utilizing the measured poses of the generic model. Hence, the different measurements are realigned to a common orientation $\mathbf{O}^*$ in the same coordinate system (Fig. 18.3c). Subsequently, the points are triangulated to a mesh, which provides normal vectors for each vertex.

For the reduction of computational cost, while keeping high accuracy it has been proven suitable to sub-sample the triangle mesh to $n = 500$ vertices, in order to create the ICP fitting model **M** (18.4) [15]. The model consists of a set of vertices

$\mathbf{a}_j$ and normal vectors $\mathbf{b}_j$ (Fig. 18.3d).

$$\mathbf{M} = (\{\mathbf{a}_1, \dots, \mathbf{a}_n\}, \{\mathbf{b}_1, \dots, \mathbf{b}_n\}),\ \mathbf{a}_j, \mathbf{b}_j \in \mathbb{R}^3,\ n = 500 \qquad (18.4)$$

18.2.3 ICP-Based Pose Estimation Using a Normal Vector Constraint

Estimation of rigid body pose usually refers to the determination of six unknown parameters, i.e. three translations and three rotations, which in the following are referred to as pose vector **t**.

$$\mathbf{t} = (t_x\ t_y\ t_z\ t_\omega\ t_\phi\ t_\kappa)^{\mathrm{T}},\ \mathbf{t}_i \in \mathbb{R} \qquad (18.5)$$

Generally, the determination of the pose from image data is an optimization problem, which is mostly solved iteratively on the basis of an error measure. Differences between the pose estimation methods arise in the definition of the error measure, the kind of utilized model and image features and the type of correspondences. In the case of captured 3-D scenes and unknown correspondences between model and world data, the Iterative Closest Point (ICP) algorithm offers opportunities for a quick and accurate solution. In general, in the ICP approach correspondences are determined between two basically n-dimensional data sets, like point clouds or geometrical descriptions, while reducing a global distance measure and approximating the pose parameters.

Accordingly, the computation of the head pose is carried out by aligning the 3-D model **M** (18.4) respectively **G** (18.3) with respect to point cloud **W** (18.1). The error function $e(\mathbf{t})$ (18.6) represents the quality of the current pose **t**. The total error results from the sum of all squared distances d_j between the model vertices $\mathbf{a}_j$ and the plane, which contains the spatially closest measuring point $\mathbf{p}_i$ in point cloud **W**. Further, this plane is oriented orthogonally to the model's normal vector $\mathbf{b}_j$ (Fig. 18.4).

$$e(\mathbf{t}) = \sum_j (d_j(\mathbf{t}))^2 \rightarrow \min,\ d_j(\mathbf{t}) = (\mathbf{a}_j(\mathbf{t}) - \mathbf{p}_i) \cdot \mathbf{b}_j, \qquad (18.6)$$

with $\mathbf{t} \in \mathbb{R}^6,\ \mathbf{a}_j, \mathbf{b}_j, \mathbf{p}_i, \in \mathbb{R}^3,\ d_j \in \mathbb{R}$.

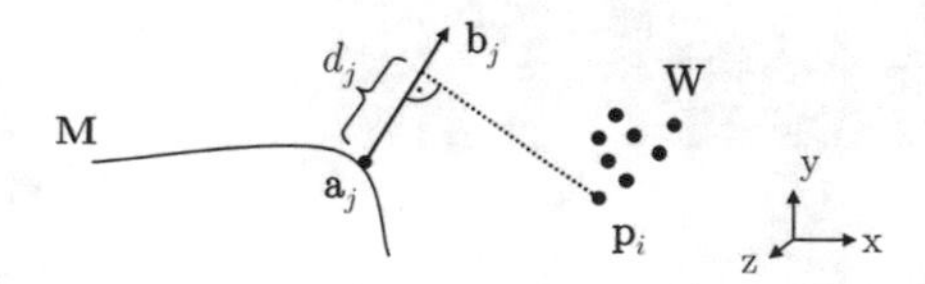

Fig. 18.4 Model fitting principle. Minimization of orthogonal model to point cloud distance d_j. Consider model **M** with vertex $\mathbf{a}_j$: searched for is the next point $\mathbf{p}_i$ in point cloud **W** lying in a plane perpendicular to $\mathbf{b}_j$

Algorithmically, the ICP method applied is as follows:

- *Initialization or reset of pose vector* $\mathbf{t}^{[0]}$ *if necessary.*
- *Let* $\mathbf{W}$ (18.1) *be a cloud of points* $\mathbf{p}_i$ *and* $\mathbf{M}$ (18.4) *an ICP model with vertices* $\mathbf{a}_j$ *and associated normals* $\mathbf{b}_j$.
- *Repeat for* $k = 1 \ldots k_{max}$ *or until convergence:*
  - *Determine a set of closest correspondences* $\mathbf{S}$
    $\mathbf{S} = \bigcup_{j=1}^{m} (\mathbf{a}_j(\mathbf{t}^k), f_{cp}(\mathbf{W}, \mathbf{a}_j(\mathbf{t}^k)))$
    with f_{cp} returning the closest point $\mathbf{p}_i$ in $\mathbf{W}$ to any point $\mathbf{a}_j$.
  - *Compute the new pose vector* $\mathbf{t}^{[k+1]}$, *which minimizes the fitting error function* $e(\mathbf{t})$ (18.6) *with respect to all pairs* $\mathbf{S}$.

In order to efficiently find the corresponding model points $\mathbf{a}_j$ and measuring points in $\mathbf{W}$ we use function f_{cp}, which applies a kd-search tree [1]. When determining the correspondence, parameter d_{max} defines the maximum allowed distance between model and target points. In this way, a robust out-of-plane rotation is ensured also for large angles, because all target points that are farther away will have no influence on the computation. In particular, we use the empirical threshold $d_{max} = 10\,\text{mm}$ (Fig. 18.5).

The optimization of pose vector $\mathbf{t}$ in error function $e(\mathbf{t})$ (18.6) is carried out iteratively on the basis of least-squares minimization. The elementary matrices for model rotation contain sine and cosine functions, which we need to linearize in order to solve the system of equations for the minimization. This is accomplished using Taylor series approximation. Then, the model vertex coordinates are differentiated with respect to the components of the pose vector. The derivatives $\partial \mathbf{a}_j / \partial \mathbf{t}$ are computed analytically, which can be done easily in the case of translations and rotations. Thus, for each 3-D model point $\mathbf{a}_j$ we can form three equations (18.7),

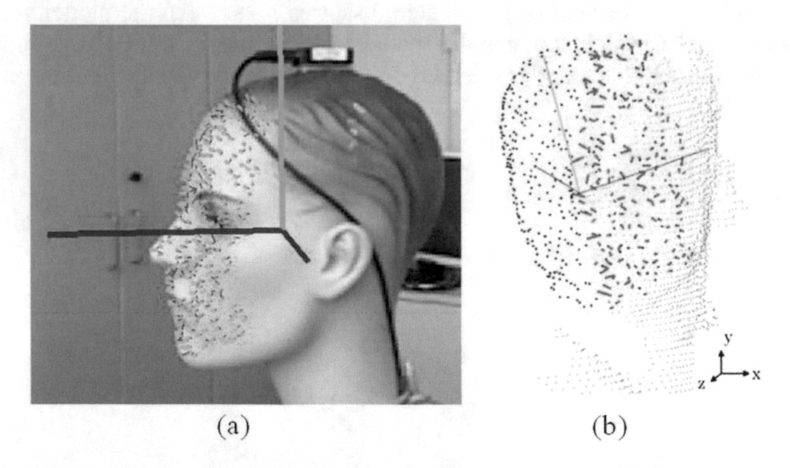

Fig. 18.5 Correspondence search. The *red lines* show associated model and measurement points. (**a**) Image with pose (*XYZ*-axes in RGB) and model in *light blue* plus correspondences in *red*; (**b**) model points of the turned-away face half exceed the maximum distance d_{max} and are no longer associated

which leads to a highly overdetermined system of equations. That itself leads to tolerance against noise and robustness of the computed pose vector.

$$\mathbf{a}_j(\mathbf{t}) + \partial \mathbf{a}_j/\partial \mathbf{t} \cdot \Delta \mathbf{t} = \mathbf{p}_i \tag{18.7}$$

The ICP approach stops if error function $e(\mathbf{t})$ goes below a threshold, or if a given number of iterations has been reached. As shown in comprehensive tests, the computed pose vector $\mathbf{t}$ accurately represents the actual orientation of the face. For the initialization and reset of the pose vector, it is assumed that the face is located in the upper half of the point cloud, which is generally the case if the camera is aligned properly. With respect to the x- and z-coordinates, we use the centroid of the point cloud, while the rotation angles are set to zero.

Also, in the beginning, and after reset, we apply $n = 20$ iterations, which leads to convergence in most cases. This can be observed through a small value of error function $e(\mathbf{t})$. After initialization, $n = 5$ iterations are applied, or less, if the error goes below threshold $f_{err} = 20$. Reset of the pose is triggered by the error function exceeding the threshold $f_{err} = 40$. That is a hint of misalignment, which can happen if the face is fully occluded. Further, reset is carried out if the number of corresponding model and target points is too low for a reliable computation, i.e. $n_{corr} < 250$. This can happen if no face is in the measurement volume.

Using these error measures, quick and robust pose estimation processing is assured. Alternatively, if it is available, one can also utilize the grayscale image corresponding to the depth map, e.g. for Haar-like feature-based face detection using Adaboost [21]. This way, one can set the initial XY-translation of the 3-D model w.r.t. the image face position, or detect if several or no faces are in the image.

18.2.4 Evaluation of Head Pose Estimation

As initially stated, our pose estimation procedure is essential for automated face analysis. Thus, in order to get a qualitative statement, we have made an evaluation on the basis of the exact tilt sensor 3DM-GX3 of the company MicroStrain, which provides ground truth rotation parameters at high accuracy (Fig. 18.6).

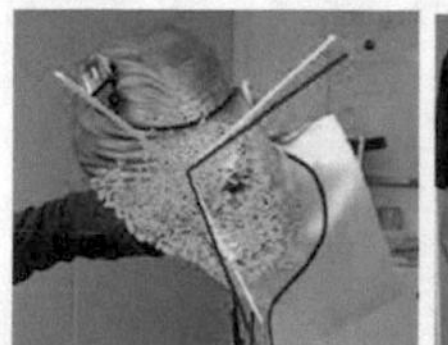
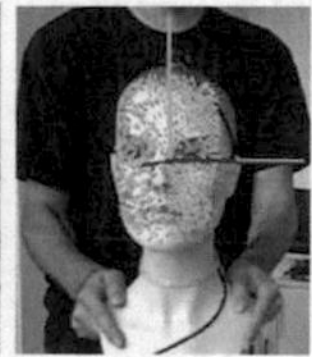
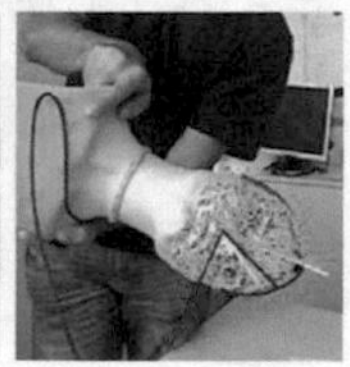
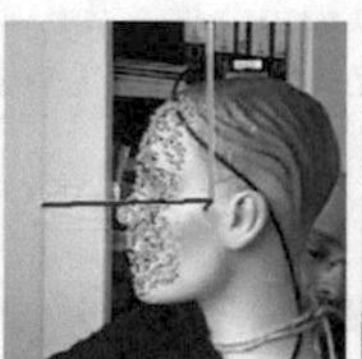
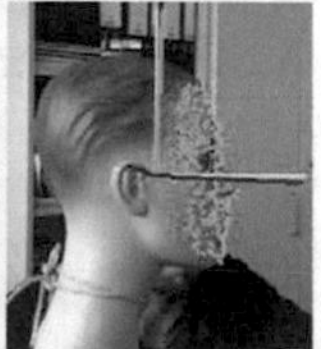

Fig. 18.6 Pose validation using a dummy head model with tilt sensor. The pose is displayed as coordinate system in RGB, the ground truth in *yellow* and the 3-D ICP model in *light blue*

Table 18.1 Evaluation: rotation and tilt sensor displacement in degrees

Rotation axis	Maximum absolute values in degrees		Ground truth displacement in degrees		
	Estimation	Tilt sensor (ground truth)	Mean μ	Std. deviation	Maximum
rx	51	50	2	2	6
ry	108	110	3	3	11
rz	115	116	2	2	4

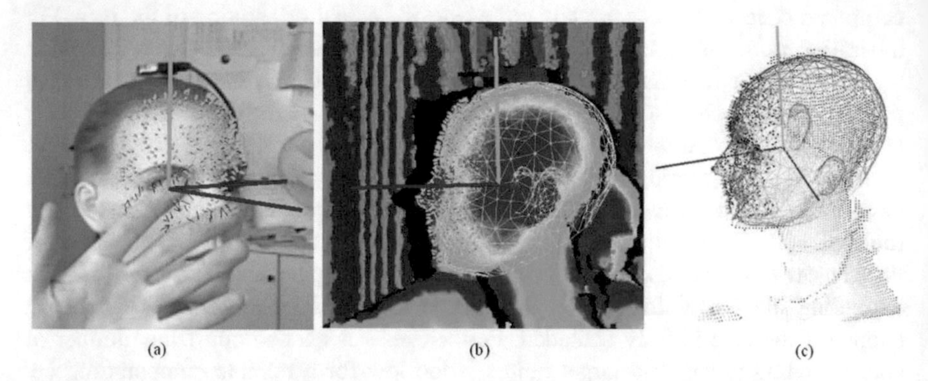

Fig. 18.7 (**a**) Occlusion tolerance, (**b**) application of an additional head model for ongoing facial feature extraction and analysis, (**c**) 3-D point cloud with head model and the determined pose

The estimated pose accuracy has been determined as the ground truth deviation in a series of 10,000 measured sample frames, including rotations of the three axes. Table 18.1 shows the resulting relevant results of the analysis. For the measured data, the tilt sensor has returned a maximum absolute rotation of $\{rx\ ry\ rz\}=\{50°\ 110°\ 116°\}$. The first two columns show the computed maximum absolute rotations, the next two the mean pose sensor displacement with corresponding standard deviations, all in degrees. The last column provides the measured maximum ground truth deviations, which have occurred at strong rotations, i.e. for yaw of more than 90°, only. For all rotation angles, the mean deviation is less than 3°.

Robust handling of occlusion is a further concern of the presented work. Due to the characteristics of the measured 3-D data, pose estimation of partially occluded heads is still possible using temporal coherence (Fig. 18.7a). That means the model is fitted step by step up to large rotation angles. The processing speed is high and reaches the maximum possible frame rate (30 Hz) of the used camera's USB port at VGA resolution on an Intel Core i7 PC.

18.2.5 Summary of Pose Estimation

The presented 3-D data-based pose estimation procedure has been shown to perform robustly and accurately in a wide range of head poses. This is achieved through the use of a model normal constraint in conjunction with an ICP algorithm. In particular, the method constitutes a solid basis for the application of 3-D head models for further analysis (Fig. 18.7b/c). Also, in experiments with real persons, we have evaluated that there is only a minimal effect on face pose accuracy related to facial expression, which is due to the fact that the pose model covers the whole of the face. This makes it an ideal basis for subsequent face analysis, which is presented in the following Sect. 18.3.

18.3 Facial Feature Processing

In image-based analysis of faces, one can discern at least three categories of commonly used features, such as holistic vs. geometric analytic, two- vs. three-dimensional, i.e. image- or volume-based and temporal vs. static features [17]. In this categorization, almost all face recognition approaches apply holistic, 2-D image-based, static features, whereas in facial expression analysis there are analytic geometric region-based and dynamic features commonly used. In our work, apart from the 3-D depth image-based pose estimation, which is done in the first step, we apply a combination of 2-D image and 3-D model data for the processing of geometric features.

18.3.1 Face Model

In the presented method, facial feature processing and evaluation are based on a geometric 3-D face model, which utilizes the Facegen Photofit routine [7]. This is a morphable model, which is adapted to a frontal face image using facial landmarks. These are quickly and reliably found using the IntraFace detector by Xiong et al. [23] in conjunction with gradient data and the active contour model algorithm of Cootes [5] (Fig. 18.8a/b).

In order to set the appropriate size of the Facegen-based model, we apply scaling in X- and Y-direction by utilizing point cloud data derived from the depth image and the ICP algorithm of Sect. 18.2 with scaling as free model parameter (Fig. 18.8c/d).

For the conducted face processing we use a rigid 3-D surface mesh description of the adapted Facegen model, which is denoted by **S** (18.8).

$$\mathbf{S} = (\{\mathbf{v}_1, \dots, \mathbf{v}_n\}, \{w_1, \dots, w_m\}), \; \mathbf{v}_i \in \mathbb{R}^3, \; w_j \in \mathbb{N}, \tag{18.8}$$

with $\mathbf{v}_i$ as mesh vertices and w_j as triangle indices.

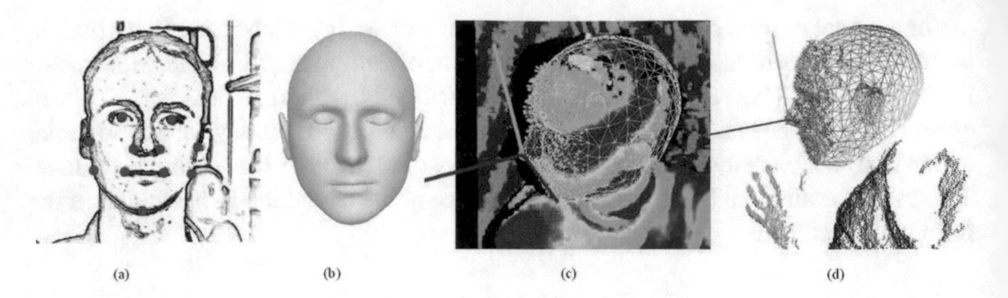

Fig. 18.8 Face model adaptation, (**a**) frontal face gradient image with detected landmark points, (**b**) reconstructed 3-D model **S**, (**c**) depth image encoded in cyclic rainbow colors with scaled model at pose, (**d**) point cloud with scaled model

18.3.2 Facial Expression-Related Features

In the presented concept, geometric features represent the facial expression at frame t through a set of distances and angles. These parameters result from an evaluation of relevant facial feature points.

18.3.2.1 Feature Points

Evaluation of feature points is a general technique in face and facial expression analysis. The Facial Animation Parameter (FAP) System [16], which was developed in the context of the MPEG-4 standard, has inspired the selection of facial points in our method. In the FAP 88 feature points were defined for the simulation of facial expression. In experiments, we found that a subset of eight relevant points suffices for the recognition of facial expression (Fig. 18.9). For this purpose we use point set $\mathbf{P}_f$ (18.9). The model-based computation of feature points requires the detection of the corresponding image points beforehand.

$$\mathbf{P}_f = \{\mathbf{p}_{le}, \mathbf{p}_{re}, \mathbf{p}_{leb}, \mathbf{p}_{reb}, \mathbf{p}_{lm}, \mathbf{p}_{rm}, \mathbf{p}_{ul}, \mathbf{p}_{ll}\}, \ \mathbf{p}_i \in \mathbb{R}^3, \tag{18.9}$$

with le/re as left and right eye, leb/reb as eyebrow and mouth points (see Fig. 18.9c).

18.3.2.2 Extraction and Transformation of Image Feature Points

For the detection of facial points in the image, we apply a Haar-like feature-based Adaboost face detector in the first step in order to find the face and constrain the facial feature search space [21]. Then we apply the IntraFace detector [23] for feature point finding. To solve non-linear least-squares (NLS) functions, which are

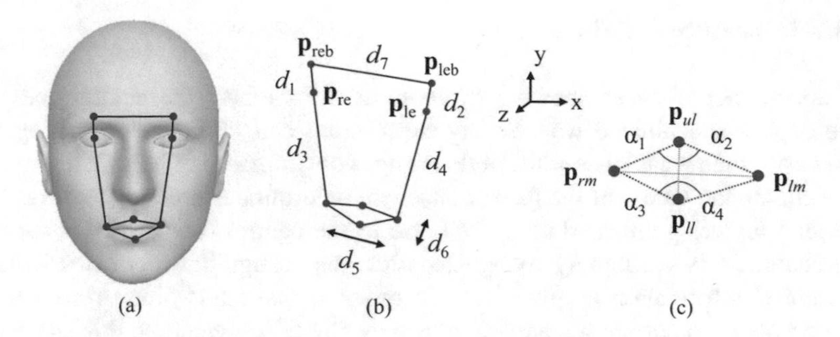

Fig. 18.9 (**a**) Face model **S** with features, (**b**) feature points **p** with distances d_i and (**c**) angles α_j, in the mouth region

taken for feature point detection, the IntraFace method uses the fast and accurate Supervised Descent Method (SDM) approach. This way facial expression relevant points are detected in the input image in a reliable and fast manner, until out-of-plane rotations up to 25°. In the following, these points are referred to as set $\mathbf{I}_f$ (18.10).

$$\mathbf{I}_f = \{\mathbf{i}_{le}, \mathbf{i}_{re}, \mathbf{i}_{leb}, \mathbf{i}_{reb}, \mathbf{i}_{lm}, \mathbf{i}_{rm}, \mathbf{i}_{ul}, \mathbf{i}_{ll}\},\ \mathbf{i}_i \in \mathbb{R}^2 \tag{18.10}$$

The 3-D transformation of image feature points requires knowledge about the underlying camera system and on the other side about the depth of the captured scene. In the case of the applied Kinect camera system, depth information is available. However, as the depth map occasionally contains artifacts, such as holes, a different and more general solution is preferred here, which can be applied to an arbitrary camera system. We infer depth information by measuring the distance d from the camera in the scene at the respective pixel raster coordinate in 3-D to the face model using a raycasting algorithm [9]. The model is oriented in the current estimated pose (Sect. 18.2). These transformation steps require information about the camera model **K**, and thus about intrinsic and extrinsic camera parameters, which are determined through calibration [12]. In this way we can easily define the transformation function k (18.11) that converts 3-D scene points **w** to image points **i** and the other way around as k^{-1} (18.12). Using this transformation the 3-D feature point set **P** (18.9) is determined.

$$\mathbf{i} = k(\mathbf{w}, \mathbf{K}), \tag{18.11}$$

with $\mathbf{i} \in \mathbb{R}^2$, $\mathbf{w} \in \mathbb{R}^3$, camera model **K**.

$$\mathbf{w} = k^{-1}(\mathbf{i}, d, \mathbf{K}), \tag{18.12}$$

with $\mathbf{i} \in \mathbb{R}^2$, $\mathbf{w} \in \mathbb{R}^3$, depth parameter $d \in \mathbb{R}$ and camera model **K**.

18.3.2.3 Feature Definition

The appearance of faces showing expressions differs from the neutral ones in a more or less accentuated way. Strong expressions can, of course, be recognized more easily than weak ones with subtle changes only. Instead of holistic approaches that evaluate the whole of the face at once, we determine expression features from the facial feature points and compare these to the neutral face, which is captured beforehand. It is commonly recognized that the recognition performs better if this neutral information is given [20]. In order to make this kind of comparison more practical, there are already strategies available to overcome this sometimes impossible initialization step, by using Point Distribution Models (PDMs), and thus, to also handle unknown faces [20].

The transition from 2-D image to 3-D facial expression features offers clear advantages, in particular, independence from the face pose. This property is utilized in the evaluation of 3-D feature point set $\mathbf{P}_f$ and the inference of the raw feature vector $\mathbf{f}$ (18.13). The feature vector for the neutral facial expression $\mathbf{f}_{neutral}$ is equivalent to $\mathbf{f}$ and kept for each subject for further processing. The vector's values are seven distances d_i (18.14) distributed across the face and four angles α_j (18.15), which contain distinct information about the current mouth shape and the facial expression as a whole (Fig. 18.9). In particular, raising and lowering of the eyebrows is captured through the parameters d_1 and d_2, mouth movements through the distance between mouth corners and eye centers d_3 and d_4. Additionally, the current mouth width, height and eyebrow distance are encoded in d_5, d_6 and d_7.

$$\mathbf{f} = (d_1 \ldots\ d_7\ \alpha_1 \ldots\ \alpha_4)^{\mathrm{T}},\ d_i, \alpha_j \in \mathbb{R},\ \mathbf{f} \in \mathbb{R}^{11}, \tag{18.13}$$

with the definition of distances d_i as

$$\begin{aligned} d_1 &= ||\mathbf{p}_{reb} - \mathbf{p}_{re}|| \quad d_2 = ||\mathbf{p}_{leb} - \mathbf{p}_{le}|| \\ d_3 &= ||\mathbf{p}_{re} - \mathbf{p}_{rm}|| \quad d_4 = ||\mathbf{p}_{le} - \mathbf{p}_{lm}|| \\ d_5 &= ||\mathbf{p}_{rm} - \mathbf{p}_{lm}|| \quad d_6 = ||\mathbf{p}_{ul} - \mathbf{p}_{ll}|| \\ d_7 &= ||\mathbf{p}_{reb} - \mathbf{p}_{leb}|| \end{aligned} \tag{18.14}$$

and angles α_j as

$$\begin{aligned} \alpha_1 &= \arccos\left(\frac{\mathbf{v_1} \cdot \mathbf{v_2}}{\|\mathbf{v_1}\| \cdot \|\mathbf{v_2}\|}\right) \quad \alpha_2 = \arccos\left(\frac{\mathbf{v_2} \cdot \mathbf{v_3}}{\|\mathbf{v_2}\| \cdot \|\mathbf{v_3}\|}\right) \\ \alpha_3 &= \arccos\left(\frac{-\mathbf{v_2} \cdot \mathbf{v_4}}{\|\mathbf{v_2}\| \cdot \|\mathbf{v_4}\|}\right) \quad \alpha_4 = \arccos\left(\frac{-\mathbf{v_2} \cdot \mathbf{v_5}}{\|\mathbf{v_2}\| \cdot \|\mathbf{v_5}\|}\right) \end{aligned} \tag{18.15}$$

with

$$\mathbf{v}_1 = \mathbf{p}_{rm} - \mathbf{p}_{ul}, \quad \mathbf{v}_2 = \mathbf{p}_{ll} - \mathbf{p}_{ul}, \quad \mathbf{v}_3 = \mathbf{p}_{lm} - \mathbf{p}_{ul},$$
$$\mathbf{v}_4 = \mathbf{p}_{rm} - \mathbf{p}_{ll}, \quad \mathbf{v}_5 = \mathbf{p}_{lm} - \mathbf{p}_{ll}, \quad \mathbf{v}_i, \mathbf{p}_j \in \mathbb{R}^3$$

18.3.2.4 Feature Normalization

In order to evaluate the facial expression captured at frame t represented by feature vector $\mathbf{f}$ (18.13), we make a comparison with the neutral face, which is provided through $\mathbf{f}_{neutral}$ as explained above. For this purpose we introduce the operator # (18.16) for component-wise division of two feature vectors **a** and **b**.

$$\mathbf{a} \,\#\, \mathbf{b} = (a_1/b_1 \; a_2/b_2 \; \ldots \; a_n/b_n)^{\mathrm{T}} \in \mathbb{R}^n, \; \mathbf{a}, \mathbf{b} \in \mathbb{R}^n \tag{18.16}$$

$$\mathbf{f}_{ratio}(t) = \mathbf{f}(t) \,\#\, \mathbf{f}_{neutral}, \; \mathbf{f}_{ratio}, \; \mathbf{f}, \; \mathbf{f}_{neutral} \in \mathbb{R}^{11}, \; t \in \mathbb{Z} \tag{18.17}$$

The ratios of $\mathbf{f}_{ratio}(t)$ (18.16) usually have large deviations between different persons and facial expressions. Thus, we apply feature normalization. For all vector components of $\mathbf{f}_{ratio}(t)$, we have determined the statistical parameters mean and standard deviation as well as the minimum and maximum, $\mathbf{c}_{min}$ and $\mathbf{c}_{max}$ (18.18), across a representative set of example data. Feature vector $\mathbf{f}(t)$ (18.19) is the normalization result, which is computed for the empirical confidence interval of 2σ.

$$\begin{aligned} \mathbf{c}_{min} &= \mu - 2\sigma, \; \mathbf{c}_{min} \in \mathbb{R}^{11} \\ \mathbf{c}_{max} &= \mu + 2\sigma, \; \mathbf{c}_{max} \in \mathbb{R}^{11} \end{aligned} \tag{18.18}$$

with μ and σ as mean and standard deviation for all vector rows.

$$\mathbf{f}(t) = (\mathbf{f}_{ratio} - \mathbf{c}_{min}) \,\#\, (\mathbf{c}_{max} - \mathbf{c}_{min}) = (\mathbf{f}_{ratio} - \mathbf{c}_{min}) \,\#\, 4\sigma, \;\; \mathbf{f} \in \mathbb{R}^{11} \tag{18.19}$$

18.4 Circumplex Model Mapping

In literature, the majority of approaches for facial expression analysis apply discrete categories, mostly emotions, pain, and sleepiness. However, often fixed categories are not optimal, as they can be mixed and ambiguous. For that reason, the approach we apply is influenced by the observation that the affect labels valence and arousal of the Circumplex Model of Affect lead to a state representation that is continuous in principle [19]. Thus, no discrete descriptions of the user state are necessary for classification.

18.4.1 Multi-Layer Perceptron Based Valence-Arousal Estimation

In order to appropriate the circumplex model, in our work we use a technical implementation of the concept from psychology. For this purpose the transformation of the 11-D feature vector to the 2-D model plane is realized by using mapping function f_{map} (18.21) (Fig. 18.10). In our implementation, the circumplex plane is spanned by six polar coordinates P_{C_i} (18.20) of the discrete emotion categories plus neutral (Fig. 18.11) [14]. Basically, this definition reflects the findings of

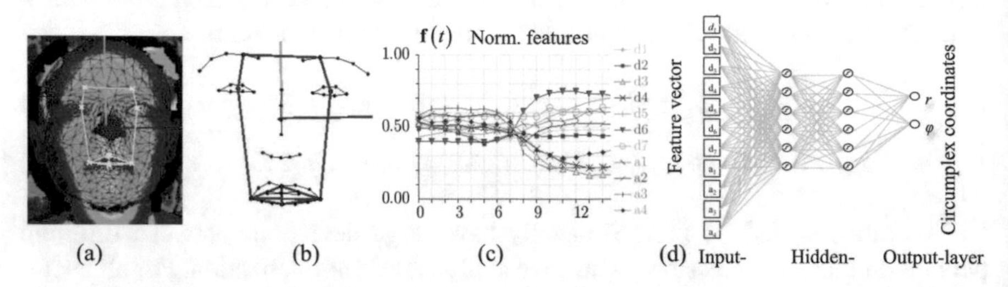

Fig. 18.10 Valence-Arousal (V-A) transformation. (**a**) Depth image with pose and overlaid feature points, (**b**) 3-D features in *blue*, (**c**) feature plot, (**d**) artificial neural network with function f_{map} of the 11-dimensional feature vector to the V-A space

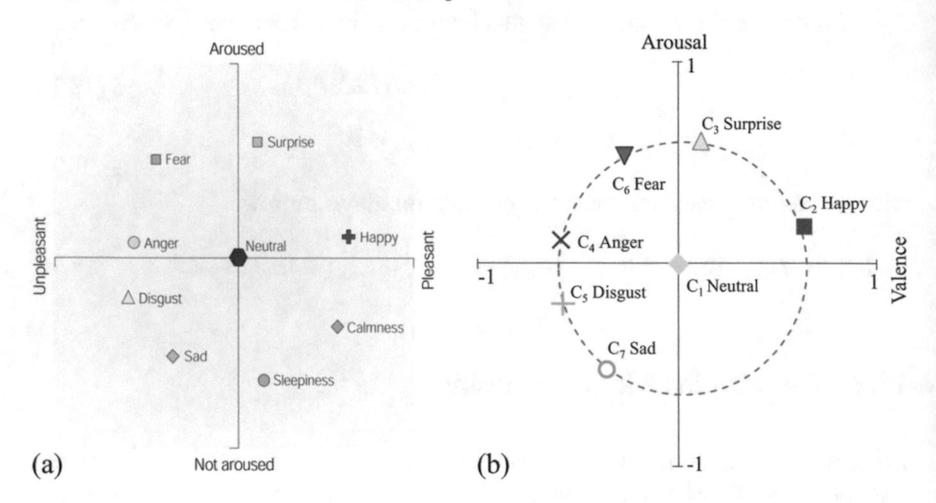

Fig. 18.11 (**a**) Circumplex model of affect as introduced by Russel (Source [4]), (**b**) technical implementation of the model with polar coordinates P_{C_i} (18.20) for the definition of the 2-D model plane

psychologist Russel [19].

$$P_{C_i}(r_{C_i};\varphi_{C_i}) \in \begin{Bmatrix} 0 & l & l & l & l & l & l \\ 0 & 10 & 85 & 170 & 200 & 125 & 240 \end{Bmatrix},\ l = 0.7,\ \varphi_{C_i}\ in\ deg,\ c_i \in 1\ldots7, \tag{18.20}$$

with the classes $C_i \in$ {Neutral, Joy, Surprise, Anger, Disgust, Fear, Sadness}, whereas radius l has been set empirically to $l = 0.7$.

We apply an artificial neural network (Multi-Layer Perceptron, MLP) [11] with a sigmoid transfer function and a backpropagation training algorithm to realize the transformation f_{map} (18.21) of a feature vector $\mathbf{f}$ at a given frame t. In particular, we apply a network with eleven input and two output neurons plus two hidden layers with six neurons each. Our hypothesis is that we can infer the 11-D to 2-D transformation based on the adapted weights of the neural network, which have been determined through supervised learning. That means, in the training phase each feature vector is assigned to the polar coordinate of its emotion class C_i (18.20), which is derived from the Circumplex Model definition [4]. Accordingly, during classification, each input vector leads to a position in the model plane, which is supposed to be at a place that corresponds to the presented emotion.

$$f_{map} : \mathbf{f}(t) \in \mathbb{R}^{11} \xrightarrow{MLP} \begin{bmatrix} V \\ A \end{bmatrix} \in \mathbb{R}^2, \tag{18.21}$$

with valence V, arousal A.

18.4.2 Dynamic Integration and Determination of Intensity

The estimation of the current affective state can be considered from the viewpoint of inverse problems. In the first stage, the current state is estimated by the evaluation and fusion of optical features using function f_{map} (18.21), leading to an observation in the 2-D V-A space. Now, the goal is to find the underlying unknown state at the current frame t, which is denoted by the variable $z(t)$. From a mathematical point of view, this inverse problem is ill-posed in Hadamard's sense since the reconstruction is potentially sensitive to noise and not guaranteed to be unambiguous. The solution is assumed to be within close distance of the observations, as measured by square norm e_{data}.

$$e_{data} = \|z(t) - f_{map}(\mathbf{f}(t))\|^2,\ z, f_{map} \rightarrow \mathbb{R}^2,\ t \in \mathbb{Z} \tag{18.22}$$

Further, the potential solution is constrained by applying operator $P(z)$, in order to achieve a smoothing of the result. The smoothness property is denoted by the first-order derivative of the desired solution, i.e. $P(z) = \dot{z}$,[1] which is scaled by the regularization parameter λ, that works as weighting constant. Taken together, the resulting energy measure is defined as the sum of e_{data} (18.22) and the smoothness constraint $\dot{z}$ defined above,

$$E(z) = \int \|z(t) - f_{map}(\mathbf{f}(t))\|^2 + \lambda \cdot \dot{z}^2(t)dt \rightarrow min. \quad (18.23)$$

For minimization of Eq. (18.23) we apply the Euler-Lagrange equation to solve the partial differential system of equations, which leads to state variable $z(t)$. The intensity level r (18.24) of the current emotion quantity is inferred from the state variable while the user state is traced over a temporal period and integrated over time.

$$z(t) = \begin{pmatrix} r \\ \beta \end{pmatrix}(t), \text{ with } r(t) = \sqrt{a^2 + v^2},\ \beta(t) = \tan^{-1}(v/a), \quad (18.24)$$

with a and v as scalar activations in the cardinal dimensions arousal and valence.

18.4.3 Evaluation

The different modules of the proposed concept have been tested with training and testing data from the BU-4DFE database [24] and exemplary online samples that were taken with a Kinect camera system. A total amount of about 18,000 data samples from seven classes according to (18.20) has been used for training and testing the neural network and the dynamic integration. The pose estimation approach presented in Sect. 18.2 was adapted in order to work with the 3-D data of the BU-database, i.e. to carry out the required 2-D/3-D transformations, image and depth data have been generated through OpenGL rendering [9] with a virtual camera and specified parameters (Fig. 18.12). Analysis has been conducted by applying feature extraction and processing plus V-A mapping to the processed BU-4DFE data (Fig. 18.13).

One motivation of this work is to overcome the use of fixed classification categories, i.e. basic emotions. However, for the evaluation of the V-A mapping, it can be reasonable to apply them. In order to get a qualitative statement about the recognition accuracy, we have analyzed the displacement $\mu(t)$ (18.25) between the calculated angle $\beta(t)$ (18.24) and the given orientation of the underlying class φ_{C_i} in the Circumplex Model's V-A space (Fig. 18.14a).

$$\mu(t) = |\beta(t) - \varphi_{C_i}| \in \mathbb{R}, \text{ see Eqs. (18.20), (18.24)} \quad (18.25)$$

[1] We use here the dot notation to denote temporal derivatives of a function with time as the independent variable.

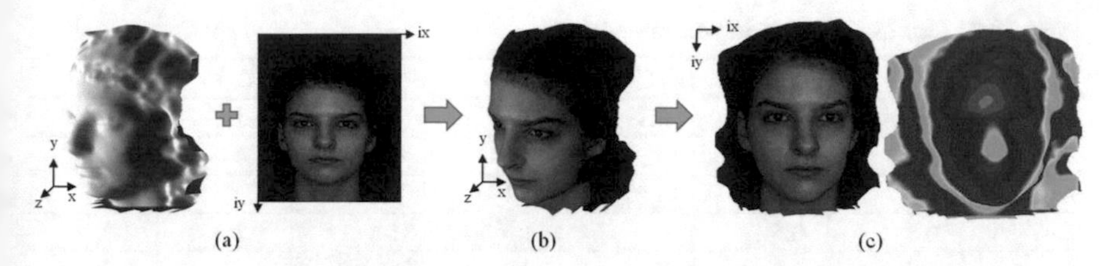

Fig. 18.12 Preprocessing of BU-4DFE data. (**a**) 3-D model and high-resolution texture image, (**b**) textured triangle mesh in 3-D, (**c**) image projection with defined camera as color and depth image

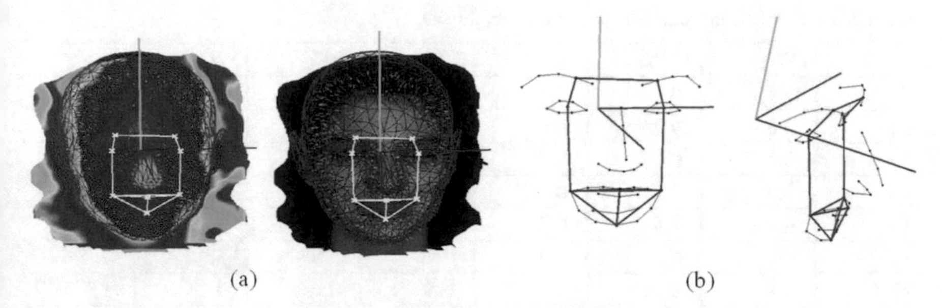

Fig. 18.13 Feature processing of a BU-sample. (**a**) Depth and color image with pose as RGB-coordinate system; further face model **S** (18.8) is shown as *blue triangle* mesh along with extracted features as *white lines*. (**b**) 3-D projection of facial features

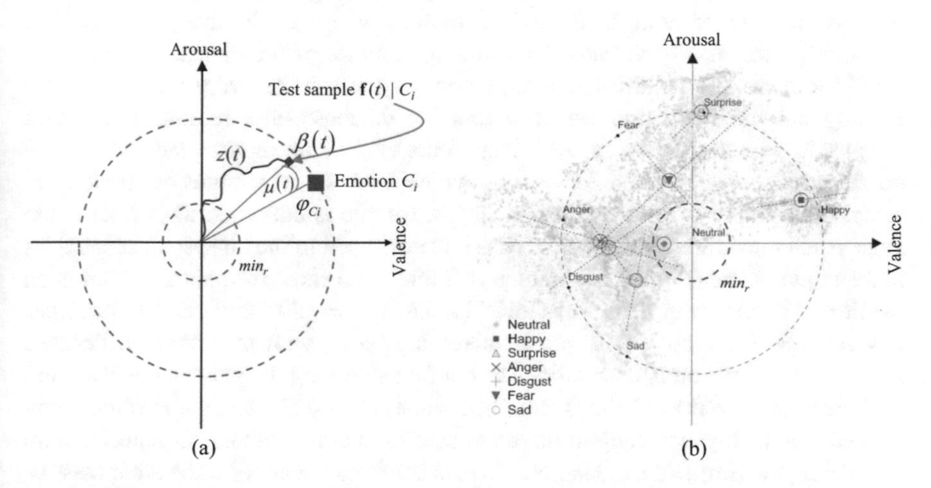

Fig. 18.14 Evaluation. (**a**) The method's accuracy for an exemplary sample $\mathbf{f}(t)$ is determined using angle $\mu(t)$, which reflects the displacement between function $z(t)$ and the given angle φ_{C_i} of the associated class C_i in V-A space. (**b**) Projection of the applied test set from the BU-4DFE database is shown in *light colors* with centroid and principal axes for each class. Data with $r(t) < min_r$ (18.24), i.e. samples with very low expression intensity, are attributed to the neutral class

Table 18.2 Confusion matrix in percent, $t_\mu = 30$

Class	$P(C_1)$	$P(C_2)$	$P(C_3)$	$P(C_4)$	$P(C_5)$	$P(C_6)$	$P(C_7)$
C_1	89.1	0	0.1	0.4	1	0.1	0.3
C_2	3.1	82.2	13.3	0	0	1.4	0
C_3	1.4	0	93.1	0.1	0.1	5.3	0
C_4	0.8	0.5	0	64.5	31.6	1.2	1.4
C_5	6.2	1.6	0.2	37.8	50.4	3.5	0.3
C_6	4.7	5	14.4	7.3	5.9	62.6	0.1
C_7	6.9	0	1.8	6.5	29.5	0.1	53.2

Table 18.3 Confusion matrix in percent, $t_\mu = 60$

Class	$P(C_1)$	$P(C_2)$	$P(C_3)$	$P(C_4)$	$P(C_5)$	$P(C_6)$	$P(C_7)$
C_1	89.1	0	0.1	0.4	1	0.1	0.3
C_2	3.0	89.0	6.6	0	0	1.4	0
C_3	1.4	0	98.4	0.1	0.1	0	0
C_4	0.8	0.5	0	65.7	31.6	0	1.4
C_5	6.2	1.6	0.2	37.8	50.7	3.5	0
C_6	4.7	5.0	1.4	0	5.9	83.0	0
C_7	8.9	0	1.8	6.5	0	0.1	82.8

The neutral class C_1 is treated in a different way, i.e. in the recognition step a sample is considered neutral if $r(t) < min_r$ according to (18.24). The threshold we have determined empirically and set to $min_r = 0.25$. The image material for the neutral class has been taken from the first frames of the database videos. Per definition there is a neutral facial expression at the start of every video. In order to carry out the evaluation we have split the database into training and testing samples in a randomized way, such that all the classes are represented equally and no training sample is taken for testing, which leads to an amount of about 1400 samples per class. In the recognition step, a sample is classified as correct if the angle μ is below threshold t_μ; otherwise it is attributed to the closest adjacent class in the model plane (Fig. 18.14). Tables 18.2 and 18.3 show the resulting confusion matrices for two empirical thresholds t_μ, i.e. $t_\mu = 30°$ and $60°$. It becomes obvious how the recognition rate is increasing along with threshold t_μ, because more samples are counted as valid. The confusion among the classes C_6(Fear) and C_3(Surprise) as well as C_7(Sad) and C_5(Disgust) clearly shows this. Further, it can be seen that the highest recognition rates occur for classes with the strongest feature distinction, i.e. Surprise and Happy, whereas confusion occurs for the other classes. The average recognition rates are 70.2 and 79.7 % for $t_\mu = 30$ and $t_\mu = 60$. Basically, these recognition rates are in accordance with category-based state-of-the-art methods for this particular database [22]. Concise inspection of the data shows that even for the human observer, the presented facial expressions cannot always be interpreted correctly. In this case, the continuous description of the user's facially expressed emotional state in the V-A space can provide more opportunities for evaluation, compared to the solely category-based recognition. Figure 18.14b shows the mapping of all used test samples. Also, here the overlapping of samples becomes obvious, in particular for the classes disgust and anger, as well as sad and surprise and fear.

Using a Kinect camera system, we have evaluated the dynamic integration with the temporal smoothness constraint. Exemplary sequences of presented basic emotions are given in Fig. 18.15. Besides extracted features, the temporally smoothed V-A space projections are shown, along with the evolution of function $z(t)$ (18.24), starting from the neutral state in the center. The examples show that the presented expressions become clearly separated, which enables evaluation in terms of valence and arousal of the emotion model.

An example of the smoothing effect of the dynamic temporal constraint is given in Fig. 18.16, which corresponds to first plot in Fig. 18.15a. Also, the underlying feature sequence plus category-based classification is shown. It demonstrates the competitive abilities of the V-A classification approach, which exemplarily shows the detection of a smile. When it comes to evaluation of intensity, the V-A approach

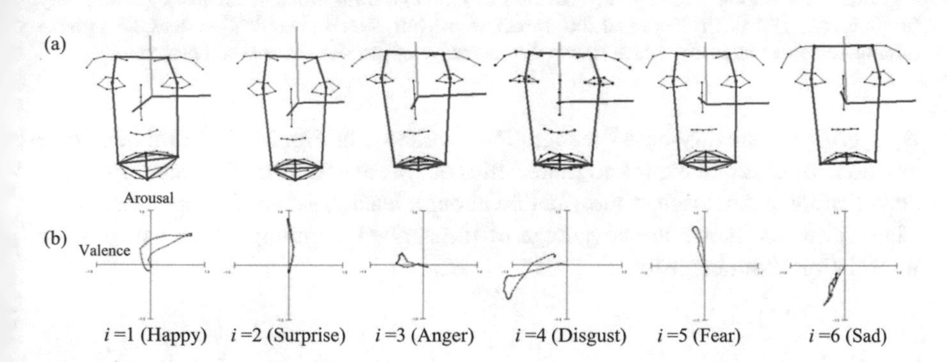

Fig. 18.15 (**a**) 3-D facial expression model with features in *blue*. (**b**) Plot of the projections of presented classic basic emotions in the V-A space using the temporal constraint

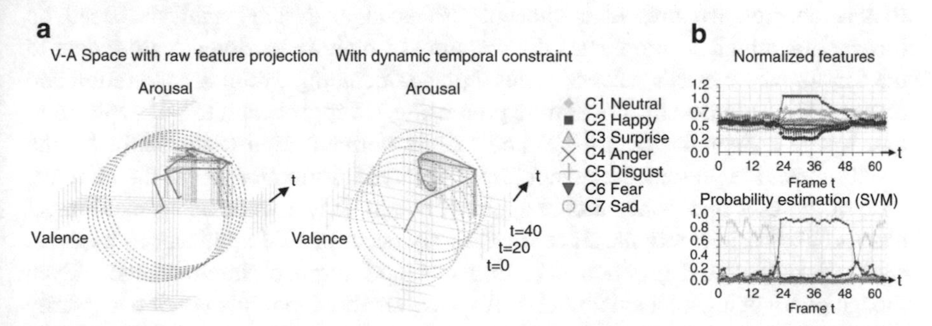

Fig. 18.16 Data processing example with a comparison between affect model and category-based classification. (**a**) Valence-Arousal mapping without and with dynamic temporal constraint, (**b**) corresponding feature vector along with category-based SVM classification

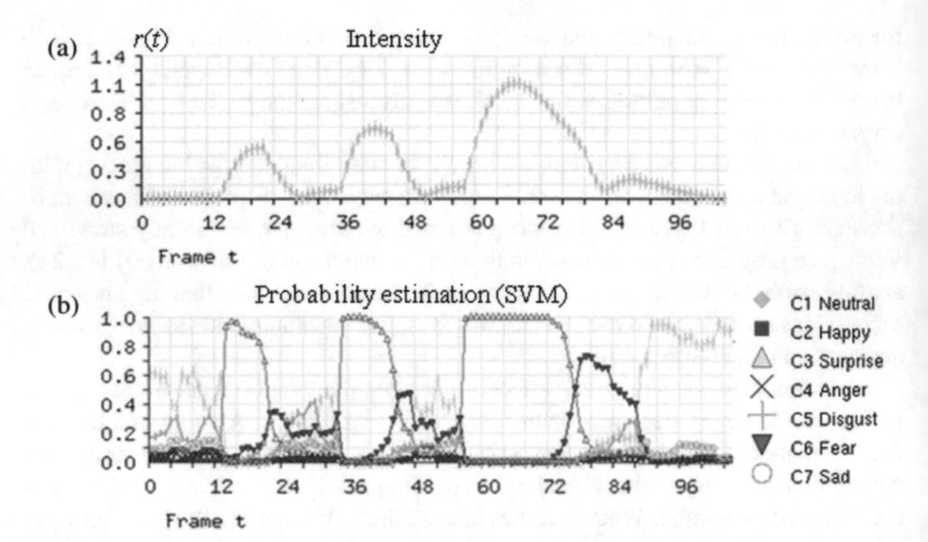

Fig. 18.17 Measuring intensity: Presentation of emotions with different intensity. (**a**) Intensity function $r(t)$ (18.24). (**b**) Note that the respective category-based classification does not provide information about intensity, which shows the advantage of the V-A space-based evaluation

is superior to category-based recognition, as shown in Fig. 18.17. In this example emotions of different expression intensities are presented. Here the category-based classification cannot reflect information about intensity, as the V-A approach does. Thus, it clearly shows the advantage of the applied mapping, which can provide information about intensity.

18.4.4 Summary and Conclusion

In this chapter, we present a concept for facial expression analysis based on image data, which accomplishes the mapping of high-dimensional feature data to the Circumplex model's valence-arousal plane, including dynamic integration and determination of intensity. For feature processing, we apply accurate 3-D depth data-based pose estimation (Sect. 18.2) and feature normalization (Sect. 18.3). As the results show, the presented concept provides more information about the affective state of the user than conventional approaches can deliver when using fixed target classes, like basic emotions, since these rarely occur in HCI applications and also can be ambiguous, e.g. when detecting fear and surprise simultaneously. Even though overlapping may exist after V-A transformation, too, this is not necessarily a bad property, and it is even unavoidable to a certain degree. Overlapping simply results from the fact that some emotions are near to each other in the affect model [19]. As such, it can be useful if the evaluation rather shows a tendency of the user's reaction, i.e. “negtive/positive/aroused or not”, than a particular discrete emotion, which is unlikely to happen. Thus, for an HCI system, this information can be more valuable, such that the presented concept has a potential impact on applicability [14].

In future work, we plan several modifications to the three main parts of our concept. First, in order to facilitate applicability, we want to generalize the adaptation of the user-specific models that are used in the processing. Next we will increase the machine’s perceptive capabilities through the use of advanced sensor technology, like NIR as well as high-speed cameras, and also apply new detection algorithms, like the one of [18], which can robustly and quickly provide a greater number of image features. Further, we also want to acquire and adapt to new application domains. Moreover, at the moment, we neglect the lower right part of the circumplex model plane. This part contains states such as sleepiness and calmness. In future work, we also want to address this quadrant, as it bears potential for vigilance recognition in medical projects, as well as sleep detection in automotive applications [13].

Acknowledgements This work was done within the Transregional Collaborative Research Centre SFB/TRR 62 “*Companion*-Technology for Cognitive Technical Systems” funded by the German Research Foundation (DFG).

References

1. Bentley, J.L.: Multidimensional binary search trees used for associative searching. Commun. ACM **18**(9), 509–517 (1975). http://doi.acm.org/10.1145/361002.361007
2. Besl, P.J., McKay, N.D.: A method for registration of 3-d shapes. IEEE Trans. Pattern Anal. Mach. Intell. **14**(2), 239–256 (1992). doi:10.1109/34.121791. http://doi.ieeecomputersociety.org/10.1109/34.121791
3. Bradski, G.: The OpenCV library. Dr. Dobb’s J. Softw. Tools **25**(11), 120–126 (2000)
4. Calder, A.J., Lawrence, A.D., Young, A.W.: Neuropsychology of fear and loathing. Nat. Rev. Neurosci. **2**(5), 352–363 (2001). doi:10.1038/35072584. http://dx.doi.org/10.1038/35072584
5. Cootes, T.F., Taylor, C.J., Cooper, D.H., Graham, J.: Active shape models - their training and application. Comput. Vis. Image Underst. **61**, 38–59 (1995)
6. Ekman, P.: Strong evidence for universals in facial expressions: a reply to Russell’s mistaken critique. Psychol. Bull. **115**, 268–287 (1994)
7. Facegen modeller. http://facegen.com/modeller.htm (June 2017)
8. Fanelli, G., Gall, J., Gool, L.J.V.: Real time head pose estimation with random regression forests. In: The 24th IEEE Conference on Computer Vision and Pattern Recognition, pp. 617–624 (2011)
9. Foley, J.D., van Dam, A., Feiner, S., Hughes, J.: Computer Graphics: Principles and Practice, 3rd edn. Addison-Wesley, Boston (2013)
10. GFMesstechnik GmbH Germany, T.: Facescan 3d. http://www.gfm3d.com (2015)
11. Haykin, S.: Neural Networks: A Comprehensive Foundation, 3rd edn. Prentice-Hall, Upper Saddle River, NJ (2008)
12. McGlone, C., Mikhail, E., Bethe, J.: Manual of Photogrammetry, 5th edn. ASPRS, ISBN: 1-57083-071-1 (2004)

13. Niese, R., Al-Hamadi, A., Panning, A., Brammen, D.G., Ebmeyer, U., Michaelis, B.: Towards pain recognition in post-operative phases using 3d-based features from video and support vector machines. Int. J. Digit. Content Technol. Appl. **3**(4), 21–33 (2009)
14. Niese, R., Al-Hamadi, A., Heuer, M., Michaelis, B., Matuszewski, B.: Machine vision based recognition of emotions using the circumplex model of affect. In: International Conference on Multimedia Technology (ICMT), pp. 6424–6427 (2011)
15. Niese, R., Werner, P., Al-Hamadi, A.: Accurate, fast and robust realtime face pose estimation using kinect camera. In: IEEE SMC International Conference, pp. 487–490 (2013)
16. Pandzic, I.S., Forchheimer, R.: MPEG-4 Facial Animation: The Standard, Implementation and Applications, 1st edn. Wiley, New York (2002). ISBN: 0-470-84465-5
17. Pantic, M., Pentland, A., Nijholt, A., Huang, T.S.: Human computing and machine understanding of human behavior: a survey. In: Artificial Intelligence for Human Computing, ICMI, pp. 47–71 (2007)
18. Ren, S., Cao, X., Wei, Y., Sun, J.: Face alignment at 3000 FPS via regressing local binary features. In: 2014 IEEE Conference on Computer Vision and Pattern Recognition, CVPR 2014, Columbus, OH, 23–28 June 2014, pp. 1685–1692 (2014)
19. Russell, J.A.: A circumplex model of affect. J. Pers. Soc. Psychol. **39**, 1161–1178 (1980)
20. Saeed, A., Al-Hamadi, A., Niese, R., Elzobi, M.: Frame-based facial expression recognition using geometrical features. Hindawi Adv. Hum.-Comput. Interaction (2014). doi:10.1155/2014/408953
21. Viola, P.A., Jones, M.J.: Robust real-time face detection. Int. J. Comput. Vis. **57**(2), 137–154 (2004)
22. Wang, J., Yin, L., Wei, X., Sun, Y.: 3d facial expression recognition based on primitive surface feature distribution. In: IEEE International Conference on Computer Vision and Pattern Recognition, CVPR06, pp. 1399–1406 (2006)
23. Xiong, X., De la Torre, F.: Supervised descent method and its applications to face alignment. In: 2013 IEEE Conference on Computer Vision and Pattern Recognition, pp. 532–539 (2013)
24. Yin, L., Chen, X., Sun, Y., Worm, T., Reale, M.: A high-resolution 3d dynamic facial expression database. In: IEEE International Conference on Automatic Face and Gesture Recognition (FG'08), pp. 1–6 (2008)

Chapter 19
Multimodal Affect Recognition in the Context of Human-Computer Interaction for *Companion*-Systems

Friedhelm Schwenker, Ronald Böck, Martin Schels, Sascha Meudt, Ingo Siegert, Michael Glodek, Markus Kächele, Miriam Schmidt-Wack, Patrick Thiam, Andreas Wendemuth, and Gerald Krell

Abstract In general, humans interact with each other using multiple modalities. The main channels are speech, facial expressions, and gesture. But also bio-physiological data such as biopotentials can convey valuable information which can be used to interpret the communication in a dedicated way. A Companion-System can use these modalities to perform an efficient human-computer interaction (HCI). To do so, the multiple sources need to be analyzed and combined in technical systems. However, so far only few studies have been published dealing with the fusion of three or even more such modalities. This chapter addresses the necessary processing steps in the development of a multimodal system applying fusion approaches.

F. Schwenker (✉) • M. Schels • S. Meudt • M. Glodek • M. Kächele • M. Schmidt-Wack • P. Thiam
Institute for Neural Information Processing, University of Ulm, 89069 Ulm, Germany
e-mail: friedhelm.schwenker@uni-ulm.de; martin.schels@uni-ulm.de; sascha.meudt@uni-ulm.de; michael.glodek@uni-ulm.de; markus.kaechele@uni-ulm.de; miriam.schmidt-wack@uni-ulm.de; patrick.thiam@uni-ulm.de

R. Böck • I. Siegert
Cognitive Systems Group, Institute for Information Technology and Communications, Otto von Guericke University, PO Box 4120, 39106 Magdeburg, Germany
e-mail: ronald.boeck@ovgu.de; ingo.siegert@ovgu.de

A. Wendemuth
Cognitive Systems Group, Institute for Information Technology and Communications, Otto von Guericke University, PO Box 4120, 39106 Magdeburg, Germany

Center for Behavioral Brain Sciences, 39118 Magdeburg, Germany
e-mail: andreas.wendemuth@ovgu.de

G. Krell
Technical Computer Science Group, Institute for Information Technology and Communications, Otto von Guericke University, PO Box 4120, 39106 Magdeburg, Germany
e-mail: gerald.krell@ovgu.de

S. Biundo, A. Wendemuth (eds.), *Companion Technology*, Cognitive Technologies,
DOI 10.1007/978-3-319-43665-4_19

ATLAS and ikannotate are presented which are designed for the pre-analyzing of multimodal data streams and the labeling of relevant parts. ATLAS allows us to display raw data, extracted features and even outputs of pre-trained classifier modules. Further, the tool integrates annotation, transcription and an active learning module. Ikannotate can be directly used for transcription and guided step-wise emotional annotation of multimodal data. The tool includes the three mainly used annotation paradigms, namely the basic emotions, the Geneva emotion wheel and the self-assessment manikins (SAMs). Furthermore, annotators using ikannotate can assign an uncertainty to samples.

Classifier architectures need to realize a fusion system in which the multiple modalities are combined. A large number of machine learning approaches were evaluated, such as data, feature, score and decision-level fusion schemes, but also temporal fusion architectures and partially supervised learning.

The proposed methods are evaluated on either multimodal benchmark corpora or on the datasets of the Transregional Collaborative Research Centre SFB/TRR 62, i.e. Last Minute Corpus and the EmoRec Dataset. Furthermore, we present results which were achieved in international challenges.

19.1 Introduction

Successful human-computer interaction (HCI) requires that the computer be able to consider and fulfill different sub-tasks [8] such as perceptual, actuatoric, and cognitive functionalities. In this chapter, we focus on the perceptual sub-system of the computer where pattern recognition methods and machine learning technologies are utilized to perceive the user and to infer the user's affective state [18] which is reflected and represented by perceptible user emotions.

Computers are endowed with various types of sensors to achieve multimodal recognition of affects. These sensors can comprise cameras and microphones to perceive the user's activities in front of the system; laser scanners for localization and tracking of the user; more complex sensors such as eye-tracking devices for detailed gaze analysis; or sensors to observe the user's bio-potentials, e.g. skin conductance (SCR), respiration, electro-cardiogram (ECG), or electroencephalography (EEG). Collecting the raw data in real-time is just the first part of the overall process. In the next step, task-relevant patterns must be identified to enable the system to perform appropriate actions. This process of transferring raw data into a number of classes or categories is known as *pattern recognition*. The idea of pattern recognition is to follow the principle of *learning by example*, utilizing general machine learning algorithms to design classifiers.

Considering the achievements in multimodal disposition recognition we emphasize four main hypotheses being considered in this chapter:

1. The training of multimodal classifiers can benefit from datasets annotated using all available modalities.

2. The training of multimodal classifiers can be conducted in a semi-automatic way without a complete labeling of the training material.
3. Classification performance can be improved by considering the temporal evolution of the observed features.
4. Recognition performance can be improved by applying a multimodal classification approach including a temporal fusion.

Basics in Pattern Recognition In HCI scenarios, technical systems have to combine information from different modalities. This is usually achieved by so-called multiple classifier systems (MCSs) which integrate several classifiers to solve a specific classification problem. The main goal is to obtain a combined output that provides a more accurate and robust classification. In a typical MCS scenario, a complex high-dimensional classification problem is decomposed into smaller sub-problems for which improved solutions can be achieved.

In MCS, it is assumed that the raw data X originates from an underlying source, but each classifier receives different subsets of X, e.g. X is applied to multiple types of feature extractors $F_1, \ldots, F_N$ computing multiple views $F_1(X), \ldots, F_N(X)$ of the same raw input data X. Feature vectors $F_j(X)$ are used as the input to the jth classifier, computing an estimate y_j of the class membership of $F_j(X)$. This output y_j might be a crisp class label or a vector of class memberships, e.g. estimates of posterior probabilities. Based on the multiple classifier outputs $y_1, \ldots, y_N$, the combiner produces the final decision y. Combiners can be grouped into fixed transformations of the classifier outputs $y_1, \ldots, y_N$ and trainable mappings. Examples of fixed combining rules are *voting*, *(weighted) averaging* and *multiplying*. By means of an additional optimization procedure, trainable mappings can be realized using the classifier decisions as the inputs to a classifier which performs the final combination. Popular members of this group are artificial neural networks, decision templates and support vector machines [28].

Training a classifier based on vector-valued data can be achieved by computing gradients of error functions with respect to the parameters of some predefined classifier model, such as a multilayer feed forward neural network or a kernel machine. Usually, the raw data comes as a continuous stream of data, e.g. video, audio streams, or waveforms of bio-potentials, and for some tasks the temporal structure of the data might be of importance. Classifier training based on sequential data is much more complex. The structure of the underlying classifier model must be able to process input sequences of different lengths, in particular. In contrast to constant length vectors, sequences may have different lengths even when they represent the same class, e.g. spoken words in speech recognition. Classifier models that often come into play in this context are hidden Markov models (HMMs) and recurrent neural networks (RNNs).

19.2 Machine Learning Framework for Emotion Recognition

A quite important channel of communication between humans is speech, which further allows the transfer of emotional states via content, prosody or paralinguistic cues, which results also in audio-based emotion classification. Another important channel is given by visual perception, which is used by humans to convey their emotional states using facial expressions or body poses. Video-based emotion recognition focuses mainly on the extraction and recognition of emotional information from facial expressions. Attempts have been made to classify emotional states from body and head gestures, as well as from the combination of different visual modalities, such as facial expressions and body gestures which were captured by two separate cameras [16]. Furthermore, emotion recognition can be based on psycho-physiological measurements, e.g. SCR, respiration, ECG, electromyography, or EEG. In contrast to speech, gestures and facial expressions, psycho-physiological measurements are directly generated in the (human) autonomic nervous system and cannot be imitated [24].

The MCS approach is very promising for improving the system's overall classification performance. The individual classifier outputs of the classifier ensemble, which is based on different feature views or modalities, need to be accurate and diverse. While high accuracy is an obvious requirement for members of the ensemble, the concept of classifier diversity is less intuitive to grasp. Members of an ensemble can be regarded as being diverse if the corresponding classifiers disagree on a set of misclassified data [50]. In our work on multimodal emotion recognition, ensemble members have been trained on various types of features extracted mainly from the user's voice and the facial region (e.g. fundamental frequency, Mel-Frequency Cepstral Coefficients (MFCCs), modulation spectrum from the audio signal and form and motion patterns from the video channel) [10, 44]. Besides these more external physical expressions, human emotions (initially studied in human-human interactions) consist of feelings, thoughts and many other types of internal (physiological) processes. Therefore, measuring physiological parameters, such as skin conductivity, heart rate, respiration, or brain activity from EEG, is the first step to studying the automatic recognition of these internal emotional states. The numerical evaluation showed that MCS using fixed and trainable fusion mappings applied to multimodal emotional data can outperform unimodal classifiers. Even in unimodal applications the overall recognition performance increases in many cases by combining outputs of multiple classifiers trained on different features views [39, 50].

Research in facial expression and that in speech-based emotion recognition [35] are usually performed independently from each other. However, in almost all practical applications, people speak and exhibit facial expressions at the same time, and consequently both modalities should be used to perform robust affect recognition. Therefore, multimodal, and in particular audio-visual, emotion recognition has been emerging as a fruitful research topic in recent times [56]. Approaches applying MCS to the classification of human emotions are presented in [6, 42, 45, 49, 58].

19.3 Data Acquisition and Benchmark Data

19.3.1 Survey of Relevant Benchmark Data Sets

In the beginning of affect recognition and, especially emotion or disposition recognition, data sets with acted material were mainly used to create controlled conditions and a setting that allows automated recognition which is not further impeded by difficulties in the extraction process. This results in high quality-data and, further, provides a kind of ground truth in the assigned labels. Such ground truth is important for the validation of recognition systems. For this, the participants of the recording were either actors or naïve speakers who were asked to react in a specific manner. Therefore, a predefined situation is created fixing intended labels by design. In general, it can be assumed that acted material is quite expressive in the shown affects [1]. Such a way of generating corpora was quite similar throughout all modalities (cf. e.g. [59]) since the focus could be set on the development of suitable classifiers and methods. Prominent examples of acted corpora include the Cohn-Kanade facial expression dataset [23], the Berlin database of emotional speech [4] and the Eight-Emotion Sentics Data for biophysiology [17]. To foster the evolution in the various fields, several researchers proposed and conducted challenges. The most prominent are the AVEC challenges—providing audio-visual data sets for various sub-tasks in the emotion recognition, usually concerned with near real-life situations—[56] and the (so-called) EmotiW challenges [7] that emulate a challenging *in-the-wild* setting using emotional snippets from movies.

On the other hand, the question arises: Why shall we consider data sets containing naïve material? For automatic emotion recognition from speech, Batliner et al. [1] discussed the importance of real-life material in detail. This is of interest since it can be assumed that dispositions or affects are expressed in a subtle manner, especially in real-life interactions. Therefore, the research community has to push towards non-acted corpora. In [1] three types are distinguished between based on emotional classes, namely acted, read, and real-life. Besides the characteristic of the data recording, the contained dispositional or affective classes have to be considered (cf. e.g. [55]). The novel shift towards more real-life scenarios is considered in both corpora recorded by the authors and various other research groups.

Prominent naturalistic corpora include the PIT corpus [54], which is conducted as a computer-assisted multi-party dialog, and the RECOLA corpus [38], in which pairs of participants are collaborating to solve a survival task in a Wizard-of-Oz setting. Another notable corpus is the MAHNOB-HCI [53] dataset. The unique part about this data collection is that a multitude of modalities has been recorded: Audio, video, bio-physiology but also EEG and eye gaze, have been recorded.

Besides the mentioned data sets, a strong focus of this work lies on two corpora providing naturalistic interactions with a technical system: the Last Minute Corpus (LMC) (cf. Chap. 13) and the EmoRec corpus [58].

19.3.2 Annotation of Emotional Data

To conduct the emotional labelling, the annotators should be supported by a tool assisting them. Several tools exist to support the literal transcription, for instance EXMARaLDA [46] or Folker [45], but for emotional labelling such tools are rare. For content analysis of videos, the tool Anvil [25] can be used.

19.3.2.1 ATLAS

The freely available ATLAS annotation, labelling and data investigation tool developed at the University of Ulm [31] was designed to assign blocked labels to affective material. In the current version of the ongoing project, it is extended to support fuzzy and fully continuous labelling techniques. Contrary to most other annotation tools, ATLAS is not limited to a maximum number of data streams or specific annotation paradigms. It is possible to depict various recorded raw data like audio, video or, in general, digital sampled values (e.g. bio-physiological signals). Additional extracted features, crisp and probabilistic results of classification procedures and mixed types of labels can be displayed in order to support researchers in obtaining a better understanding of their data, algorithms and results. Synchronous playback of all streams and information is possible.

The presentation complexity of the UI is adaptive to the user's needs. That means that an expert researcher is able to visualize a large amount of detailed information and investigate it at once (see Fig. 19.1), while complex details can also be hidden from unexperienced users to prevent them from being confused. ATLAS is platform-independent and it provides an interface to many common data formats, like MATLAB files. In the case of large datasets a client-server-based distributed annotation structure is implemented, in order to divide computational cost and to give the possibility to annotate with multiple raters at the same time. The annotation supports generic predefined structures that can be tailored to the researchers' needs.

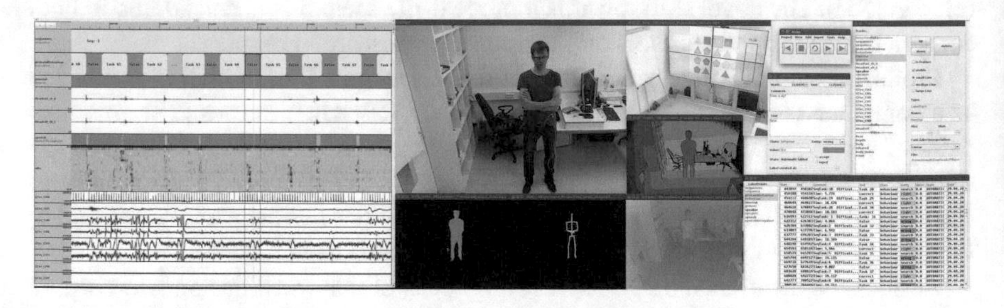

Fig. 19.1 Overview of the adaptive ATLAS UI (expert mode shown). Labels, acoustic properties and physiological raw data are depicted on the *left*. Video and infrared and depth information coupled with their corresponding extracted body data are shown in the *center*. On the *right*, some additional control and detail information windows are presented

Thus ATLAS is not restricted to specific existing emotional models. This leads to the fact that ATLAS is also usable outside the core affective computing community.

Finally, ATLAS includes active learning techniques to provide assistance to the researchers and raters. Externally extracted features can be combined with annotated labels in order to train a classifier. This classifier then suggests additional labels of instances which seem to be most certain, which can be either accepted or rejected by the rater. This information is added to the training information of the next iteration steps. Active learning and semi-supervised learning techniques can improve annotation speed dramatically. As a practicable application of this active learning approach a speaker segmentation tool is available to segment silence and different voices in a quick way.

19.3.2.2 ikannotate

The tool ***i***nterdisciplinary ***k***nowledge-based ***anno***tation ***t***ool for ***a***ided ***t***ranscription of ***e***motions (*ikannotate*) [2] is hosted at the Otto von Guericke University Magdeburg (cf. http://ikannotate.cognitive-systems-magdeburg.de). The particular focus of *ikannotate* is the support of a literal transcription enhanced with phonetic annotations and an emotional labelling using different methods. Both steps are pursued in one tool that provides a more convenient way of data processing. Further, both tools, *ikannotate* and ATLAS, complement each other.

As stated in [2], *ikannotate*'s first release (in 2011) was focused on audio material only. Besides audio processing, the latest version integrates modules which allow a labelling based on visual information, as well. Additionally, several support functions are implemented in *ikannotate*, helping to enrich the labelling or the post-processing. These are, for instance, tagging of dispositionally colored words in the utterance, assigning of uncertainty levels for the labelling, and modules for post-processing like feature extraction.

Enhanced Literal Transcription As it is known from [51], literal transcription is reasonably done on the utterance level since emotions and dispositions change slowly and, thus, one utterance covers one affective state. Therefore, *ikannotate* uses utterances as basic units in transcription. Generally, transcription is done by well-trained experts using tools like Folker [45] or common text editors. Enriching the transcript with prosodic or phonetic annotation usually demands expert knowledge since the utilized annotation paradigms are quite complex. *ikannotate* combines transcription and annotation, and further allows even non-experts to handle audio material properly since the annotator is supported by the tool in both steps.

Necessary information for transcription like start and end times are set by click events. The dialogue structure can be examined by corresponding tabs, and the text input is focused on the current utterance. The annotation is supported by click events as well. Thus, internal complexity of transcription methods in terms of symbols is avoided and annotation is made possible for non-expert users. In *ikannotate* the Gesprächsanalytisches Transkriptionssystem (dialogue analytic transcription system) is implemented (cf. [2]) including the corresponding features.

Emotional Labelling Another important step in the pre-processing of audio material is the assignment of emotional labels. For this the labeller is assisted by *ikannotate* as well. The tool supports three emotional labelling paradigms: (1) list of emotions (particular emotional phrases are combined in various lists as discussed in Chap. 21), (2) Geneva Wheel of Emotions (GEW) as proposed by Scherer [41], and (3) Self-Assessment Manikins (SAMs) according to [29].

19.4 Context-Aware Temporal Information Fusion Architectures for Multimodal Affect Recognition

Modern fusion architectures for affect recognition have to implement numerous features to take into account the characteristics of emotions. Information about emotions can be gathered from different origins such as multimodality, temporality or the context [15]. Emotions are inherently conveyed by humans using multiple channels [36] which complement each other. The most prominent ones are the auditory and visual channels. Furthermore, emotional analysis can be grouped into categories of different temporal granularities, like expression, attitude, mood, and trait [5]. Thus, affective recognition results have to be temporally combined using a suitable fusion technique. In most cases, and especially in the context of HCI, emotions are related to events or entities in the world [43]. The recognition of this relation is not only crucial to enhance the performance of the classifier system, but it is also of central importance for the further processing in the *Companion*-System. Therefore, affective fusion architectures should incorporate additional user or environmental context.

19.4.1 Fusion of Time-Windowed Features

Audio and video provide a less invasive way to obtain user data for the estimation of affective user states compared to directly user-attached or implanted biometric sensors. However, information on speech, facial expressions, and hand and body gestures [27, 37, 44] is often superimposed by noise and signals unrelated to the affective state to be detected. For instance, facial expression detectors have to cope with problems when a subject turns away, when feature extraction is hampered by wearing glasses, or when mouth movement caused by spoken utterances overlays facial expression. In addition, the target class of the affective user state may be characterized by a vast variety of facial appearances, which makes affective state recognition out of facial expression even more complicated.

Since audio and video are functions over time, we obtain time series of features and intermediate classifier decisions. It is obvious that temporal dependencies between different states exist and that the previous user states have an influence on the current state. We exploited these dependencies by considering the dynamic properties of features [33, 44].

Our investigations showed that linear classifiers often outperform non-linear classifiers [11, 37], since more restrictive classifier functions can be more robust against noise and overfitting in case of training data shortage. Using ensembles of classifiers and exploiting the temporal characteristics of features improved the results significantly. While ensemble learning approaches helped to capture the variety of the target class [3, 12], a time window of features has been applied to consider the dynamics of affective states [37].

The proposed approaches have been examined on the LAST MINUTE corpus (cf. Chap. 13) providing non-acted data of a Wizard-of-Oz setting. In this example, two selected affective user states—namely the normal (Baseline) and the stressed (Challenge) user states—had to be classified by analysing video data of the face. The screenplay of the LAST MINUTE experiment defines the time periods of induced affective state classes to be detected. An extra annotation of ground truth is therefore not required for training data generation because it is directly given by the start and end instants of the events specified in the screenplay. We must be aware that the subject can only be assumed to be in the desired affective state. In reality we try to detect the event according to the screenplay of the experiment and not necessarily the actual affective state.

Figure 19.2 (left) shows the time series of facial measurements creating feature channels for classification. In this case, 13 normalized geometric distances between significant facial points and additionally the eye blink frequency have been collected in a temporal window as input for a linear classifier. The classifier weights for the time series of feature data have been determined in a similar way as for matched

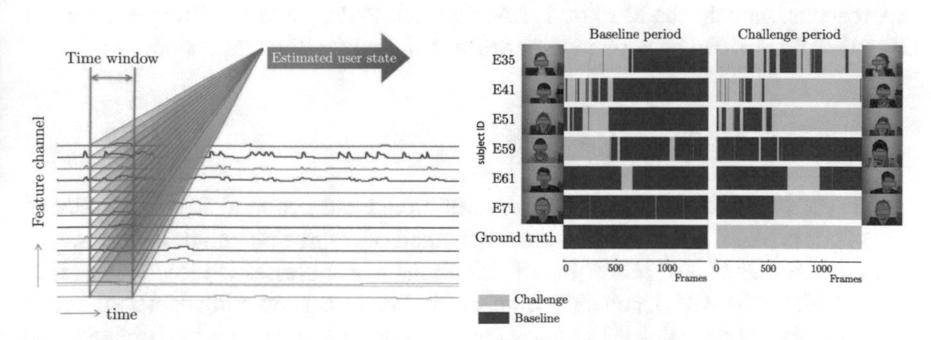

Fig. 19.2 *Left:* Time channels of features are collected in a time window providing the input for user state estimation, in this case 13 geometric features and the eye blink frequency in a time window of 0.6 s. *Right:* Output of linear classification (leave-one-out cross-validation) for six (anonymized) subjects in Baseline (*dark*) and Challenge (*bright*) periods and the ground truth defined by the screenplay of the experiment. Subject ID-related accuracies: E35: 61%, E41: 85%, E51: 77%, E59: 33%, E61: 57%, E71: 80%

filters or deconvolution [26, 34]. Applying a simple threshold to the filter output gives the decision of this linear two-class classifier. The length of the time window, which has been determined for best classification results using leave-one-out cross-validation, comprises 15 frames (0.6 s). A detailed evaluation of the influence of the window size is given in [37].

Figure 19.2 (right) shows the classification results over time for six selected subjects together with the screenplay-defined ground truth values. The recognition accuracy ranges from 33 to 80% for the individual subjects. Depending on such factors as age, biological gender and individual temper, but also on how the subject is used to communicate with a technical system, the facial expressiveness obviously varies a lot by the individual subject. This also holds for the ability of a classifier to detect a user state out of just facial features. Additionally, the unrestricted setting in this scenario does not ensure that the subject is really in the desired affective state (ground truth). Nevertheless, Fig. 19.2 (right) shows that the affective state recognition is capable of distinguishing the given user states to some extent even without creating ensembles of individuals.

An overall classification accuracy of 66% has been calculated in this example, which is quite vague on its own, but may be considered as a typical value for non-acted experimental data. A combination with other modalities is therefore one way to aim for higher confidence in the detected user state.

19.4.2 Temporal Multimodal Fusion Architectures

In the multimodal recognition of affective user states in real-world scenarios, decisions from multiple sources have to be combined. These sources can become inoperative, e.g. due to sensor malfunctions or a missing signal. This issue can be addressed by making use of the temporal process of classifier decisions. Two approaches, namely the Markov fusion network (MFN) and the Kalman filter for classifier fusion, can be applied to perform temporal multimodal fusion.

19.4.2.1 Markov Fusion Networks

The MFN is a probabilistic model for multimodal and temporal fusion introduced in [14]. It is based on a Markov network defined over three potential functions.

A number of $M \times T$ probability distributions over I classes is provided as input to the MFN, where M denotes the number of classifiers generating decisions and T is the number of time steps. The classifiers provide a class distribution for each time step based on each modality. The class probability distribution $m = 1, \dots, M$ at time step $t \in \mathscr{L}_m$ is a vector $\mathbf{x}_{mt} \in [0, 1]^I$ summing up to 1. Since a classifier decision might be temporally missing, the set $\mathscr{L}_m$ contains only the time steps in which class probability distributions of the classifier m are available. Assuming, without loss of generality, that the probability distributions are available for all time steps, the

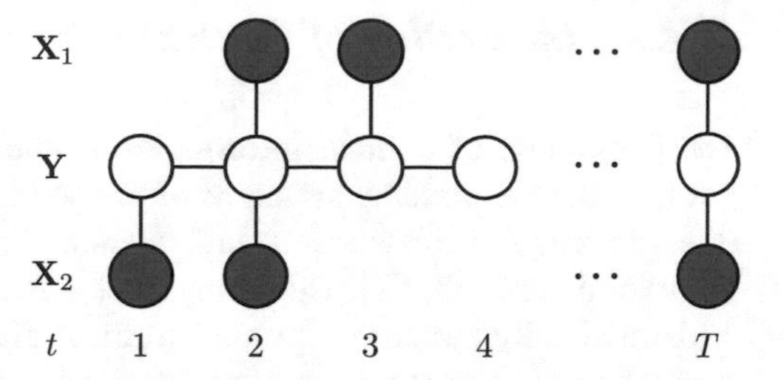

Fig. 19.3 Graphical model of the MFN. The sequence of combined estimates $\mathbf{y}_t$ is influenced by the available decisions $\mathbf{X}_{mt}$ of the source m and $t \in \mathscr{L}_m$ and adjacent combined estimates $\mathbf{y}_{t-1}$ and $\mathbf{y}_{t+1}$

MFN integrates the classifier predictions $\mathbf{X}_m \in [0, 1]^{I \times T}$ to a combined estimate $\mathbf{Y} \in [0, 1]^{I \times T}$ by making use of two main objectives. The first objective states that the combined estimated probability distributions will be similar to the provided class probability distributions. The second objective states that the estimated probability distributions are similar in the temporal proximity.

At first, the MFN defining function enforces the estimates to be similar to the observed class probability distributions. The second objective is implemented by temporally connecting the estimates in a Markov chain. The MFN reconstructs regions without classifier decisions by propagating information along the Markov chain.

Figure 19.3 depicts the graphical model of an exemplary MFN, which integrates two sequences of classifier decisions $\mathbf{X}_1$ and $\mathbf{X}_2$ to a combined estimate $\mathbf{Y}$. The classifier decisions are connected to the corresponding estimates in each time step. Whenever a class distribution is unavailable, the input node and the connecting link are omitted. The estimates themselves are temporally connected by the Markov chain.

19.4.2.2 Kalman Filter Architectures

The Kalman filter for classifier fusion operates on the same input data as the MFN [11]. However, the studied implementation was restricted to a two-class classification problem.

The conventional Kalman filter [22] is a well-known algorithm to enhance the quality of noisy measurements over time. It is commonly applied in the field of navigation and object tracking. Instead of calculating a rather simple average of measurements, the Kalman filter explicitly models the measurement noise. The modeled uncertainty can significantly enhance the quality of tracking. The Kalman filter itself is closely related to the HMM, but uses a Markov chain of continuous latent variables.

In [11] the Kalman filter was first applied to classifier decision fusion over time and was extended in order to handle missing classifier decisions. The Kalman filter for classifier fusion approach was tested on the AVEC 2013 challenge data [20] which is discussed in Sect. 19.5.1.

19.4.3 Integration of Context Information

The integration of context information is challenging for machine learning [15]. For this, the hierarchical structure of temporal patterns with different complexities is exploited. For instance, social signals can be decomposed into short-term behavioral cues [43, 57]. However, since no dataset has been recorded so far with a hierarchically structured ground truth of affective states, basic research on the integration of context information was conducted in the field of activity recognition. Several approaches have been proposed, like conditioned hidden Markov model (CHMM), unidirectional layered architecture (ULA), and HMM/CHMM using graph probability densities (HMM/CHMM-GPD).

The CHMM is an extension of the classic HMM, in which the hidden states are influenced by an additional sequence of causes. These causes can be provided by an external classifier decision which serves as additional context information [13].

CHMM can be studied as part of the ULA [13], in which each layer recognizes different classes with increasing complexities. The lowermost layer operates on features derived from the sensors and recognizes basic entities. The subsequent layers operate on the output of the preceding layers, which is given by their probabilistic class predictions. User preferences were recognized in the uppermost layer using a dynamic Markov logic network (DMLN) which models context information in the form of probabilistic logical rules. Studies on the baseline (cf. Sect. 19.4.1) investigate the propagation of context information down to the lowermost layer in order to influence the CHMM.

19.5 Multimodal Affect Recognition Results

19.5.1 Public Benchmark

AVEC Results The two sub-challenges of the 2013/2014 edition of the AVEC challenge comprise the two-/three-dimensional continuous affect sub-challenge and the discrete depression sub-challenge. The dataset contains 150 audio-visual recordings of participants of a clinical study in an inquiry-response cycle in front of a consumer notebook. The task was to estimate the continuous label traces for a test set of 50 videos. Performance was measured using (the magnitude of) Pearson's correlation coefficient, averaged over the test videos. For the 2013 edition of the challenge, a recognition system was developed that combines the input of a hierarchical classifier (consisting of multiple individual SVR and MLP classifiers) for the video modality with a diverse set of audio features using a Kalman filter (see Sect. 19.4.2.2) [20]. The results can be seen in Fig. 19.4 (left). For the 2014 edition, a slightly different approach has been taken. Based on the annotated trajectories, prototypical label traces were created using PCA and SVR to highlight difficulties in the annotation process (i.e. arbitrary starting point and subsequent transient phases

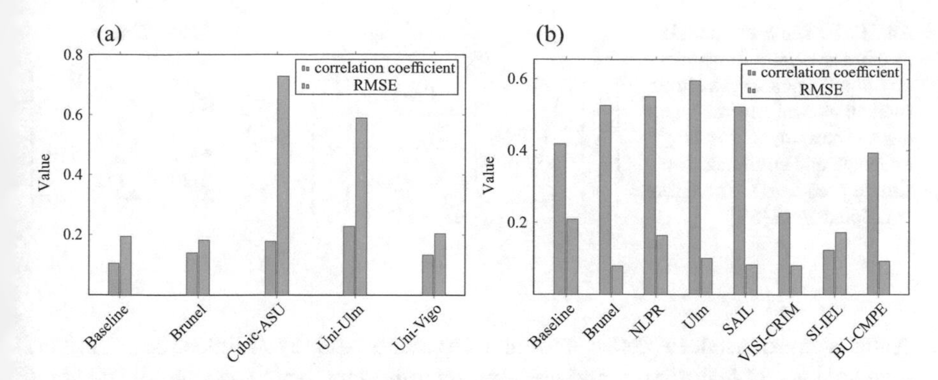

Fig. 19.4 Average correlation coefficient and RMSE of the affect sub-challenge of the 2013 and 2014 editions of the AVEC challenges. Source: http://sspnet.eu/avec2013/ and http://sspnet.eu/avec2014/. (**a**) Challenge results of AVEC 2013. (**b**) Challenge results of AVEC 2014

in the beginning) and of the performance measures. Combined with a clustering to reveal participant groups, the approach led to superior results and the win of the affect sub-challenge. For details the reader is referred to Fig. 19.4 and [21].

EmotiW Results The Emotion Recognition in the Wild challenge [7] focuses on audio-visual emotion recognition from movie snippets extracted from feature films. The snippets offer unconstrained movement, difficult lighting and speech overlapped with music and background noise. To tackle the challenge, the authors presented an approach that combined basic audio features with feature selection and achieved, using only a single modality, competitive results in the 2013 edition of the challenge [32]. Building on this success, in the follow-up challenge the application of enhanced auto-correlation features for the recognition of emotions from speech was proposed [30].

19.5.2 Results

Last Minute Corpus The LMC material has the advantage that a fusion of multimodal classifications can be combined in the context of more subtle dispositions like "concentration" and "thinking". For this, novel characteristics such as self-touching and eye-blink frequency can be observed. These approaches are pursued based on the main idea presented in Fig. 19.5 focussing on two situations, namely baseline (BSL) and weight limit barrier (WLB) [9].

In [52] these novel characteristics are investigated with a focus on the analysis of facial expressions. For the generation of a visual-based classifier the following features are considered: mouth deformations, eye-blink, eyebrow movement, and the general movement of the head (global) as the most prominent and reliably detectable features for the face as used in [27, 52]. On the one hand, common

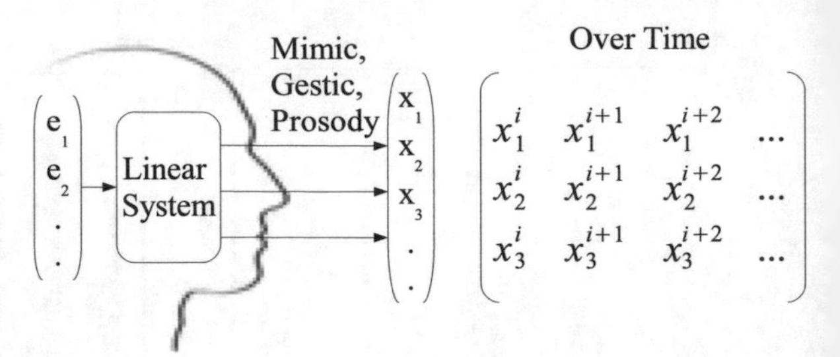

Fig. 19.5 The user, modeled as a linear system, transmits emotional signals via various modalities over several time steps. These inputs can be collected and combined for further processing. The figure is adapted from [37]

features were used to detect mouth movements and eye-blinks [52]. Further, extracted visual features of hand gestures are important, from a psychological point of view, especially "self-touch" and "no-self-touch" when the subject touches his face. Therefore, using skin color and connected component analysis, the overlap of hand and face regions can be detected. For analysing purposes, we investigated 13 subjects of the LMC for visual classification due to different illumination, head positions, and occlusions. According to [37, 52] the processing is as follows: A time window with the size of ten frames (0.4 s) is considered instead of a decision based on individual frames, as it is assumed that the bodily response, which is reflected in changes of the features, is short-time stable.

The visual classification is further combined with an acoustic evaluation. Therefore, an automatic classification system was trained on the material of the same 13 subjects. For evaluation, we applied a leave-one-subject-out (LOSO) strategy and used HMMs/GMMs with common MFCCs with delta and acceleration. This results in an overall mean of the weighted accuracy of 76.01% (std. dev. 6.45%) for the two-class problem.

Based on these classifications a fusion was conducted. For this, an MFN with the following parameters was used: W = 1000, kf = 0.5, kp = 4, kg = 4. The uni-modal classifiers are the facial expressions, gestural analysis, and the acoustic classifier. As pointed out in [52], each modality possesses its own distinct characteristic distribution of decisions over time. The recognition of the emotional state based on facial expressions requires the subject's face to be in the view of the camera. However, in case the subject turns away, a decision may become infeasible. A similar problem occurs in prosodic analysis since it can be performed only if the subject produces an utterance. In the given setting, the decisions derived from the gestural analysis are even more demanding, because they only give evidence for the class WLB. The classifier based on facial expression provides decision probabilities for all frames, the acoustic analysis only for 15.9% of the frames and a gestural analysis only for 9% of the frames. The overall average accuracy is 85.29% (std. dev. 14.22%).

EmoRec The experimental validation can be divided into uni-modal and multi-modal approaches since audio, video, and bio-physiology of the user were recorded. Grounded on the experimental design, a small set of representative experimental sequences (ESs) is selected for classification. ES 2—the experimental part linked to "positive pleasure, low arousal, high dominance" feeling of the user—and ES 5—referring to "negative pleasure, high arousal, low dominance"—are selected based on their location as opposing octants in the VAD space. For a more detailed view of the experimental setting, we refer you to [58].

The video modality is pre-processed as follows: First, face detection is done based on Viola and Jones' boosted Haar cascade. Salient facial points are detected using a constraint local model followed by an alignment procedure based on selected points. The selected features are: optical flow, motion histograms, pyramids of histograms of oriented gradients and local binary patterns.

Using the diverse information captured by the different feature sets, different fusion methods are employed. First, an ensemble of Support Vector Machines (SVMs) with softmax output was trained on bootstrapped subsets of the training data for each of the four feature sets. The results of each ensemble were aggregated by a trainable combiner in the form of a multilayer perceptron. To compensate for different time resolutions of the features, a common reference time window of 2 s is used to integrate the per-channel decisions. Finally, in another trainable fusion mapping the estimates of the individual channels are aggregated into the final decision. Using this fusion scheme, a final accuracy of 69.2% can be achieved. For an overview of the results (including other fusion methods), the reader is referred to Fig. 19.6. More details are given in [19] as these results were obtained only on a subset of 11 people.

Further results are obtained by analyzing the bio-physiological channels. Since the recorded modalities are inherently different, each channel has to be individually preprocessed. For example, to extract information from the blood volume pulse, first the heartbeats in the form of so-called QRS complexes have to be located. After

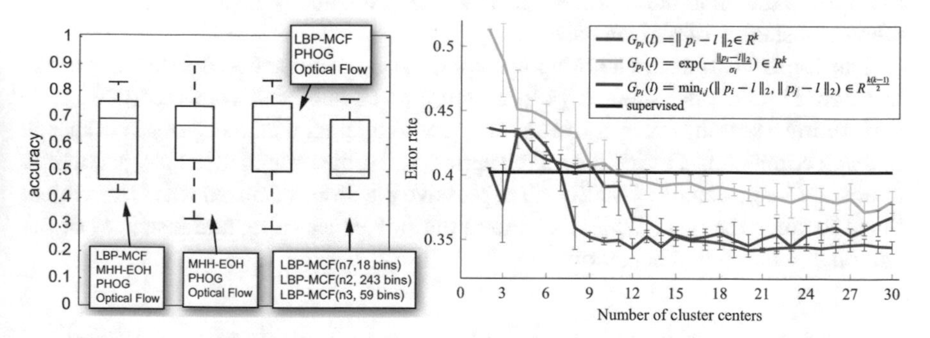

Fig. 19.6 *Left*: The final results obtained with the fusion architecture with different settings. The highest median classification rate was 69.2%, achieved with a combination of all channels. Other channel selections also performed relatively well. Fusing results of the same channel with different settings (bin and neighbourhood size) did not result in an improvement. *Right*: Classification error with standard deviation computed over ten runs. Different cluster techniques are used to augment the classification process. The supervised reference is given as a *black line*. The unsupervised techniques outperform the supervised approach given enough model complexity (here: number of cluster centers)

additional filtering and detrending, different features are computed for each channel on suitable time windows. For the heart rate, well-known statistical features such as the standard deviation, RMSSD, or pNN50 are applied. Additionally, the non-linear features that approximate entropy, recurrence rate and dimensions of the ellipse in a Poincaré plot are calculated. Finally, three features based on the power spectrum density are computed.

The experimental results based on the individual channels suggest that robust recognition is only possible to a certain extent, i.e. the recognition rate is only slightly above chance level (for details, the reader is referred to [40]). To alleviate this problem, the following means are introduced: A fusion step is introduced to combine the individual bio-physiological channels and, additionally, a technique from the field of semi-supervised learning is used to incorporate the additional data provided by the remaining experimental sequences (i.e. ES 1,3,4,6). In the procedure, an unsupervised pre-processing step is used as transformation into a data-driven representation. In Fig. 19.6, the results are summarized. The fusion itself improves the result to an error rate of about 40%. In addition, the unsupervised pre-processing step further lowers the error rate to about 35%.

19.5.3 Active Learning in an HCI Scenario

Plausible annotation of multimodal HCI data is an enormous problem due to the fact of the time-consuming and sensitive annotation process. Furthermore, emotional reactions are often very sparse, resulting in a large annotation overhead to gather the interesting moments of a recording. Active learning techniques provide methods to improve the annotation processes since the annotator is asked to label only the relevant instances of a given dataset.

The approach of active learning was applied on an interaction data set, described in Chap. 12 and published in [48]. A number of subjects were recorded while performing a search task on a screen. They could interact with the system via speech or touch commands. During the search period, the subjects tended to react just a little or not at all emotionally. Usually, all expressive reactions occurred when the subject failed to solve the task. Table 19.1 shows the imbalance of feature instances of the neutral and emotional behaviors.

Table 19.1 Number of neutral and emotional feature instances, enlisted for each participant

ID	12	15	17	23	26	30	Sum.
#Neutral	3387	3365	4149	3406	3409	4489	22,205
#Emotion	108	135	66	208	429	665	1661
Responsiveness	Good	Good	Moderate	Good	Moderate	Moderate	

The last column shows the estimation of the participant's emotional expressiveness

For feature extraction, the facial region was located and extracted within each single frame of a video sequence. Subsequently, each face was divided into a fixed number of non-overlapping blocks. The *Local Binary Pattern* operator on *Three Orthogonal Planes (LBP-TOP)* [60] was applied to each cuboid consisting of each block of facial region for an entire sequence to generate the corresponding histogram of descriptors. These histograms were concatenated to form the feature vector for each sequence.

The feature vectors were utilized as input for a *One Class Support Vector Machine (OCSVM)* [47]. It is an extension of the binary SVM, defining a decision function that takes the value $+1$ in a small region capturing most of the instances, and -1 elsewhere. The instances of the target class are mapped into a Hilbert space H and subsequently separated from the origin by a hyperplane with maximum margin. Consequently, data objects are classified either as outliers or as belonging to the target class. The target class was designated as neutral, because of the extreme imbalance in the data; therefore the outliers are the sparse emotional moments.

The OCSVM was used for an active learning approach. The initial adjustment of the classifier pooled all data points. The first step presented the data points, which were explained least by the model (depending on their distance to the hyperplane), to an expert to be labeled. The labeling information was utilized to improve the classifier's performance. Both steps were applied in a loop several times.

In each iteration, the ten worst data points (outliers) were presented to an annotator. In the first iteration steps, most of the outliers represented emotional moments and were labeled accordingly. During the further course, the number of data points differing from neutral decreased. After 50 iterations (500 data points) the process was stopped. Figure 19.7 shows this behavior for six participants.

Moreover, a closer look shows that 82.41% of the emotional moments of participant 12 were identified by labeling just 14% of the entire dataset. The same observations can be made for participant 15: 74.07% of the outliers were identified by labeling about 14.3% of the entire dataset.

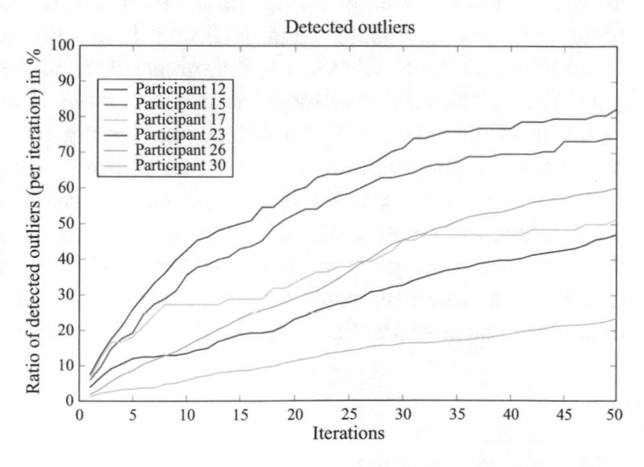

Fig. 19.7 Number of data points labeled as emotional moments in relation to all emotional moments in the data set for each iteration step. As can be seen, about 82.41% of emotional events could be detected for participant 12 by labeling just 14% of the data

In a subsequent experiment a binary SVM was trained on the 500 labeled data points and achieved a g-mean value[1] of about 0.8. To compare this performance with common methods, the whole data set was manually labeled and classified with a binary SVM, trained on all data points. The accomplished g-mean was almost the same. The reason for this is that most of the support vectors lie within the labeled 500 data points and the remaining points do not contribute to the decision hyperplane. Hence, the proposed active learning approach generates an effective classification model while labeling just a small portion of the dataset.

19.6 Conclusion

The multimodal recognition of user affects is a challenging issue which could be faced by the combination of various modalities. Besides the combination of suitable modalities and features, effort should be applied to a stringent development and handling of recognition architectures. In this chapter, we presented an overview of our work done in the context of affect recognition based on multiple sources. Especially, the use of active learning approaches and Markov fusion networks provides an improvement in the processing of naturalistic, affectively afflicted material. The achieved results are either based on corpora recorded in the authors' research groups or on publicly available benchmark data sets. For both categories remarkable results were obtained, in particular in the benchmark challenges (cf. Sect. 19.5.1). Based on such achievements, we processed the internally recorded data sets—namely LMC (cf. Chap. 13) and EmoRec [58]—showing that the discussed recognizer architectures can be applied in naturalistic scenarios. In particular, the use of temporal and contextual information for multimodal fusion improved the classification (cf. Hypotheses 2 and 4).

Besides the classification issue, the preprocessing of data should also be approached multimodally (cf. Hypothesis 3). Therefore, we introduced two annotation tools assisting during the annotation and labelling. *ikannotate* mainly focuses on the annotation and labelling based on audio and video streams. In addition, ATLAS allows a synchronous handling of audio, video, and biophysiological data. Furthermore, it applies active learning techniques to assist in the labelling process. From the active learning perspective, we can conclude that the multimodal classifiers can be established without completely annotated material. The starting point is a small subset which provides reasonable class information. Iteratively, a classifier can be trained while simultaneously annotating the material (cf. Hypothesis 1). Finally, we briefly discussed that data which is recorded synchronously can improve the performance of the classification in terms of multimodal investigations. As is elaborated on in Chap. 22, there are several

[1]The g-mean was chosen because of the strong imbalance between the two classes.

ways of establishing synchronous recordings. For real-world applications, we refer to Chap. 22.

Acknowledgements We thank our highly regarded deceased colleague and friend Prof. Dr. Bernd Michaelis who contributed to the SFB on various topics and provided well-informed suggestions. This work was done within the Transregional Collaborative Research Centre SFB/TRR 62 "*Companion*-Technology for Cognitive Technical Systems" funded by the German Research Foundation (DFG).

References

1. Batliner, A., Fischer, K., Huber, R., Spiker, J., Nöth, E.: Desperately seeking emotions: Actors, wizards and human beings. In: Proceedings of the ISCA Workshop on Speech and Emotion: A Conceptual Framework for Research, pp. 195–200 (2000)
2. Böck, R., Siegert, I., Haase, M., Lange, J., Wendemuth, A.: ikannotate - a tool for labelling, transcription, and annotation of emotionally coloured speech. In: D'Mello, S., Graesser, A., Schuller, B., Martin, J.C. (eds.) Proceedings of ACII. Lecture Notes on Computer Science, vol. 6974, pp. 25–34. Springer, Berlin (2011)
3. Breiman, L.: Bagging predictors. Mach. Learn. **24**(2), 123–140 (1996)
4. Burkhardt, F., Paeschke, A., Rolfes, M., Sendlmeier, W., Weiss, B.: A database of German emotional speech. In: Proceedings of Interspeech 2005, pp. 1517–1520 (2005)
5. Cowie, R., Douglas-Cowie, E., Tsapatsoulis, N., Votsis, G., Kollias, S., Fellenz, W., Taylor, J.: Emotion recognition in human-computer interaction. IEEE Signal Process. Mag. **18**(1), 32–80 (2001)
6. Devillers, L., Vidrascu, L., Lamel, L.: Challenges in real-life emotion annotation and machine learning based detection. Neural Netw. **18**(4), 407–422 (2005)
7. Dhall, A., Goecke, R., Joshi, J., Sikka, K., Gedeon, T.: Emotion recognition in the wild challenge 2014: baseline, data and protocol. In: Proceedings of ICMI, pp. 461–466. ACM, New York (2014)
8. Dix, A., Finlay, J., Abowd, G., Beale, R.: Human-computer Interaction. Prentice-Hall, Upper Saddle River, NJ (1997)
9. Frommer, J., Michaelis, B., Rösner, D., Wendemuth, A., Friesen, R., Haase, M., Kunze, M., Andrich, R., Lange, J., Panning, A., Siegert, I.: Towards emotion and affect detection in the multimodal last minute corpus. In: Calzolari, N., Choukri, K., Declerck, T., Doğan, M.U., Maegaard, B., Mariani, J., Odijk, J., Piperidis, S. (eds.) Proceedings of LREC. ELRA, Paris (2012)
10. Glodek, M., Tschechne, S., Layher, G., Schels, M., Brosch, T., Scherer, S., Kächele, M., Schmidt, M., Neumann, H., Palm, G., Schwenker, F.: Multiple classifier systems for the classification of audio-visual emotional states. In: D'Mello, S., Graesser, A., Schuller, B., Martin, J.C. (eds.) Proceedings of ACII - Part II, Lecture Notes on Computer Science, vol. 6975, pp. 359–368. Springer, Berlin (2011)
11. Glodek, M., Reuter, S., Schels, M., Dietmayer, K., Schwenker, F.: Kalman filter based classifier fusion for affective state recognition. In: Zhou, Z.H., Roli, F., Kittler, J. (eds.) Multiple Classifier Systems (MCS). Lecture Notes on Computer Science, vol. 7872, pp. 85–94. Springer, Berlin (2013)
12. Glodek, M., Schels, M., Schwenker, F.: Ensemble Gaussian mixture models for probability density estimation. Comput. Stat. **27**(1), 127–138 (2013)
13. Glodek, M., Geier, T., Biundo, S., Palm, G.: A layered architecture for probabilistic complex pattern recognition to detect user preferences. J. Biol. Inspired Cognitive Archit. **9**, 46–56 (2014)

14. Glodek, M., Schels, M., Schwenker, F., Palm, G.: Combination of sequential class distributions from multiple channels using Markov fusion networks. J. Multimodal User Interfaces **8**(3), 257–272 (2014)
15. Glodek, M., Honold, F., Geier, T., Krell, G., Nothdurft, F., Reuter, S., Schüssel, F., Hörnle, T., Dietmayer, K., Minker, W., Biundo, S., Weber, M., Palm, G., Schwenker, F.: Fusion paradigms in cognitive technical systems for human-computer interaction. Neurocomputing **161**, 17–37 (2015)
16. Gunes, H., Piccardi, M.: Bi-modal emotion recognition from expressive face and body gestures. J. Netw. Comput. Appl. **30**(4), 1334–1345 (2007)
17. Healey, J.: Wearable and automotive systems for affect recognition from physiology. Ph.D. thesis, MIT (2000)
18. Hudlicka, E.: To feel or not to feel: The role of affect in human-computer interaction. Int. J. Hum.-Comput. Stud. **59**(1-2), 1–32 (2003)
19. Kächele, M., Schwenker, F.: Cascaded fusion of dynamic, spatial, and textural feature sets for person-independent facial emotion recognition. In: Proceedings of ICPR, pp. 4660–4665 (2014)
20. Kächele, M., Glodek, M., Zharkov, D., Meudt, S., Schwenker, F.: Fusion of audio-visual features using hierarchical classifier systems for the recognition of affective states and the state of depression. In: De Marsico, M., Tabbone, A., Fred, A. (eds.) Proceedings of ICPRAM, pp. 671–678. SciTePress, Setúbal (2014)
21. Kächele, M., Schels, M., Schwenker, F.: Inferring depression and affect from application dependent meta knowledge. In: Proceedings of the 4th International Workshop on Audio/Visual Emotion Challenge, AVEC '14, pp. 41–48. ACM, New York (2014)
22. Kalman, R.E.: A new approach to linear filtering and prediction problems. J. Fluids Eng. **82**(1), 35–45 (1960)
23. Kanade, T., Cohn, J., Tian, Y.: Comprehensive database for facial expression analysis. In: Automatic Face and Gesture Recognition, 2000, pp. 46–53 (2000)
24. Kim, K., Bang, S., Kim, S.: Emotion recognition system using short-term monitoring of physiological signals. Med. Biol. Eng. Comput. **42**(3), 419–427 (2004)
25. Kipp, M.: Anvil - a generic annotation tool for multimodal dialogue. In: INTERSPEECH-2001, Aalborg, Denmark, pp. 1367–1370 (2001)
26. Krell, G., Niese, R., Al-Hamadi, A., Michaelis, B.: Suppression of uncertainties at emotional transitions — facial mimics recognition in video with 3-D model. In: Richard, P., Braz, J. (eds.) Proceedings of the International Conference on Computer Vision Theory and Applications (VISAPP), vol. 2, pp. 537–542 (2010)
27. Krell, G., Glodek, M., Panning, A., Siegert, I., Michaelis, B., Wendemuth, A., Schwenker, F.: Fusion of fragmentary classifier decisions for affective state recognition. In: MPRSS, Lecture Notes on Artificial Intelligence, vol. 7742, pp. 116–130. Springer, Berlin (2012)
28. Kuncheva, L.: Combining Pattern Classifiers: Methods and Algorithms. Wiley, New York (2004)
29. Lang, P.J.: Behavioral Treatment and Bio-Behavioral Assessment: Computer Applications, pp. 119–137. Ablex Publishing, New York (1980)
30. Meudt, S., Schwenker, F.: Enhanced autocorrelation in real world emotion recognition. In: Proceedings of the 16th International Conference on Multimodal Interaction, ICMI '14, pp. 502–507. ACM, New York (2014)
31. Meudt, S., Bigalke, L., Schwenker, F.: Atlas – an annotation tool for HCI data utilizing machine learning methods. In: International Conference on Affective and Pleasurable Design (APD'12), pp. 5347–5352 (2012)
32. Meudt, S., Zharkov, D., Kächele, M., Schwenker, F.: Multi classifier systems and forward backward feature selection algorithms to classify emotional coloured speech. In: International Conference on Multimodal Interaction, ICMI 2013, pp. 551–556. ACM, New York (2013)

33. Niese, R., Al-Hamadi, A., Heuer, M., Michaelis, B., Matuszewski, B.: Machine vision based recognition of emotions using the circumplex model of affect. In: Proceedings of the International Conference on Multimedia Technology (ICMT), pp. 6424–6427. IEEE, New York (2011)
34. North, D.O.: An analysis of the factors which determine signal/noise discrimination in pulsed-carrier systems. Proc. IEEE **51**(7), 1016–1027 (1963)
35. Oudeyer, P.: The production and recognition of emotions in speech: features and algorithms. Int. J. Hum.-Comput. Stud. **59**(1-2), 157–183 (2003)
36. Palm, G., Glodek, M.: Towards emotion recognition in human computer interaction. In: Esposito, A., Squartini, S., Palm, G. (eds.) Neural Nets and Surroundings, vol. 19, pp. 323–336. Springer, Berlin (2013)
37. Panning, A., Siegert, I., Al-Hamadi, A., Wendemuth, A., Rösner, D., Frommer, J., Krell, G., Michaelis, B.: Multimodal affect recognition in spontaneous HCI environment. In: 2012 IEEE International Conference on Signal Processing, Communication and Computing, pp. 430–435. IEEE, New York (2012)
38. Ringeval, F., Sonderegger, A., Sauer, J., Lalanne, D.: Introducing the RECOLA multimodal corpus of remote collaborative and affective interactions. In: 2013 10th IEEE International Conference and Workshops on Automatic Face and Gesture Recognition (FG), pp. 1–8 (2013)
39. Schels, M., Scherer, S., Glodek, M., Kestler, H., Palm, G., Schwenker, F.: On the discovery of events in EEG data utilizing information fusion. Comput. Stat. **28**(1), 5–18 (2013)
40. Schels, M., Kächele, M., Glodek, M., Hrabal, D., Walter, S., Schwenker, F.: Using unlabeled data to improve classification of emotional states in human computer interaction. J. Multimodal User Interfaces **8**(1), 5–16 (2014)
41. Scherer, K.R.: What are emotions? and how can they be measured? Soc. Sci. Inf. **44**, 695–729 (2005)
42. Scherer, S., Schwenker, F., Palm, G.: Classifier fusion for emotion recognition from speech. In: Advanced Intelligent Environments, pp. 95–117. Springer, Boston (2009)
43. Scherer, S., Glodek, M., Layher, G., Schels, M., Schmidt, M., Brosch, T., Tschechne, S., Schwenker, F., Neumann, H., Palm, G.: A generic framework for the inference of user states in human computer interaction: how patterns of low level behavioral cues support complex user states in HCI. J. Multimodal User Interfaces **6**(3–4), 117–141 (2012)
44. Scherer, S., Glodek, M., Schwenker, F., Campbell, N., Palm, G.: Spotting laughter in natural multiparty conversations: a comparison of automatic online and offline approaches using audiovisual data. ACM Trans. Interactive Intell. Syst. **2**(1), 4:1–4:31 (2012)
45. Schmidt, T., Schütte, W.: FOLKER: an annotation tool for efficient transcription of natural, multi-party interaction. In: Proceedings of the 7th International Conference on Language Resources and Evaluation (2010)
46. Schmidt, T., Wörner, K.: EXMARaLDA – Creating, analysing and sharing spoken language corpora for pragmatic research. Pragmatics **19**, 565–582 (2009)
47. Schölkopf, B., Williamson, R.C., Smola, A.J., Shawe-Taylor, J., Platt, J.C.: Support vector method for novelty detection. In: NIPS, vol. 12, pp. 582–588 (1999)
48. Schüssel, F., Honold, F., Weber, M., Schmidt, M., Bubalo, N., Huckauf, A.: Multimodal interaction history and its use in error detection and recovery. In: Proceedings of the 16th ACM International Conference on Multimodal Interaction (ICMI'14), pp. 164–171. ACM, New York (2014)
49. Schwenker, F., Scherer, S., Magdi, Y.M., Palm, G.: The GMM-SVM supervector approach for the recognition of the emotional status from speech. In: ICANN (1), Lecture Notes on Computer Science, vol. 5768, pp. 894–903. Springer, Berlin (2009)
50. Schwenker, F., Scherer, S., Schmidt, M., Schels, M., Glodek, M.: Multiple classifier systems for the recognition of human emotions. In: Multiple Classifier Systems, Lecture Notes on Computer Science, vol. 5997, pp. 315–324. Springer, Berlin (2010)

51. Sezgin, M.C., Gunsel, B., Kurt, G.: Perceptual audio features for emotion detection. EURASIP J. Audio Speech Music Process. **2012**, 1–21 (2012)
52. Siegert, I., Glodek, M., Krell, G.: Using speaker group dependent modelling to improve fusion of fragmentary classifier decisions. In: Proceedings of the International IEEE Conference on Cybernetics (CYBCONF), pp. 132–137. IEEE, New York (2013)
53. Soleymani, M., Lichtenauer, J., Pun, T., Pantic, M.: A multimodal database for affect recognition and implicit tagging. IEEE Trans. Affect. Comput. **3**, 42–55 (2012).
54. Strauß, P.M., Hoffmann, H., Minker, W., Neumann, H., Palm, G., Scherer, S., Schwenker, F., Traue, H., Walter, W., Weidenbacher, U.: Wizard-of-oz data collection for perception and interaction in multi-user environments. In: Proceedings of LREC, pp. 2014–2017 (2006)
55. Traue, H.C., Ohl, F., Brechmann, A., Schwenker, F., Kessler, H., Limbrecht, K., Hoffman, H., Scherer, S., Kotzyba, M., Scheck, A., Walter, S.: A framework for emotions and dispositions in man-companion interaction. In: Rojc, M., Campbell, N. (eds.) Converbal Synchrony in Human-Machine Interaction, pp. 98–140. CRC Press, Boca Raton (2013)
56. Valstar, M., Schuller, B., Smith, K., Almaev, T., Eyben, F., Krajewski, J., Cowie, R., Pantic, M.: AVEC 2014: 3d dimensional affect and depression recognition challenge. In: Proceedings of ACM MM, AVEC '14, pp. 3–10. ACM, New York (2014)
57. Vinciarelli, A., Pantic, M., Bourlard, H., Pentland, A.: Social signal processing: state-of-the-art and future perspectives of an emerging domain. In: Proceedings of the International ACM Conference on Multimedia (MM), pp. 1061–1070. ACM, New York, NY (2008)
58. Walter, S., Scherer, S., Schels, M., Glodek, M., Hrabal, D., Schmidt, M., Böck, R., Limbrecht, K., Traue, H.C., Schwenker, F.: Multimodal emotion classification in naturalistic user behavior. In: Jacko, J.A. (ed.) Proceedings of the 14th International Conference on Human Computer Interaction (HCI'11), Lecture Notes on Computer Science, vol. 6763, pp. 603–611. Springer, Berlin (2011)
59. Zeng, Z., Pantic, M., Roisman, G.I., Huang, T.S.: A survey of affect recognition methods: audio, visual, and spontaneous expressions. IEEE Trans. Pattern Anal. Mach. Intell. **31**(1), 39–58 (2009)
60. Zhao, G., Pietikainen, M.: Dynamic texture recognition using local binary patterns with an application to facial expressions. IEEE Trans. Pattern Anal. Mach. Intell. **29**(6), 915–928 (2007)

Chapter 20
Emotion Recognition from Speech

Andreas Wendemuth, Bogdan Vlasenko, Ingo Siegert, Ronald Böck, Friedhelm Schwenker, and Günther Palm

Abstract Spoken language is one of the main interaction patterns in human-human as well as in natural, companion-like human-machine interactions. Speech conveys content, but also emotions and interaction patterns determining the nature and quality of the user's relationship to his counterpart. Hence, we consider emotion recognition from speech in the wider sense of application in Companion-systems. This requires a dedicated annotation process to label emotions and to describe their temporal evolution in view of a proper regulation and control of a system's reaction. This problem is peculiar for naturalistic interactions, where the emotional labels are no longer a priori given. This calls for generating and measuring of a reliable ground truth, where the measurement is closely related to the usage of appropriate emotional features and classification techniques. Further, acted and naturalistic spoken data has to be available in operational form (corpora) for the development of emotion classification; we address the difficulties arising from the variety of these data sources. Speaker clustering and speaker adaptation will as well improve the emotional modeling. Additionally, a combination of the acoustical affective evaluation and the interpretation of non-verbal interaction patterns will lead to a better understanding of and reaction to user-specific emotional behavior.

A. Wendemuth (✉)
Cognitive Systems Group, Otto von Guericke University, PF-4120, 39016 Magdeburg, Germany

Center for Behavioral Brain Sciences, 39118 Magdeburg, Germany
e-mail: andreas.wendemuth@ovgu.de

B. Vlasenko • I. Siegert • R. Böck
Cognitive Systems Group, Otto von Guericke University, PF-4120, 39016 Magdeburg, Germany
e-mail: bodgan.vlasenko@ovgu.de; ingo.siegert@ovgu.de; ronald.boeck@ovgu.de

F. Schwenker • G. Palm
Institute for Neural Information Processing, University of Ulm, 89069 Ulm, Germany
e-mail: friedhelm.schwenker@uni-ulm.de; guenther.palm@uni-ulm.de

S. Biundo, A. Wendemuth (eds.), *Companion Technology*, Cognitive Technologies,
DOI 10.1007/978-3-319-43665-4_20

20.1 Introduction

Human-machine interaction (HCI) has recently received increased attention. Besides making the operation of technical systems as simple as possible, a main goal is to enable a natural interaction with the user. However, today's speech-based operation still seems artificial, as only the content of the speech is evaluated. The way in which something is said remains unconsidered, although it is well known that also emotions are important to communicate successfully. "*Companion*-Systems" aim to fill this gap by adapting to the user's individual skills, preferences and emotions to be able to recognize, interpret and respond to emotional utterances appropriately (cf. Chap. 1).

A first prerequisite for emotion recognition is the availability of data for training and testing classifiers. In [43] it is pointed out that for a speech-based emotion recognition a rather straightforward engineering approach is usually used: "we take what we get, and so far, performance has been the decisive criterion". This very practical approach allows the evaluation of various feature extraction and classification methods. But to regulate and control a system's reaction towards a user adequately, the interpretation of the data labels can no longer be neglected. This problem has been addressed by many researchers in the community (cf. [4, 54, 63]), but a proper solution has not appeared yet.

Automatic emotion recognition is treated here as a branch of pattern recognition. It is data-driven—insights are gathered from sampled data, as it is difficult to rely on empirical evidence from emotion psychology: there is no universal emotion representation. This problem is arising for naturalistic[1] interactions. One has to rely on the emotional annotation of data, as the emotional labels are not given a priori. Furthermore, there are barely two datasets using the same emotional terms. Thus, for cross-corpora analyses researchers have to, for instance, combine different emotional classes to arrive at common labels.

Another issue that has only been rarely investigated for emotion recognition is speaker(-group) adaptation, to improve the emotional modeling. Although a model adaptation towards a specific group of speakers or an individual speaker has been used to improve automatic speech recognition [25], it has only rarely been used for emotion recognition. Additionally, these studies are conducted on databases of simulated affects only. Thus, there is no proof that these methods are suitable for natural interactions as well.

Furthermore, speech contains more than just content and emotions. It includes also "interaction patterns" determining the nature and quality of the user's relationship to his counterpart. These cues are sensed and interpreted by the humans, often without conscious awareness, but greatly influence the perception and interpretation of messages and situations. Thus a combination of the acoustical affective evaluation

[1]The term "naturalistic" is used to clarify the fact that a computer system always is a conversational partner less powerful than a human and thus HCI cannot be a natural interaction.

and the interpretation of these interaction patterns could enhance the prediction power for the process of naturalistic interactions.

From these considerations we derive the following research questions, which will be addressed in the following:

- How do we generate a reliable ground truth for emotion recognition from speech?
- How do we adapt models for user-specific emotional behavior?
- How do we combine acoustic and linguistic features for an improved emotion recognition?

In the next section we sketch common methods for speech-based emotion recognition and the utilized datasets. The novel and specialized methods which we used for the analysis of the research questions will be introduced in the later sections.

20.2 Methods for Acoustic Emotion Analysis

Speech-based emotion classifiers used in recent research publications include a broad variety [55] of different classification methods. There are two predominant emotion classification paradigms: frame-level dynamic modeling by means of Hidden Markov Models (HMMs) and turn-level static modeling [42]. In frame-level analysis, the speech data is divided into short frames of about 15–25 ms, where the human vocal apparatus generates a stable short-term spectrum; the extracted features are called segmental features or low-level features. For turn-level analyses, the whole turn, mostly the speaker's utterance, is investigated, and supra-segmental features are used to describe long-term acoustic characteristics (cf. [3]). It has to be taken into account that turn-level modeling in comparison with frame-level modeling does not provide good flexibility for modeling emotional intensity variability within a turn. But most emotional datasets provide an emotion annotation on the turn level, and generating a reliable annotation on the frame level is not as feasible.

Among dynamic modeling techniques, hidden Markov models are dominant. But so is "bag-of-frames" technique for multi-instance learning [45]. Further, dynamic Bayesian network architectures [29] could help to combine acoustic features on different time levels, such as the supra-segmental prosodic level, and spectral features on a frame-level basis. Regarding static modeling, the list of possible classification techniques seems endless: multi-layer perceptrons or other types of neural networks, Bayes classifiers, Bayesian decision networks, random forests, Gaussian mixture models (GMMs), decision trees, k-nearest neighbor distance classifiers, and support vector machines (SVM) are applied most often (cf. [24]).

Also, a selection of ensemble methods has been used, such as bagging, boosting, multi-boosting, and stacking with and without confidence scores. Newly developing approaches are long-short-term-memory recurrent neural networks, hidden conditional random fields, and tandem GMMs with support vector machines. They could further become more popular in the near future.

An important question which is investigated in the research community is about the choice and length of the best emotional unit. Unfortunately, this question has not been answered to date. Depending on the material and experimental setup, the emotion classification performance of sub-turn entries—automatically extracted quasi-stationary segments as well as manually marked syllables—falls behind models trained on turn-levels or sub-turn units that are related to the phonetic content clearly outperformed turn-level models; [3]. Lee et al. presented an acceptable acoustic-based emotion recognition performance using phoneme-class-dependent HMM classifiers with short-term spectral acoustic features [28]. The authors reached a classification performance of 76.1% for their four-class recognition problem, but used very expressive acted emotional data.

Still, most of the aforementioned phonetic pattern-dependent emotion classification techniques used forced alignment or manual annotation for the extraction of the phoneme borders. Just a few techniques faced real-life conditions by using automatic speech recognition (ASR) engines for the generation of phoneme alignments. Current ASR techniques, however, are not able to provide phoneme alignment on affective speech samples in quality comparable to that of manual phonetic transcription or alignment obtained with forced alignment. In order to exploit the advantages of ASR techniques and to meet real-life conditions, a phoneme-level emotion processing technique should use modified ASR methods for the phoneme time alignment. To be able to obtain the best possible phoneme alignment within real-world development tasks, we used an ASR system with acoustic models adapted on emotional speech samples.

For our experiments we applied a low-level feature modeling on a frame level for acoustic emotion recognition. The HMMs with Gaussian mixture models (GMMs) have been used for this purpose. Three different segments can be used for dynamic analysis: *utterance*, *chunk,* and *phoneme* (cf. [3]). We applied utterance- and phoneme-level analysis for our experiments. It is also possible to classify emotions with an average formant's value extracted from vowel segments. The phoneme boundaries' estimation is based on a *forced alignment*, provided by the Hidden Markov Toolkit (HTK) [62]. Within our experiments we use a simplified version of a BAS SAMPA with a set of 39 phonemes (18 vowels and 21 consonants). A list of emotion-indicative vowels with their corresponding instance number is given in [57].

The time evolution of emotions is another central question. We apply a state transition model for this purpose. Instead of the standard ASR task to deduce the most likely word sequence hypothesis Ω_k from a given acoustic vector sequence $\mathbf{O}$ of M observations $\mathbf{o}$, we recognize the speaker's emotional state. This is solved with standard Bayes' ASR recognition criteria, with a different argument interpretation: $P(\mathbf{O}|\Omega)$ is called the emotion acoustic model, $P(\Omega)$ is the prior user-behavior information and Ω is one of all system-known emotions.

20.3 Utilized Datasets of Emotional Speech

In this section, we shortly describe all datasets used in our experiments to be reported later. A broader overview of emotional speech databases can be found in the following survey articles [36] as well as [55]. Some important details are given in Table 20.1.

The Berlin Database of Emotional Speech (emoDB) [11] is one of the most common emotional acoustic databases. This corpus contains studio-recorded emotionally neutral German sentences for seven affective states and contains 494 phrases with a total length up to 20 min. The content is pre-defined and spoken by ten (five male, five female) actors. The age of the actors is in the range of 21–35.

The Vera am Mittag audio-visual emotional speech database (VAM) [22] contains spontaneous and unscripted discussions between two to five persons from a German talk show. The labeling uses Self-Assessment Manikins (cf. [34]). The recordings cover low and high expressive emotional speech, due to the nature of the origin as TV talk-show. This database contains 947 sentences derived with a total length of 47 min. The age of participants ranges from 16 to 58 years.

The LAST MINUTE CORPUS (LMC) (cf. [37, 38]) contains synchronous audio and video recordings in a so-called Wizard-of-Oz (WoZ) experiment including 130 participants with nearly 56 h. For our experiments, we selected those 79 participants with best signal-to-noise ratio, having 31 min of audio material. During the dialogue critical events are induced that could lead to a dialogue-break-off [20]. We focus on two key events: *baseline* (BSL) and *weight limit barrier* (WLB). A detailed description can be found in Chaps. 13 and 14.

The EmoRec corpus (EmoRec) [60] simulates a natural verbal human-computer interaction, also implemented as a WoZ experiment. The design of the trainer followed the principle of the popular game "Concentration". The procedure of emotion induction included differentiated experimental sequences during which the user passed through specific valence/arousal/dominance (PAD) [10] octants in a

Table 20.1 Overview of selected emotional speech corpora

Name	Emotions	HH:MM	# Speaker
Acted emotions			
emoDB [11]	Anger boredom disgust fear happiness neutral sadness	00:22	10
Excerpts of human-human interaction			
VAM [22]	Values of arousal and valence	00:48	47
UAH [12]	Anger boredom doubt neutral	02:30	60
Naturalistic interaction			
SAL [32]	Continuous traces	10:00	20
EmoRec [60]	Four quadrants of valence-arousal space	33:00	100
LMC [38]	Four dialogue barriers	56:00	130
EmoGest [5]	Happy, neutral, sad	12:00	32

controlled fashion. In our experiment, we investigate only ES-2, which is assumed to be positive, and ES-5, which is negative.

The UAH emotional speech corpus (UAH) [12] contains 85 dialogues from a telephone-based information system spoken in Andalusian dialect from 60 different users. They used four emotional terms to discern emotions. The annotation process was conducted by nine labelers assessing complete utterances.

The Belfast Sensitive Artificial Listener corpus (SAL) is built from emotionally colored multimodal conversations. With four different operator behaviors, the scenario is designed to evoke emotional reactions. To obtain annotations, trace-style continuous ratings were made on five core dimensions (valence, activation, power, expectation, overall emotional intensity) utilizing FEELTRACE [15]. The number of labelers varied between two and six.

The EmoGest corpus (EmoGest) [5] consists of audio and video recordings as well as Kinect data based on a linguistic experiment which consisted of two experimental phases: a musical emotion induction procedure and a gesture-eliciting task in dialogical interaction with a confederate partner. In total, the corpus provides roughly 12 h of multimodal material from 32 participants.

20.4 Ground Truth, Adaptation and Non-verbals in Emotion Recognition

20.4.1 Generating and Measuring a Reliable Ground Truth for Emotion Recognition from Speech

As stated in Sect. 20.1, finding appropriate emotional labels for spoken expressions in naturalistic interactions is a challenging issue. Thus, besides the support of the labelers with suitable emotional annotation tools like *ikannotate* [7] or ATLAS [33] (cf. Chap. 19), valid and well-founded emotion-labeling methods have also to be utilized (cf. [14, 21]).

The research community started with rather small datasets containing acted, studio-recorded, non-interactional, high-quality emotions like in emoDB [11] which are related to the early days of speech corpora generation. The next step in data collection was the emotional inducement. Emotional stimuli were presented to or induced in a subject, whose reactions were recorded [55]. For this, in (HCI) a Wizard-of-Oz setup is often used, providing optimal conditions to influence the recorded user (cf. e.g. Chap. 13). Therefore, the data can be directly used to analyze human reactions while interacting with a technical system [59, 60]. Further, an overview on emotional classes in selected corpora is given in Table 20.1.

Various emotional labels are used in different corpora, situations, tasks, and setups [12, 35]. The emotion recognition community is aware of these difficulties and in [47, 50] a comparative study is conducted. The authors stated that for naturalistic interactions the emotion labels are generated by a quite complex and

Table 20.2 Qualitative assessment of labeling qualities for different labeling methods on a 5-item scale in the range of −− to ++

Method	Usability	Emotion coverage	Label reproducibility
Basic emotions	++	−−	+
GEW	−	++	++
SAM	+	+	−−

++: valid to high degree, −−: not valid at all

time-consuming annotation process. Such an annotation should cover the full range of observed emotions and, further, be proper for the labeling process.

Which labels are needed for labeling emotions in speech? To answer this question, three mainly used emotion assessment methods were compared (cf. [46]), namely Basic Emotions [17], Geneva Emotion Wheel (GEW) [39], and self-assessment manikins (SAMs) [22], and additionally questionnaires assessing the methods were applied. Comparative results are shown in Table 20.2. The main results can be summarized as follows (cf. [46]): Basic Emotions are not sufficient for emotional labeling of spontaneous speech since more variations are observable than covered by Basic Emotions. SAMs are able to cover these variations. On the other hand, labelers have to identify the three values of valence, arousal and dominance and, further, non-trained labelers have difficulties identifying valence or dominance only from speech. Siegert et al. suggest using GEW, which provides a mapping of Basic Emotions into a subset of the (valence-)arousal-dominance space and thus a possible clustering (cf. [46]). Labelers could cover nearly all variations with GEW, which can become quite complex, as the annotator has to chose 1 of 17 emotion families, each with five graduations of intensity. Though evaluated labeling tools exist, the selection of a proper labeling method highly depends on the established system and on the scenario. Although this comparison is not complete, as several other emotion assessment methods and tools exist, it gives a first advice which methods to prefer.

To assess the annotation and the appropriateness of the methods themselves, measures which allow a statement concerning the correctness of the found phenomena and, thus, the reliability, should be investigated. Since reliability values are usually low regarding naturalistic speech, a new interpretation is needed (cf. [50]). For this purpose, the inter-rater-reliability (IRR) is a good measure where the general ideas are presented in [2]. In particular, inter-rater reliability determines the extent to which two or more coders obtain the same result when measuring a certain object, called agreement.

A good reliability measure must fulfill the demands of stability, reproducibility, and accuracy, where reproducibility is the strongest demand. To calculate the IRR, mostly a kappa-like statistic is used. Siegert et al. [50] decided on Krippendorff's alpha since this is generally more applicable than the κ statistics, but the same scheme to interpret the values can be used. Further, it has the advantage of incorporating several distance metrics (ordinal, nominal, ...) for the utilized labels.

Table 20.3 Comparison of different agreement interpretations of kappa-like coefficients utilized in content analysis

Agreement	Landis and Koch [27]	Altmann [1]	Fleiss [19]	Krippendorff [26]
Poor	>0	>0.2	>0.4	>0.66
Slight	0.0–0.2	–	–	–
Fair	0.2–0.4	0.2–0.4	–	–
Moderate/Good	0.4–0.6	0.4–0.6	0.4–0.75	0.67–0.8
Substantial	0.6–0.8	0.6–0.8	–	–
Excellent/Very good	>0.8	>0.8	>0.75	>0.8

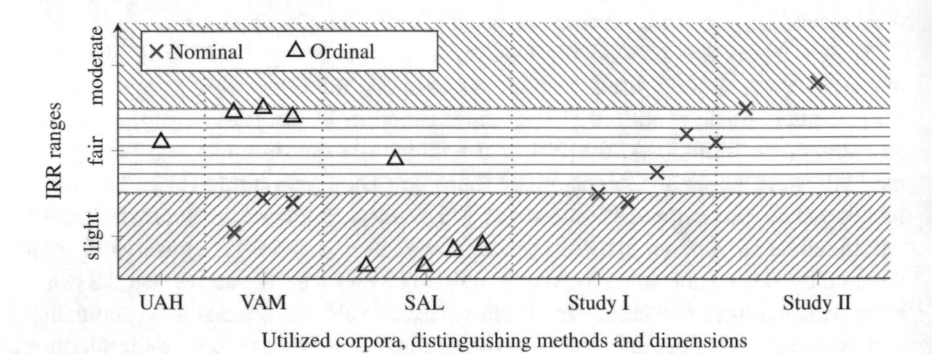

Fig. 20.1 Compilation of IIRs reported in [50], plotted against the agreement interpretation by Landis and Koch [27]

Table 20.3 presents the interpretation schema for IRR, which is widely accepted. Comparing the IRRs for the presented corpora, we notice that for all annotation methods and types of material, the reported reliabilities are far away from the values regarded as reliable. Even well-known and widely used corpora like VAM and SAL reveal a low inter-rater agreement. In particular, nominal alpha is between 0.01 and 0.34. Also, an ordinal metric increased alpha only up to 0.48 at best. Both cases are interpreted as a slight to fair reliability (cf. Table 20.3). Furthermore, comparing the three different annotation methods shows that the methods themselves only have a small impact on the reliability value. Even the use of additional information did not increase the reliability (cf. [50]). The quite low IRR values are due to the subjective nature of emotion perception and emphasize the need for well-trained labelers.

In [50] four corpora, namely UAH, VAM, SAL, and LMC (cf. Table 20.1), are investigated in terms of IRR, considering the previously mentioned annotation paradigms. Furthermore, two approaches are presented to increase the reliability values on LMC. At first, both audio and video recordings of the interaction (denoted as Study I in Fig. 20.1) as well as the natural time order of the interaction (denoted as Study II in Fig. 20.1) were used to increase the reliability. Also, training the annotators with preselected material avoided the Kappa paradoxes [13, 18] and, further, improved the IRR.

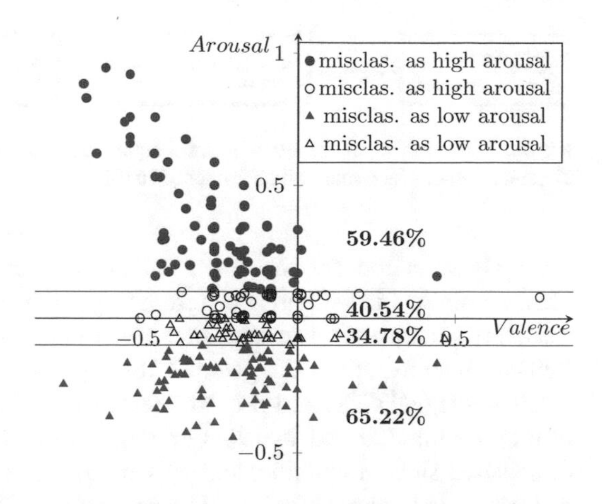

Fig. 20.2 Misclassified emotional instances for a two-class emotion classification task. *Filled markers* represent instances with arousal level in ranges $[-1, -0.1)$ and $(0.1, 1]$. The plot is adapted from [56]

During the annotation, human labelers tend to a subjective view on the material. This results in a variance in the agreement of samples into classes. Utilizing learning methods, such variance can be identified and, further, a proper number of classes can be derived. In several studies, subspaces of the valence-arousal-dominance (VAD) space were considered (cf. e.g. [43]). In [56] a two-class emotion classification was applied to suggest a grouping of emotional samples in a valence-arousal space. The optimal classification performance on emoDB was obtained with 31 GMMs. To establish a cross-corpus grouping of emotions the classifier was tested on VAM, which results in three classes in terms of the valence-arousal space, namely high-arousal, neutral, and low-arousal (cf. [56]). The applied measure is the number of misclassified emotional instances in VAM utilizing the emoDB models. As shown in Fig. 20.2 about 40.54% of misrecognized low-arousal instances and about 34.78% of misrecognized high-arousal samples are located in the range (−0.1, 0.1) in the arousal dimension. Therefore, a third class which covers the emotionally neutral arousal instances should be introduced. This approach improves the recognition performance in a densely populated arousal subspace by about 2.7% absolute.

Up to this point, we have discussed whether and how the annotation method influences the reliability of the labeling. Further, we proposed a grouping of emotions based on cross-corpora evaluation of emoDB and VAM which provides an approach to handle observed data in the beginning of an interaction (cf. [56]). On the other hand, as stated in [9], the annotation process as such is of interest and, thus, should be considered to be objectified.

We know that emotions are usually expressed by multiple modalities like speech, facial expressions, and gestures. Thus, an annotation system, which will work in a semi-automatic fashion, can rely on a large amount of data that argues for an efficient way in the annotation. Therefore, Böck et al. [9] presented an approach towards semi-automatic annotation in a multimodal environment. The main idea is as follows: Automatic audio analyses are used to identify relevant affective sequences which are aligned with the corresponding video material. This indicates

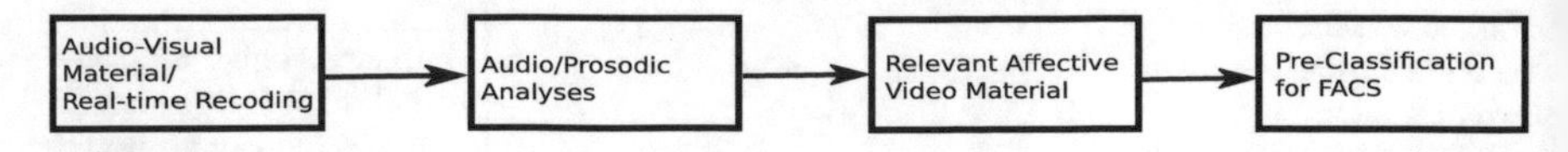

Fig. 20.3 Workflow to establish a semi-automatic annotation based on an audio-based pre-classification of video material as proposed in [9]

a pre-classification for the facial annotation (cf. Fig. 20.3). Since each person shows emotions in (slightly) different ways, the utilized audio features should be relatively general, so that a wide range of domains and audio conditions are covered. Suitable features are investigated in [6, 31], resulting in Mel-Frequency-Cepstral-Coefficients (MFCCs) and prosodic features. A GMM-based identification method is proposed and tested in [9]. For this, an objective way of annotation can be established since a classification system does not tend to interpret user reactions differently in several situations. Further, the approach reduces the manual effort as human annotators are just asked to label debatable sequences. This approach can also be used to preselect emotional material for the improvement of IRR.

20.4.2 User-Specific Adaptation for Speech-Based Emotion Recognition

An issue that has only been rarely investigated is the user-specific adaptation for speech-based emotion recognition to improve the corresponding emotional modeling [52]. For speech-based emotion recognition, an adaptation onto the speaker's biological gender has been used, for instance, in [16]. The authors utilized a gender-specific UBM-MAP approach to increase the recognition performance, but did not apply their methods on several corpora. In [58] a gender differentiation is used to improve the automatic emotion recognition from speech. The authors achieved an absolute difference of approx. 3% between the usage of correct and automatically recognized gender information.

All these publications only investigate the rather obvious gender-dependency. No other factors such as, for instance, age are considered, although it is known that age has an impact on both, the vocal characteristics (cf. [23]) and the emotional response (cf. [30]). It has not been investigated whether the previously mentioned improvements are dependent on the utilized material, as most of these studies are conducted on databases of acted emotions. Thus, a proof of whether these methods are suitable for natural interactions is still missing.

Therefore, in our experiments (cf. [48, 52]), we investigated whether a combination of age and gender can further improve the classification performance. As the analyses were conducted on different corpora, comparatively general statements can be derived. Being able to compare our results on emoDB and VAM, we used the two-class emotional set generated by Schuller et al. [42]. They defined combinations of emotional classes to cluster into `low arousal (A-)` and `high arousal`

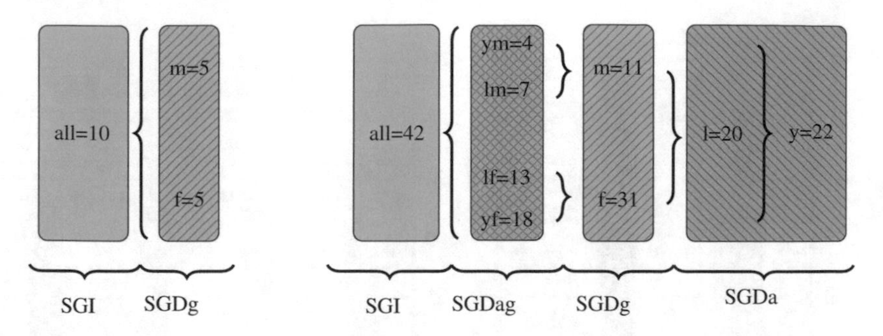

Fig. 20.4 Distribution of speaker groupings and their abbreviations on emoDB (*left*) and VAM (*right*). *SGI* speaker group independent, *SGD* speaker group dependent, *a* age, *g* gender

(A+) for several datasets (cf. Sect. 20.3). The different age-gender groupings together with the number of corresponding speakers are depicted in Fig. 20.4. The combination of both grouping factors led to sub-groups according to the speakers' gender: **m**ale vs. **f**emale; according to their age: mid-***l***ife vs. **y**oung speakers; and combinations of both: i.e. **y**oung **m**ale speakers.

The following acoustic characteristics are utilized as features: 12 mel-frequency cepstral coefficients, zeroth cepstral coefficient, fundamental frequency, and energy. The Δ and $\Delta\Delta$ regression coefficients of all features are used to include contextual information. As channel normalization technique, relative spectral (RASTA)-filtering is applied. GMMs with 120 mixture components utilizing four iteration steps are used as classifiers. For validation we use a leave-one-speaker-out (LOSO) strategy. As performance measure, the unweighted average recall (UAR) is applied. The UAR indicated the averaged recall taking into account the recognition results for each class independently. More details about the parameter optimization can be found in [52]. Afterwards, we performed the experiments on the SGI set as well as the SGDa, SGDg, and SGDag sets. For this, the subjects are grouped according to their age and gender in order to train the corresponding classifiers in a LOSO manner. To allow a comparison between all speaker groupings, we combined the different results. For instance, the results for each male and female speaker are merged to obtain the overall result for the SGDg set. This result can be directly compared with results gained on the SGI set. The outcome is shown in Fig. 20.5.

The SGD results on a two-class problem are outperforming the classification result of 96.8% from [42]. In comparison to the SGI results the SGD classifiers achieved an absolute improvement of approx. 4%. This improvement is significant ($F = 4.48238$, $p = 0.0281$). Utilizing VAM allows us to examine a grouping of the speakers on different age ranges, namely young adults (y) and mid-life adults (l). These groupings comprise the speakers' age (SGDa), the speakers' gender (SGDg), and the combination of both (SGDag); see Fig. 20.5. For all three combinations, a substantial improvement was achieved in comparison to the BSL classification (SGI). Unfortunately, SGDa and SGDag achieve lower results than the classification using only the gender characteristic. This is mostly caused by the

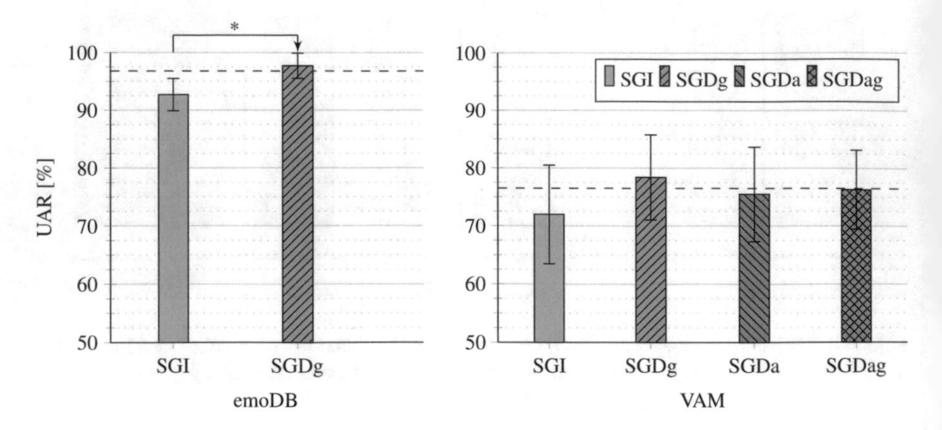

Fig. 20.5 UARs in percent for the two-class problem on emoDB and VAM comparing SGI and SGDg utilizing LOSO validation on different feature sets. For comparison, the best results from [42] are marked with a *dashed line*. The *star* denotes the significance level: * ($p < 0.05$)

declined performance for the m group. It has to be further investigated whether this can be attributed to the small amount of material available or to the fact that the present acoustical differences within the mid-life adults are larger than those in the young adults' group (cf. [52]).

If we take into account datasets with high differences in the age grouping, we can observe that the SGDag grouping outperforms the results of the other models. More details about this investigation can be found in Chap. 14. In general, it can be stated that for the investigated groups of young adults and mid-life adults the gender grouping is the dominant factor. Taking also corpora into account, where a high age difference can be observed, a combination of age and gender groups is needed.

In addition to the speaker-dependent characteristics like age and gender, we also see that a combination of several sources of information could provide a gain in the classification performance of emotions (cf. Chap. 19). Therefore, we conducted another study (cf. [8]). In this experiment, we prove the performance of chosen modalities in emotion recognition. The most prominent modalities are speech, facial expression, and, as they are speaker-dependent, biopsychological characteristics like skin conductance, heart rate, etc. In this case, a comparison of intraindividual against interindividual classification of emotions is highly recommended; cf. Table 20.4. Intraindividual analysis means that only the data of a particular speaker is analyzed and used for both training and test. In contrast, interindividual analysis directly compares the performance of a classifier training on all material except this of a certain person against the material of this particular speaker (test material). For the experiments, we need a corpus which provides material for all three modalities and, thus, allows a comparative study. Therefore, the EmoRec data set was used (cf. Sect. 20.3). Indeed, these evaluations are necessary to investigate the effect of personalization. Using biopsychological features for classification, a calibration of the features is necessary since the intraindividual characteristics of each speaker

Table 20.4 Recognition rates on the EmoRec Corpus for intraindividual and interindividual classification of ES-2 vs. ES-5 in percent with HTK, taken from [8]

Subject	*1*	*2*	*3*	*4*	*5*	*6*	*7*	*8*	*9*	*10*
ES-2 (**intra**individual)	50.0	50.0	38.5	50.0	50.0	65.0	68.4	71.4	100.0	81.3
ES-5 (**intra**individual)	66.7	75.0	75.0	70.8	84.6	70.0	84.0	100.0	100.0	63.2
ES-2 (**inter**individual)	84.0	66.7	88.5	74.2	82.1	86.2	73.1	72.4	90.5	50.0
ES-5 (**inter**individual)	33.3	36.5	3.8	34.0	22.6	24.3	26.4	86.7	13.3	63.2
Subject	*11*	*12*	*13*	*14*	*15*	*16*	*17*	*18*	*19*	*20*
ES-2 (**intra**individual)	100.0	100.0	80.0	43.9	68.8	58.5	40.0	40.6	70.0	61.1
ES-5 (**intra**individual)	93.3	72.5	55.8	59.1	68.8	83.8	65.0	81.6	68.0	59.1
ES-2 (**inter**individual)	21.4	28.1	40.6	14.3	55.2	31.2	97.0	79.3	10.7	54.5
ES-5 (**inter**individual)	83.8	85.0	75.0	97.7	66.7	71.4	9.3	16.7	73.9	80.0

influence the features as such. This means that for a particular person the heart rate corresponding to an emotion can be higher rather in a calm situation than for another person. Unfortunately, the calibration is difficult due to the unknown baseline of emotional behavior for each user; this means obtaining a value for each feature which represents a neutral emotion. Therefore, the idea is to use a classifier based on other modalities to provide an emotional rating which can be used to calibrate the biopsychological observations. In the experiment (cf. [8]), two emotional situations—for short positive and negative (for details, cf., e.g., [8])—were distinguished that are given by the design of the data recording scenario (cf. [60]). For this, no external annotation was conducted.

In the case of audio classification we found that intraindividual and interindividual systems provided good results. Since we compared our analyses to multimodal investigations (cf. [8]), we can state: Biopsychological classifiers, for instance neural networks (cf. [61]), in particular, if they are calibrated (details cf. [60]), showed, in general, recognition accuracies of more than 75%. Comparable results were gained by video analysis. Given these results, a personalization process for a technical system should be based on interindividual approaches, first using material which is clearly detected to adapt the system, and finally adapting classifiers which are adjusted to a certain user. This is guiding us to several issues: Which modality will influence the phases of the system adaptation? Which is the best combination of modalities in information fusion? Such aspects are discussed in Chap. 19.

20.4.3 Combination of Acoustic and Linguistic Features for Emotion Recognition

In this section, we compare emotion recognition of acoustic features containing both spectral and prosodic information combined with the linguistic bag-of-words features.

Several research groups have used two feature sets defined in the following (cf. Table 20.5): The first set consists of 12 MFCCs and the zeroth coefficient with corresponding Delta and Acceleration coefficients. The first three formants, the corresponding bandwidths, intensity, jitter, and pitch were accompanied into the acoustic feature vector. The second acoustic feature set is a subset of the first and is focused on spectral features only, namely MFCC (cf. MFCC_0_D_A in Table 20.5). Both feature sets have recently been heavily applied in the research community (cf. [4, 41]). Furthermore, they are also applied in classification of naturalistic and/or spontaneous emotion recognition from speech (cf. e.g. [9]).

The GMMs provided by HTK [62] were trained for all feature sets given in Table 20.5. Notice that for each emotional class, in particular, happy and sad, a separate GMM was generated, representing the characteristics of the class, and a final decision was established by passing through the models using the Viterbi decoding algorithm. The classifiers with nine (cf. [9]), 81 (cf. [42]), and 120 (cf. [52]) Gaussian mixtures gained the best performance on the EmoGest corpus.

Table 20.6 shows that the achieved classification performance has two maxima. Employing a LOSO strategy, the ability to classify emotions on the speaker-independent level can be shown. In particular, remarkable results were obtained on the full feature set (i.e., MFCC_0_D_A_F_B_I_J_P) with 0.785 recall (cf. Table 20.6). For the classifier with 120 Gaussian mixtures the variance has also

Table 20.5 The two feature sets (with total number of features) used in the experiments with the corresponding applied features

Feature set	Number	Applied features
MFCC_0_D_A	39	MFCC, zeroth cepstral coefficient, delta, acceleration
MFCC_0_D_A_F_B_I_J_P	48	MFCC_0_D_A, formants 1–3, bandwidths, intensity, jitter, pitch

Table 20.6 Experimental results for the two acoustic features sets (cf. Table 20.5 on page 422) employing GMMs with different numbers of mixtures (#mix)

Feature set	#mix	Recall	Variance
MFCC_0_D_A	9	0.822	0.375
	81	0.826	0.380
	120	0.822	0.375
MFCC_0_D_A_F_B_I_J_P	9	0.755	0.324
	81	0.782	0.327
	120	0.785	0.322

Recall with corresponding variance is given

its minimum. Since we oriented ourselves on the experiment presented in [52], we have not tested more than 120 mixtures, which was the maximum number utilized.

One determines that the spectral feature set outperforms the combined feature set. The recall rates are more than (absolute) 4% higher, staying with comparable variance rates. We conclude that spectral features can better distinguish or cover the two emotions sad and happy. This leads to the discussion about which acoustic features are the most meaningful ones for emotions [6, 10, 40]. In [44], the authors state that, usually, optimal feature sets are highly dependent on the evaluated dataset.

A further application of feature selection is to automatically find potential significant dialog turns. For the recognition of WLB events we analyze acoustic, linguistic and non-linguistic content with a static classification setup. We investigated several feature sets: a full set consisting of 54 acoustic and 124 linguistic features, two reduced sets focusing on the most relevant features, and four sets corresponding to a different combination of linguistic and acoustic features. The proposed classification techniques were evaluated on the LAST MINUTE corpus. The dataset provides emotional instances for WLB and BSL classes. An unweighted average recall of 0.83 was achieved by using the full feature set.

Using Bag-of-words (BoW) features usually leads to a high-dimensional sparse feature vector, as only a limited vocabulary is used in the utterances we are interested in. We pre-selected only those words which appear in at least three turns over the database. Finally, this resulted in 1242 BoW features. Therefore we employed a feature ranking method to select the most informative ones. Using WEKA, we employed an information gain attribute evaluator in conjunction with a feature ranker as the search method. The top 100 BoW features selected were added to the combined (acoustic, linguistic) feature set described above, giving us 278 features in total. The resulting feature vector is referenced hereinafter as *full*. In addition to this, we constructed two further feature vectors, denoted by *top 100* and *top 50*, corresponding to the top 100 and top 50 meaningful features from the full set.

In Table 20.7 our experimental results are presented. The evaluation has been conducted using tenfold speaker-independent cross-validation. Classification results are slightly better for the BSL class, which reflects the grouped material belonging to the barrier called BSL, denoting the experiment's part where the first excitement has been gone (cf. [20]). In terms of the feature selection, the top 100 features provide only 1–2% absolute improvement compared to the full feature set; however, the differences are not significant.

Among 50 (from *top 100* to *top 50*) reduced features, 45 belong to BoW. Such reduction is accompanied with a severe performance degradation, which suggests that BoW features contribute much to the final recognition result.

Therefore, these results do not answer the question of the development of an optimal feature set, which goes beyond the scope of the current research.

Table 20.7 Classification performance for speaker-independent evaluation on the LAST MINUTE corpus

Feature set	Full 178	Top 100	Top 50
General results			
Weighted avg recall	0.86	0.86	0.81
UAR	0.83	0.84	0.76
Weighted F-score	0.86	0.86	0.80
Class WLB			
Precision	0.77	0.79	0.72
Recall	0.77	0.77	0.62
F-score	0.77	0.78	0.67
Class BSL			
Precision	0.90	0.89	0.84
Recall	0.89	0.91	0.89
F-score	0.90	0.90	0.86

Full 178, Top 100, Top 50 feature set

20.5 Conclusion and Outlook

We have considered emotions and other interaction patterns in spoken language, determining the nature and quality of the users' interaction with a machine in the wider sense of application in *Companion*-Systems. Acted and naturalistic spoken data in operational form (corpora) have been used for the development of emotion classification; we have addressed the difficulties arising from the variety of these data sources. A dedicated annotation process to label emotions and to describe their temporal evolution has been considered, and its quality and reliability have been evaluated. The novelty here is the consideration of naturalistic interactions, where the emotional labels are no longer a priori given. We have further investigated appropriate emotional features and classification techniques. Optimized feature sets with combined spectral, prosodic and Bag-of-Words components have been derived. Speaker clustering and speaker adaptation have been shown to improve the emotional modeling, where detailed analyses on gender and age dependencies have been given.

In summary, it can be said that acoustical emotion recognition techniques have been considerably advanced and tailored for naturalistic environments, approaching a *Companion*-like scenario.

In the future, emotional personalization will be the next stage, following the presented speaker grouping. Also, the further incorporation of established methods in psychology will improve the labeling and annotation process towards higher reliability and ease-of-use for annotators. A combination of the acoustical affective evaluation and the interpretation of non-verbal interaction patterns will lead to a better understanding of and reaction to user-specific emotional behavior (cf. [49, 51, 53]). The meaning and usage of acoustic emotion, disposition and intention in the general human-machine context will be further investigated.

Acknowledgements This work was done within the Transregional Collaborative Research Centre SFB/TRR 62 "*Companion*-Technology for Cognitive Technical Systems" funded by the German Research Foundation (DFG).

References

1. Altman, D.G.: Practical Statistics for Medical Research. Chapman & Hall, London (1991)
2. Artstein, R., Poesio, M.: Inter-coder agreement for computational linguistics. Comput. Linguist. **34**, 555–596 (2008)
3. Batliner, A., Seppi, D., Steidl, S., Schuller, B.: Segmenting into adequate units for automatic recognition of emotion-related episodes: a speech-based approach. Adv. Hum. Comput. Interact. **2010**, 15 (2010)
4. Batliner, A., Steidl, S., Schuller, B., Seppi, D., Vogt, T., Wagner, J., Devillers, L., Vidrascu, L., Aharonson, V., Kessous, L., Amir, N.: Whodunnit – searching for the most important feature types signalling emotion-related user states in speech. Comput. Speech Lang. **25**, 4–28 (2011)
5. Bergmann, K., Böck, R., Jaecks, P.: Emogest: investigating the impact of emotions on spontaneous co-speech gestures. In: Proceedings of the Workshop on Multimodal Corpora 2014, pp. 13–16. LREC, Reykjavik (2014)
6. Böck, R., Hübner, D., Wendemuth, A.: Determining optimal signal features and parameters for HMM-based emotion classification. In: Proceedings of the 15th IEEE MELECON, Valletta, Malta, pp. 1586–1590 (2010)
7. Böck, R., Siegert, I., Vlasenko, B., Wendemuth, A., Haase, M., Lange, J.: A processing tool for emotionally coloured speech. In: Proceedings of the 2011 IEEE ICME, p. s.p, Barcelona (2011)
8. Böck, R., Limbrecht, K., Walter, S., Hrabal, D., Traue, H., Glüge, S., Wendemuth, A.: Intraindividual and interindividual multimodal emotion analyses in human-machine-interaction. In: Proceedings of the IEEE CogSIMA, New Orleans, pp. 59–64 (2012)
9. Böck, R., Limbrecht-Ecklundt, K., Siegert, I., Walter, S., Wendemuth, A.: Audio-based pre-classification for semi-automatic facial expression coding. In: Kurosu, M. (ed.) Human-Computer Interaction. Towards Intelligent and Implicit Interaction. Lecture Notes in Computer Science, vol. 8008, pp. 301–309. Springer, Berlin/Heidelberg (2013)
10. Böck, R., Bergmann, K., Jaecks, P.: Disposition recognition from spontaneous speech towards a combination with co-speech gestures. In: Böck, R., Bonin, F., Campbell, N., Poppe, R. (eds.) Multimodal Analyses Enabling Artificial Agents in Human-Machine Interaction. Lecture Notes in Artificial Intelligence, vol. 8757, pp. 57–66. Springer, Cham (2015)
11. Burkhardt, F., Paeschke, A., Rolfes, M., Sendlmeier, W., Weiss, B.: A database of German emotional speech. In: Proceedings of the INTERSPEECH-2005, Lisbon, pp. 1517–1520 (2005)
12. Callejas, Z., López-Cózar, R.: Influence of contextual information in emotion annotation for spoken dialogue systems. Speech Comm. **50**, 416–433 (2008)
13. Cicchetti, D., Feinstein, A.: High agreement but low kappa: II. Resolving the paradoxes. J. Clin. Epidemiol. **43**, 551–558 (1990)
14. Cowie, R., Cornelius, R.R.: Describing the emotional states that are expressed in speech. Speech Comm. **40**, 5–32 (2003)
15. Cowie, R., Douglas-Cowie, E., Savvidou, S., McMahon, E., Sawey, M., Schröder, M.: FEELTRACE: an instrument for recording perceived emotion in real time. In: Proceedings of the SpeechEmotion-2000, Newcastle, pp. 19–24 (2000)
16. Dobrišek, S., Gajšek, R., Mihelič, F., Pavešić, N., Štruc, V.: Towards efficient multi-modal emotion recognition. Int. J. Adv. Robot. Syst. **10**, 1–10 (2013)
17. Ekman, P.: Are there basic emotions? Psychol. Rev. **99**, 550–553 (1992)

18. Feinstein, A., Cicchetti, D.: High agreement but low kappa: I. The problems of two paradoxes. J. Clin. Epidemiol. **43**, 543–549 (1990)
19. Fleiss, J.: Measuring nominal scale agreement among many raters. Psychol. Bull. **76**, 378–382 (1971)
20. Frommer, J., Rösner, D., Haase, M., Lange, J., Friesen, R., Otto, M.: Detection and Avoidance of Failures in Dialogues – Wizard of Oz Experiment Operator's Manual. Pabst Science Publishers, Lengerich (2012)
21. Grimm, M., Kroschel, K., Mower, E., Narayanan, S.: Primitives-based evaluation and estimation of emotions in speech. Speech Comm. **49**, 787–800 (2007)
22. Grimm, M., Kroschel, K., Narayanan, S.: The Vera am Mittag German audio-visual emotional speech database. In: Proceedings of the 2008 IEEE ICME, Hannover, pp. 865–868 (2008)
23. Harrington, J., Palethorpe, S., Watson, C.: Age-related changes in fundamental frequency and formants: a longitudinal study of four speakers. In: Proceedings of the INTERSPEECH-2007, Antwerp, vol. 2, pp. 1081–1084 (2007)
24. Iliou, T., Anagnostopoulos, C.N.: Comparison of different classifiers for emotion recognition. In: Proceedings of the Panhellenic Conference on Informatics, pp. 102–106 (2009)
25. Kelly, F., Harte, N.: Effects of long-term ageing on speaker verification. In: Vielhauer, C., Dittmann, J., Drygajlo, A., Juul, N., Fairhurst, M. (eds.) Biometrics and ID Management. Lecture Notes in Computer Science, vol. 6583, pp. 113–124. Springer, Berlin/Heidelberg (2011)
26. Krippendorff, K.: Content Analysis: An Introduction to Its Methodology, 3rd edn. SAGE, Thousand Oaks (2012)
27. Landis, J.R., Koch, G.G.: The measurement of observer agreement for categorical data. Biometrics **33**, 159–174 (1977)
28. Lee, C.M., Yildirim, S., Bulut, M., Kazemzadeh, A., Busso, C., Deng, Z., Lee, S., Narayanan, S.: Emotion recognition based on phoneme classes. In: Proceedings of the INTERSPEECH 2004, Jeju Island, pp. 889–892 (2004)
29. Lee, C., Busso, C., Lee, S., Narayanan, S.: Modeling mutual influence of interlocutor emotion states in dyadic spoken interactions. In: Proceedings of the INTERSPEECH 2009, pp. 1983–1986 (2009)
30. Lipovčan, L., Prizmić, Z., Franc, R.: Age and gender differences in affect regulation strategies. Drustvena istrazivanja: J. Gen. Soc. Issues **18**, 1075–1088 (2009)
31. Maganti, H.K., Scherer, S., Palm, G.: A novel feature for emotion recognition in voice based applications. In: Affective Computing and Intelligent Interaction, pp. 710–711. Springer, Berlin/Heidelberg (2007)
32. McKeown, G., Valstar, M., Cowie, R., Pantic, M., Schroder, M.: The SEMAINE database: annotated multimodal records of emotionally colored conversations between a person and a limited agent. IEEE Trans. Affect. Comput. **3**, 5–17 (2012)
33. Meudt, S., Bigalke, L., Schwenker, F.: ATLAS – an annotation tool for HCI data utilizing machine learning methods. In: Proceedings of the 1st APD, San Francisco, pp. 5347–5352 (2012)
34. Morris, J.D.: SAM: the self-assessment manikin an efficient cross-cultural measurement of emotional response. J. Adv. Res. **35**, 63–68 (1995)
35. Palm, G., Glodek, M.: Towards emotion recognition in human computer interaction. In: Neural nets and surroundings, pp. 323–336. Springer, Berlin/Heidelberg (2013)
36. Pittermann, J., Pittermann, A., Minker, W.: Handling Emotions in Human-Computer Dialogues. Springer, Amsterdam (2010)
37. Prylipko, D., Rösner, D., Siegert, I., Günther, S., Friesen, R., Haase, M., Vlasenko, B., Wendemuth, A.: Analysis of significant dialog events in realistic human–computer interaction. J. Multimodal User Interfaces **8**, 75–86 (2014)
38. Rösner, D., Frommer, J., Friesen, R., Haase, M., Lange, J., Otto, M.: LAST MINUTE: a multimodal corpus of speech-based user-companion interactions. In: Proceedings of the 8th LREC, Istanbul, pp. 96–103 (2012)

39. Scherer, K.R.: Unconscious Processes in Emotion: The Bulk of the Iceberg, pp. 312–334. Guilford Press, New York (2005)
40. Scherer, S., Kane, J., Gobl, C., Schwenker, F.: Investigating fuzzy-input fuzzy-output support vector machines for robust voice quality classification. Comput. Speech Lang. **27**(1), 263–287 (2013)
41. Schuller, B., Steidl, S., Batliner, A.: The INTERSPEECH 2009 emotion challenge. In: Proceedings of the INTERSPEECH-2009, Brighton, pp. 312–315 (2009)
42. Schuller, B., Vlasenko, B., Eyben, F., Rigoll, G., Wendemuth, A.: Acoustic emotion recognition: a benchmark comparison of performances. In: Proceedings of the IEEE ASRU-2009, Merano, pp. 552–557 (2009)
43. Schuller, B., Batliner, A., Steidl, S., Seppi, D.: Recognising realistic emotions and affect in speech: State of the art and lessons learnt from the first challenge. Speech Comm. **53**, 1062–1087 (2011)
44. Schuller, B., Valstar, M., Eyben, F., McKeown, G., Cowie, R., Pantic, M.: AVEC 2011–the first international audio/visual emotion challenge. In: D'Mello, S., Graesser, A., Schuller, B., Martin, J.C. (eds.) Affective Computing and Intelligent Interaction. Lecture Notes in Computer Science, vol. 6975, pp. 415–424. Springer, Berlin/Heidelberg (2011)
45. Shami, M., Verhelst, W.: Automatic classification of emotions in speech using multi-corpora approaches. In: Proceedings of the 2nd IEEE Signal Processing Symposium, Antwerp, pp. 3–6 (2006)
46. Siegert, I., Böck, R., Philippou-Hübner, D., Vlasenko, B., Wendemuth, A.: Appropriate emotional labeling of non-acted speech using basic emotions, Geneva emotion wheel and self assessment manikins. In: Proceedings of the 2011 IEEE ICME, p. s.p, Barcelona (2011)
47. Siegert, I., Böck, R., Wendemuth, A.: The influence of context knowledge for multi-modal affective annotation. In: Kurosu, M. (ed.) Human-Computer Interaction. Towards Intelligent and Implicit Interaction. Lecture Notes in Computer Science , vol. 8008, pp. 381–390. Springer, Berlin/Heidelberg (2013)
48. Siegert, I., Glodek, M., Panning, A., Krell, G., Schwenker, F., Al-Hamadi, A., Wendemuth, A.: Using speaker group dependent modelling to improve fusion of fragmentary classifier decisions. In: Proceedings of 2013 IEEE CYBCONF, Lausanne, pp. 132–137 (2013)
49. Siegert, I., Hartmann, K., Philippou-Hübner, D., Wendemuth, A.: Human behaviour in HCI: complex emotion detection through sparse speech features. In: Salah, A., Hung, H., Aran, O., Gunes, H. (eds.) Human Behavior Understanding. Lecture Notes in Computer Science, vol. 8212, pp. 246–257. Springer, Berlin/Heidelberg (2013)
50. Siegert, I., Böck, R., Wendemuth, A.: Inter-rater reliability for emotion annotation in human-computer interaction – comparison and methodological improvements. J. Multimodal User Interfaces **8**, 17–28 (2014)
51. Siegert, I., Haase, M., Prylipko, D., Wendemuth, A.: Discourse particles and user characteristics in naturalistic human-computer interaction. In: Kurosu, M. (ed.) Human-Computer Interaction. Advanced Interaction Modalities and Techniques. Lecture Notes in Computer Science , vol. 8511, pp. 492–501. Springer, Berlin/Heidelberg (2014)
52. Siegert, I., Philippou-Hübner, D., Hartmann, K., Böck, R., Wendemuth, A.: Investigation of speaker group-dependent modelling for recognition of affective states from speech. Cogn. Comput. **6**(4), 892–913 (2014)
53. Siegert, I., Prylipko, D., Hartmann, K., Böck, R., Wendemuth, A.: Investigating the form-function-relation of the discourse particle "hm" in a naturalistic human-computer interaction. In: Bassis, S., Esposito, A., Morabito, F. (eds.) Recent Advances of Neural Network Models and Applications. Smart Innovation, Systems and Technologies, vol. 26, pp. 387–394. Springer, Berlin/Heidelberg (2014)
54. Strauß, P.M., Hoffmann, H., Minker, W., Neumann, H., Palm, G., Scherer, S., Schwenker, F., Traue, H., Walter, W., Weidenbacher, U.: Wizard-of-oz data collection for perception and interaction in multi-user environments. In: International Conference on Language Resources and Evaluation (LREC) (2006)

55. Ververidis, D., Kotropoulos, C.: Emotional speech recognition: resources, features, and methods. Speech Comm. **48**, 1162–1181 (2006)
56. Vlasenko, B., Wendemuth, A.: Location of an emotionally neutral region in valence-arousal space. Two-class vs. three-class cross corpora emotion recognition evaluations. In: Proceedings of 2014 IEEE ICME (2014)
57. Vlasenko, B., Philippou-Hübner, D., Prylipko, D., Böck, R., Siegert, I., Wendemuth, A.: Vowels formants analysis allows straightforward detection of high arousal emotions. In: Proceedings of 2011 IEEE ICME, Barcelona (2011)
58. Vogt, T., André, E.: Improving automatic emotion recognition from speech via gender differentiation. In: Proceedings of the 5th LREC, p. s.p, Genoa (2006)
59. Wahlster, W. (ed.): SmartKom: Foundations of Multimodal Dialogue Systems. Springer, Heidelberg/Berlin (2006)
60. Walter, S., Scherer, S., Schels, M., Glodek, M., Hrabal, D., Schmidt, M., Böck, R., Limbrecht, K., Traue, H., Schwenker, F.: Multimodal emotion classification in naturalistic user behavior. In: Jacko, J. (ed.) Human-Computer Interaction. Towards Mobile and Intelligent Interaction Environments. Lecture Notes in Computer Science, vol. 6763, pp. 603–611. Springer, Berlin/Heidelberg (2011)
61. Walter, S., Kim, J., Hrabal, D., Crawcour, S., Kessler, H., Traue, H.: Transsituational individual-specific biopsychological classification of emotions. IEEE Trans. Syst. Man Cybern. Syst. Hum. **43**(4), 988–995 (2013)
62. Young, S., Evermann, G., Gales, M., Hasin, T., Kershaw, D., Liu, X., Moore, G., Odell, J., Ollason, D., Povey, D., Valtchev, V., Woodland, P.: The HTK Book (for HTK Version 3.4). Engineering Department, Cambridge University, Cambridge (2009)
63. Zeng, Z., Pantic, M., Roisman, G.I., Huang, T.S.: A survey of affect recognition methods: audio, visual, and spontaneous expressions. IEEE Trans. Pattern Anal. Mach. Intell. **31**, 39–58 (2009)

Chapter 21
Modeling Emotions in Simulated Computer-Mediated Human-Human Interactions in a Virtual Game Environment

Andreas Scheck, Holger Hoffmann, Harald C. Traue, and Henrik Kessler

Abstract Emotions form a major part of humans' day-to-day lives, especially in the areas of communication and interaction with others. They modify our gesture or facial expression and therefore serve as an additional communication channel. Furthermore, they have an impact on decision-making. This has two possible implications for computer science in the field of human-computer-interaction. First, computers should be able to adequately recognize and model human emotions if they genuinely want to help users in applied fields of human-human interactions. Second, a reliable and valid computer model of users' emotions is the basis of effective implementations for human-computer interaction, with the computer thus being able to adapt to users' emotions flexibly in any given application.

From an empirical point of view, though, computerized recognition of human emotions still lacks substantial reliability and validity. In our opinion there are two main reasons for this shortcoming. First, triggers of emotional responses, i.e. eliciting situations, are typically complex in nature and thus difficult to predict or even assess once apparent. Second, the emotional response itself is a complex reaction involving subjects' individual learning history, appraisal, preparedness, bodily reactions, and so forth. Both factors make it difficult for any algorithm to recognize real-life emotions.

In a venture to approach this problem, the main goal of our study is to test an implementation of a computer model (COMPLEX) that predicts user emotions in a simulated human-human interaction. The prediction is supported by an elaborate appraisal model of emotions and the assessment of user bodily reactions, facial expression and speech. This article will give an overview of the theoretical background, the practical implementation of our new approach and first results of an empirical validation.

A. Scheck (✉) • H. Hoffmann • H.C. Traue
Medical Psychology, University of Ulm, Frauensteige 6, 89075 Ulm, Germany
e-mail: andreas.scheck@uni-ulm.de; holger.hoffmann@uni-ulm.de; harald.traue@uni-ulm.de

H. Kessler
University Clinic for Psychosomatic Medicine and Psychotherapy, Bochum, Germany
e-mail: henrik.kessler@ruhr-uni-bochum.de

S. Biundo, A. Wendemuth (eds.), *Companion Technology*, Cognitive Technologies,
DOI 10.1007/978-3-319-43665-4_21

21.1 Introduction

Emotions form a major part of humans' day-to-day lives, especially in the areas of communication and interaction with others. Human-computer interactions on the other hand often lack the emotional aspects of communication. We suggest that in the field of human-computer interaction computers should be able to adequately recognize and model human emotions if they genuinely want to help users and create an effective framework of communication between user and computer. If, for instance, the use is continuously stressed while working with an application, that application could measure stress levels and suggest a break or switch its background colors to a more relaxing tone. Provided with algorithms to model human emotions, the computer thus could be able to adapt to users' emotions flexibly in a given application. From an empirical point of view, though, computerized recognition of human emotions still lacks substantial reliability and validity. In our opinion there are two main reasons for this shortcoming. First, situations eliciting emotional responses in real life ("triggers") are typically complex in nature and thus difficult to predict or even assess once apparent. Second, the emotional response itself is a complex reaction involving, besides other factors, a subject's individual learning history, appraisal, preparedness and bodily reactions. These complex factors make it difficult for any algorithm to recognize or track real-life emotions. In a venture to approach this problem, the main goal of our study is to test an implementation that predicts user emotions in a simulated human-human interaction on an individual basis. The prediction is supported by an elaborate appraisal model of emotions developed by Ortony, Clore and Collins (OCC) [10] and the assessment of user bodily reactions, facial expression and speech. This article will give an overview of the theoretical background and the implementation of our new approach and report findings of our first empirical study testing the model.

The novel aspects of our approach should be outlined as follows. Tackling the first problem, the complexity of triggers, we simulated a scenario of human-human interaction in a computer game with well-defined situations supposed to elicit emotional responses under experimental conditions. Additionally, using OCC as an appraisal-based algorithm to "generate" (i.e. predict) emotions gives us a powerful tool to do justice to the complexity of triggers. Considering the second problem, the complexity of emotional responses, an individual calibration procedure takes place for every user before starting the game. This calibration provides an individual mapping of OCC emotions into the three-dimensional space of the PAD model (pleasure, arousal, dominance [9]), in order to handle and transfer emotional states from one model to the other. Finally, measuring physiological responses (skin conductance, heart rate, respiration), facial expressions and speech parameters helps to disambiguate situations where multiple emotions could be predicted only considering OCC derivations.

For the empirical validation, subjects play a competitive multiplayer-game in which a subject is told to play against other real humans being connected via the Internet In reality these other players are controlled by a Wizard-of-Oz following a protocol to elicit certain emotions within the subject. After individual calibration (mapping of OCC emotions into subjective PAD space) subjects are frequently asked to rate their current emotional state in the PAD space throughout the game. The proximity of this rating to the OCC emotion/PAD position predicted by our algorithm serves as a measure for its validity.

21.2 Emotion Theories

In order to better understand the choices we made to implement our model and experimental setting, a brief overview over psychological emotion theories should be provided. Basically, there are two major types of psychological emotion models: categorical and dimensional models.

Categorical models conceptualize emotions to be single, discrete entities. Typically, these entities can be directly mapped to subjective experience, distinct facial expressions or neural representations. The most common and widespread one was developed by Ekman [3–5]. This model suggests that there are the six basic emotions: anger, disgust, fear, joy, sadness and surprise.

Dimensional models of emotion, on the other hand, typically conceptualize emotions as being a composite of various dimensions reflecting distinct aspects of the emotion itself. The most prominent model uses three dimensions: pleasure, arousal and dominance (PAD) [9]. All emotions can be localized in a space created by those three dimensions, i.e. can be understood as a mixture of a certain amount of pleasure, arousal and dominance. From a technical point of view, the advantage of treating emotion as lying in one continuous space is that it is not just easier to use for technically representing emotions, but it also allows us to represent emotions that might not even have a name. While pleasure and arousal are rather self-explanatory in both models, dominance in regard to an emotion is a measure of experienced control of or influence on the situation. For example, this distinguishes between the emotions fear (low control) and anger (high control). Both are states of negative pleasure and high arousal, but not feeling in control is what separates fear from anger, where the agent would at least believe it has potential influence.

For the prediction of user emotions according to eliciting situations (triggers), we chose the (categorical) OCC model (3) and for the measurement of users' physiological reactions we chose the (dimensional) PAD model. Finally, we provided a mapping algorithm to transform emotions from the OCC model into the PAD model.

21.3 Computational Models of Emotion

A detailed overview of existing computational models of emotion is given in [8].

21.3.1 The OCC Model of Emotions

This model is briefly described as it provides an essential foundation of our approach presented in this book. Having been used in various scientific and practical applications of affective computing, OCC provides a sound algorithm to model emotions [6]. The basis of the model is that emotions are reactions to the attributes of objects, to events or to actions (see Fig. 21.1). Objects, events and actions are evaluated in an appraisal process based on specific criteria and result in multiple emotions of different intensities. Figure 21.1 provides an overview of the OCC appraisal process approach. Appraising the aspects of objects requires the agent to have attitudes (tastes or preferences) in order to decide whether the object is appealing or not. Since our game scenario does not include specified objects, this part of OCC is omitted in our analysis. Events, or rather consequences of events, are appraised by analyzing their impact on the agent's goals. This determines the desirability of events. The degree of desirability depends on the distance of how much closer to vs. further away from achieving a goal an event will bring the agent. Finally, the criterion used to appraise the actions of agents is their praiseworthiness,

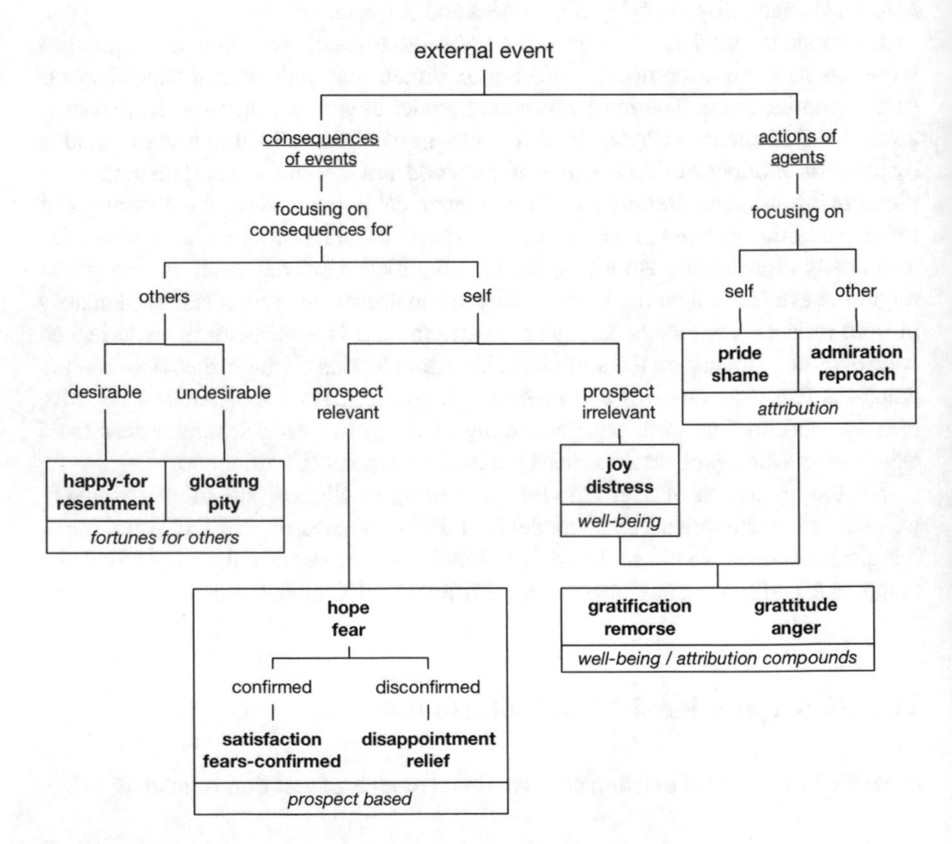

Fig. 21.1 The OCC appraisal process implemented in COMPLEX. Note that the branch *aspects of objects* is neglected here

which is based on the agent's standards. In general, praiseworthy actions cause pride and blameworthy actions cause shame, if the agent himself is the one acting. When the actions of other agents are appraised, the emotions triggered are admiration or reproach. The OCC model with its appraisal-driven prediction of emotions is an adequate framework for our game scenario because eliciting situations of OCC (e.g. actions of others) form an important part of interactions.

21.4 Implementation of the COMPLEX Model

This chapter describes the COMPLEX framework in as much detail as possible. First of all, COMPLEX can be regarded as a plugin module for *Companion*-Systems for being aware of the emotional state of a user interacting with the system. It is up to the *Companion*-System how this information is used in terms of adapting the dialog strategy the modifying the facial expression, the language or and gesture of an artificial avatar to achieve a more believable behavior.

In Fig. 21.2 an overview schema is depicted showing all the important modules of COMPLEX. In the subsequent sections they will be described in more detail. But in advance, a brief overview.

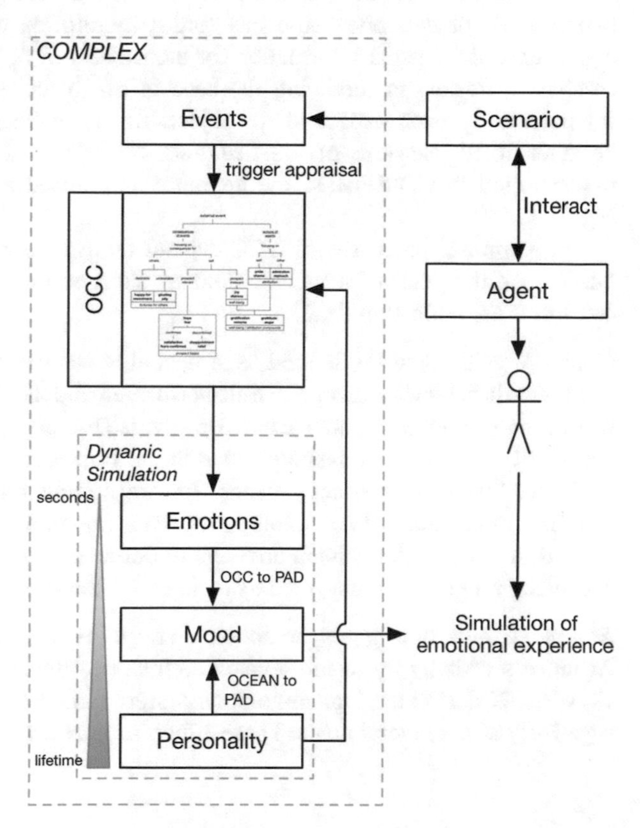

Fig. 21.2 COMPLEX. Events are processed into emotions by the OCC implementation. The mood is distracted from its current position and pulled back to the default mood determined from personality

Having a look at Fig. 21.2, there are two major parts in COMPLEX. On the one hand there is the emotion generator module based on OCC. It is triggered by events created during a human-computer interaction within a scenario. Furthermore, there is the dynamic simulation module. Its task is to continuously (in terms of time) model the emotional experience within the PAD space to achieve a continuously available emotional state. The idea is that the emotional state is represented by the mood, which is influenced by generated incoming emotions and also by the personality. In case of an incoming emotion (short in its duration), the mood is instantly distracted from its current PAD position. As long as there are active emotions, the mood will not change; otherwise it return to a default value which is determined by the OCEAN personality traits. The personality traits also influence the appraisal process of the OCC implementation.

As input, COMPLEX takes events and at any time it can be queried for the current emotional state of the agent (the mood).

21.5 Emotion Generator Framework (OCC)

The emotion generator framework of COMPLEX is based on the OCC appraisal process except for the branch of *aspects of objects*. This appraisal process requires two aspects: objects one could face and a monitoring or sensory system which could sense the needed information for the aspects of objects. Also, an increasing complexity regarding modeling the aspects of objects should not be neglected: for potentially each aspect of an object, the appealingness information has to be determined. Because of the fact that this appraisal branch is currently not implemented in COMPLEX, the appraisal is reduced to the one as depicted in Fig. 21.1.

The appraisal processes of OCC depend on available actions and their consequences on the goals of a specific domain. First of all will be clarified how these constructs are defined in detail.

Goals A goal could be defined by a desirable state in the future and all efforts aim towards achieving this goal. Self-performed and also ones performed by other persons can either be conducive but also adverse for one's goals. But a goal does not necessarily have to be a desirable state in the future. It could also be a behavioral goal, like "try to help other persons" or "only focus on your own needs". Such goals are more similar to a ongoing processes than to a finite state in the future. In the end, it is up to the domain architect to define appropriate goals for generating. Technically a goal is represented by an id and a name (see Listing 21.1).

Actions Similarly to goals, an action is simply represented by an id and a name. Again, it's actually up to the domain architect to model the actions available for COMPLEX during the human-computer interaction. Each action can positively or negatively affect several goals. These affects are named consequences.

```
goal.0.id=go_not_into_jail
goal.0.name=Do not go into jail
goal.0.default.desirability=0.8

action.0.id=betray
action.0.name=betray friend
action.0.default.praiseworthiness=-0.6

//the consequences of action "betray"
action.consequence.0.goal=go_not_into_jail
action.consequence.0.affectedGroup=ORIGINATOR
action.consequence.0.affect=1

action.consequence.1.goal=go_not_into_jail
action.consequence.1.affectedGroup=AFFECTED
action.consequence.1.affect=-1

action.consequence.2.goal=go_not_into_jail
action.consequence.2.affectedGroup=OTHERS
action.consequence.2.affect=1
```

Listing 21.1 Definition of actions

Consequences Now that goals and actions are defined, the definition of consequences can be introduced to connect them together: an action that can have an impact on one or more goals. The impact can be conducive or prejudice for a goal. Thus, it can have a positive or a negative effect. A complete example for a simple domain can be seen in Listing 21.1

Going into details of Listing 21.1, a goal is represented by a name and an id. The default desirability is a default value for this goal. It can be overwritten by an individual value. Next, a single action is described, again by an id and a name. The praiseworthiness value is a default value which also can be overwritten by an individual one. Regarding the consequences, the **.goal* property is a readable name for a goal and the **.affect* tells whether the action is conducive (+1) or aversive (−1) for a goal. The **.affectedGroup* property tells the OCC appraisal process which agents are actually affected by a consequence. This is because actions in a systems with multiple agent are typically caused by one agent and are directed to another one. For sure, this must not be the case and, therefore, COMPLEX distinguishes between three groups: *AFFECTED*, *ORIGINATOR* and *OTHERS*.

The Appraisal Imagining the scenario described by the listing above: three persons committed a crime. All of them were arrested and separately interrogated. Their goal is to not go to jail. To reach this goal, they could keep quiet and hope for an "innocent until guilty proven" verdict. But they actually could also try to betray each other and hope that the other one will go to jail and not he. So there is the action *betray* and the goal *go_not_to_ jail* described in Listing 21.1. Further, imagining that suspect A betrays B: regarding *Consequences of events → for self*; possible emotions for suspect A: *joy* because *go_not_to_ jail* was positively affected; possible emotions for suspect B: *distress* because *go_not_to_jail* was negatively affected. Suspect C seems more like a spectator and no emotions are generated.

Regarding the appraisal for *Consequences of events → for others*, COMPLEX generates emotions including the relationship between the originator and the affected suspect. Assume that the relationship between A (the betrayer) and C was negative. C would feel *resentment* because A's goal *go_not_to_jail* was positively affected. Further, assuming that the relationship between B (the betrayed suspect) and C is positive, *pity* would be generated because of the positive relationship and the fact that the betrayed suspect's goal *go_not_to_jail* was negatively affected. Tables 21.1 and 21.2 again show the rules for the consequences of the events' appraisal branch.

Regarding the appraisal branch **actions of agents**, COMPLEX would generate emotions according to the standards of an agent (praiseworthiness). Because the action is not judged to be praiseworthy for each suspect ($pr_{betray} < 0$), A would feel *shame* for himself independently of whether suspect A likes B or not. Quite similar is the appraisal of all other agents B and C. They also appraise the action in regard to their own standards independently of the relationship. Assuming that betraying another person is negative, COMPLEX would generate the emotion *reproach* for suspects B and C in regard to the betrayer A. Tables 21.1, 21.2 and 21.3 describe the appraisal rules with their dependent variables (Table 21.4).

Relationship Changes The relationship $rel(A, B)$ between two agents is defined by a value within $[-1, 1]$. This means that an agent A perceives an agent B as a friend (positive value) or an enemy (negative value). Initially, COMPLEX assumes a value of 0 which is neutral. Note that $rel(A, B) = 0.5$ does not imply $rel(B, A) = 0.5$. With

Table 21.1 Emotion-generating rules for *consequences of events → for others*

Desirability ∗ effect	Relationship between agents	Emotion
(0, 1]	(0, 1]	Happy-for
(0, 1]	[−1, 0)	Resentment
[−1, 0)	(0, 1]	Pity
[−1, 0)	[−1, 0)	Gloating
[−1, 1] (any)	0	–

If there is no relationship, no emotions will be generated in regard to some other agent. The values for desirability and relationship range from −1 to 1. The value for effect is either 1 or −1

Table 21.2 Emotion-generating rules for *consequences of events → for self*

Desirability ∗ effect	Emotion
(0, 1]	Joy
[−1, 0)	Distress
0	–

If there is no relationship, no emotions will be generated in regard to some other agent

Table 21.3 Emotion-generating rules for *Actions of agents → others*

Praiseworthiness of action	Emotion
(0, 1]	Admiration
[−1, 0)	Reproach
[−1, 1] (any)	–

Note: the praiseworthiness value is not defined by the agent who performed the action. It's always the judged value of the agent the appraisal process is running for (the spectator or the affected agent)

Table 21.4 Emotion-generating rules for *consequences of events → for self*

Praiseworthiness of action	Emotion
(0, 1]	Pride
[−1, 0)	Shame
0	–

Table 21.5 Generated emotions when suspect A betrays suspect B

Agent	Emotions	Cause
A	Joy	Not go to jail
	Pity	B has to go to jail; positive relationship
B	Distress	Go to jail
	Reproach	Go to jail; positive relationship
C	Joy	Not go to jail
	Pity	Towards B; B goes to jail
	Reproach	Towards A; B goes to jail
	Happy-for	Towards A; not go to jail

In consequence, B has to go to jail. Note that multiple emotions are a consequence of one single action

each action being performed, the relationships between agents potentially change, depending on whether an action is conducive or aversive to a goal. The change by itself is described by:

$$rel_{new}(A,B) = (rel_{old}(A,B) + (des(Con) * effect * extraversion))/2 \qquad (21.1)$$

Having a look at Eq. (21.1), the direction and power of a relationship change due to an action depend on the desirability ($des(Cons) \in [-1, 1]$) of a consequence and its effect (−1 or 1) on the goal. Therefore a relationship will increase if (a) an undesirable consequence has a negative effect on a goal or (b) a desirable consequence has a positive effect on a goal. Assuming that the goal *go_not_to_ jail* would be highly undesirable for the betrayed person B, the betrayed person would consider A as friend because his action *betray* has a negative impact on the goal not going to jail and it is undesirable. If only one of the variables desirability and effect is negative, $rel(A, B)$ will decrease. The power of the change is additionally weakened by the extraversion trait ($\in [0, 1]$), suggesting that this trait determines the tendency of a person to connect to other persons.

Table 21.5 provides an overview of all emotions generated by COMPLEX based on the above example with the three suspects A (betrayer), B (betrayed one) and C.

21.6 Mapping OCC Emotions to PAD

The dynamic simulation module of COMPLEX completely handles emotions in the PAD space. Therefore the multiple emotions (Table 21.5) must be mapped and merged into this space. Hoffmann et al. created an empirical basis for a mapping of (among other emotional terms) the 22 OCC emotions into the PAD space [7]. Table 21.6 shows the mapping results, but also the mapping of ALMA suggested by Gebhard [6]. For some emotions like *gloating*, the coordinates are not even in the same octant. Since the mapping values of ALMA don't seem to be empirically well-founded, the data of Hoffman et al. will be used for a default mapping. "Default" because these mappings should be overridden by individual values.

Table 21.6 Mapping of OCC Emotions into three-dimensional VAD space

	Hoffmann			ALMA		
Emotion	*P*	*A*	*D*	*P*	*A*	*D*
Hope	0.22	0.28	−0.23	0.2	0.2	−0.1
Resentment	−0.52	0.00	0.03	−0.2	−0.3	−0.2
Reproach	−0.41	0.47	0.50	−0.3	−0.1	0.4
Fear	−0.74	0.47	−0.62	−0.64	0.6	−0.43
Hate	−0.52	0.00	0.28	−0.6	0.6	0.3
Fears confirmed	−0.74	0.42	−0.52	−0.5	−0.3	−0.7
Gratitude	0.69	−0.09	0.05	0.4	0.2	−0.3
Disappointment	−0.64	−0.17	−0.41	−0.3	0.1	−0.4
Relief	0.73	−0.24	0.06	0.2	−0.3	0.4
Gratification	0.39	−0.18	0.41	0.6	0.5	0.4
Pity	−0.27	−0.24	0.24	−0.4	−0.2	−0.5
Admiration	0.49	−0.19	0.05	0.5	0.3	−0.2
Remorse	−0.42	−0.01	−0.35	−0.3	0.1	−0.6
Gloating	0.08	0.11	0.44	0.3	−0.3	−0.1
Joy	0.82	0.43	0.55	0.4	0.2	0.1
Happy for	0.75	0.17	0.37	0.4	0.2	0.2
Shame	−0.66	0.05	−0.63	−0.3	0.1	−0.6
Pride	0.72	0.20	0.57	0.4	0.3	0.3
Distress	−0.75	−0.31	−0.47	−0.4	−0.2	−0.5
Anger	−0.62	0.59	0.23	−0.51	0.59	0.25
Satisfaction	0.65	−0.42	0.35	0.3	−0.2	0.4
Love	0.80	0.14	0.30	0.3	0.1	0.2

On the left: mapping provided by Hoffmann et al. [7]. On the right: mapping provided by ALMA [6]. *P* Pleasure, *A* Arousal, *D* Dominance

21.6.1 Conclusion

So far, COMPEX is able to generate emotions when actions are performed. Therefore the actions and their consequences must be defined and modelled by an domain-architect. For generating the emotions, COMPEX uses an implementation of the OCC framework which is an appraisal-based theory. Because the appraisal of a single action typically results in more than one emotion, COMPLEX maps the emotions into the dimensional PAD space and merges them in this space. Subsequently will be explained how this merged output of the appraisal process impacts the mood.

21.7 Dynamic Simulation Module

In the dynamic simulation of COMPLEX, three different affective concepts in regard to time are involved: short-term emotions, mid-term mood, which represents the current emotional state and is distracted by short-term emotions, and finally the long-term personality, which is assumed not to change at all on adults. Figure 21.2 shows the interplay between those three concepts.

21.7.1 The Influence of Short-Term Emotions on Mood

There are two aspects which have to be modeled in regard to the influence of short-term emotions generated by the OCC framework on the mood. First of all, the direction and the strength (a vector), emotions distract the current mood from its current PAD position has to be modeled. Second, even if emotions are supposed to be very short-lasting, the duration for which emotions are active has to be defined because as long as there are any active emotions, the current mood will stay at its current PAD position.

Figure 21.3 depicts step by step how generated OCC emotions affect the current mood. The generated OCC emotions are mapped into the PAD space using the information of the mapping table (Table 21.6). Because there are potentially multiple emotions, COMPLEX merges them into one virtual emotion by simply using the mean value in the PAD space. This virtual emotion will be called emotion center E_C. The vector $\mathbf{e}$ formed by $\overline{(0,0,0)E_C}$ defines the direction; the current mood M_{curr} will be distracted from its current position to its new position M_{new}. The strength is additionally affected by the user's neuroticism personality trait P_{neur}:

$$M_{new} = M_{curr} + \mathbf{e} * P_{neur} \tag{21.2}$$

The emotion center can be seen as an attractor which moves the current mood towards itself. It serves as input for a function pushing the current mood away from

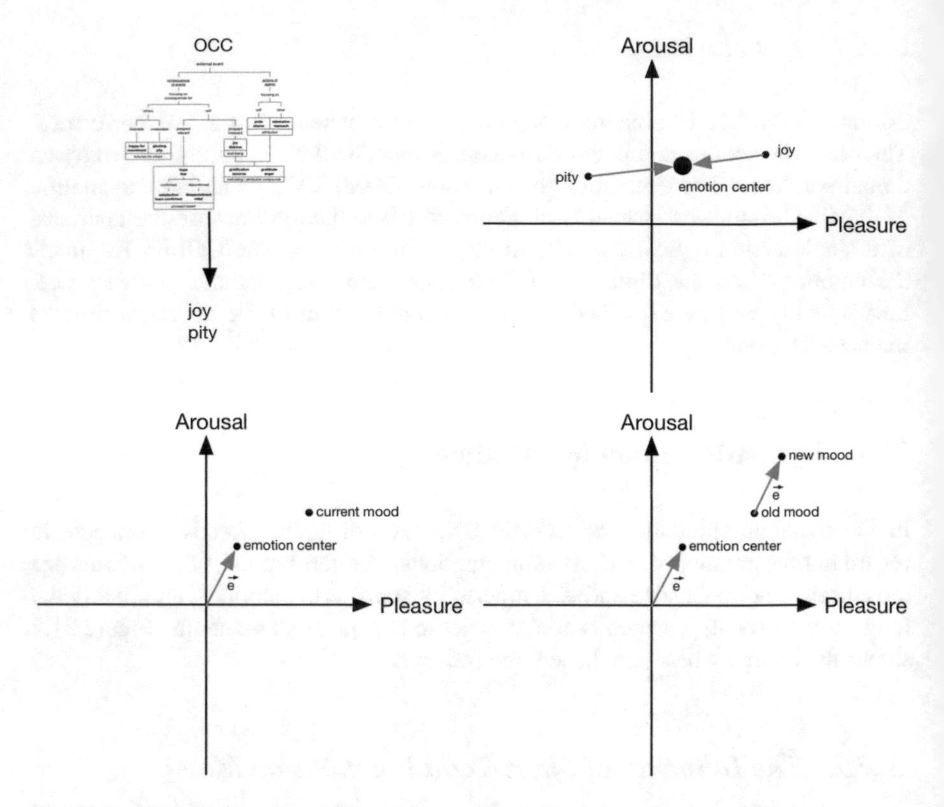

Fig. 21.3 Affect of short-term OCC emotions (*upper-left*) on the current mood: The emotions are mapped into the PAD space and an emotion center is calculated (the mean, *upper-right*). With the emotion center, a vector **e** is formed (*lower-left*) which pushes the mood into its direction (*lower-right*)

itself if it is located between the current mood and (0, 0, 0) and attracting the current mood to itself if the current mood is between the emotion center and (0, 0, 0). As Eq. (21.2) shows, the absolute value by which the current mood is moved from its current position is affected by the neuroticism personalty trait P_{neur}. Because its value ranges from 0 to 1, P_{neur} decreases the absolute value. The less the value of P_{neur} is (the more emotionally stable a person is), the less the user's mood will be changed by events and vice versa.

Decay Functions As mentioned, the current mood will always try to move back to the default mood as soon as there is no more active emotion left. Two aspects have to be taken into account: at what time an emotion is activated and how fast will it become inactive (decay). Regarding the first one, COMPLEX assumes an emotion to be active as long its intensity is above a given threshold (the intensity of an emotion ranges from 0 to 1). Therefore, a decay function has to be found which decreases the intensity of an emotion over time. Answering the question about how fast emotions decay in their intensity is an impossible task because of being number of dependencies not the assessable. To name just a few of them: the decay of an emotion will differ from emotion to emotion. E.g. *fear* will potentially not decay while the reason for the emotion is still present. But as soon this reason disappears, it will maybe decay quite fast or be replaced by another emotion like *relief*. Other emotions like *surprise* caused by a startle reflex will be very high in intensity but will also decay quite fast. Second, decay functions potentially differ inter-individually. Persons with a high emotional stability (which could be related to their neuroticism value) may differ in their decay functions compared to persons with less emotional stability. Third: it is not possible to exactly (in terms of high resolution in time) measure the decay function of a particular emotion of an individual, but this is approximated using fast responding physiological channels such as electroencephalography (EEG) or electromyography (EMG). These statements may appear like an apology, but they also give hints about why previous works on especially this topic (decay of emotions) do not explain why particular decay functions where chosen: it is not possible to find a scientifically founded explanation. Reilly [11] suggests different functions for different emotions:

- The default decay function suggests that the intensity continuously is reduced by 1 per tick $f(x) = I - x$ (linear)
- For startle emotions a function is suggested which rapidly reduces the intensity: $f(x) = I/2^x$
- As third decay function for emotions like *fear* or *hope*, the intensity of an emotion does not change as long as its reason is present. Afterwards it linearly decreases 1 per tick.

The three decay functions suggested above, the definition of a "tick" should be mentioned because this plays an important role when defining the time which is needed an emotion needs to "disappear". In Reily's case this seems to be not possible, because a "tick" is not necessarily related to time. In his round-robin scenario a tick is one cycle.

For this reason, another function is used in the implementation of the dynamic simulation module of COMPLEX:

$$I(t) = \cos(\frac{\pi}{2 * D_E} * t) \tag{21.3}$$

D_E defines the time in milliseconds an emotion needs to decay to 0. The idea is that the decay is modeled by a cosine function between 0 and $\frac{\pi}{2}$. Within this span, its value first decays slowly starting with $\cos(0) = 1$ and continuously faster heads to 0 with $\cos(\frac{\pi}{2}) = 0$. This decay function is based on the one defined by Becker [1, p. 67], who in turn based his function on the basis of the behavior of a spiral spring: $I(t) = I * \cos(\sqrt{\frac{d}{m}} * t)$ (m=mass, d=spring constant).

All the mentioned functions are also depicted in Fig. 21.4.

The last function, $I_3(t)$, in Fig. 21.4 is used both for decaying the fallback to the default mood if there is no active emotion left and also for decaying single

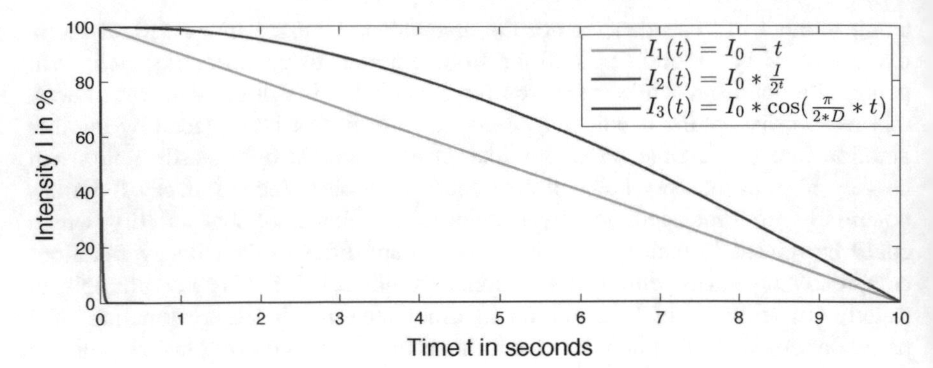

Fig. 21.4 Note that a tick is not related to time. Therefore it's not correct to compare it with other decay functions in this diagram. $f_1(t)$ models a decay function which reduces the intensity by half each tick. $f_2(t)$ linearly reduces the intensity. $f_3(t)$, used by COMPLEX, is based on a cos function between 0 and $\frac{\pi}{2}$

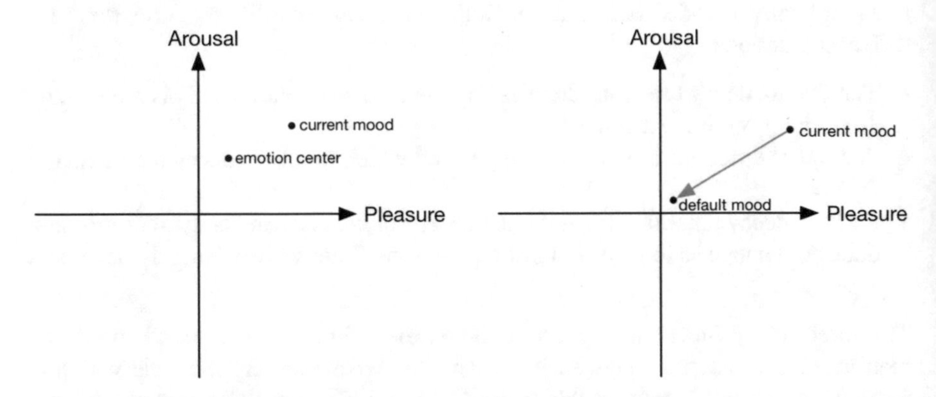

Fig. 21.5 Mood fallback—as long as there is an active OCC emotion, the generated emotion center will also stay active, and therefore the current mood stays at its position (*left picture*). As soon as the intensity of all emotion decays below a threshold, the current mood will fall back to the default mood. This transition back to the default mood is describe by in $f_3(t) = \cos(\frac{\pi}{2*D_E} * t)$ in Fig. 21.4

short-term emotions. Regarding the definition of an "active emotion", in COMPLEX it is assumed that emotion is active as long its intensity is above a threshold of 20%. Afterwards the emotion center will disappear and the mood will move back the default mood according to function $f_3(t)$. This is also depicted in Fig. 21.5.

For sure, other OCC emotions can be generated by subsequent actions with their own emotions centers before all the emotions of the previous action become inactive, and this makes it possible that the same action again and again generating the same emotion center pushes the user further in an octant of the PAD space.

Regarding the time the current mood needs to return to the default mood, COMPLEX assumes 20 min [6]. For short-term emotions, 20 s are assumed necessary for the intensity to become zero. For the scenario (computer game) described in the subsequent chapter it is not really important if the decay time of an emotion is 20 s or 5 s because the density of the actions is assumed to be high enough that the mood won't have any chance to return to the default mood.

21.7.2 The Influence of Personality

A user's personalty is acquired by a ten item questionnaire and stored by the five OCEAN variables (big fives). The basic idea is that the work as screws in the COMPLEX framework and tune different aspects to adapt the COMPLEX framework to a single individual user. In the current implementation the influence of the big fives is limited to the following aspects.

- The neuroticism value affects how much the current mood is distracted by generated emotions as a consequence of an event. If E_C is the virtual emotion center calculated by generated OCC emotions in order of an event and M is the current mood, the mood change is described by the following formulas:

$$Pleasure_{new} = Pleasure_{old} + M_P * P_{Neuroticism} \tag{21.4}$$

$$Arousal_{new} = Arousal_{old} + M_A * P_{Neuroticism} \tag{21.5}$$

$$Dominance_{new} = Dominance_{old} + M_D * P_{Neuroticism} \tag{21.6}$$

- The extraversion value affects the perception of events performed by other agents and in consequence how fast an agent perceives other agents as friends or enemies. The relationship between two agents A and B is represented by $R(A, B) \in [-1, 1]$. If D is the desirability of the event E, the relationship change caused by an event is described as follows:

$$R_{new}(A, B) = \frac{R_{old}(A, B) + D(E) * P_{Extraversion}}{2} \tag{21.7}$$

  The relationship itself influences the OCC emotion generator framework focusing on the emotions in regard to other agents (*admiration, reproach, happy-for, resentment*, etc.)

In the current implementation, these are the only two OCEAN variables affecting the model. But having in mind that individual values for praiseworthiness of actions and desirability of goals for COMPLEX are postulated, the individualization to a specific subject is sufficient.

There are additional other possibilities for OCEAN variables affecting the model in the literature e.g. Egges et al. propose in [2, p. 458] even if there seem to be a lack of arguments for their proposals.

21.8 The Scenario: Emowars

The choice of the scenario was mainly influenced by two aspects: first of all, the OCC model is designed to generate emotions for multiple agents in parallel. This is emphasized by Fig. 21.2 in the distinction between *others* and *self* branches. Therefore a multi-agent scenario was chosen. Second, high intensity emotions should be elicited by the scenario, and this led to the decision for a multiplayer computer game subsequently called Emowars. Designed as a strategic game, its four players—each the owner of an island with a set of fields—try to survive by choosing the right strategy that is a mixture of aggressiveness and supportiveness. Losing all fields means game over for the player (Fig. 21.6). On each player's turn, he can choose from five actions with different consequences. Obviously, these actions serve as input for COMPLEX (the events):

1. Attacking with cannon: fires a single shot destroying one field.
2. Attacking with artillery: fires six shots at a time. The number of destroyed fields ranges from zero to six, given by chance. On average, three fields are destroyed.
3. building shield: the shield withstands one shot and another one by chance.
4. building artillery: artillery must be built before it can be used.
5. building field: one field is added to the player's island.

Regarding his winning supportive strategy, a player can perform the actions 3–5 for himself but also for other players with the idea that if two players help each other and form a team, they have a better chance to survive. For sure, at some time they have to break their collaboration to win the game. Having a look at the goals serving

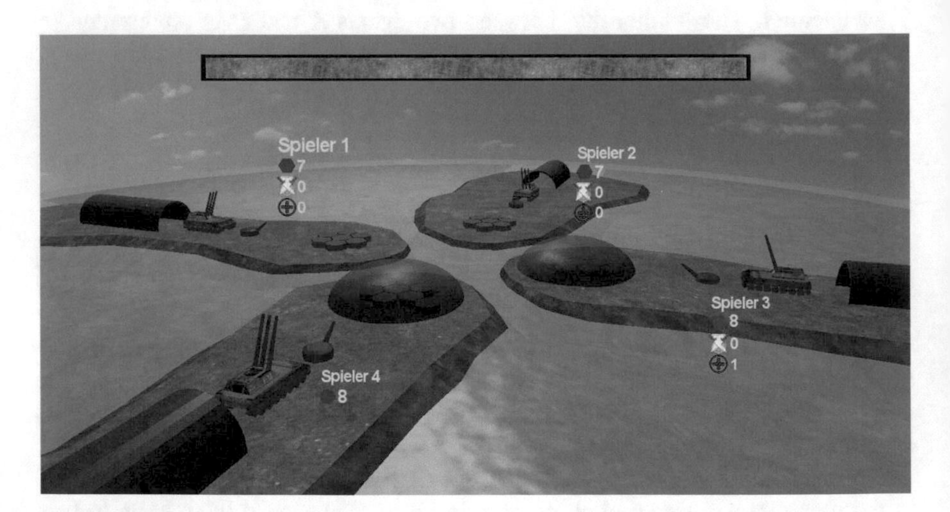

Fig. 21.6 Emowars—User Interface—one Island per player consisting of: a set of fields (*blue hexagons*), a cannon (single shot per use), artillery (six shots per use), a shield (*blue hemisphere*), user-specific information (remaining fields, number of attacks, number of supportive actions to determine between friend and enemy)

Table 21.7 Currently modeled consequences (impact of actions on goals)

Action	Group in focus	Impact on goals
Cannon attack	AFFECTED	Raise fields(−1)
Cannon attack	ORIGINATOR	Aggressive (1)
Artillery attack	AFFECTED	Raise fields(−1)
Artillery attack	ORIGINATOR	Aggressive (1)
Build field	AFFECTED	Raise fields (1)
Build artillery	AFFECTED	Aggressive (1)
Build artillery	ORIGINATOR	Raise fields (−1)
Build artillery	OTHER	Raise fields (−1)
Build shield	AFFECTED	Raise fields (1)

as input for the COMPLEX model, there are currently only two goals defined: first of all it is assumed that a player always increases the number of his fields, leading to his winning the game: *raise fields*. Second, an abstract goal is formulated regarding the chosen strategy: be *aggressive*. With these two goals, the consequences listed in Table 21.7 can be formulated (positive or negative impact of actions on goals):

Game Rules Regarding the rules of the game: each player starts with the same number of fields. Going counter-clockwise, each player can perform two actions with a limit of 15 s to decide which action to choose. Exceeding this limit means that the opportunity for the action is missed. Afterwards the next player comes to play. If a player loses all his fields, he's out of the game, and the whole game ends if only one player is left.

The Playbook To have maximum control on the affective state of a player, the experiment was designed as a Wizard of Oz (WoZ) scenario, which means that three of the players were controlled by the experimenter and only one human player participated at about time. At the very beginning of each experiment, the subject was instructed the procedure and was told especially that the other players were also being currently instructed in different rooms and buildings, Therefore a Skype session was shown to the subject (with muted audio). After being instructed, the subjects were seated in a chair facing two screens: one screen for continuously assessing his current affective state and the other showing the game (Fig. 21.6). When seated, the subject was wired to a NeXus 32 system for acquiring the physiological parameters EEG (corrugator supercilii and zygomaticus), skin conductance level (SCL), respiration rate (RSP) and blood volume pressure (BVP). Furthermore, three video channels (frontal-left and frontal-right—both 45° and frontal) and one audio channel were recorded. After the wiring was done, the subject had to map all OCC emotions into the three-dimensional Pleasure-Arousal-Dominance (PAD) space. This mapping data was fed after the whole procedure into COMPLEX. As soon as the mapping was done another previous Skype session was played and the experimenter asked the experimenters on the other side if they were ready, and the recorded session gave an answer the subject could hear. This was done to reach maximum believability that the subject was playing against real human players to get a better emotional response during the game. To further incentivise the subject to take the experiment seriously, he was offered additional 20 € when he won the game. During the game itself, the subject was alone in the laboratory giving commands to Emowars by voice.

There were three to five rounds depending on the time, and the experimenter tried to ensure the following sequence:

- Round 1: The subject wins and starts a collaboration with another player.
- Round 2: The subject hardly wins and his friend is attacked by the other two players and drops out.
- Round 3: The subject prematurely drops out and gets no help from his friend.
- Round 4 and subsequent: The subject wins again.

In summary, the experimenter ensures that the subject wins and receives the reward. During the game, the experimenter paused the game from time to time and the subject assessed his affective state on the other screen in the PAD space. These assessments were afterwards used to validate the output of the COMPLEX model.

In the last step, the subject had to fill out questionnaires for acquiring

- personality (NEO fife-factor personality inventory)
- praiseworthiness of the actions
- importance of goals

The last two items were also fed into the COMPLEX model for validating its output. Afterwards, the reward was handed out and the experiment ended. The whole procedure took nearly 2 h.

21.9 Results and Discussion

There are a lot of variables for COMPLEX, which makes it hard to evaluate the data of the 20 recorded subjects. Each subject was continuously asked to asses his current affective state within the PAD space displayed on a separate screen. The average duration of each session was about 40 min. The wizard decided when to ask the subject for a self-assessment, which was about every 2 min depending on whether the wizard was examining interesting situations. The wizard tried to reduce interruptions due to these self-assessments as much as possible and tried not to pull him out of the situation too much. On average there were about 20 samples for one subject. Figures 21.7 and 21.8 shows the raw data of one subject. The red lines show the output of the COMPLEX framework and the blue ones the self-assessing data of the subject. For the pleasure (Fig. 21.7) dimension both lines match fairly well until sample 12 and after. Quite similarly for the arousal dimension (Fig. 21.8): both lines seem to match quite well, but at sample 17 and 20 the subject assessed the arousal dimension lower than the COMPLEX framework simulated. The dominance dimension isn't mentioned here because the subjects didn't take the time during playing the game for assessing this dimension. The dominance was visualized by

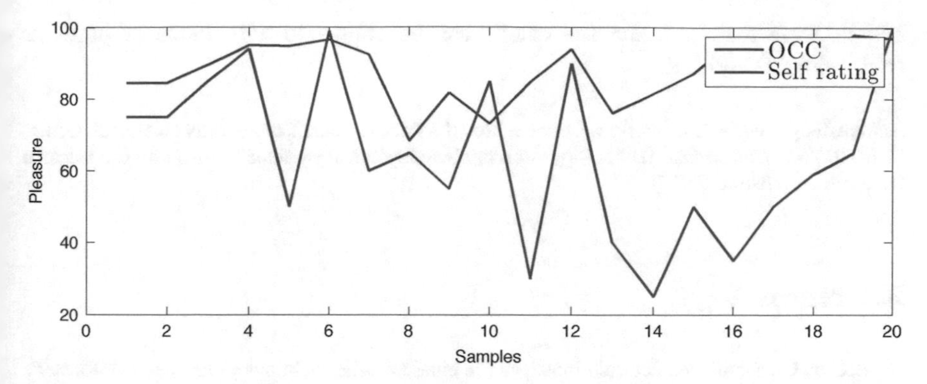

Fig. 21.7 Comparison of self-assessed and simulated pleasure of a single subject

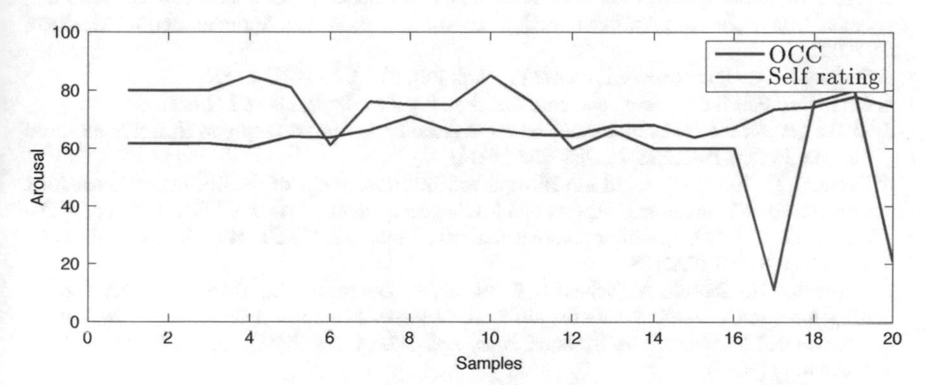

Fig. 21.8 Comparison of self-assessed and simulated arousal of a single subject

a ball pushed into the back for low dominance and pulled to the front for high dominance using the mouse wheel.

There were other subjects, the simulations didn't work at all and there are a lot of potential causes: the continuous self-assessments caused a distraction from the current situations to the subjects. A video of the subjects' faces were transmitted to the wizard's room and as soon they were asked for a self-assessment they looked more and more annoyed as soon as the game was paused and the prompt "Please assess your current condition" appeared. Another source of error could be found in the user-specific configuration of COMPLEX regarding the praiseworthiness of actions and the desirability of goals. Also the implementation of the decay functions for emotions and mood could be an issue, because they are only approximations and do not fully represent the reality. And last but not least, also the goals and consequences were not modeled correctly, and more goals were needed. The more goals and consequences, the more the emotions that would be potentially generated by a single action.

For improving the simulation of COMPLEX, most of the parameters can be adapted and afterwards each session can be replayed because also the performed actions were recorded, and the results can be compared with those of previous settings of COMPLEX.

Acknowledgements This work was done within the Transregional Collaborative Research Centre SFB/TRR 62 "*Companion*-Technology for Cognitive Technical Systems" funded by the German Research Foundation (DFG).

References

1. Becker, C.: Simulation der emotionsdynamik eines künstlichen humanoiden agenten. Master's thesis, Bielefeld (2003)
2. Egges, A., Kshirsagar, S., Magnenat-Thalmann, N.: A model for personality and emotion simulation. Lecture Notes in Artificial Intelligence, vol. 1, pp. 453–461. Springer, Berlin/Heidelberg (2003)
3. Ekman, P.: Are there basic emotions? Psychol. Rev. **99**(3), 550–553 (1992)
4. Ekman, P.: Facial expression and emotion. Am. Psychol. **48**(4), 384–92 (1993)
5. Ekman, P.: Strong evidence for universals in facial expressions: a reply to Russell's mistaken critique. Pychol. Bull. **115**(2), 268–287 (1994)
6. Gebhard, P.: Alma: a layered model of affect. In: Proceedings of the 4th International Joint Conference on Autonomous Agents and Multiagent Systems, AAMAS '05, pp. 29–36. ACM, New York, NY (2005). doi:http://doi.acm.org/10.1145/1082473.1082478. http://doi.acm.org/10.1145/1082473.1082478
7. Hoffmann, H., Scheck, A., Schuster, T., Walter, S., Limbrecht, K., Traue, H.C., Kessler, H.: Mapping discrete emotions into the dimensional space: an empirical approach. In: 2012 IEEE International Conference on Systems, Man, and Cybernetics (SMC), pp. 3316–3320. IEEE, Piscataway (2012)
8. Marsella, S., Gratch, J., Petta, P.: Computational models of emotion. In: Blueprint for Affective Computing. Series in Affective Science, pp. 21–46. Oxford University Press, Oxford (2010)
9. Mehrabian, A.: Pleasure-arousal-dominance: a general framework for describing and measuring individual differences in temperament. Curr. Psychol. **14**(4), 261–292 (1996)
10. Ortony, A., Clore, G.L., Collins A.: The Cognitive Structure of Emotions. Cambridge University Press, Cambridge (1988)
11. Reilly, W.S.N.: Believable social and emotional agents. Master's thesis, Carnegie Mellon University Pittsburgh (1996)

Chapter 22
Companion-Systems: A Reference Architecture

Thilo Hörnle, Michael Tornow, Frank Honold, Reinhard Schwegler, Ralph Heinemann, Susanne Biundo, and Andreas Wendemuth

Abstract *Companion*-Technology for cognitive technical systems consists of a multitude of components that implement different properties. A primary point is the architecture which is responsible for the interoperability of all components. It defines the capabilities of the systems crucially. For research concerning the requirements and effects of the architecture, several demonstration scenarios were developed. Each of these demonstration scenarios focuses on some aspects of a *Companion*-System. For the implementation a middleware concept was used, having the capability to realize the major part of the *Companion*-Systems. Currently the system architecture takes up only a minor property in projects which are working on related research topics. For the description of an architecture representing the major part of possible *Companion*-Systems, the demonstration scenarios are studied with regard to their system structure and the constituting components. A monolithic architecture enables a simple system design and fast direct connections between the components, such as: sensors with their processing and fusion components, knowledge bases, planning components, dialog systems and interaction components. Herein, only a limited number of possible *Companion*-Systems can be represented. In a principled approach, a dynamic architecture, capable of including new components during run time, is able to represent almost all *Companion*-Systems. Furthermore, an approach for enhancing the architecture is introduced.

T. Hörnle (✉) • R. Schwegler • S. Biundo
Institute of Artificial Intelligence, Ulm University, 89069 Ulm, Germany
e-mail: thilo.hoernle@uni-ulm.de; reinhard.schwegler@uni-ulm.de; susanne.biundo@uni-ulm.de

M. Tornow • R. Heinemann • A. Wendemuth
Institute of Information Technology and Communications, Otto-von-Guericke University, 39016 Magdeburg, Germany
e-mail: michael.tornow@ovgu.de; ralph.heinemann@ovgu.de; andreas.wendemuth@ovgu.de

F. Honold
Institute of Media Informatics, Ulm University, 89069 Ulm, Germany
e-mail: frank.honold@uni-ulm.de

S. Biundo, A. Wendemuth (eds.), *Companion Technology*, Cognitive Technologies,
DOI 10.1007/978-3-319-43665-4_22

22.1 Introduction

Future technical systems will use *Companion*-Technology to increase their usability and extend their functionality. By adding the feasibility to adapt specific system components to the individual user in any given context of use, the system could increase its efficiency and cooperativeness by assisting the user with his everyday tasks. *Companion*-Systems should be highly available in an individual way and they should always act on behalf of their users, striving for a maximum degree of reliability (cf. Chap. 1). By these properties *Companion*-Systems differ from contemporary technical systems. To offer these properties, some new capabilities need to be implemented as shown in Fig. 22.1. In that way, existing applications, services, or technical systems can be wrapped by specialized modules to form a *Companion*-System. The required modules from Fig. 22.1 can be classified as follows:

Recognition: These modules encompass diverse sensors for emotion and intention recognition. In addition, ongoing changes in the environment are also recognized to allow adaptive reasoning and responses of other modules.

Knowledge: A central knowledge base (KB) infers and provides knowledge based on the recognizers' inputs. Since sensor data may be afflicted with uncertainty, the KB provides probability distributions along with the requested values.

Planning, Reasoning, Decision: The process of adaptive output generation decides about task planning, dialog strategy and user interface (UI) configuration. Each involved reasoning process infers its results with the use of the knowledge as provided by the KB.

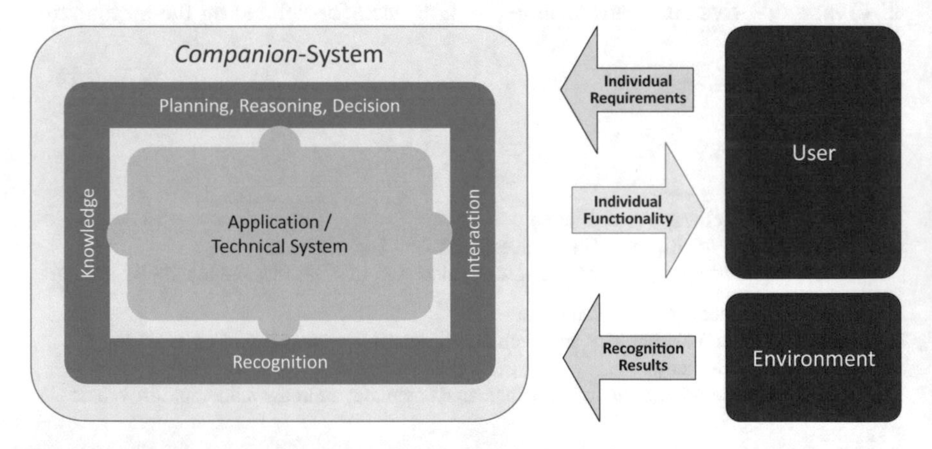

Fig. 22.1 Abstract integration of the *Companion*-Technology in a technical system to generate a specific *Companion*-System (cf. [4]). A basic technical system is wrapped by different aspects of *Companion*-Technology. The added concepts relate to the four areas of recognition, knowledge acquisition, planning and reasoning, as well as interaction

Interaction: Modules for adaptive dialog and interaction management along with different input and output devices are responsible for providing a suitable UI. With that, the user can perform either implicit or explicit interaction.

A *Companion*-System is a kind of evolution of cognitive technical systems, as it combines miscellaneous cognitive properties implemented in different modules. The application itself as well as the linked process of human-computer interaction (HCI) need a perception of the emotion and the intention of the user as well as of the situation. According to Knappmeyer et al. [15], context data can be collected via physical sensors, logical sensors, virtual sensors, and user profiles. A context-provisioning middleware can be used to analyze and aggregate such data. Decision processes of application and service logic are able to access the so-aggregated data [15]. *Companion*-Technology utilizes all these sensor classes to comprehend the current situation of the user and the environment. This knowledge in combination with the system's internal state is what we call the context of use (CoU). In our knowledge-centered approach, the KB acts as the one provider for any information that is related to the CoU.

To gain easy access to the complex sensor data from the different sensors, hierarchical fusion concepts for combining data are necessary. Fusion can be performed on different levels [5, 7]. Based on Caschera et al. [5], this chapter addresses fusion approaches on the data-level and feature-level (also known as early and late or semantic fusion). Fusion on the application-level as mentioned by Glodek et al. [7] takes place in diverse modules (application, planning, reasoning, decision) and does not have a direct impact on the architecture.

The range of applications in the domain of *Companion*-Systems is multifaceted. In order to cover a variety of possible applications we focus on the following three exemplary settings as possible implementations for *Companion*-Systems: (1) an adaptive *Wiring Assistant* for setting up a home cinema system (see Chap. 24), (2) an adaptive and emotion-aware *Ticket Vending Machine* (see Chap. 25) and (3) a multi-sensor setup for data acquisition (see Sect. 22.3.2).

The first setting realizes a cooperative system, where the focus is on the adaptivity of modules for planning, dialog management and multimodal interaction management. Thereby, the process of output generation is realized as a pipeline as described by Honold et al. [12]. A central KB provides the involved modules with knowledge required for reasoning. The second setting focuses on multimodal data-acquisition and includes fusion concepts on different levels. This setting makes use of diverse sensors and addresses fusion of temporal and spatial events. The third setting tackles the challenge of almost real-time multimodal data acquisition. Hence, this setting requires an infrastructure that allows exact timing and fast processing under conditions of chronological synchronism.

The remainder of this chapter analyzes the requirements of these three heterogeneous exemplary settings with regard to the architecture in order to come up with a dynamic reference architecture for future *Companion*-Systems.

22.2 Requirements and Middleware

Evolving a technical system to a *Companion*-System requires some additional functionalities. These functionalities originate from different modules and research domains. To facilitate interdisciplinary research and to ensure flexibility, *Companion*-Technology with inter-modular dependencies should not be realized in a monolithic manner. Spreading desired functionality over several modules in a distributed system comes with further advantages. In that way, modules can be shared and exchanged to increase availability and maintainability.

To handle the complexity that results from the interplay of different modules and various sensors, we recommend the use of a dedicated middleware concept for inter-process communication within such a distributed system. In addition, the employed middleware shall facilitate the use of different operation systems and programming languages, and shall further allow offering application programming interfaces (APIs) for any module.

The following Sect. 22.2.1 focuses on the impact of derived requirements from *Companion*-Technology on the architecture. The use of a suitable middleware concept is discussed in the subsequent Sect. 22.2.2.

22.2.1 Impact of Requirements on the Architecture

A *Companion*-System's architecture encompasses different modules, which add desired properties to some basic technical system or service. In that way, applied *Companion*-Technology shall originate, add, and combine characteristics like "competence, individuality, adaptability, availability, cooperativeness and trustworthiness" [4]. To realize these characteristics, a basic application can be wrapped by several different *Companion*-Technology components, as illustrated in Fig. 22.1. Some of these modules lead to specific requirements for the architecture to fulfill their function (e.g. the availability of linked modules of reference, or a minimum infrastructural data throughput with respect to a desired quality of service). Therefore a complex and inter-component architecture evolves, whereas every subsystem inherits its own architecture. For easy maintenance, the use of coherent concepts is an important aspect; furthermore, privacy and security issues have to be covered by the architectural approach as well.

One major aspect of *Companion*-Technology is user-adaptive behavior. For this reason, a user-specific KB is necessary to support the involved reasoning processes and the application. Only with sufficient knowledge about the user is a technical system able to act and decide as a true *Companion* in a satisfying and user-intended way. Although eligible, this property often raises concerns of privacy, reliability, and trustworthiness. From this it follows that the architectural communication approach shall at least support an encryption mechanism to ensure secure inter-process communication. Another reasonable approach might be to virtually split the KB and to swap user-related data to one of the user's personal devices (e.g. his smartphone). A third aspect addresses the process of output generation to ensure a UI that respects specific privacy guidelines. This could be realized by utilizing an additional *privacy decision engine*, as motivated by Schaub [23].

Besides the mentioned modules, the applied architectural concept itself could help to realize the aforementioned characteristics. Most architectural approaches so far are rather rigid. Once a system is set up, the architecture remains static without any changes during runtime. In that regard, an architecture-specific managing component could help to dynamically configure the architecture's topology on demand, even during runtime. Such an ability would lead to a dynamic, knowledge-based architecture, which allows us to individually integrate required and available modules.

The analysis of the *Wiring Assistant* setting (see Chap. 24 and [2, 3, 12]) shows that several modules act as processing pipeline (e.g. output generation via planner, dialog management (DM), fission, and interface components). Despite this dependency, all of these components refer to the central KB as a host for necessary data. If one of these modules stops working, the whole process chain is affected by that. To meet the requirement of continuous availability, the architecture shall provide two additional mechanisms. First, it shall allow us to detect the status of each participating module, and second, it shall offer a possibility to dynamically add and remove modules to the overall architecture. With such adaptability, envisaged but stalled modules can be replaced by other available modules that offer the same functionality. This could be realized for the architecture's modules in the same way as already implemented with the available input and output devices, as applied in the *Wiring Assistant* setting. There, the fission module adaptively decides about the configuration of the utilized end devices.

A system's competence depends on its knowledge. *Companion*-Systems shall be able to gain and infer necessary knowledge to some extent by means of recognition and cognition. This requires different sensors and fusion mechanisms to derive and extract desired information for further reasoning. Thereby, the architecture shall allow us to easily integrate all kind of sensors, even dynamically at runtime. This concept of a *Self-Organizing Sensor Network* is explained more in detail in Sect. 22.4.

Knowledge about the time of data acquisition is of great importance for the semantically correct fusion of such sensor data [5]. Fusion of old data might result in misinterpretation. There are several methods for taking care of timing, e.g. by synchronizing data acquisition and data processing or by adding time stamps to recorded and processed data. Both concepts support time-related processing of data within the fusion process. *Companion*-Systems require data fusion on different levels e.g. raw sensor data, semantic features or logical data (see Sects. 22.3.1 and 22.3.3).

The system needs the ability to be customized to the actual application by configuring a set of required modules. Due to the number of components and their function-specific requirements, it is hardly possible to implement a *Companion*-System in a monolithic way. Since the components are implemented as independent modules, the overall system needs a standardized way of communication, preferably via network. This is the topic of the next section.

22.2.2 *Middleware*

Companion-Technology can be implemented in different ways, using various technologies. According to Knappmeyer et al. [15], a middleware concept can be used to "overcome problems of heterogeneity and hide complexity". As almost every device can handle network communication, this would be a good and simple opportunity to connect all components to each other, whether they operate on the same or on different computers. To realize a flexible and extensible middleware, a client-server protocol would be best. Furthermore, this central server component would be able to record all the communication for the debugging process, and take care that the communication is still working if one component is disabled.

For reducing development effort, the middleware from the message broker system-based *SEMAINE* project [24] was used and adapted. It is set up using the well-known and tested ActiveMQ[1] server from the Apache foundation utilizing the OpenWire[2] communication protocol. This communication protocol is an improved version of the Java Messaging Service (JMS), which also offers implementations for programming languages beyond Java. The OpenWire protocol uses topics or direct connections between the communicating components. Furthermore, the *SEMAINE* implementation offers password-protected communication to ensure privacy.

As shown in Fig. 22.2, the *SEMAINE* middleware is another layer between the OpenWire protocol and the application layer. This allows abstraction and simplifies the implementation of modules on the application layer. The communication is organized by the so-called *System Manager*, which is a Java application that offers visualization concepts of the distributed system's topology, the senders and receivers of messages along with the topics, as well as the sent messages. The UI of this singleton-like central instance can support debugging processes and can perform on-demand logging of communication items and communication-specific events.

To support the integration of small ubiquitous sensors, other message broker concepts have to be taken into account. Therefore we added an adapter for the MQTT-protocol.[3] This can also be used to connect Android devices as described by Glodek et al. [7].

[1]Apache ActiveMQ: An open source messaging and integration patterns server—http://activemq.apache.org [accessed: 2015-12-18].

[2]Apache ActiveMQ—OpenWire: A cross-language wire protocol—http://activemq.apache.org/openwire.html [accessed: 2015-12-18].

[3]Message Queue Telemetry Transport (MQTT) is a machine-to-machine connectivity protocol for the internet of things—http://www.mqtt.org/ [accessed: 2015-12-18].

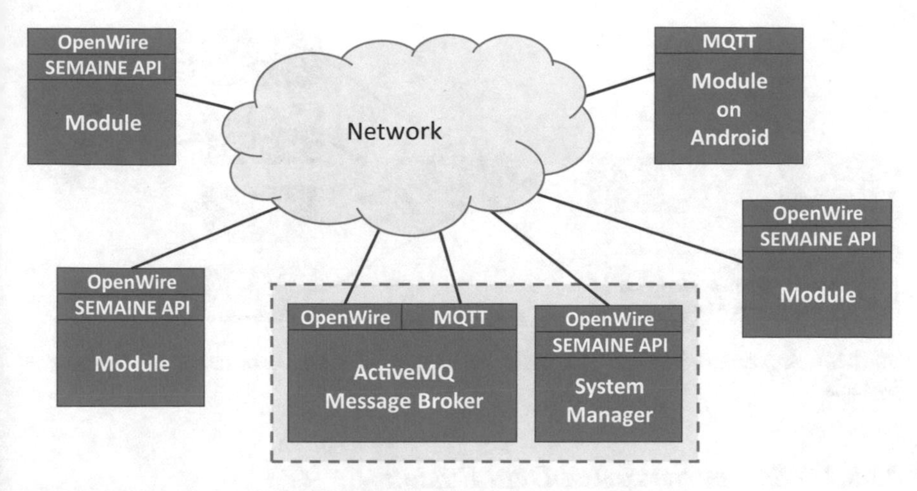

Fig. 22.2 *SEMAINE* Middleware Adoption

22.3 Deriving System Requirements from Prototypical Systems

When working on *Companion*-Technology, the problem of finding suitable architectures for those systems comes along. Several prototypical systems were developed for researching the structure of *Companion*-Systems. In this section some of the prototypical systems will be analyzed to gain information about the general problems of architectures and *Companion*-Systems. The topics of the prototypical systems are following the data flow starting with the sensors and data fusion and finishing with situation- and user-adaptive functionality. For developing and demonstrating the sensor setup and the data fusion of sensor and input signals (cf. Fig. 22.3) a prototypical system was set up, supporting the user in the task of using a *Ticket Vending Machine*, as described in Chap. 25. The architecture of this system includes sensors, a multi-level data fusion, and the application. While the actual architecture of this system is covered in Sect. 22.3.3, a general overview is given in Sects. 22.3.1 and 22.3.2. Section 22.3.5 describes the second analyzed system. It is an assistance system which helps a user to set up a home theater. It utilizes a KB and a planner to resolve problems for the user. Based on that, the system uses modules for dialog and interaction management to guide the use to the solution of said problems. Thus, both systems cover different parts of the described *Companion*-Technology. The applied concepts are compatible to each other and need to be combined according to the requirements. Both described scenarios were set up using the middleware as described in Sect. 22.2.2.

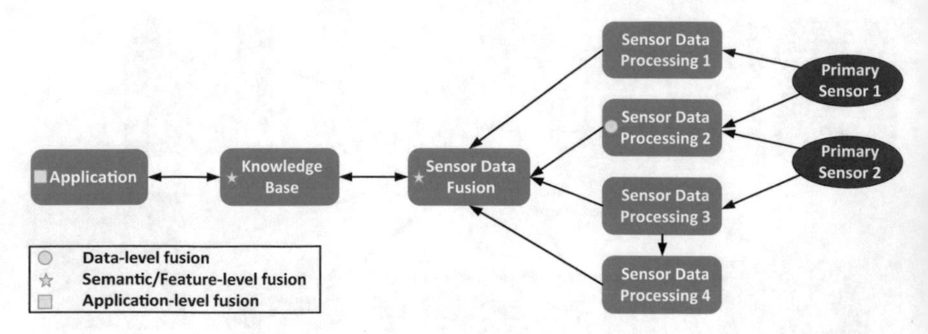

Fig. 22.3 A potential instance of a basic architecture of sensor data fusion for *Companion*-Systems

22.3.1 Sensor Setup and Data Fusion

This section focuses on the general definitions of sensor signal acquisition and sensor data fusion to give a short introduction on the topic. In complex systems like *Companion*-Systems, there are many layers between the physical sensor and the actual application, as only in very simple systems is the application directly connected to physical sensors.

There are various publications with differing definitions dealing with sensor setup and data fusion. Knappmeyer et al. [15] distinguish between physical sensors, which are in contact with the environment; virtual sensors, results of software-based algorithms; and logical sensors, like calender data.

Figure 22.3 shows the basic sensor setup and data fusion structure from a sensory perspective. In comparison to the architecture proposed by Knappmeyer et al. [15], the physical sensors are called primary sensors since they deliver direct and unprocessed information of the environment. A primary sensor is an electronic device which records data from the environment and converts it to a machine-processable digital representation (raw data). Thus the primary sensors “Primary Sensor 1” and “Primary Sensor 2” in Fig. 22.3 just record data and send it to linked sensor data processing components.

The sensor data processing is realized in software, extracting relevant information from raw data, provided by the primary sensor(s). In that manner, data is transferred to a more abstract or symbolic representation. Primary sensors and sensor data processing modules constitute virtual sensors (cf. Sect. 22.1), as in the definition given by Knappmeyer et al. [15]. Virtual sensors emit the result of simple sensor data processing, as well as data-level fusion processes.

A virtual sensor deals with data of a primary sensor, which can also be located in a mobile device. The virtual sensor’s processing probably needs to be run on a computing server due to the low computing power of most mobile devices. This case induces an additional communication layer between the primary sensor and the sensor data processing, which needs to be established. The data communication management is covered by a *Self-Organizing Sensor Network*, which can change and control the sensor setup under runtime conditions (see Sect. 22.3.4).

Several virtual sensors can share a primary sensor as exemplarily shown in Fig. 22.3. Furthermore, a virtual sensor can rely on data from sensor data processing; e.g. emotion recognition from speech [26] needs knowledge about the structure of the spoken sentence, and thus the speech recognition needs to be run first.

Data fusion for *Companion*-Systems [7] is a major topic of research for *Companion*-Technology. A wide range of fusion methods on different data abstraction levels were developed in recent years. From the architectural point of view, mainly the fusion levels are relevant, as various kinds of data can be processed in the data fusion for *Companion*-Systems. In their surveys on multimodal fusion methods Atrey et al. [1] as well as Caschera et al. [5] give an overview on the fusion levels as well as the fusion methods. They differentiate between feature- and recognition-level fusion, so called early fusion, which is performed on a very low level of abstraction. The decision-level fusion, also known as late fusion (which processes abstract data), results from sensor data processing [5].

The different levels of data fusion that are relevant for *Companion*-Systems are discussed by Glodek et al. [7]. The different fusion concepts are shown in Figs. 22.3 and 22.6. The authors distinguish between perception-level and data-level fusion, the KB- or semantic-level fusion and additionally the application-level fusion. The first level is the closest to the sensor data and is therefore dealing with raw or only slightly processed data and complies with the feature-level fusion (cf. [1]). Fusion on the feature level is usually performed in a separate fusion module, but can be required in sensor data processing as well (see Fig. 22.3: *Sensor Data Processing* 2). The data is represented on a semantic level after the sensor data processing. Hence, the fusion algorithms have to deal with more complex data.

The classes recognized on the perception level of a knowledge-based data fusion module are enriched with models created by human experts aiming at a more abstract or contradiction-free representation of data. Furthermore the temporal variation and context information are taken into account on this fusion level. On the KB level, this data is combined with context information in a second semantic data fusion, which provides information to the application. Application-level fusion is directly involved in the decision making process and combines abstract information from multiple sources as the explicit user input, the KB, the DM, and the actual application.

The primary sensors can share trigger signals for synchronization to take care of relying on a common time base (see Sect. 22.3.2). Thus the data fusion processing can be restricted to a certain time slot.

22.3.2 Concepts of Sensor Synchronization

In cases where many sensors are analyzed, it is necessary to ensure that all sensors' data ground on the same time base. The common methods to realize a synchronous recording are by using a hard-wired synchronization signal, embedding a time code, or using the network time for distributed systems. Due to the wide range of used sensors, embedding the same time code in every recorded sample is hard to realize. The hard-wired synchronization signal and the network time base were tested to gain their potential for *Companion*-Systems. From the synchronization point of view, digital sensors can be divided in value-based and frame-based devices. Value-based sensors use only one recording clock controlling the sensor data acquisition, which is usually the case for audio signal acquisition or biophysical data. Data of frame-based sensors enclose the basic raw data items plus a data structure, e.g. RGB cameras or laser rangefinders. In the latter case, the sensor synchronization can be realized by synchronizing to the frame clock or to the raw data clock.

The common approach for laboratory conditions is the hard-wired synchronization signal. A structure for a multimodal synchronization method as shown in Fig. 22.4 was developed and tested for a group of psychological experiments, e.g. for the Last Minute experiment (see Rösner et al. [22]). Such experiments focus on the psychological aspects of HCI. The aim is to build validated classifiers to gain the emotional state or disposition of the user from the different modalities. The synchronization method is set up by an asynchronous communication protocol and a hard-wired synchronous clock. In these experiments many modalities, namely four high-resolution video cameras, two 3D cameras, several sensors for biophysical data measurement (cf. [14]), and several audio channels, are used. Furthermore, the screen of the user is recorded for analyzing the user's reactions and the run of the experiment.

In these experiments the fastest recording clock was used as a master clock. As the used cameras were working with an internal pixel clock, the camera synchronization was frame-based. Thus, the 44.1 kHz audio sampling rate of the eight line in/out audio interface Yamaha Steinberg MR-X816 was used as the master clock. In recent experiments the Yamaha 01V96i was used as audio interface to improve the recorded audio quality and set the number of lost audio samples. Using this clock, the trigger signal for the cameras and stereo cameras is derived using the computer-controlled timer counter module (Measurement Computing USB4303) by dividing it by 1764 (cf. *Triggerbox* in Fig. 22.4). The resulting 25 Hz signal is fed into the trigger input of those cameras and results in images, which are synchronous to the audio clock.

This method cannot be used directly for the biophysical measurement system "Nexus32" and the screen recorder. Both devices use built-in clocks and offer no opportunity to feed them with an external signal. The Nexus32 is able to record a signal that is derived from the audio clock as well; thus its data can be reconstructed close to the desired synchronous state. The screen recorder is recording with the clock of the graphic card, which cannot be fed by any other signal. The only opportunity here is to record an audio stream with the computer.

Due to the required computing power for recording the different modalities, the recording system is set up as a distributed system, using an asynchronous communication of a topic-based message broker system. The different programs can be divided into the recorders of the different modalities and the management

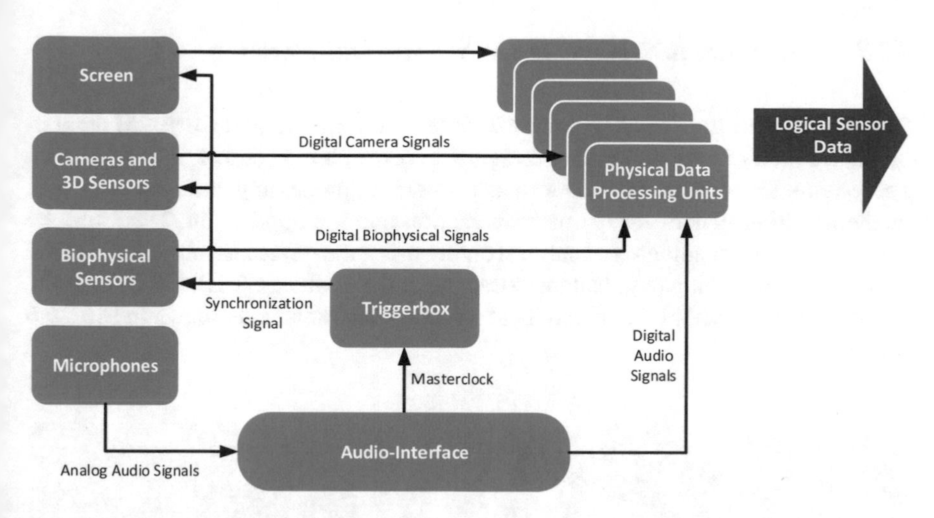

Fig. 22.4 Sensor synchronization structure

system. The management system is able to control the available recorders and start a recording with a specified name. By starting the recording process, all recorders are set in an armed state via the asynchronous message broker system. To start the synchronous trigger signal it sends a code word via USB to the USB4303 device. Along with all audio signals, the reference/trigger signal gets recorded and enables the synchronization of visual, 3D, and audio channels.

The advantage of this system is that all signals which are recorded with a sampling rate derived from the base clock are perfectly aligned. The technical overhead for a compact sensor phalanx covering multiple sensor principles is acceptable, but gets inefficient for a large sensor array distributed over a wide area.

The setup is flexible by design, but only under laboratory conditions. For a real-world scenario it is a rather static and monolithic approach, requiring hard-wired sensor setups and it synchronizes only the acquisition of a data recording. By synchronizing the sensor data processing, the effort will pay off, but the effort for a distributed system is very high. There are only very limited possibilities to correct the missing or misaligned data in an online processing. Thus, this synchronization method is mainly capable of recording a dataset which will be processed offline.

A middleware-oriented synchronization method is used in recording settings in which a flexible setup with sensors without external inputs was required. The synchronization is realized by using *SEMAINE* network time. This enables the synchronous recording of sensors like webcams, Kinect, etc., which does not allow hard-wired synchronization. Due to the network delay, the accuracy is limited to the range of milliseconds, which is appropriate for HCI interactions [25]. Advanced HCI capabilities are one of the main objectives of *Companion*-Systems; hence the system needs to meet at least weak real-time conditions.

22.3.3 Architecture Aspects for Sensor Data Fusion

The setting with the *Ticket Vending Machine* (see Fig. 22.5 and Chap. 25) demonstrates an interactive HCI system, designed to research the cooperation of different subsystems for multimodal sensor data acquisition, processing and fusion, as well as the resulting benefit for the user interaction and the application. Therefore the primary sensors' readings are analyzed online using the developed algorithms in the sensor data processing (e.g. feature extraction and classification). Based on that, the results are combined in the hierarchical data fusion structure, as shown in Fig. 22.6 (cf. [25]).

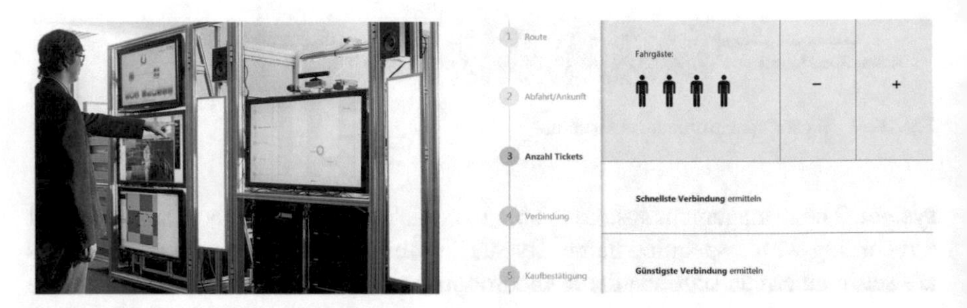

Fig. 22.5 Setup and application of the prototype system for multimodal sensor data acquisition, processing, and fusion

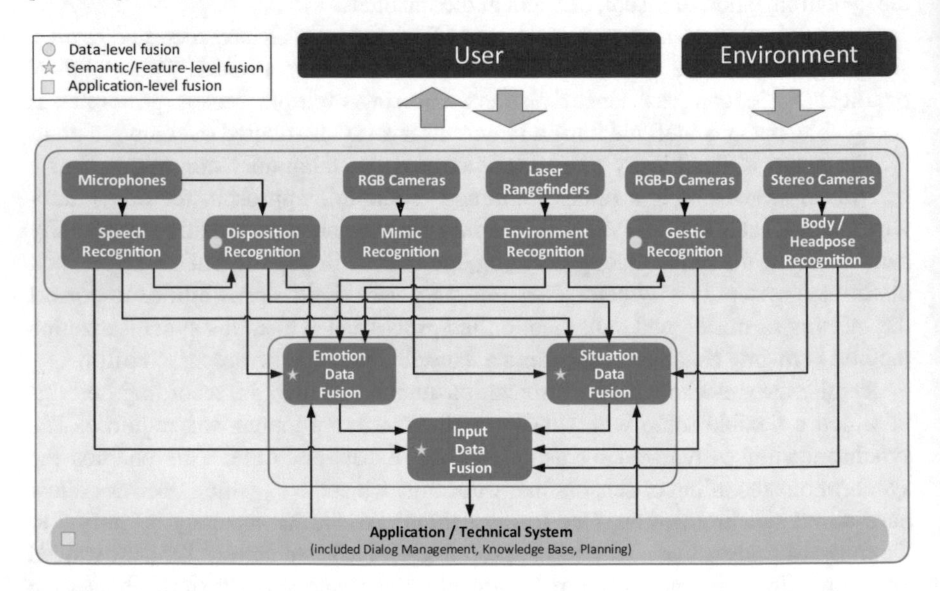

Fig. 22.6 Architecture of the system for multimodal sensor data acquisition, processing, and fusion

The sensor phalanx of the system consists of a high-resolution camera (AVT Pike F145C) for mimic analysis [18, 19], a Point Grey Bumblebee 2 stereo camera attached to a pose detection [17], a RGB-D camera (Kinect V1) for gesture recognition [8], a laser rangefinder array containing two SIC LD-MRS 4-layer laser scanners for environment recognition [21], and a head-mounted radio microphone (Sennheiser HSP2 via EW100) for speech and disposition recognition [6, 26]. In the following, only the virtual sensor results will be used, as the physical sensors are directly connected to their sensor data processing components. Those virtual sensor results are sent via the *SEMAINE* message broker system to different data fusion processes.

The levels of data fusion introduced in Sect. 22.3.1, are implemented in separate components with different requirements with regard to the data processing. The application-level fusion is localized in the input fusion component, which is reading the smart input modalities directly, e.g. the touch events, the speech and the gesture recognition results. Furthermore, it processes the results of the emotional fusion, which analyzes the state of the user and the environment fusion, which processes all data available regarding the surroundings.

The fusion of emotion-related sensor data uses mainly mimic and prosody for its decisions; thus the system is able to determine the emotional state of the user. Hence the HCI can react properly if the user is happy, neutral, or unhappy. Information from the sensors is transformed from raw data to a complex semantic level by the preceding fusion levels and is combined in the input fusion component.

As the environmental fusion is being fed by the laser rangefinders, the pose and gesture recognizers gain information to support the task of the user. In that way, an approaching person initializes the system, which responds with a greeting. According to the user's approach velocity, the system decides to start in either *standard* or *compact* dialog mode. Due to reduced complexity, the compact mode is more suitable for a precipitant user. Furthermore, the environment fusion determines the probable number of persons traveling with the user via the laser scanners and the RGB-D camera. The body pose gained from the stereo camera and the Kinect provides important information to the HCI, whether the user is currently corresponding with the system or someone else. Therefore the pose and gesture recognition data is used to extract information whether the user is looking towards the system or not, or if he is probably on the phone. This information is used to minimize the misconception of the speech recognition. The information transferred to the application represents the essence of the complete multimodal sensor phalanx. The environmental fusion results are used as an additional input for the subsequent input fusion component. The results of the emotional fusion function as a control input for the HCI and allow implicit interaction. Due to the data fusion, the system can cover a missing sensor and keep up the overall functionality.

All components are connected indirectly via the message broker system *SEMAINE*, controlled by a central management system, which is able to start up all required software components of the distributed system. All logical sensor data is processed online and the results are immediately sent to one of the fusion processes, but not directly to the application. The acquisition of all sensors apart from the Kinect is synchronized using the hard-wired method described in Sect. 22.3.2.

22.3.4 Self-Organizing Sensor Network

The sensor phalanx of the prototyped *Ticked Vending Machine* is set up as a distributed sensor network, mostly sending abstract information via the *SEMAINE* protocol. But including sensors of mobile devices exceeds the computation power of those. Therefore the data processing needs to be relocated in some situations. The signal processing is commonly implemented in software executable on every available processor. By introducing a client-server model for sensor data processing, computationally demanding processing parts can be transferred to any computation center. This allows *Companion*-Systems even to be used with smart phones or tablets. The idea of the *Self-Organizing Sensor Network* is to act as a subsystem of a *Companion*-System. It realizes and controls the required sensor data streams and distributes them according to the computation capabilities. Control the sensor data streams includes a network time-based synchronization method to relate the sensor readings to a common time base.

A generalizing abstraction is given in Fig. 22.7, where the client only contains the physical sensor and a preprocessing unit. A transfer layer is distributing the sensor data to several signal processing units if required. The required sensor synchronization is realized by time stamps derived from a high-precision network time, which gives the required accuracy in the time domain. Several signal processing units can use the data of a certain sensor, e.g. prosody extraction and speech recognition. The transfer layer can realize connections with n inputs and m outputs and is able to realize a complex sensor network. The *Self-Organizing Sensor Network* enables the dynamization of the sensor setup (see Sect. 22.4).

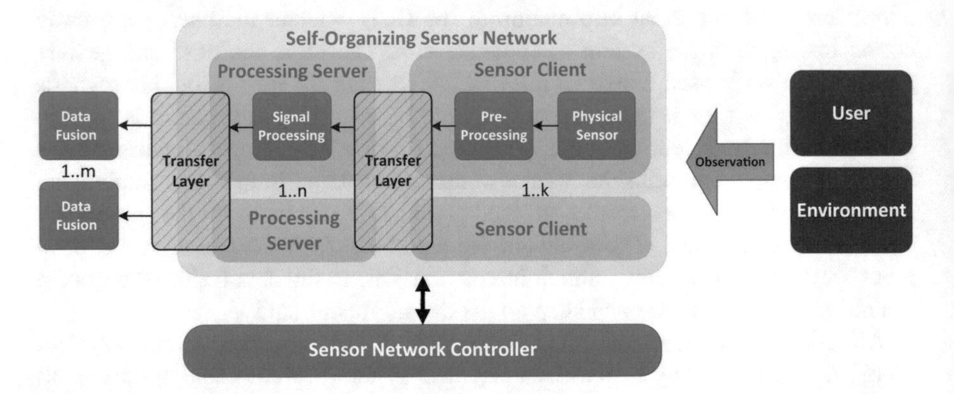

Fig. 22.7 The smallest version of a *Self-Organizing Sensor Network*. The *Transfer Layer* can have n inputs and m outputs. Based on that it is possible to build a complex sensor network as described in Fig. 22.9

22.3.5 Situation- and User-Adaptive System Functionality

The *Wiring Assistant* setting [2, 3, 12] implements an assistance system for wiring a complex home theater (see Chap. 24). The architecture (see Fig. 22.8) and the used components are domain-independent; thus a variety of application areas can be implemented using this system structure. The domain-specific knowledge is stored in different areas of the KB. It manages all necessary knowledge for the working system. Every system component has a connection to the centralized KB. The data structures in the KB are separated in static and time-stamp-based information from the user, the environment, and the system. The knowledge-based system integrates substantial planning-capabilities (Plan Generation, Execution, Explanation and Repair) with an adaptive multimodal user interface (Dialog and Interaction Management). This *Wiring Assistant* system is only equipped with a minimal sensor setup, being sufficient for demonstrating the capabilities for an assistance system like this. The laser rangefinder is used for user-localization. Data from the used touch screen sensors in combination with data from the laser rangefinder is used for user identification. The integration of a comprehensive sensor-setting has already been implemented in the *Ticket Vending Machine* setting (see Chap. 25 and Sect. 22.3.3) and is used in that configuration. All components of the system communicate via messages through specific channels of the message-based middleware. By a user requesting assistance of the application, the assistance process is initialized by passing the user given task dedicated planning problem to the plan generation component. It generates a solution, e.g. a wiring plan, which describes how to do a complete wiring of the devices of the respective home theater system. This solution is sent to the plan explanation and plan execution components. The plan execution component monitors the system, identifying the next plan step

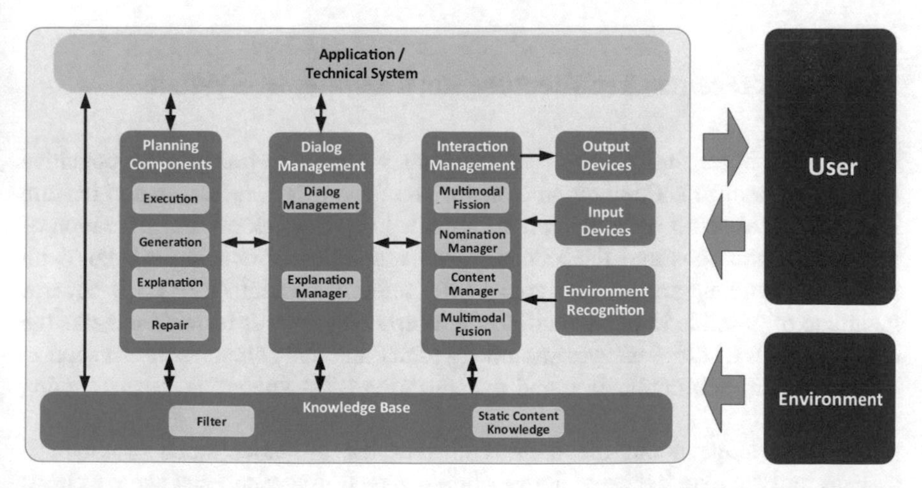

Fig. 22.8 Abstract view of the architecture of the prototyped *Wiring Assistant* for wiring a home theater, based on the architectural approach as presented by Honold et al. [11]

in dependency of the current state of the wiring process and sends it to the dialog management.

The dialog management component generates the facility of presentation for every activated plan step. To reach the goals of each plan step, the dialog management identifies the most successful dialog goal by combining the actual plan step with the associated hierarchical dialog model. The goal-oriented and adaptive dialog structure is dissected and step-wise forwarded to the interaction management.

In that way, every dialog step is transmitted to the interaction management, where the fission reasons about the currently valid UI configuration. In the present system, the UI can consist of multiple devices (see Chap. 10). Based on the fission's process, the most reasonable input and output devices are in charge of rendering their dedicated parts of the UI. The user's inputs can be made explicitly over several devices (e.g. touch or speech input) or can be interpreted as implicit input via the environment sensors (e.g. when moving from one spot to another). To get a distinct input, all input channels are fused in the input fusion module.

If the user has obscurities about why the system reacts in a certain way, he can ask the system at any time for an explanation of the currently presented plan step. The dialog management gets the user's demand through the interaction management and requests an explanation for the current step from the plan explanation component. The derived explanation is generated from a formal one, which shows the necessity of the step for the overall user task (see Chaps. 5 and 24). It is sent back to the dialog management, which starts the presentation process of the answer upon the user's demand.

In some cases, the plan step does not work (e.g. if a required cable is broken and the system does not know this yet). The plan execution component can be told about the problem via user interaction and initializes plan repair. The plan generation module derives a new plan with the modified requirements.

22.4 A Reference Architecture for *Companion*-Systems

Figure 22.1 shows that *Companion*-Technology allows to transform a cognitive technical system to a *Companion*-System. The benefit of a reference architecture is a methodical setup for *Companion*-Systems [13]. It supports the generation of new implementations and the extension of capabilities of existing systems with regard to future upgrades. The goal of the reference architecture is to cover a multitude of possible implementations for a certain system. It further increases the compatibilities of different systems among each other and concludes in the support of inter-system-communication and dynamization (also known as loose coupling (cf. [16])).

The knowledge about the architectures of the aforementioned prototypical systems and the experience of the synchronous recording setup can serve as basis for a common reference architecture. To manage the dynamic architecture and its components, an additional control structure is necessary. The architecture's

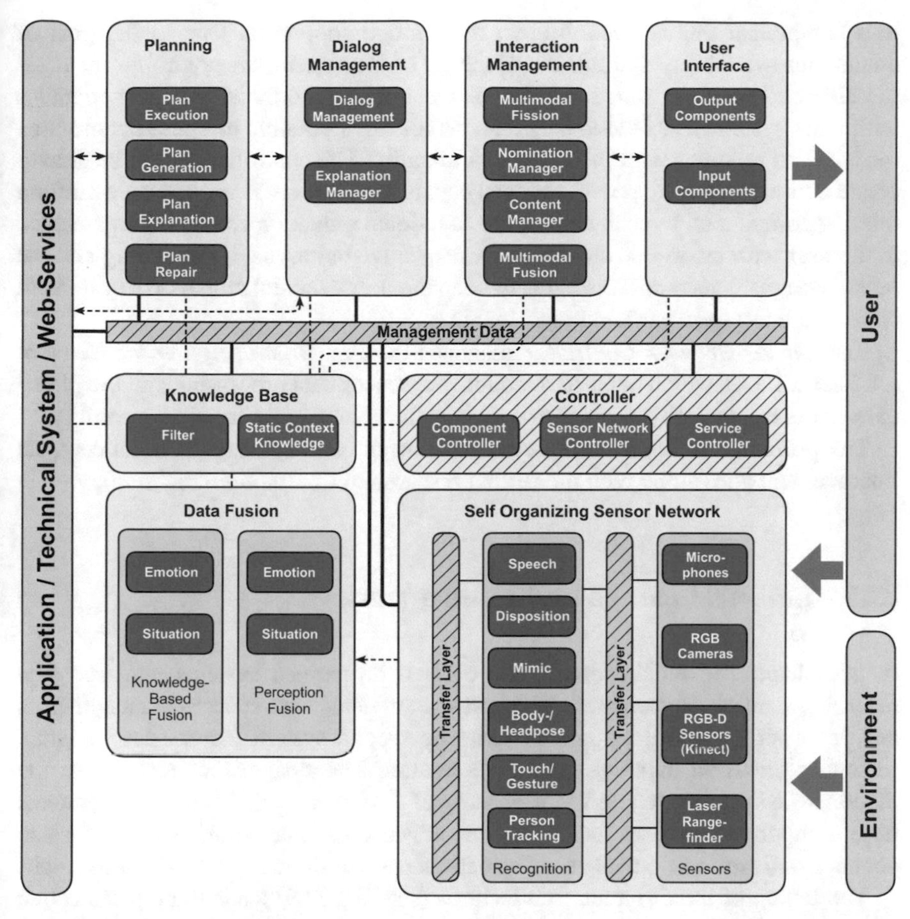

Fig. 22.9 A reference architecture for *Companion*-Systems with *Controller* components and *Self-Organizing Sensor Network*

central component is the *Controller* (see Fig. 22.9). The controller consists of three components: the *Component Controller*, the *Sensor Network Controller*, and the *Service Controller*. Each controller focuses on different aspects of the overall system.

The *Service Controller* acts as the administrative interface for other components (e.g. management of requests to several controller components). It also controls the communication between the linked system modules like the *SEMAINE* system controller (cf. [24]).

All components in the system are administrated by the central *Component Controller*. It has knowledge about the configurations of possible and desired *Companion*-Systems. Using said knowledge, the controller can identify the necessary components for a specific and desired implementation. The controller accesses the current state of all available components, which enables the controller to react to a component failure according to the configuration data. Due to the goal of continuous availability and full functionality of the system, the controller manages the distribution of the functionalities between the appropriate, available modules within the system. For redundancy reasons it is also possible to operate more than the required modules with the same functionality. The controller should be able to create a functional system with reduced capabilities if there is no replacement for a missing component. In such a case, the functional reduction can be communicated to the user with an ad hoc creation of a respective output using the dialog and the interaction management. This helps to keep the user's mental model up to date with respect to the system's offered functionality.

The *Sensor Network Controller* manages the *Self-Organizing Sensor Network* (cf. Sect. 22.3.4), which helps to decouple the sensor data processing and the physical sensors dynamically in order to optimize the system's observation capability.

The presented reference architecture is currently established as a concept. Its improvement goes along with the further research of *Companion*-Technology.

22.5 Lessons Learned and Future Work

In this chapter an architectural view on two prototyped systems and one data recording scenario is presented. The first system deals with sensor data acquisition and processing, as well as data fusion. The second system focuses on planning, reasoning, decision making, and user adaption. The system architectures of the described systems cover different aspects of *Companion*-Systems. Both systems have components of data fusion and an advanced human-computer interface in common and are consequently sharing certain parts of the architectural components.

The fusion of the different architectures (see Fig. 22.9) leads to an architecture which is close to being referential for *Companion*-Systems. The resulting architecture is still static and by now limited to a certain range of use cases. A future goal is to increase the adaptivity of the architecture to reach a wider range of applications in technical systems.

The definition of the communication structure among all necessary components of a *Companion*-System is an important task in architecture development. A standardized interface for *Companion*-Systems would be helpful for an easy integration with existing and future projects. Furthermore, such an architecture needs to cover aspects of the component interaction such as: loose coupling, service discovery, data communication management, synchronization, security, and privacy. To realize dynamization, the concept of loose coupling of modules can be applied as described by Krafzig et al. [16]. This concept is closely related with the idea of a Service-Oriented Architecture (SOA), where logical modules can be combined on demand based on a common service description language. The required modules could be discovered from a repository as described by Papazoglou et al. [20]. If necessary, specific and service-related UI elements can be realized as described by Honold et al. [9]. Privacy issues can be addressed using Schaub's

PriMA[4] approach [23] along with the process of modality arbitration as described by Honold et al. [10].

Mobile devices, such as smart phones and tablet-PCs, are probable target systems for the main interaction with the user. They contain sensors, recording a user's voice, face and position, and carry private data as well. Especially when using an individual, specific KB in insecure networks, it is a goal for the future to consider and integrate issues of security and privacy. The KB needs to contain information about several areas, which may be required to be secured against each other. Mobile devices are usually wireless components with very limited computing power, which will be part of the distributed system used to generate a *Companion*-System. Therefore no hard sensor data acquisition synchronization can be established. The synchronization needs to be realized via network-based communication.

A distributed system can be configured dynamically for different users' demands by setting up the *Companion*-System using existing hardware components as sensors and interaction devices in the current environment of any user. It is possible to add modules of the architecture—like KB or data fusion—solely using the network communication. Then, the time constraints regarding the data fusion are important for the design of the system architecture. Finally, the benefit of *Companion*-Technology could increase if one running *Companion*-System could connect to and communicate with other *Companion*-Systems around.

Acknowledgements This work was supported by the Transregional Collaborative Research Centre SFB/TRR 62 "*Companion*-Technology for Cognitive Technical Systems" which is funded by the German Research Foundation (DFG). The authors thank the following colleagues for their invaluable support (in alphabetical order): Pascal Bercher, Peter Kurzok, Andreas Meinecke, Bernd Schattenberg, and Felix Schüssel.

References

1. Atrey, P.K., Hossain, M.A., El Saddik, A., Kankanhalli, M.S.: Multimodal fusion for multimedia analysis: a survey. Multimed. Syst. **16**(6), 345–379 (2010). doi:10.1007/s00530-010-0182-0
2. Bercher, P., Biundo, S., Geier, T., Hoernle, T., Nothdurft, F., Richter, F., Schattenberg, B.: Plan, repair, execute, explain – how planning helps to assemble your home theater. In: Proceedings of the 24th International Conference on Automated Planning and Scheduling (ICAPS 2014), pp. 386–394. AAAI Press, Palo Alto (2014)
3. Bercher, P., Richter, F., Hörnle, T., Geier, T., Höller, D., Behnke, G., Nothdurft, F., Honold, F., Minker, W., Weber, M., Biundo, S.: A planning-based assistance system for setting up a home theater. In: Proceedings of the 29th National Conference on Artificial Intelligence (AAAI 2015), pp. 4264–4265. AAAI Press, Palo Alto (2015)
4. Biundo, S., Wendemuth, A.: Companion-technology for cognitive technical systems. KI – Künstl. Intell. (2015). doi:10.1007/s13218-015-0414-8

[4]PriMA—privacy-aware modality adaptation.

5. Caschera, M.C., D'Ulizia, A., Ferri, F., Grifoni, P.: Multimodal systems: an excursus of the main research questions. In: Ciuciu, I., Panetto, H., Debruyne, C., Aubry, A., Bollen, P., Valencia-García, R., Mishra, A., Fensel, A., Ferri, F. (eds.) On the Move to Meaningful Internet Systems: OTM 2015 Workshops. Lecture Notes in Computer Science, vol. 9416, pp. 546–558. Springer, Cham (2015). doi:10.1007/978-3-319-26138-6_59
6. Glodek, M., Tschechne, S., Layher, G., Schels, M., Brosch, T., Scherer, S., Kächele, M., Schmidt, M., Neumann, H., Palm, G., Schwenker, F.: Multiple classifier systems for the classification of audio-visual emotional states. In: D'Mello, S., Graesser, A., Schuller, B., Martin, J.C. (eds.) Affective Computing and Intelligent Interaction. Lecture Notes in Computer Science, vol. 6975, pp. 359–368. Springer, Berlin/Heidelberg (2011). doi:10.1007/978-3-642-24571-8_47
7. Glodek, M., Honold, F., Geier, T., Krell, G., Nothdurft, F., Reuter, S., Schüssel, F., Hörnle, T., Dietmayer, K., Minker, W., Biundo, S., Weber, M., Palm, G., Schwenker, F.: Fusion paradigms in cognitive technical systems for human–computer interaction. Neurocomputing **161**(0), 17–37 (2015). doi:10.1016/j.neucom.2015.01.076
8. Handrich, S., Al-Hamadi, A.: Multi hypotheses based object tracking in HCI environments. In: 2012 19th IEEE International Conference on Image Processing (ICIP), pp. 1981–1984 (2012). doi:10.1109/ICIP.2012.6467276
9. Honold, F., Poguntke, M., Schüssel, F., Weber, M.: Adaptive dialogue management and UIDL-based interactive applications. In: Proceedings of the International Workshop on Software Support for User Interface Description Language (UIDL 2011). Thales Research and Technology France, Paris (2011)
10. Honold, F., Schüssel, F., Weber, M.: Adaptive probabilistic fission for multimodal systems. In: Proceedings of the 24th Australian Computer-Human Interaction Conference, OzCHI '12, pp. 222–231. ACM, New York (2012). doi:10.1145/2414536.2414575
11. Honold, F., Schüssel, F., Weber, M., Nothdurft, F., Bertrand, G., Minker, W.: Context models for adaptive dialogs and multimodal interaction. In: 2013 9th International Conference on Intelligent Environments (IE), pp. 57–64. IEEE, Piscataway (2013). doi:10.1109/IE.2013.54
12. Honold, F., Bercher, P., Richter, F., Nothdurft, F., Geier, T., Barth, R., Hörnle, T., Schüssel, F., Reuter, S., Rau, M., Bertrand, G., Seegebarth, B., Kurzok, P., Schattenberg, B., Minker, W., Weber, M., Biundo, S.: *Companion*-technology: towards user- and situation-adaptive functionality of technical systems. In: 2014 10th International Conference on Intelligent Environments (IE), pp. 378–381. IEEE, Piscataway (2014). doi:10.1109/IE.2014.60
13. Hörnle, T., Tornow, M.: Reference architecture approach for companion-systems. Presented on the 1st International Symposium on Companion-Technology (2015)
14. Hrabal, D., Kohrs, C., Brechmann, A., Tan, J.W., Rukavina, S., Traue, H.: Physiological effects of delayed system response time on skin conductance. In: Schwenker, F., Scherer, S., Morency, L.P. (eds.) Multimodal Pattern Recognition of Social Signals in Human-Computer-Interaction. Lecture Notes in Computer Science, vol. 7742, pp. 52–62. Springer, Cham (2013). doi:10.1007/978-3-642-37081-6_7
15. Knappmeyer, M., Kiani, S., Reetz, E., Baker, N., Tonjes, R.: Survey of context provisioning middleware. IEEE Commun. Surv. Tutorials **15**(3), 1492–1519 (2013). doi:10.1109/SURV.2013.010413.00207
16. Krafzig, D., Banke, K., Slama, D.: Enterprise SOA: Service-Oriented Architecture Best Practices. The Coad Series. Prentice Hall Professional Technical Reference, Indianapolis, IN (2005)
17. Layher, G., Liebau, H., Niese, R., Al-Hamadi, A., Michaelis, B., Neumann, H.: Robust stereoscopic head pose estimation in human-computer interaction and a unified evaluation framework. In: Maino, G., Foresti, G. (eds.) Image Analysis and Processing – ICIAP 2011. Lecture Notes in Computer Science, vol. 6978, pp. 227–236. Springer, Berlin/Heidelberg (2011). doi:10.1007/978-3-642-24085-0_24
18. Niese, R., Al-Hamadi, A., Panning, A., Michaelis, B.: Emotion recognition based on 2D-3D facial feature extraction from color image sequences. J. Multimed. **5**(5), 488–500 (2010)

19. Panning, A., Al-Hamadi, A., Michaelis, B., Neumann, H.: Colored and anchored active shape models for tracking and form description of the facial features under image-specific disturbances. In: 2010 5th International Symposium on I/V Communications and Mobile Network (ISVC), pp. 1–4. IEEE, Piscataway (2010)
20. Papazoglou, M.P., Traverso, P., Dustdar, S., Leymann, F.: Service-oriented computing: state of the art and research challenges. Computer **40**(11), 38–45 (2007)
21. Reuter, S., Dietmayer, K.: Pedestrian tracking using random finite sets. In: 2011 Proceedings of the 14th International Conference on Information Fusion (FUSION), pp. 1–8 (2011)
22. Rösner, D., Frommer, J., Friesen, R., Haase, M., Lange, J., Otto, M.: LAST MINUTE: a multimodal corpus of speech-based user-companion interactions. In: LREC, pp. 2559–2566 (2012)
23. Schaub, F.M.: Dynamic privacy adaptation in ubiquitous computing. Dissertation, Universität Ulm. Fakultät für Ingenieurwissenschaften und Informatik (2014)
24. Schröder, M.: The SEMAINE API: towards a standards-based framework for building emotion-oriented systems. Adv. Hum. Comput. Interact. **2010**, 21 (2010). doi:10.1155/2010/319406
25. Sharma, R., Pavlovic, V., Huang, T.: Toward multimodal human-computer interface. Proc. IEEE **86**(5), 853–869 (1998). doi:10.1109/5.664275
26. Siegert, I., Hartmann, K., Philippou-Hübner, D., Wendemuth, A.: Human behaviour in HCI: complex emotion detection through sparse speech features. In: Salah, A., Hung, H., Aran, O., Gunes, H. (eds.) Human Behavior Understanding, Lecture Notes in Computer Science, vol. 8212, pp. 246–257. Springer, Berlin/Heidelberg (2013). doi:10.1007/978-3-319-02714-2_21

Chapter 23
Investigation of an Augmented Reality-Based Machine Operator Assistance-System

Frerk Saxen, Anne Köpsel, Simon Adler, Rüdiger Mecke, Ayoub Al-Hamadi, Johannes Tümler, and Anke Huckauf

Abstract In this work we propose three applications towards an augmented reality-based machine operator assistance system. The application context is worker training in motor vehicle production. The assistance system visualizes information relevant to any particular procedure directly at the workplace. Mobile display devices in combination with augmented reality (AR) technologies present situational information. Head-mounted displays (HMD) can be used in industrial environments when workers need to have both hands free. Such systems augment the user's field of view with visual information relevant to a particular job. The potentials of HMDs are well known and their capabilities have been demonstrated in different application scenarios. Nonetheless, many systems are not user-friendly and may lead to rejection or prejudice among users. The need for research on user-related aspects as well as methods of intuitive user interaction arose early but has not been met until now. Therefore, a robust prototypical system was developed, modified and validated. We present image-based methods for robust recognition of static and dynamic hand gestures in real time. These methods are used for intuitive interaction with the mobile assistance system. The selection of gestures (e.g., static vs. dynamic) and devices is based on psychological findings and ensured by experimental studies.

F. Saxen (✉) • A. Al-Hamadi
IIKT, Otto von Guericke University Magdeburg, Magdeburg, Germany
e-mail: Frerk.Saxen@ovgu.de; Ayoub.Al-Hamadi@ovgu.de

A. Köpsel • A. Huckauf
General Psychology, Ulm University, 89081 Ulm, Germany
e-mail: Anne.Koepsel@uni-ulm.de; Anke.Huckauf@uni-ulm.de

S. Adler • R. Mecke
Fraunhofer IFF, Magdeburg, Germany
e-mail: Simon.Adler@iff.fraunhofer.de; Ruediger.Mecke@iff.fraunhofer.de

J. Tümler
Volkswagen AG, Wolfsburg, Germany
e-mail: Johannes.Tuemler@volkswagen.de

S. Biundo, A. Wendemuth (eds.), *Companion Technology*, Cognitive Technologies,
DOI 10.1007/978-3-319-43665-4_23

23.1 Introduction

Current production techniques aim at including modern augmented reality technologies in order to improve the work flow by assisting the workers. Although optical characteristics of augmented reality technologies are well elaborated on, they still suffer from impeded possibilities of interaction with users. Since the noise during production is too large to enable speech-based interaction, we applied gesture-based interaction. Technological challenges were the automated recognition of body parts and of body postures in front of a complex and dynamic background with a moving camera. For human-machine-interaction, the questions of necessary commands and usable gestures were of utmost importance. As application scenario, the assembly of the electrical box (e-box) from the transporter T5 was chosen. The e-box mainly contains wires, fuses and relays. This scenario requires carefully installing various assemblies (e.g. cables) in a fixed order at defined places, partly with blocked view.

23.2 Gesture Selection

The given scenario of workers being assisted by augmented reality head-mounted devices leads to several demands for control gestures:

- Since several procedures require tool usage, gestures must be executed unimanually.
- Some workers wear gloves during work. Hence, gestures must be executable and recognizable even with gloves.
- The primary goal of an augmented reality HMD lies in displaying working instructions.

Thus, assisting functions should primarily focus on navigation aid. In a first step, based on an analysis of the planned working scenario, important commands were identified. These were: on/off, help, back, forward, zoom in, zoom out, scroll up, scroll down. All eight gestures were realized in two language-based gesture systems (American sign language, Edge Write), and two frequently used associative gesture systems (static pointing, wipe gestures). Examples for a gesture for the different systems are shown in Fig. 23.1. In an experimental series, we investigated the usability of these four kinds of gestures for the eight basic commands. As described

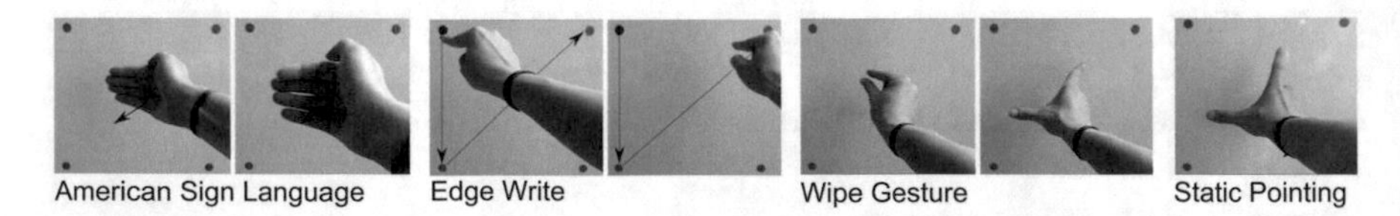

Fig. 23.1 Examples for the four different gesture systems. The meaning of the shown gesture is zoom in

in [4], understanding the commands without further explanations or training is better for the associative gestures than for language-based gesture systems. Hence, associative gesture systems might be interpreted as more intuitive. However, after a short training, respective advantages declined. In gesture production, there were slight advantages for language-based gestures. However, the users' acceptance equaled the very first performances before training. That is, associative gesture systems were liked more than language-based gesture systems.

23.3 Activating Control Elements

The described application scenario needs the user to activate control elements which are displaced in the HMD in order to interact with the assistance system. An example is the navigation within the design step of an assembly sequence, which can be activated by positioning the hand in the sensitive region (displayed arrow symbols in the HMD). For that, no specific pointing gestures need to be executed. It is enough if hand regions occupy the sensitive regions (location based gesture).

Resource-efficient methods are used to allow portability to mobile devices. First, image regions are identified that contain the user's hand. This is done by classifying pixels in the RGB color space. Thresholds and rules are applied that distinguish between the color values of the user's hand (skin color or work glove) from the background. Empirical studies have shown that the detection rate can be increased if the RGB and the HSV color space are used in combination. The control element is activated if the number of pixels that are identified to belong to the user's hand relative to the size of the sensitive region exceed a threshold. Unfortunately this method is error-prone if the detection is used on single images only. Considering the time context improves the robustness significantly. We do not allow the activation of the control element until the coverage of the sensitive region exceeds the threshold for a certain amount of time (see Fig. 23.2). The activation status is visualized in

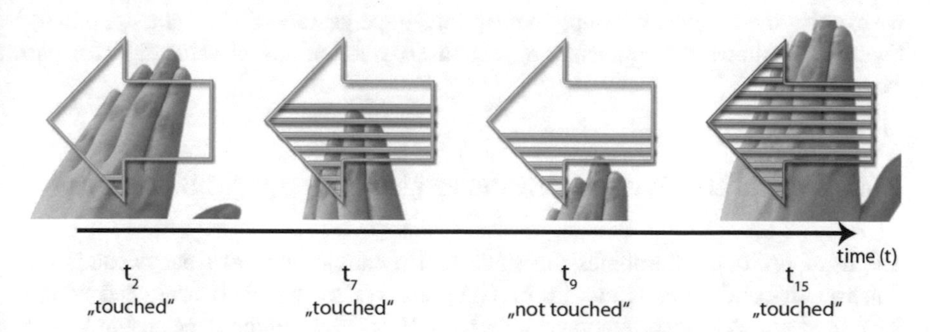

Fig. 23.2 Control element in HMD is activated over time by touching. The element is touched (t_1) and fills up with time (t_7). Withdrawing the hand drops the level indicator (t_9). If the element is re-touched it fills up again and reaches its maximum after a certain amount of time ($t_1 5$) and triggers a system function

the HMD via level indicator at the control element. If the coverage goes below the threshold, the level indicator drops and the control element starts to get deactivated. The setup for the activation/deactivation status (approx. 1 s for activation, approx. 0.5 s for deactivation) is based upon user studies [6].

23.4 Detecting Dynamic Hand Gestures

The assistance-based system should also detect dynamic hand gestures next to the location-based gestures to activate control elements. The dynamic gesture detection system is based on a pixel-based segmentation step, whereas histograms of color values in the IHLS color space [3] are used (trained on the ECU dataset [5]). Background segmentation algorithms cannot be used without modelling the 3D environment because the camera is mounted on the HMD, which moves with the user's perspective. Nevertheless, the estimation of the camera motion is possible for gathering the hand movement if we can assume that the background is static while the user performs hand gestures. Our studies have shown that this assumption often holds because the user needs to concentrate on the interaction with the HMD, which is very difficult during head movements. To estimate the camera movement we calculate the mean and variance of the optical flow of two consecutive frames. Image borders are taken more into consideration to minimize the effect of hand gestures on the compensation of the camera motion. The estimated camera motion is then subtracted from the optical flow to identify the hand gestures. This approach comes with uncertainties but provides robust hand detection along with the skin segmentation. The feature calculation is based on tracking the hand during the gesture. At each frame we calculate the hand centroid (x_t, y_t) based on the combination of the skin segmentation and the optical flow. We concatenate the discrete angles (grouped in eight bins) between the centroids of two consecutive frames for T frames $(\phi_t), (\phi_{t+1}), \ldots, (\phi_{t+T})$ for the classification. The length T of the feature vector depends on the gesture. With a Hidden Markov Model (HMM) we classify the feature. We implemented four wipe gestures (left, right, up, down). Figure 23.3 shows the segmentation, feature extraction and classification for one example.

23.5 Detecting Dynamic Gestures Using Temporal Signatures

The third prototype combines the static and dynamic approach but expands the functionality and embeds this into a complete system which is evaluated within the user study. Beside the static gestures from Sect. 23.3 and the wipe gestures from Sect. 23.4 we added EdgeWrite to the repertoire of gestures. EdgeWrite gestures are quickly executable and easily distinguishable from natural hand gestures. Wipe gestures on the other hand sometimes appear without the intention of any system interaction.

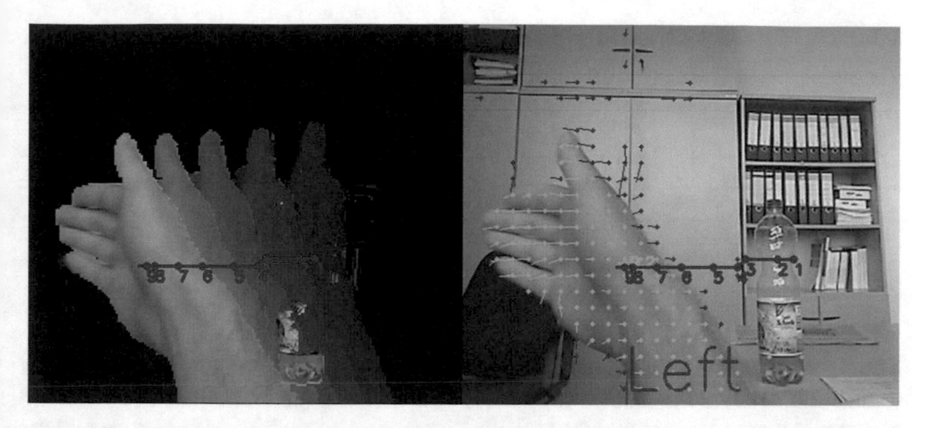

Fig. 23.3 Dynamic gesture recognition: Segmentation shown as motion history image along with tracked hand centroids (*left*). Motion compensated optical flow with trajectories and classification (*right*)

The skin segmentation, as suggested in Sect. 23.4, is not applicable because the users often use gloves in our application scenario. In contrast to the previous approach, we detect dynamic gestures by identifying a temporal signature without detecting the hand explicitly. This allows the user to vary the hand pose, wear gloves, and handle tools while executing the gesture. This approach works for EdgeWrite as well as for wipe gestures.

In this application we assume that the background scene changes relatively little while performing a gesture. This has stood the test and was confirmed in user studies. The detection of dynamic gestures is done in three steps: feature extraction, feature reduction, and classification.

To train the gesture detection system we created a dataset with ten subjects containing eight different gestures of three gesture systems. Each subject performed each gesture five to ten times. We marked the start and the end of each gesture manually.

23.5.1 Feature Extraction

To extract the global motion, we calculate the forward and backward optical flow on a regular grid between the last two frames and discard the 25% least consistent vectors to eliminate outliers. Consistent vectors have a low Euclidean distance between the forward and the backward optical flow [2].

We then calculate local histograms of the optical flow orientation (using eight bins). The histograms are of equal size and span across the whole image. We performed cross-validations to optimize the histogram and bin size for each gesture, and most gestures perform best using 3×3 histograms with eight bins (see Table 23.1). The frame-based feature contains $3 \times 3 \times 8 = 72$ dimensions for the most gestures (e.g. all wipe gestures and most EdgeWrite gestures).

Table 23.1 Optimal parameter set for each EdgeWrite gesture

Gesture	1	2	3	4	5	6	7	8
# Local histograms (hor. × vert.)	3 × 3	3 × 3	3 × 4	3 × 3	3 × 3	3 × 4	3 × 3	3 × 3
# Bins	8	8	8	8	8	8	8	8
# Parts	4	4	5	4	4	5	4	4

23.5.2 *Feature Reduction*

The time to perform a gesture varies between gestures and subjects but also varies within subjects. In our dataset we labeled the start and the end of each gesture and calculated three time periods for each gesture to cope with different variations of the same gesture. The first period represents the median, the second and third represent the lower and upper quantile. We observed that the features from quickly performed gestures are not just the aggregation of features from slowly performed examples. Therefore we trained a separate classifier for each time period.

We now aggregate the frame-based features (72 dimensions) until the time period has been exceeded. This results in a 72×N matrix, whereas N represents the number of frames captured during the time period. Because consecutive frames contain similar flow histograms and most gestures can be separated into four or five combinations of linear motions, we divide the matrix into four or five parts (depending on the gesture), average each part individually and concatenate them back together (resulting in a 72×4 matrix or a 72×5 matrix respectively). This descriptor is then used for classification. All parameters for each gesture have been optimized using cross-validation within the training set to avoid overfitting.

23.5.3 *Classification*

Our dataset provides gestures from ten subjects. We sampled the data from five subjects as our training set. To evaluate the parameters (e.g. the histogram size) we perform a leave-one-subject-out cross-validation within the training set. The final training is done with all five subjects using the previously determined optimal parameters. The optimal parameters are shown in Table 23.1.

We use a linear Support Vector Machine (SVM) to classify each gesture and time period at each given frame individually. Thus, at each frame we classify by evaluating the highest SVM score (distance to the hyperplane). Because consecutive features are similar due to the sampling, we average the scores (for each gesture and time period separately) from the last 250 ms (a few frames only) to reduce the number of false positives. If all SVM scores are negative, no gesture has been detected. This is by far the most frequent result, because during the usage of our system, gestures are not performed that frequently and the gesture should only be detected at its end.

23.6 Overall System

For the evaluation with users, we needed an integration of the different approaches in the central application. The aim was not only a technical validation but also to analyze and to compare the solutions with subjects. The scope of this system is to provide different systems of interaction, a protocol of important data during the interaction, a graphical user interface (GUI), and a flexible way of presenting different scenarios. To validate the gestures, interactions must be performed for a specific aim. For this purpose, a specific task is done by each subject. For each gesture system a tutorial was realized in which the subject can practice the interaction. The evaluation scenario provides the assistance in an easy assembly process. With gestures the user can navigate within the instructions, fading out the GUI, or increase/decrease the size of the instructional video content.

The developed application is used to evaluate different gestures and gesture systems. The investigations of interactions are focused, where subjects perform straight-forward work steps. On the other hand, the prototype shall represent the procedure in a selected automotive application scenario (e.g. e-box; see Fig. 23.4) to investigate the acceptance of gestures in this application context. The flexible

Fig. 23.4 User with HMD in the application scenario e-box. Fraunhofer IFF, Dirk Mahler

design of the application allows different concepts of interactions and scenarios to develop the user interfaces and evaluate the interactions.

During the development we focused on different topics, which are explained more elaborately in the next sections.

23.6.1 Interaction

In our system five different interaction systems are integrated in one application. The software design was realized in a modular concept to allow the easy integration of new interaction systems. The implementation is completely separated from the user interface. The developed prototype can be used as a platform to systematically evaluate and compare different interaction systems.

The integrated interaction systems are the three EdgeWrite, wipe, and static gestures, plus two conventional methods for comparison: Mouse and Button. The interaction with buttons is based on a system which is dominant in older head-mounted mobile devices. Future head-mounted devices allow the interaction with simple touch gestures. Until now these systems have not been part of industrial pilot studies. Elder systems and systems used in pilot studies allow the interaction with few buttons, which are part of the mobile device. Simple inputs (forward, backward) can be achieved efficiently, but complex inputs are not possible. These buttons can be used along with head-mounted devices while standing. The interaction with a mouse, which provides more possibilities, needs a tray. The gesture-based interaction works completely without any utilities.

23.6.2 System Properties

The given scenario addresses users in a specific working domain. The evaluation focuses on the interaction and not on the knowledge of the user about the specific task. The prototype has to allow different scenarios with partly different settings.

The required flexible configuration has a direct influence on the GUI and therefore needs an interactive adjustment. The classical configuration via static configuration files does not allow the needed flexibility. We used the script language LUA [1] to adopt the requirements on the fly with a command window. The configuration of the application is done via scripts that allow high-level programming languages.

This flexibility is important especially for users of different working domains. The script language allows the adjustment of GUI elements and central functions during run-time, such that a direct adaptation on the user needs can be achieved together with experts of the working domain. Thus, results from evaluations can be adopted directly.

With an optical-see-through system (OST) the user sees not the camera image but the real environment through a half-transparent display, where overlays can be presented. The camera itself is mounted on the HMD but has an offset and a different aperture angle than the real sight of the user. Thus, a calibration is necessary between display overlay and user sight. In this process the regions in the camera image are identified that match with the sight of the user (mapping).

23.6.3 Graphical User Interface/Visual Feedback

The 3D visualization of the GUI for head-mounted devices was implemented with OpenGL. The data transfer to the HMD provides a side-by-side view to allow the perception of virtual 3D elements. By calibrating the user interface with the user's left and right eye the depth perception of virtual objects can be improved. The camera is monocular though. This can lead to discrepancies during the execution of gestures in relation to virtual elements. Studies have shown that displaying these virtual elements on only one eye can decrease the discrepancies if we calibrate on the user's dominant eye.

23.6.4 Subject-Dependent Protocol of Interactions for Evaluation

For the evaluation of the gestures, the system needs to collect data. The name of the subject is automatically ciphered to provide anonymity and a unique log name. The log file of each subject contains metadata of the subject and the interactions with the system. We log the time stamps at the start and the end of each interaction.

23.7 Results and Evaluation

23.7.1 Quantitative Results

We conducted two quantitative evaluations. First we trained a very conservative model (which was used in the user study). We tested 53 videos with 346 wipe gestures in total (approx. 87 examples per gesture). For EdgeWrite, 103 videos contain 625 gesture sequences in total (approx. 78 examples per gesture). Figure 23.5 provides the classification rates for this first quantitative evaluation. Fifty-eight gesture sequences have been correctly detected for the first wipe gesture. This corresponds to a detection rate of 65.2%. The confusion matrix also shows that all 45,265 non-gesture sequences have been correctly classified as non-gestures. Also

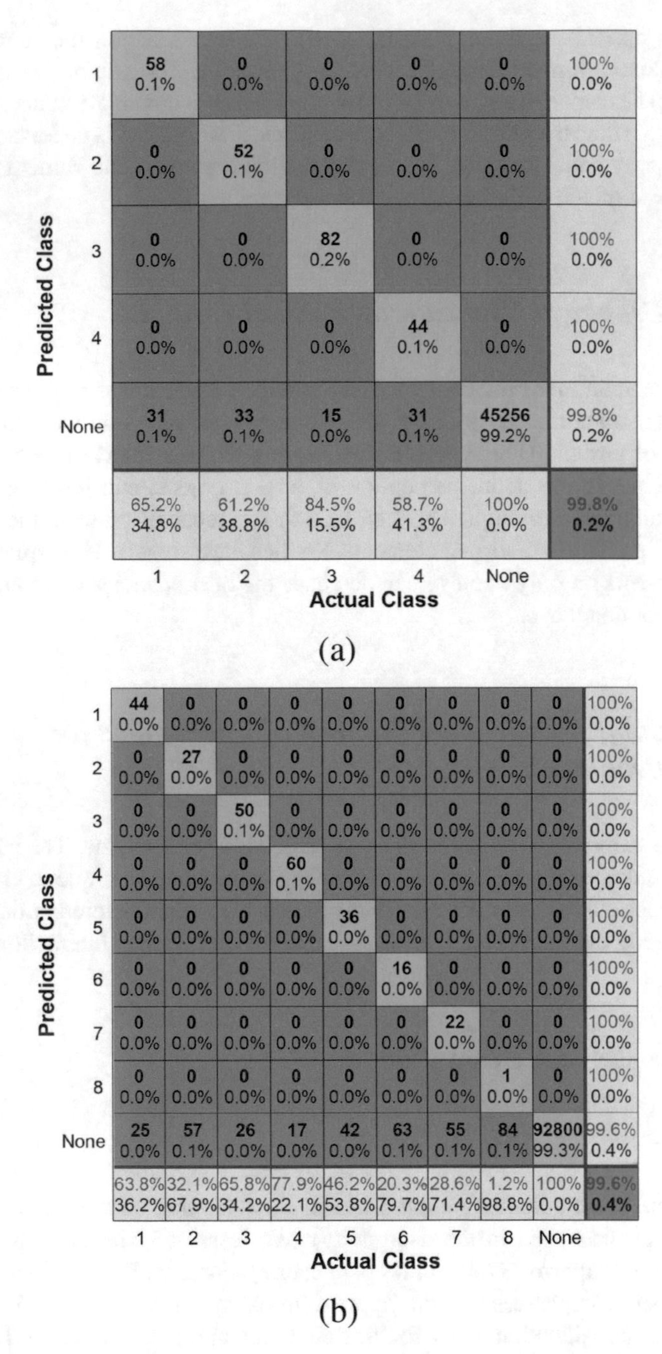

Fig. 23.5 Confusion matrix for the first quantitative evaluation of Wipe and EdgeWrite. (**a**) Wipe gesture confusion matrix. (**b**) EdgeWrite confusion matrix

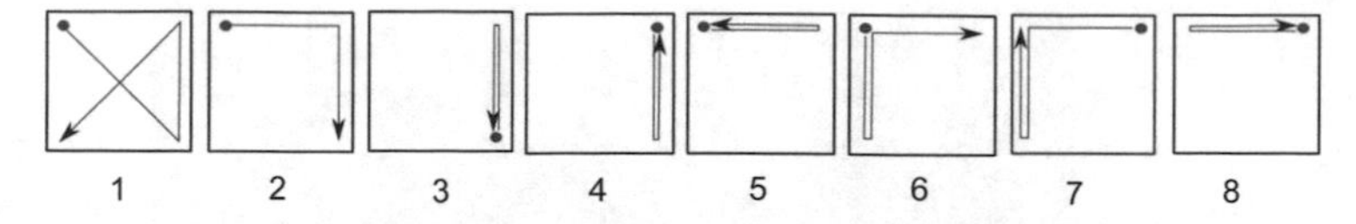

Fig. 23.6 Symbolic representation of the gestures from the user's perspective

for the EdgeWrite gestures, all 92,800 sequences have been correctly classified as non-gestures. Figure 23.6 illustrates the execution of the evaluated gestures.

In the second quantitative evaluation, we trained a model that increased the detection rate significantly for the cost of a few false alarms (see Fig. 23.7). Unfortunately, this model could not be tested in the evaluation scenario. In the second evaluation we increased the dataset size by one subject. We tested 71 videos containing 454 gesture sequences in total. For the wipe gestures, 105 gesture sequences have been classified correctly, six have been misclassified to gesture 2, and one to gesture 4. This corresponds to a detection rate of 93.8%. Of all 46,431 negative examples, one was classified as gesture 4 (false alarm). This corresponds to one false alarm every 26 min.

23.7.2 *Qualitative Results*

The prototype was finally evaluated. Here, static pointing (as an example of a very easy gesture system, for users as well as for system recognition) as well as Edge Write (as an example of a rather difficult gesture system for both users and the technical system) were compared to a standard control via buttons. Since participants were acquired at the university, the e-box production process was transferred to an abstract scenario. Twenty-one users first performed a practice session with the respective gesture system. Afterwards, they had to construct a plane using material handed out to them. Each production step was presented as written text as well as a picture on the head-mounted display (see Fig. 23.8). Users could navigate within the instruction while constructing the plane. All users were able to build a plane within 10 min. The effectivity was highest for button control followed by static pointing and lowest for EdgeWrite. At times for gesture execution, static pointing was, with about 2766 ms, faster than EdgeWrite with 3069 ms. Interestingly, the variability which is known to affect the reliability of the system gave an advantage to EdgeWrite. Subjective ratings of usability showed no significant differences between the gesture systems.

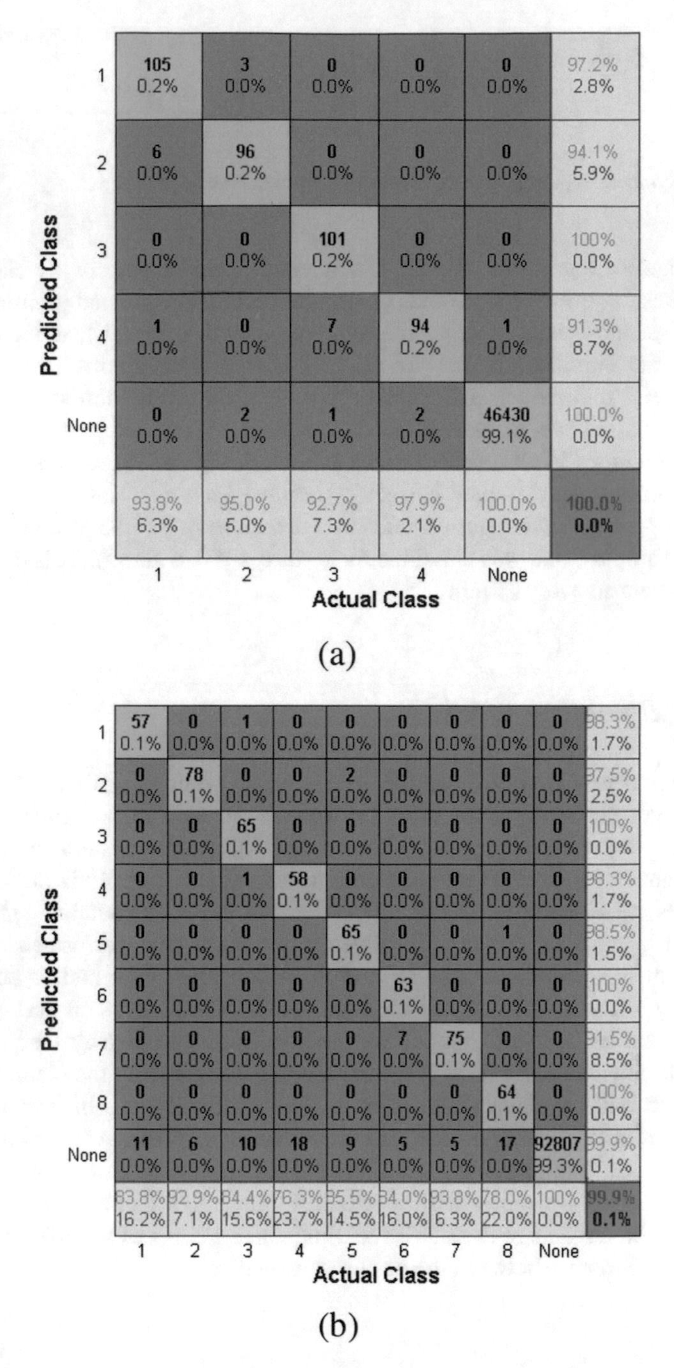

Fig. 23.7 Confusion matrix for the second quantitative evaluation of Wipe and EdgeWrite. (**a**) Wipe gesture confusion matrix. (**b**) EdgeWrite confusion matrix

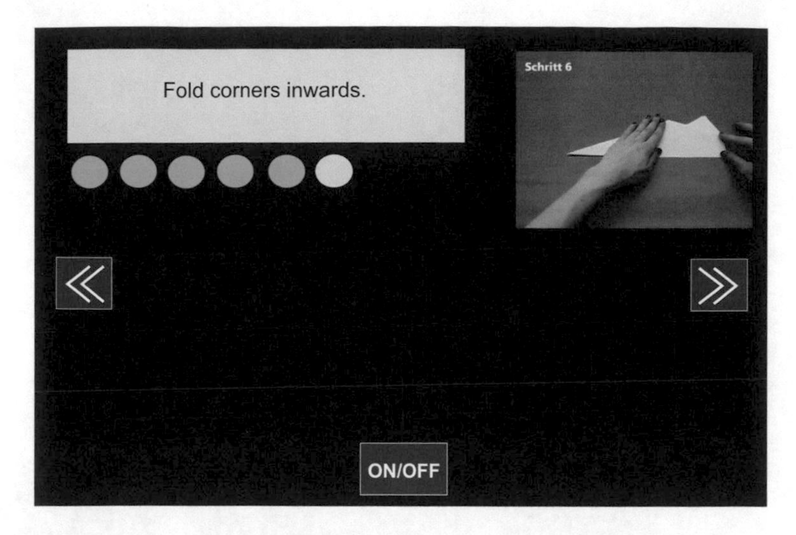

Fig. 23.8 The content seen through the Vuzix 1200XLD by the participant

23.8 Conclusion

We have developed a natural gesture-based interaction between users and an assistive system implemented on a head-mounted augmented reality device and demonstrated its applicability. Also the applicant, Volkswagen AG, sees a huge potential in such a system and will pursue respective technological solutions. When gesture-based assisting technologies will be implemented in the production cycle cannot be estimated so far.

Acknowledgements This work was done within the Transregional Collaborative Research Centre SFB/TRR 62 "*Companion*-Technology for Cognitive Technical Systems" funded by the German Research Foundation (DFG).

References

1. Ierusalimschy, R., Celes, W., de Figueiredo, L.H.: Lua (1993). http://www.lua.org/
2. Kalal, Z., Mikolajczyk, K., Matas, J.: Tracking-learning-detection. IEEE Trans. Pattern Anal. Mach. Intell. **6**(1), 1409–1422 (2010)
3. Khan, R., Hanbury, A., Stöttinger, J., Bais, A.: Color based skin classification. Pattern Recogn. Lett. **33**, 157–163 (2012). doi:10.1016/j.patrec.2011.09.032
4. Köpsel, A., Huckauf, A.: Evaluation of static and dynamic freehand gestures in device control. In: Proceedings of the Tilburg Gesture Research Meeting (2013)
5. Phung, S.L., Bouzerdoum, A., Chai, D.: Skin segmentation using color pixel classification: analysis and comparison. IEEE Trans. Pattern Anal. Mach. Intell. **27**(1), 148–154 (2005)
6. Wobbrock, J.O., Myers, B.A.: Edgewrite: a new text entry technique designed for stability. Technical report, Institute for Software Research, Carnegie Mellon University (2005)

Chapter 24
Advanced User Assistance for Setting Up a Home Theater

Pascal Bercher, Felix Richter, Thilo Hörnle, Thomas Geier, Daniel Höller, Gregor Behnke, Florian Nielsen, Frank Honold, Felix Schüssel, Stephan Reuter, Wolfgang Minker, Michael Weber, Klaus Dietmayer, and Susanne Biundo

Abstract In many situations of daily life, such as in educational, work-related, or social contexts, one can observe an increasing demand for intelligent assistance systems. In this chapter, we show how such assistance can be provided in a wide range of application scenarios—based on the integration of user-centered planning with advanced dialog and interaction management capabilities. Our approach is demonstrated by a system that assists a user in the task of setting up a complex home theater. The theater consists of several hi-fi devices that need to be connected with each other using the available cables and adapters. In particular for technically inexperienced users, the task is quite challenging due to the high number of different ports of the devices and because the used cables might not be known to the user. Support is provided by presenting a detailed sequence of instructions that solves the task.

24.1 Introduction

In many situations of daily life, such as in educational, work-related, or social contexts, one can observe an increasing demand for intelligent assistance systems. *Companion*-Technology can be used to design and implement systems that provide intelligent assistance in a wide range of application areas. For an introduction to *Companion*-Technology we refer you to Chap. 1, and for a survey to the article by Biundo et al. [6]. Here, we demonstrate a prototypical *Companion*-System

P. Bercher (✉) • F. Richter • T. Hörnle • T. Geier • D. Höller • G. Behnke • F. Nielsen •
F. Honold • F. Schüssel • S. Reuter • W. Minker • M. Weber • K. Dietmayer • S. Biundo
Faculty of Engineering, Computer Science and Psychology, Ulm University, Ulm, Germany
e-mail: pascal.bercher@uni-ulm.de; felix.richter@alumni.uni-ulm.de; thilo.hoernle@uni-ulm.de; thomas.geier@alumni.uni-ulm.de; daniel.hoeller@uni-ulm.de; gregor.behnke@uni-ulm.de; florian.nothdurft@alumni.uni-ulm.de; frank.honold@uni-ulm.de; felix.schuessel@uni-ulm.de; stephan.reuter@uni-ulm.de; wolfgang.minker@uni-ulm.de; michael.weber@uni-ulm.de; klaus.dietmayer@uni-ulm.de; susanne.biundo@uni-ulm.de

S. Biundo, A. Wendemuth (eds.), *Companion Technology*, Cognitive Technologies,
DOI 10.1007/978-3-319-43665-4_24

that provides intelligent assistance with the task of setting up a complex home theater [1, 2]. The respective system is also presented in a short video [11].

The home theater consists of several hi-fi devices that need to be connected with each other using the available cables and adapters. In particular for technically inexperienced users, the task is quite challenging due to the high number of different ports of the devices, and because the user might not be familiar with the different kinds of available cables. Support is provided by presenting a detailed sequence of individualized instructions that describe how to solve the task. In the case of unexpected changes in the environment, for example if a cable turns out to be damaged, the system adapts to the new situation and presents an alternative solution. It can further adapt its multimodal user interface to changes in the context of use (CoU) and is able to provide explanations about the presented instructions. Within the SFB/TRR 62 [4], this *Companion*-Technology has been implemented and demonstrated in a prototype system that integrates advanced user-centered planning capabilities (cf. Chap. 5) with advanced dialog and user interaction capabilities (cf. Chaps. 9 and 10, respectively) to ensure an individual, convenient, and highly flexible interaction with the system.

24.2 Application Scenario: The Home Assembly Task

Our system is capable of assisting a human user in the task of setting up his or her home theater. The home theater consists of several hi-fi devices which need to be connected with each other using different types of cables and adapters, such that all components correctly work together.

In a fully general assistance system of the depicted application domain, the currently available hi-fi devices and cables and adapters had to be selected by the user. For demonstration purposes, we fixed one specific setting, where the following devices are given: a television for visual output, an audio/video receiver with attached boxes for audio output (video signals can be looped through that receiver, though), a Blu-ray player, and a satellite receiver. For these devices to be fully functional, the television needs to receive the video signals of the Blu-ray player and the satellite receiver, and the audio/video receiver needs to receive their audio signals.

24.3 System Description

The proposed assistance system uses the technical specification of the available hardware and automatically generates a course of action that solves the assembly task. That is, we modeled the assembly task as a hybrid planning task,[1] the solution

[1]The hybrid planning model is deterministic. A non-deterministic variant of the planning task (a hierarchical POMDP model) is considered in Chap. 6.

to which is then presented step-by-step to the user in an adequate way to enable a natural interaction with the system. The respective system is a knowledge-based system that implements a generic architecture (cf. Chaps. 10 and 22) that can be used as a basis for the creation of intelligent assistance systems in a wide range of application scenarios [1, 9]. It comprises different modules, which are described in the sequel.

Knowledge Base The knowledge base serves as a central hub for any domain and user information.[2] It integrates information from various sources, such as data coming from sensors—usually preprocessed through machine learning algorithms—or explicit user input originating from system components responsible for human-computer interaction.

Besides storing information central to the application domain, the main purpose of the knowledge base is the integration of sensory data. Our system can be configured in such a way that—except for touch or sound sensors to gain user input—it works without use of further sensors. However, we can also use laser sensors to track the user's position [7, 18, 19] in order to determine the most appropriate output modality [10]. To cope with uncertainty and to facilitate the concise representation of large graphical models, the knowledge base uses Markov Logic [20] as modeling language. Since the planning components of the system are based on a deterministic model, they require special treatment. To make the probabilistic view of the knowledge base compatible with this deterministic representation, we consider the most probable world state according to the current beliefs as the actual world state.

Planning, Dialog, and Interaction: An Interactive Cycle We modeled the assembly task as a *hybrid planning problem*, as it enables us to provide advanced user assistance [5]. For a detailed discussion about the user-centered planning capabilities that it integrates and further arguments about why it is well suited for planning with or for humans, we refer you to Chap. 5.

Each presented instruction is a single planning action that specifies where the specific ends of the cables and adapters have to be plugged in. Before any instruction is presented to the user, a plan must be found that solves the specified problem. In hybrid planning, plans are partially ordered sequences of actions. Actions have a name (in our example scenario, there are only *plugIn* actions) and a sequence of parameters that specify the involved devices and their ports that are used to connect the devices with each other. For instance,

$$\textit{plugIn}(\text{SCART-CABLE}, \text{AUDIO-PORT}_1, \text{AV-RECEIVER}, \text{AUDIO-PORT}_2) \tag{24.1}$$

[2]The instances of the user and environment model can be loaded from a pre-defined source and altered at runtime by updating their respective properties. These two context model instances can be displayed and manipulated on additional devices, e.g., via a touch-enabled tablet computer or an interactive surface using tangible interaction.

depicts an action describing that the audio end of the SCART cable should be connected with a specific audio port of the audio/video receiver. Since plans are only partially ordered, but actions need to be presented one after another, one first needs to find an action linearization that is plausible to the user [8]. After such a sequence is found, they need to be communicated step-wise in an adequate way.

For this, actions are passed on one-by-one to the dialog management (cf. Chap. 9). Here, every received action is mapped to a dialog model [1, 14] specifying several so-called dialog goals that need to be achieved. Every dialog goal may achieve some or all of the required effects defined by the action. The dialog goals may be structured in a hierarchical way: actions may be decomposed into several user-adapted sequences of so-called dialog steps suitable for the individual user (e.g., novice users receive more detailed instructions distributed to several dialog steps instead of a single one). These dialog steps are modality-independent representations referencing the content to be communicated to the user, where modality means a combination of interaction language plus device component. Verbal intelligence can also be addressed [16]. This allows for user-adaptive selections of the most appropriate dialog steps. The resulting sequence implements those effects of the given action that need to be achieved in a user interaction. Graphical approaches can assist a domain expert in the modeling process of dialog models [3].

Each dialog step is passed on to the fission component (cf. Chap. 10). The fission is responsible for modality arbitration of the respective dialog step [9]. Each of the referenced information items within such a dialog step can be realized with the use of different interaction languages such as video, automatic speech recognition, or text. For instance, depending on the current CoU, it might be inappropriate for the system to read instructions aloud, but appropriate to present them using written text or graphics, instead. Further, the fission decides which information fragment is realized via which device component. One example for a possible output of a dialog step is presented in Fig. 24.1. In this case, the respective action (cf. Eq. (24.1)) is reflected using only one single dialog step. In an empirical evaluation we studied adaptation rules for the fission to improve user interaction [21].

At any point in time during interaction with the system, the system receives user input. This input might be implicit (like the position of the user) or explicit (such as pointing gestures for selection purposes). Since the user often interacts in a multimodal fashion (such as pointing and using speech at the same time), these interactions need to be fused to a coherent system input [22]. Sensor failures or fuzziness can be limited or even resolved when incorporating the interaction history of the user [23]. To resolve remaining conflicts and uncertainty, the system can initiate additional questions to the user [12].

Assuming the instructions are correctly carried out, all components of the home theater will be correctly connected with each other. Note that we do not have sensors to detect execution failures. Those need to be reported by the user manually. So, at any point in time during interaction with the system, the user may state execution errors. For instance, in case one of the cables breaks down, he may state so using either speech or text input (as for example: “The cable is broken!”). In such a case,

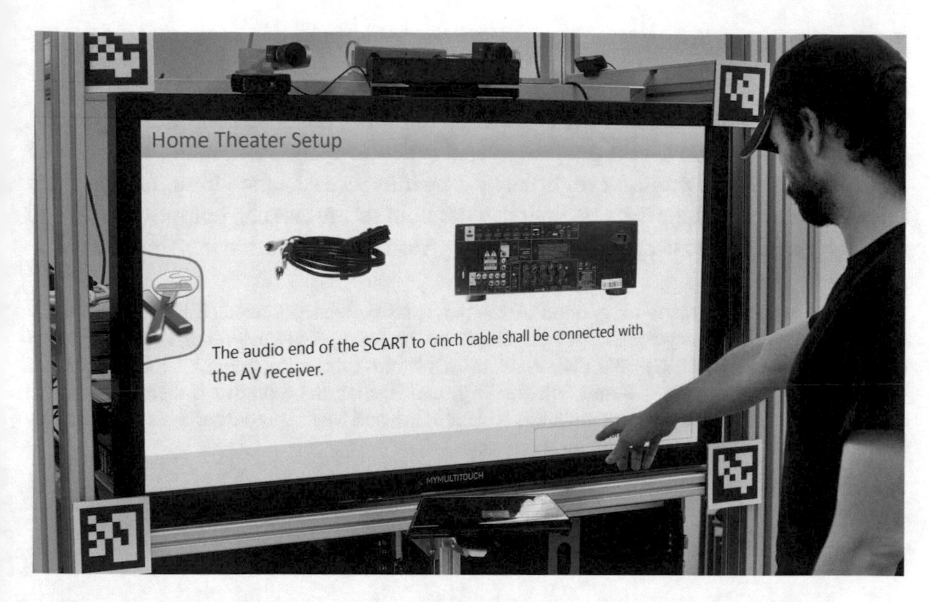

Fig. 24.1 The user interacts with the system. It displays the instruction representing the action given in Eq. (24.1). The respective port of the audio/video receiver depicted is flashing in *red*

the system marks the respective cable as unusable and initiates plan repair to find a repaired solution to the problem that does not use that specific cable [1]. The plan repair technique could also handle other unexpected failures (such as unusable ports, for example), but these are currently not handled by the system.

In particular, when execution errors occur and a plan needs to be changed, questions might arise. The user might wonder, for example, what purpose a presented instruction serves (as he or she might be asked to use a different type of cable than before the execution failure, for instance). Such a question, i.e., the demanded justification for a specific instruction, can be answered by the plan explanation component [1, 24]. It derives a sequence of arguments that explains the purpose of the instruction in question. In an empirical evaluation we studied the benefit of such plan explanations in the given scenario and how the assembly task assistance system is perceived in general [1]. In a different scenario, an empirical evaluation was conducted to study how providing justification and transparency explanations may improve the user-experience in unexpected situations (e.g., the change of plans) [17]. In addition, the system can explain declarative facts (such as technical details about a specific cable) and may decide when to present these (declarative) explanations depending on the current user knowledge and the currently presented instruction [13, 15, 16].

24.4 Conclusion

We presented the integration and interaction of dialog and interaction management, probabilistic reasoning, and planning capabilities in a single system. That system realizes several *Companion* properties: it provides its assistance functionality in an individual manner, adapts its behavior in the light of unforeseen complications, and actively maintains its user's trust by offering various explanation capabilities.

Acknowledgements This work is done within the Transregional Collaborative Research Centre SFB/TRR 62 "*Companion*-Technology for Cognitive Technical Systems" funded by the German Research Foundation (DFG). We also want to thank our former colleagues Gregor Bertrand, Peter Kurzok, Louisa Pragst, Bernd Schattenberg, and Bastian Seegebarth for their work on the demonstrated system. We also acknowledge Roland Barth and Matthias Rau for their contributions in producing the video [11] explaining the system.

References

1. Bercher, P., Biundo, S., Geier, T., Hörnle, T., Nothdurft, F., Richter, F., Schattenberg, B.: Plan, repair, execute, explain - how planning helps to assemble your home theater. In: Proceedings of the 24th International Conference on Automated Planning and Scheduling (ICAPS), pp. 386–394. AAAI Press, Palo Alto (2014)
2. Bercher, P., Richter, F., Hörnle, T., Geier, T., Höller, D., Behnke, G., Nothdurft, F., Honold, F., Minker, W., Weber, M., Biundo, S.: A planning-based assistance system for setting up a home theater. In: Proceedings of the 29th National Conference on AI (AAAI), pp. 4264–4265. AAAI Press, Palo Alto (2015)
3. Bertrand, G., Nothdurft, F., Honold, F., Schüssel, F.: CALIGRAPHI - creation of adaptive dialogues using a graphical interface. In: 35th Annual 2011 IEEE Computer Software and Applications Conference (COMPSAC), pp. 393–400. IEEE, Piscataway (2011)
4. Biundo, S., Wendemuth, A.: Companion-technology for cognitive technical systems. Künstl. Intell. **30**(1), 71–75 (2016). doi:10.1007/s13218-015-0414-8
5. Biundo, S., Bercher, P., Geier, T., Müller, F., Schattenberg, B.: Advanced user assistance based on AI planning. Cogn. Syst. Res. **12**(3–4), 219–236 (2011). Special Issue on Complex Cognition
6. Biundo, S., Höller, D., Schattenberg, B., Bercher, P.: Companion-technology: an overview. Künstl. Intell. **30**(1), 11–20 (2016). doi:10.1007/s13218-015-0419-3
7. Geier, T., Reuter, S., Dietmayer, K., Biundo, S.: Track-person association using a first-order probabilistic model. In: Proceedings of the 24th IEEE International Conference on Tools with AI (ICTAI 2012), pp. 844–851. IEEE, Piscataway (2012)
8. Höller, D., Bercher, P., Richter, F., Schiller, M., Geier, T., Biundo, S.: Finding user-friendly linearizations of partially ordered plans. In: 28th PuK Workshop "Planen, Scheduling und Konfigurieren, Entwerfen" (2014)
9. Honold, F., Schüssel, F., Weber, M.: Adaptive probabilistic fission for multimodal systems. In: Proceedings of the 24th Australian Computer-Human Interaction Conference, OzCHI '12, pp. 222–231. ACM, New York (2012)
10. Honold, F., Schüssel, F., Weber, M., Nothdurft, F., Bertrand, G., Minker, W.: Context models for adaptive dialogs and multimodal interaction. In: 9th International Conference on Intelligent Environments, pp. 57–64. IEEE, Piscataway (2013)

11. Honold, F., Bercher, P., Richter, F., Nothdurft, F., Geier, T., Barth, R., Hörnle, T., Schüssel, F., Reuter, S., Rau, M., Bertrand, G., Seegebarth, B., Kurzok, P., Schattenberg, B., Minker, W., Weber, M., Biundo, S.: Companion-technology: towards user- and situation-adaptive functionality of technical systems. In: 10th International Conference on Intelligent Environments, pp. 378–381. IEEE, Piscataway (2014). http://companion.informatik.uni-ulm.de/ie2014/companion-system.mp4
12. Honold, F., Schüssel, F., Weber, M.: The automated interplay of multimodal fission and fusion in adaptive HCI. In: 10th International Conference on Intelligent Environments, pp. 170–177 (2014)
13. Nothdurft, F., Minker, W.: Using multimodal resources for explanation approaches in intelligent systems. In: Proceedings of the Eight International Conference on Language Resources and Evaluation (LREC), pp. 411–415. European Language Resources Association (ELRA), Paris (2012)
14. Nothdurft, F., Bertrand, G., Heinroth, T., Minker, W.: GEEDI - guards for emotional and explanatory dialogues. In: 6th International Conference on Intelligent Environments, pp. 90–95. IEEE, Piscataway (2010)
15. Nothdurft, F., Honold, F., Kurzok, P.: Using explanations for runtime dialogue adaptation. In: Proceedings of the 14th ACM International Conference on Multimodal Interaction, pp. 63–64. ACM, New York (2012)
16. Nothdurft, F., Honold, F., Zablotskaya, K., Diab, A., Minker, W.: Application of verbal intelligence in dialog systems for multimodal interaction. In: 10th International Conference on Intelligent Environments, pp. 361–364. IEEE, Piscataway (2014)
17. Nothdurft, F., Richter, F., Minker, W.: Probabilistic human-computer trust handling. In: Proceedings of the 15th Annual Meeting of the Special Interest Group on Discourse and Dialogue (SIGDIAL), pp. 51–59. Association for Computational Linguistics, Stroudsburg (2014)
18. Reuter, S., Dietmayer, K.: Pedestrian tracking using random finite sets. In: Proceedings of the 14th International Conference on Information Fusion, pp. 1–8. IEEE, Piscataway (2011)
19. Reuter, S., Wilking, B., Wiest, J., Munz, M., Dietmayer, K.: Real-time multi-object tracking using random finite sets. IEEE Trans. Aerosp. Electron. Syst. **49**(4), 2666–2678 (2013)
20. Richardson, M., Domingos, P.: Markov logic networks. Mach. Learn. **62**(1–2), 107–136 (2006)
21. Schüssel, F., Honold, F., Weber, M.: Influencing factors on multimodal interaction during selection tasks. J. Multimodal User Interfaces **7**(4), 299–310 (2013)
22. Schüssel, F., Honold, F., Weber, M.: Using the transferable belief model for multimodal input fusion in companion systems. In: Schwenker, F., Scherer, S., Morency, L.P. (eds.) Multimodal Pattern Recognition of Social Signals in Human-Computer-Interaction, vol. 7742, pp. 100–115. Springer, Berlin/Heidelberg (2013)
23. Schüssel, F., Honold, F., Schmidt, M., Bubalo, N., Huckauf, A., Weber, M.: Multimodal interaction history and its use in error detection and recovery. In: Proceedings of the 16th International Conference on Multimodal Interaction, ICMI '14, pp. 164–171. ACM, New York (2014)
24. Seegebarth, B., Müller, F., Schattenberg, B., Biundo, S.: Making hybrid plans more clear to human users – a formal approach for generating sound explanations. In: Proceedings of the 22nd International Conference on Automated Planning and Scheduling (ICAPS), pp. 225–233. AAAI Press, Palo Alto (2012)

Chapter 25
Multi-modal Information Processing in *Companion*-Systems: A Ticket Purchase System

Ingo Siegert, Felix Schüssel, Miriam Schmidt, Stephan Reuter, Sascha Meudt, Georg Layher, Gerald Krell , Thilo Hörnle, Sebastian Handrich, Ayoub Al-Hamadi, Klaus Dietmayer, Heiko Neumann, Günther Palm, Friedhelm Schwenker, and Andreas Wendemuth

Abstract We demonstrate a successful multimodal dynamic human-computer interaction (HCI) in which the system adapts to the current situation and the user's state is provided using the scenario of purchasing a train ticket. This scenario demonstrates that Companion Systems are facing the challenge of analyzing and interpreting explicit and implicit observations obtained from sensors under changing environmental conditions. In a dedicated experimental setup, a wide range of sensors was used to capture the situative context and the user, comprising video and audio capturing devices, laser scanners, a touch screen, and a depth sensor. Explicit signals describe a user's direct interaction with the system, such as interaction gestures, speech and touch input. Implicit signals are not directly addressed to the system; they comprise the user's situative context, his or her gesture, speech, body pose, facial expressions and prosody. Both multimodally fused explicit signals and interpreted information from implicit signals steer the application component, which was kept deliberately robust. The application offers stepwise dialogs gathering the most relevant information for purchasing a train

---

All authors contributed equally.

I. Siegert (✉) • G. Krell • S. Handrich • A. Al-Hamadi • A. Wendemuth
Otto von Guericke University Magdeburg, 39016 Magdeburg, Germany
e-mail: ingo.siegert@ovgu.de; gerald.krell@ovgu.de; sebastian.handrich@ovgu.de; ayoub.al-hamadi@ovgu.de; andreas.wendemuth@ovgu.de

F. Schüssel • M. Schmidt • S. Reuter • S. Meudt • G. Layher • T. Hörnle • K. Dietmayer • H. Neumann • G. Palm • F. Schwenker
Ulm University, 89081 Ulm, Germany
e-mail: felix.schuessel@uni-ulm.de; miriam.k.schmidt@uni-ulm.de; stephan.reuter@uni-ulm.de; sascha.meudt@uni-ulm.de; georg.layher@uni-ulm.de; thilo.hoernle@uni-ulm.de; klaus.dietmayer@uni-ulm.de; heiko.neumann@uni-ulm.de; guenther.palm@uni-ulm.de; friedhelm.schwenker@uni-ulm.de

S. Biundo, A. Wendemuth (eds.), *Companion Technology*, Cognitive Technologies,
DOI 10.1007/978-3-319-43665-4_25

ticket, where the dialog steps are sensitive and adaptable within the processing time to the interpreted signals and data. We further highlight the system's potential for a fast-track ticket purchase when several pieces of information indicate a hurried user.

A video of the complete scenario in German language is available at: http://www.uni-ulm.de/en/in/sfb-transregio-62/pr-and-press/videos.html

25.1 Introduction

Companion-Systems are faced with the challenge of analyzing and interpreting observations obtained from sensors under changing environmental conditions. We demonstrate a successful multimodal dynamic human–computer interaction in which the system adapts to the current situation by implementing multimodal information processing and the user's state is provided using the scenario of purchasing a train ticket.[1] The technical setup defines the available sensors and may change depending on the location of the user. The environmental conditions parametrize the interpretation of the observations from the sensors and constrain the reliability of the information processing. The inferred knowledge about the user's state and the context of use finally enables the system to adapt not only its functionality, but also the way of interacting with the user in terms of available in- and output modalities.

An experimental platform shown in Fig. 25.1 was equipped with a wide range of sensors to capture the situative context and the user. The sensors comprise video and audio capturing devices, laser scanners, a touch screen, and a depth-sensing camera. The information processing is depicted in Fig. 25.2 and was realized by multiple components. Some of them retrieve data from sensors while others are pure software components, depicting a complex conceptual information flow. The components are categorized according to explicit and implicit user signals. Explicit signals describe commands performed by the user with the intention to interact with the system, such as interaction gestures, speech and touch input. Implicit signals are not directly addressed to the system but nevertheless contain a rich set of relevant information. These signals comprise the user's situative context, his or her gesture, speech, body pose, facial expressions and prosody.

While explicit user signals are directly fed into the input fusion component, sending signals to the ticket application, implicit signals are first combined and further abstracted within a dedicated data fusion component. The architecture and the communication middleware of the underlying system represent a specific instance of the generic *Companion* architecture as described in Chap. 22. The planning and dialog management tasks are realized within the application component. The same applies to the storage about the user and his or her preferences using a knowledge base component. The application offers stepwise dialogs gathering the most relevant

[1]A video of the complete scenario in German language is available at: http://www.uni-ulm.de/en/in/sfb-transregio-62/pr-and-press/videos.html.

Fig. 25.1 A user interacting with the ticket purchase system. While the application is shown on the *right screen* in front of the user, the *left screens* visualize internal system states and signal processing details. Various sensors can be seen mounted on the rack

information for purchasing a train ticket, where the dialog steps are sensitive to the interpreted signals and data. The dialog flow can be automatically adapted within the processing time.

The following section describes a complete run through of a normal ticket purchase with details on aspects of signal processing, information fusion, user adaption and interaction. The final section highlights the possibilities of a fast-track ticket purchase when several pieces of information indicate a hurried user.

25.2 User- and Situation-Adaptive Ticket Purchase

The ticket purchase starts with a user approaching the system. The standard process requires the specification of the destination, travel time, number of tickets and train connection. Leaving the device finally marks the end of a normal purchase.

Approaching the Device: The activation of the ticket purchase system is triggered based on the environment perception system and the head and body pose recognition. The environment perception system estimates the locations of all persons in the proximity using two laser-range-finders and the multi-object

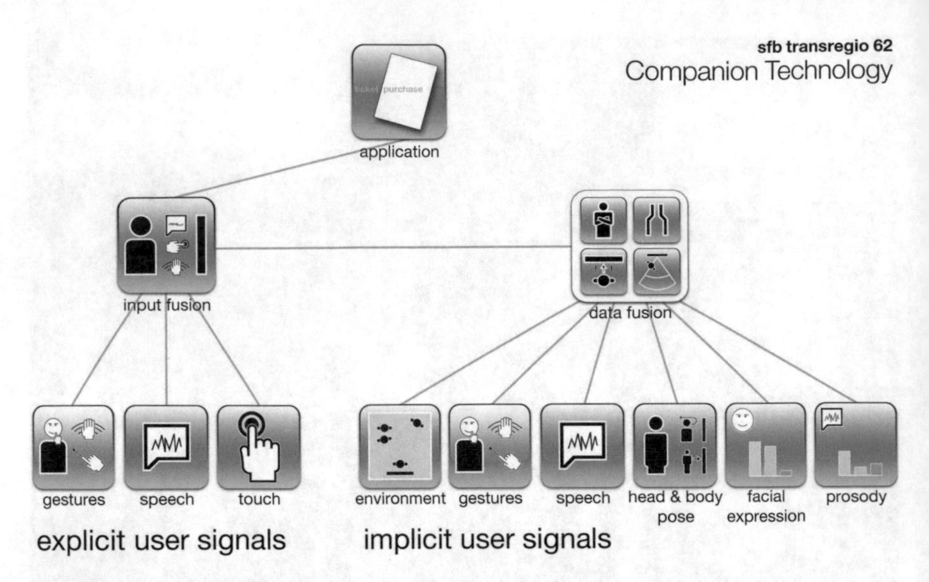

Fig. 25.2 Bottom-up information flow in the train ticket purchase system. Explicit user signals (i.e. gestures, speech and touch inputs) are directly combined in the input fusion and sent to the application. Implicit signals stem from various sensors. Laser scanners observe the user's environment; video capturing devices gather implicit gestures, facial expressions, as well as the head and body pose; audio devices analyze the speech and nonverbal signals. In the data fusion component, implicit signals are combined into high-level information about the user's state, such as disagreement with the system behavior, attentiveness to the system, or hastiness

tracking algorithms introduced in Chap. 15. All tracked persons heading towards the ticket purchase system are considered as potential users. In order to prevent an activation of the system due to passers-by, the data fusion software component combines the states of the potential users with the results of the head and body pose estimation (cf. Chap. 17). Hence, only users who are approaching and facing the system will trigger the beginning of a ticket purchase. After a new user has been detected, the application starts using a range-dependent fade-over from the stand-by screen to the welcome screen of the purchase process. The position of the active user in the calibrated coordinate system is transferred to other software components to prevent the system from confusing the active user with non-active users.

Destination Selection: The next step in the ticket purchase process is the selection of the travel destination. Given that the user is already known to the system,[2] individual contextual information is provided, e.g. the user's most frequent travel destinations are automatically suggested on a graphical map. The user selects a destination by performing a touch input on the displayed map or by specifying the destination via direct speech input.

[2]This can be elicited, e.g., by authorized data transfer from the user's mobile device.

Time Selection: The third step comprises the selection of the date and time. Again, the system displays a personalized dialog of the user's schedule given that the user is already known. The dialog allows the user to select a time slot for the trip. The *Companion*-System is aware that the interaction takes place in a public area and, therefore, observes the predefined privacy policy. Hence, the system will display only whether a specific slot is already blocked or not.
The travel time selection is conducted using both speech and gestural input. While it is more convenient to specify the date and time using speech, e.g. "I want to travel at 8 am on Wednesday", the browsing through the calendar is performed more naturally with the gestural input, e.g. by using the "wiping to the left" gesture to get to the next week. However, the complete functionality is provided by both modalities, e.g. a specific time slot is selected by holding a pointing gesture for a few seconds. The screen coordinate of the pointing direction is computed in two ways. If the gesture recognition system recognizes the user's arm as being outstretched, the line from the head to the hand is extended until it intersects with the screen. When the user's hand is recognized as being close to the user's body, the pointing direction is adjusted by local hand movements. Additionally, a graphical feedback is presented on the screen to indicate the location the user is pointing at. The speech and gestural inputs are recognized independently and integrated within the input fusion (see Fig. 25.2). The systems further allow combined and relative inputs such as pointing on a specific time and uttering the speech command "This time" to perform a fast confirmation of the selected time. In this case the input fusion does not wait, since it can take advantage of the explicit speech command.

Ticket Number Selection: The system further performs an adaption to the current context. This is exemplified by the automatic pre-selection of the number of tickets. The system has initially observed with the help of the multi-object tracking algorithm whether the user detached himself from a group of other persons. The group is defined by spatial regions and common trajectories in the past. The tracking algorithm provides the group size such that the application is able to automatically suggest buying either a single ticket or tickets for all people in the group.

Interruption: Another issue a speech-controlled technical system has to deal with is distinguishing between user commands intended to control the system and other unrelated utterances. Unrelated utterances can often be denoted as "off-talk" in human–computer interaction (cf. Chap. 20). As long as the content of this off-talk is different from system commands, this differentiation can be purely based on the speech content in the speech recognizer itself. But in situations where the off-talk contains the same phrases, e.g. the user conducts clarifying dialogs with his or her co-passengers or agrees upon the journey via mobile phone, this assumption cannot be made. In this case the decision whether the user utterances are intended to control the system or not can only be made using additional modalities. Two types of off-talk events are recognized by the system: (1) turning away from the system and (2) talking to somebody over the phone.

In the first case, the system recognizes the pose of the active user. As long as the pose is not directed towards the system, the output of the speech recognizer will be discarded until the user turns again towards the system.

In the second case, the system interprets both the recognized speech and the gesture. It is assumed that in order to make a phone call the mobile phone will be moved to the user's ear and typical phrases of receiving or initiating a call are uttered. The self-touch of the user's ear is detected by the gesture recognizer, while the speech recognizer detects the greeting at the beginning phone conversation. Both events are recognized independently in their respective software components and passed over to the data fusion component. If both events occur within the same short period of time, they are detected as off-talk. The off-talk disables the speech recognizer and the speech synthesizer as long as the gesture of the user does not change. Once the phone conversation ends, the system enables both the speech recognizer and the speech synthesizer again.

The off-talk is a welcome example of an implicit user signal. The *Companion*-System has to make an active decision which interferes with the ongoing interaction although no direct command is given.

Connection Selection: After the required information is gathered, the system seeks train connections which suit the user's preferences known from the knowledge base. Ideally, a suitable connection can be provided and the user can successfully complete the ticket purchasing process. However, in our demonstration we will assume that no suitable train connection exists and the system can only approximately match the known user preferences, i.e. reservations are possible and there is a low number of changes between trains. The system shows connections in which the user can make a reservation, but unfortunately has to change trains very often.

In this phase of interaction, the video and audio data are analyzed to capture the user's emotional states, which will serve as an implicit input. The goal is to recognize whether the user shows a facial expression or performs an utterance which indicates that he or she is not satisfied with the pre-selection. The emotional state, i.e. positive and negative valence, is recognized by software components for each modality independently. The recognition using the video data analyzes the facial expressions on the basis of features derived from geometric distances measured in the face of the active user (e.g. mouth width/height, eye-brow distance); see Chap. 18. The recognition using the audio channel starts by extracting mel frequency cepstral coefficients which are then classified using a probabilistic support vector machine (cf. Chap. 20). The outputs of the audio- and video-based recognitions are then combined in the data fusion component using a Markov fusion network which is able to deal with temporally fragmented intermediate classification inputs (cf. Chap. 19).

In case a negative reaction is recognized, this information is sent to the input fusion module (for the connections see Fig. 25.2), which triggers the application component in order to ask the user if the pre-selection should be adapted. The application then expands the list of train connections such that the user is able to choose a connection which is the most acceptable and to continue by paying the tickets.

Leaving the Device: After the purchase process, the system remains active as long as the user does not turn away from the system. The end of the interaction is triggered by the environment perception system and the head and body pose, i.e. the system only returns to stand-by mode if the distance of the user exceeds a certain threshold and the user is no longer facing the system.

25.3 Hurried User

The ticket purchase system supports two basic interaction modes: the normal ticket purchase mode and a quick purchase for users in a hurry.

The selection of the actual interaction mode (normal vs. quick purchase) is done automatically based on an analysis of the approaching speed of the user to the ticket purchase system. In case the approach speed exceeds a given threshold, the system decides for the quick purchase mode, it double-checks by asking the user at the beginning of the interaction whether he or she is really in a hurry to avoid misinterpretations of the approaching speed.

In quick purchase mode, the user has less choices during the purchase to achieve a shorter overall interaction time. The system automatically makes choices by considering the available information in order to omit the corresponding queries. For instance, the system sets the number of tickets to be purchased equal to the size of the group of people the active user has detached from. This number is derived based on the data of the laser range sensors. Since the system knows that the user is in a hurry, it proposes taking a train that leaves close to the current time. The current train station is set for departure such that only the destination needs to be selected, e.g. via speech command. If there is an ambiguity, the system resolves it by presenting additional user queries. Finally, the user has to confirm the pre-selected train connection to complete the purchase.

25.4 Conclusion

In this chapter, we showed how several key components of *Companion* technology have been exemplarily integrated into a prototypical ticket purchase system. Besides the possibility of using several input modalities, the proposed system adapts its behavior to the current situation and interprets implicit input data. The adaptation to the current situation is demonstrated by the interruption handling, the ticket number selection and the automatic switching between standard and hurried mode. Examples for implicit input data are the interpretation of the facial expressions and voice. Deliberately, the potentials of planning, data interpretation, and dialog components have been kept low in this system. They are demonstrated in Chap. 24, for example. The main focus of this demonstrator was to elaborate on the multimodal signal processing capabilities developed for a *Companion*-System.

Acknowledgements The authors thank the following colleagues for their invaluable support in realizing the scenario (in alphabetical order): Michael Glodek, Ralph Heinemann, Peter Kurzok, Sebastian Mechelke, Andreas Meinecke, Axel Panning, and Johannes Voss.

This work was done within the Transregional Collaborative Research Centre SFB/TRR 62 "*Companion*-Technology for Cognitive Technical Systems" funded by the German Research Foundation (DFG).

Printed in the United States
By Bookmasters